Lecture Notes in Mathematics

Edited by A. Dold and B. Eckmann

Subseries: USSR

Adviser: L.D. Faddeev, Leningr

1289

Yu. I. Manin (Ed.)

K-Theory, Arithmetic and Geometry

Seminar, Moscow University, 1984–1986

Springer-Verlag

Berlin Heidelberg New York London Paris Tokyo

Editor

Yuri I. Manin
Steklov Mathematical Institute
Vavilova 42, 117 966 Moscow, GSP-1, USSR

ISBN 3-540-18571-2 Springer-Verlag Berlin Heidelberg New York
ISBN 0-387-18571-2 Springer-Verlag New York Berlin Heidelberg

Printed in Germany

Printing and binding: Druckhaus Beltz, Hemsbach/Bergstr.
2146/3140-543210

PREFACE

This volume contains a collection of articles that originated from lectures given at the Manin Seminar at Moscow University, in 1984-1986. One of the principal motivations of the seminar was a collective desire to understand various ramifications of motivic and K-theoretic thinking in modern geometry and arithmetics. The final product is a volume of research papers reflecting the individual tastes of contributors. We hope however that it retains some internal coherence that is difficult to verbalize.

Всего, что знал еще Евгений,
Пересказать мне недосуг ...
(А. С. Пушкин, Евгений Онегин, I.VIII)

A.A.Bèilinson
Yu.I.Manin
V.V.Schechtman

February 1987

TABLE OF CONTENTS

Height pairing between algebraic cycles

A.A. Beilinson

Introduction

Let X be a smooth projective variety over Q; assume that its L-functions $L(H^j(X), s)$ satisfy the standard analytic continuation conjectures. Let $CH^i(X)$ be the group of codimension i cycles on X modulo rational equivalence, and $Ch^i(X)^o \subset CH^i(X)$ be the subgroup of cycles homologous to zero on $X(\mathbb{C})$; in particular $CH^1(X)^\circ = Pic^\circ(X)(\mathbb{Q})$. The conjecture of Birch and Swinnerton-Dyer claims that at $s = 1$ the function $L(H^1(X), s)$ has zero of order rk $CH^1(X)^\circ$ with the leading coefficient equal to the determinant of Neron-Tate canonical height pairing multiplied by the period matrix determinant up to some rational multiple (we do not need its exact value in what follows). As for the other L-functions, Swinnerton-Dyer conjectured [20] that the function $L(H^{2i-1}(X), s)$ has at s=i (=the middle of the critical strip) the zero of order rk $CH^i(X)^\circ$. The aim of this note is to define the canonical height pairing between $CH^i(X)^\circ$ and $CH^{\dim X+1-i}(X)^\circ$ that coincides with the Neron-Tate one for i=1 and whose determinant multiplied by the period matrix determinant should conjecturally be equal up to a rational multiple (of the nature I cannot imagine) to the leading coefficient of $L(H^{2i-1}(X), s)$ at s=i *). This pairing should also occure in Riemann-Roch type theorems a la Arakelov-Faltings (see [15]).

---

*) In fact our height pairing is defined on a certain subgroup of $CH^{\cdot}(X)^\circ$; under the very plausible (:=of rank of evidence far higher then B-SwD) local conjectures this subgroups should coincide with the whole $CH^{\cdot}(X)^\circ$.

The paper goes as follows. To motivate the basic construction, we begin with the simpler geometric case: here our base field is a field $k(C)$ of rational functions on a smooth projective curve C. Then the height pairing $\langle , \rangle$ comes from the global Poincaré duality on ℓ-adic cohomology. We may compute $\langle , \rangle$ in terms of local data round the points of C : if a_1, a_2 are cycles with disjoint supports that are homologous to zero on $X \otimes \overline{k(C)}$ and $v \in C$ is a closed point then the local link index $\langle a_1, a_2 \rangle_v$ is defined, and we have

$$(*) \qquad \langle a_1, a_2 \rangle = \sum_{v \in C} \langle a_1, a_2 \rangle_v$$

In the arithmetic situation, when the base field is a number field, the global construction fails due to the lack of appropriate cohomology theory. But we may still define the local indices $\langle , \rangle_v$ numbered by the places of the base field, and then use (*) as the definition of $\langle , \rangle$. These indices are defined using ℓ-adic cohomology for non-archimedean v and using the absolute Hodge-Deligne cohomology (see [2], [3], [19]) for archimedean ones; in case of pairing between divisors and zero cycles they are just Neron's quasifunctions ([17], [13], [22])

We also consider the intersection pairings. In the geometric case this is just the usual intersection pairing between the cycles of complementary dimensions on the regular scheme X_C proper over C. In the arithmetic case the role of X_C plays the A-variety $X = (X_{\mathbb{Z}}, \omega)$: where $X_{\mathbb{Z}}$ is a regular scheme projective over Spec $\mathbb{Z}$ and ω is a Kahler (1,1)-form on $X_{\mathbb{Z}} \otimes \mathbb{R}$ (see [15]). We define the corresponding Chow groups $CH^{\cdot}(X)$ and the $\mathbb{R}$-valued intersection pairing between $CH^i(X)$ and $CH^{\dim X - i}(X)$. This construction was independently found by H. Gillet and Ch. Soulé [10].

The final § contains some conjectures and motivic speculations about algebraic cycles, heights, L-functions and absolute cohomology groups.

The different construction of height pairing was proposed by S. Bloch [6]; I hope that our pairings coincide.

I would like to thank S. Bloch, P. Deligne, Yu. Manin, V. Schechtman and Ch. Soulé for stimulating ideas and interest.

§1. Height pairing in geometric case

(global construction)

In this § k will be an algebraically closed field, and ℓ will be a prime different from char k.

1.0. First recall some basic facts about the intersections. Let Y be a smooth projective scheme over k of dimension N + 1. The intersection of cycles defines the ring structure on Chow group $CH^{\bullet}(Y)$. This, together with an obvious trace map $CH^{N+1}(Y) \to \mathbb{Q}$, determines the intersection pairing $(,) : CH^{i}(Y) \otimes CH^{N+1-i}(Y) \to \mathbb{Q}$. We also have étale cohomology ring and the trace $H^{2(N+1)}(Y, \mathbb{Q}_\ell(N+1)) \to \mathbb{Q}_\ell$. The class map $cl : CH^{\bullet}(Y) \to H^{2\bullet}(Y, \mathbb{Q}_\ell(\bullet))$ is compatible with these structures, so we may compute (,) using ℓ-adic cohomology classes. In particular, (,) factors through $\overline{CH}^{\bullet}(Y) := Im(CH^{\bullet}(Y) \to H^{2\bullet}(Y, \mathbb{Q}_\ell(\bullet)))$. One hopes that the following standard conjectures hold

- the obvious map $\overline{CH}^{\bullet}(Y) \otimes \mathbb{Q}_\ell \to H^{2\bullet}(Y, \mathbb{Q}_\ell(\bullet))$ should be injective, so $\overline{CH}^{\bullet}(Y) \otimes \mathbb{Q}$ should be finite-dimensional $\mathbb{Q}$-vector spaces (this is obviously true in char 0)
- (,) should be non-degenerate on $\overline{CH}^{\bullet}(Y) \otimes \mathbb{Q}$
- $\overline{CH}^{\bullet} \otimes \mathbb{Q}$ and (,) should satisfy hard Lefschetz and Hodge-index theorems.
- Finally if k is an algebraic closure of a finite field, then one should have $CH^{\bullet} \otimes \mathbb{Q} = \overline{CH}^{\bullet} \otimes \mathbb{Q}$. More precisely, for an Y_0/F_q, $Y_0 \otimes k = Y$, the group $CH^{\bullet}(Y_0) \otimes \mathbb{Q}_\ell$ should coincide with invariants of Frobenius in $H^{2\bullet}(Y, \mathbb{Q}_\ell(\bullet))$ (Tate's conjecture).

1.1. The height pairing arises in a slightly different situation. Fix a smooth projective irreducible curve C over k; let $\eta \in C$ be its generic point, and $\bar{\eta}/\eta$ be some geometric generic point. Now let X be a smooth projective N-dimensional η-scheme, and $X_{\bar{\eta}} := X \times_{\eta} \bar{\eta}$ be its geometric fiber. Choose some projective scheme $X_C \xrightarrow{\pi} C$ over C with the generic fiber X; for an open $U \subset C$ put $X_U = \pi^{-1}(U)$. Now take $j : U \hookrightarrow C$ s.t. U is affine and $\pi|_U$ is smooth. Put $H^{\bullet}_{!*}(X, \mathbb{Q}_\ell(*)) := Im(H^{\bullet}_c(X_U, \mathbb{Q}_\ell(*)) \to H^{\bullet}(X_U, \mathbb{Q}_\ell(*)))$; the Poincaré duality on X_U induces the perfect pairing between $H^{\bullet}_{!*}(X, \mathbb{Q}_\ell(*))$ and $H^{2N+2-\bullet}_{!*}(X, \mathbb{Q}_\ell(N+1-*))$. We'll see in a moment that this groups and pairing depend only on X itself (and not on the choice of particular model X_C/C). To do this consider the smooth sheaf $R^{\bullet-1}\pi|_{U*}(\mathbb{Q}_\ell)$ on U and its middle extension $\mathcal{J}^{\bullet} :=$ $j_*R^{\bullet-1}\pi|_{U*}(\mathbb{Q}_\ell)$ on C. The Poincare duality along the fibers of π defines the perfect pairing $\langle,\rangle : \mathcal{J}^{\bullet}\otimes\mathcal{J}^{2N+2-\bullet} \longrightarrow \mathbb{Q}_\ell(-N)$ in $D^b(C,\mathbb{Q}_\ell)$ and so the same-noted perfect duality $\langle,\rangle : H^1(C,\mathcal{J}^{\bullet})\otimes H^1(C,\mathcal{J}^{2N+2-\bullet}) \longrightarrow \mathbb{Q}_\ell(-N-1)$. Clearly both $\mathcal{J}^{\bullet}$ and $\langle,\rangle$ depend only on X (and not on X_C).

Lemma 1.1.1. We have $H^{\bullet}_{!*}(X,\mathbb{Q}_\ell(*)) = H^1(C,\mathcal{J}^{\bullet}(*))$ and $\langle,\rangle$ coincides with the duality on $H_{!*}$ induced by the pairing between $H_c(X_U)$ and $H(X_U)$.

Proof. I'll prove only the first statement since the second is immediate. Since U is affine, the Leray spectral sequence for π degenerates and reduces to two-step filtrations on $H^{\bullet}_c(X_U)$ and $H^{\bullet}(X_U)$ with factors $Gr_{-2}(H^{\bullet}_c) = H^2_c(U, R^{\bullet-2}\pi_{U*}(\mathbb{Q}_\ell))$, $Gr_{-1}(H^{\bullet}_c) = H^1_c(U, R^{\bullet-1}\pi_*(\mathbb{Q}_\ell))$ and $Gr_{-1}(H^{\bullet}) = H^1(U, R^{\bullet-1}\pi_*(\mathbb{Q}_\ell))$, $Gr_0(H^{\bullet}) = H^0(U, R\,\pi^{\bullet}_*(\mathbb{Q}_\ell))$. So $H^{\bullet}_{!*}(X,\mathbb{Q}_\ell) = \mathrm{Im}\,(H^1_c(U, R^{\bullet-1}\pi_*(\mathbb{Q}_\ell)) \longrightarrow H^1(U, R^{\bullet-1}\pi_*(\mathbb{Q}_\ell))) = H^1(C,\mathcal{J}^{\bullet})$, q.e.d. ■

For a moment put $H^{\bullet}(X,\mathbb{Q}_\ell) := \varinjlim H^{\bullet}(X_U,\mathbb{Q}_\ell) = \varinjlim H^{\bullet}(U, R\pi_*(\mathbb{Q}_\ell))$, $H^{\bullet}(X,\mathbb{Q}_\ell)^0 := \mathrm{Ker}(H^{\bullet}(X,\mathbb{Q}_\ell) \longrightarrow H^{\bullet}(X_{\bar{\eta}},\mathbb{Q}_\ell)) = \varinjlim H^1(U,R^{\bullet-1}\pi_*(\mathbb{Q}_\ell))$. Clearly $H^{\bullet}_{!*}(X)\subset H^{\bullet}(X)^0\subset H^{\bullet}(X)$ and this groups depend on X only. Now consider the cycles on X; we have the class map $cl : CH^{\bullet}(X) \longrightarrow H^{2\bullet}(X,\mathbb{Q}_\ell(\bullet))$, put $CH^{\bullet}(X)^0 := cl^{-1}(H^{2\bullet}(X,\mathbb{Q}_\ell(\bullet))^0)$ to be the subgroup of cycles whose intersection with generic geometric fiber is homologous to zero.

Key-lemma 1.1.2. One has $cl(CH^{\bullet}(X)^0)\subset H^{2\bullet}_{!*}(X,\mathbb{Q}_\ell(\bullet))$.

Proof. The problem is local round the points of C (and exists only at points of bad reduction). So let η_v be the generic point of a henselisation of C at some closed point, $\bar{\eta}_v$ be a separable closure of η_v, and $X_{\eta_v} = X\times\eta_v$, $X_{\bar{\eta}_v} = X\times\bar{\eta}_v$. We have to show that whenever $a\in CH^{\bullet}(X_{\eta_v})$ is a cycle s.t. $cl\,a\in H^{2\bullet}(X_{\eta_v},\mathbb{Q}_\ell(\bullet))^0 :=$ $\mathrm{Ker}(H^{2\bullet}(X_{\eta_v},\mathbb{Q}_\ell(\bullet)) \longrightarrow H^2(X_{\bar{\eta}_v},\mathbb{Q}_\ell(\bullet)))$ then $cl\,a$ is zero. But $H^{2\bullet}(X_{\eta_v},\mathbb{Q}_\ell(\bullet))^0 = H^{2\bullet-1}(X_{\bar{\eta}_v},\mathbb{Q}_\ell(\bullet-1))_{\mathrm{Gal}\,\bar{\eta}_v/\eta_v}$. According to [9] th.1.8.4. the weights on this group are ≥ 1. Since the class of an algebraic cycle has weight zero, we are done. ■

Now, by lemma, we may define the height pairing $\langle,\rangle: CH^{i}(X)^{o} \times CH^{N+1-i}(X)^{o} \longrightarrow \mathbb{Q}_{\ell}$ by means of class map followed by the duality between $H^{\cdot}_{!*}$.

One may conjecture that this pairing is $\mathbb{Q}$-valued and independent of ℓ ; of course this is obviously true in char 0, and also true in any char if $i = 1$ (for the discussion of this see §2). The variant of standard conjectures claims that $\overline{CH}^{\cdot}(X)^{o} := \mathrm{Im}\,(CH^{\cdot}(X)^{o} \otimes \mathbb{Q} \longrightarrow H^{2\cdot}_{!*}(X, \mathbb{Q}_{\ell}(\cdot)))$ should be finite-dimensional $\mathbb{Q}$-vector space, that this groups satisfy hard Lefschetz, the form $\langle,\rangle$ should be non-degenerate on them and satisfy Hodge-index theorem. If k = algebraic closure of finite field, then one should have $CH^{\cdot}(X)^{o} \otimes \mathbb{Q} = \overline{CH}^{\cdot}(X)^{o}$.

Problem. Consider the case $\dim \eta/k > 1$.

1.2. The both types of pairings - the intersection and the height one - are related as follows. Suppose that we have $\pi : X_{C} \longrightarrow C$ s.t. X_{C} is regular projective with generic fiber X. Define $CH^{\cdot}(X_{C})^{o} := \mathrm{Ker}(CH^{\cdot}(X_{C}) \longrightarrow H^{0}(C, R^{2\cdot}\pi_{*}(\mathbb{Q}_{\ell}(\cdot)))$ to be the subgroup of cycles whose intersection with any geometric fiber of π is homologous to zero. The restriction map $CH^{\cdot}(X_{C}) \longrightarrow CH^{\cdot}(X)$, $a_{C} \longmapsto a$ maps $CH^{\cdot}(X_{C})^{o}$ into $CH^{\cdot}(X)^{o}$ and we have an obvious

Lemma 1.2.1. For any $a_{1C} \in CH^{\cdot}(X_{C})^{o}$, $a_{2C} \in CH^{N+1-\cdot}(X_{C})^{o}$ one has $(a_{1C}, a_{2C}) = \langle a_{1}, a_{2}\rangle$ ∎

One may suppose that the image of $CH(X\,)^{o}$ under the restriction map coincides with $\overline{CH}^{\cdot}(X)^{o}$; see §2 for details.

§2. Local indices over non-archimedean places

In this § we'll define the local pairing between cycles, and will show how to decompose the global pairings of the previous § into the sum of local ones.

In what follows C_{v} will be any strictly henselian trait with the generic point η_{v}, the special point s_{v}, and an algebraic generic geometric point $\bar{\eta}_{v}$; X_{v} will be a smooth projective scheme over η_{v} of dimension N, and $X_{\bar{v}} := X_{v} \times_{\eta_{v}} \bar{\eta}_{v}$ will be its geometric fiber; ℓ will be a prime $\neq$ char s.

2.0. Let me start with the local intersection pairing. Let $X_{C_{v}}$ be a regular projective scheme over C_{v} with the generic fiber X_{v} and the special one X_{s}. If $a_{C_{v}}$ is any cycle on $X_{C_{v}}$ of codimension d then one has its classes $\widetilde{cl}\,(a_{C_{v}})$ in cohomology groups with supports:

the universal one $\widetilde{cl}_M(a_{C_v}) \in H^{2d}_{M\ \mathrm{supp}\,a_{C_v}}(X_{C_v}, \mathbb{Q}(d)) =$ $H^{2d}_M(X_{C_v}, X_{C_v} \setminus \mathrm{supp}\,a_{C_v}, \mathbb{Q}(d))$ and the ℓ-adic one $\widetilde{cl}_{\mathbb{Q}_\ell}(a_{C_v}) \in$ $H^{2d}_{\mathrm{supp}\,a_{C_v}}(X_{C_v}, \mathbb{Q}_\ell(d))$; clearly $cl_M \longmapsto cl_{\mathbb{Q}_\ell}$ under canonical map (see [18], [3]).

Now let $a_{1C_v} \in Z^{d_1}(X_{C_v})$, $a_{2C_v} \in Z^{d_2}(X_{C_v})$ be two cycles on X_{C_v} of supports Y_{1C_v}, Y_{2C_v} respectively such that $d_1 + d_2 = N + 1$ and $Y_{1v} \cap Y_{2v} := Y_{1C_v} \cap Y_{2C_v} \cap X_v = \emptyset$. Define the intersection index $(a_{1C_v}, a_{2C_v})_v \in \mathbb{Q}$ to be the image of $\widetilde{cl}_M(a_{1C_v}) \cup \widetilde{cl}_M(a_{2C_v}) \in$ $H^{2N+2}_{M\ Y_1 \cap Y_2}(X_{C_v}, \mathbb{Q}(N+1))$ by $H^{2N+2}_{M\ X_s}(X_{C_v}, \mathbb{Q}(N+1)) \xrightarrow{\mathrm{Tr}_\pi} H^1_{M\ s_v}(C_v, \mathbb{Q}(1)) = \mathbb{Q}$. We may replace here H_M by ℓ-adic cohomology; since $H_M \longrightarrow H_{\mathbb{Q}_\ell}$ commutes with any canonical map, we'll get the same answer.

If we are in a global geometric situation 1.2, then for any closed point s_v of C we may consider the henselisation C_v of C at s_v and thus get our local situation. If a_{1C}, a_{2C} are cycles on X_C s.t. a_1, a_2 have disjoint supports on X, then for any s_v we get local intersection index $(a_{1C}, a_{2C})_v$ and clearly one has

Lemma 2.0.1. $(a_{1C}, a_{2C}) = \sum (a_{1C}, a_{2C})_v$ ∎

2.1. Now let us consider the local components for height pairing. Let a_1, a_2, $a_i \in Z^{d_i}(X_v)$, be a cycles on X_v; put $Y_i = \mathrm{supp}\, a_i$ $U_i = X_v - Y_i$. Suppose that $d_1 + d_2 = N+1$, $Y_1 \cap Y_2 = \emptyset$ and both $cl(a_i) \in H^{2d_i}(X_v, \mathbb{Q}_\ell(d_i))$ are zero. In this situation one has link index $\langle a_1, a_2 \rangle_v \in \mathbb{Q}_\ell$. The intuitive picture for $\langle , \rangle_v$ is following: from the homotopy point of view η_v is a circle round s_v and X_v is $2N+1$-dimensional topological manifold fibered over this circle, a_i are $2d_i$-codimensional cycles on it that doesn't intersect and homologous to zero; the link index $\langle a_1, a_2 \rangle_v$ is the intersection number of a_1 and the chain that bounds a_2. Here are the number of exact definitions of $\langle a_1, a_2 \rangle_v$; the proof of their equivalence is left to the reader. In what follows $\alpha_i \in H^{2d_i-1}(U_i, \mathbb{Q}_\ell(d_i))$ are classes that bound a_i : this means that $\alpha_i \longmapsto \widetilde{cl}(a_i)$ under the boundary map $H^{2d_i-1}(U_i) \longrightarrow H^{2d_i}_{Y_i}(X_v)$.

<u>Lemma-definition 2.1.1.</u> The following definitions of $\langle a_1, a_2\rangle_v$ are equivalent.

a) $\langle a_1, a_2\rangle_v$ is the image of $\alpha_1 \cup \widetilde{cl}(a_2)$ by $H^{2N+1}_{Y_2}(U_1, \mathbb{Q}(N+1)) \longrightarrow H^{2N+1}(X_v, \mathbb{Q}_\ell(N+1)) \xrightarrow{Tr} H^1(\eta_v, \mathbb{Q}_\ell(1)) = \mathbb{Q}_\ell$

b) $\langle a_1, a_2\rangle_v$ is the image of $\alpha_1 \cup \alpha_2$ by $H^{2N}(U_1 \cap U_2, \mathbb{Q}_\ell(N+1)) \longrightarrow H^{2N+1}(X_v, \mathbb{Q}_\ell(N+1)) \xrightarrow{Tr} \mathbb{Q}_\ell$; here the first arrow comes from the Myer-Vietoris exact sequence for the covering $U_1 \cup U_2 = X_v$.

c) Choose a projective $\pi: X_{C_v} \longrightarrow C_v$ with the generic fiber X_v, $\beta_1 \in H^{2d_1}_{\pi^{-1}(s_v)\cup Y_1}(X_{C_v}, \mathbb{Q}_\ell(d_1))$ and $\beta_2 \in H^{2d_2-2N}_{\pi^{-1}(s_v)\cup Y_2}(X_{C_v}, R\pi^!\mathbb{Q}_\ell(d_2-N))$ s.t. the restriction of β_i on X_v coincide with $\widetilde{cl}(a_i)$ and the image of β_1 in $H^{2d_1}(X_{C_v}, \mathbb{Q}_\ell(d_1))$ is zero (e.g. you may take β_1 to be the boundary of α_1 in X_{C_v}). Then $\langle a_1, a_2\rangle_v$ is the image of $\beta_1 \cup \beta_2$ by $H^2_{\pi^{-1}(s_v)}(X_{C_v}, R\pi^!\mathbb{Q}_\ell(1)) \longrightarrow H^2_{s_v}(C_v, \mathbb{Q}_\ell(1)) = \mathbb{Q}_\ell$ ∎

This way we get link pairing between the cycles with disjoint supports; clearly 2.1.1. b) shows that this pairing is bilinear and symmetric. Now we are going to show that it behaves well under the action of correspondences. To do this first let us see that the above definition may be easily generalised to the case of arbitrary many cycles. Namely let $a_1, \ldots, a_n$ be cycles of codimensions d_i on X_v s.t. $\sum d_i = N+1$ and $\bigcap Y_i = \emptyset$ (here Y_i = supp a_i, $U_i = X_v \setminus Y_i$) assume that at least one of them is homologous to zero in X. Choose a non-empty subset $S \subset \{1,\ldots,n\}$ s.t. for any $j \in S$ the cycle a_j is homologous to zero, and for $a_j \in S$ some $\alpha_j \in H^{2d_j-1}(U_j, \mathbb{Q}_\ell(d_j))$ that bounds a_j. Define $\langle a_1,\ldots,a_n\rangle_v$ to be $\mathrm{Tr}\,[\partial(\bigcup_{j\in S}\alpha_j) \cup (\bigcup_{j\notin S} a_j)]$; here $\partial: H^\cdot(\bigcap_{j\in S} U_j) \longrightarrow H^{\cdot+\#S-1}(\bigcup_{j\in S} U_j)$ is the differential in the spectral sequence of the covering $\{U_j\}_{j\in S}$ of $\bigcup_{j\in S} U_j$, and

We have the following easy generalisation of 2.1.1:

<u>Lemma 2.1.2.</u> If at least two of a_i are homologous to zero, then $\langle a_1,\ldots,a_n\rangle_v$ depends on $a_1,\ldots,a_n$ only (and not on the choice of S and α_j) ∎

Clearly in this case the pairing is also bilinear and symmetric.

Now we may look at correspondences. Consider two schemes X_{1v}, X_{2v} of dimensions N_1, N_2, a cycles a_i of codimensions d_i on

X_{iv} and a cycle b of codimension d_3 on $X_{1v} \times X_{2v}$. Assume that both a_i are homologous to zero, $d_1 + d_2 + d_3 = N_1 + N_2 + 1$ and $p_1^{-1}(\text{supp } a_1) \cap \text{supp } b \cap p_2^{-1}(\text{supp } a_2) = \emptyset$ (here p_i are projections $X_{1v} \times X_{2v} \longrightarrow X_{iv}$). Then 2.1.1 implies

Lemma 2.1.3. $\langle p_1^*(a_1), b, p_2^*(a_2)\rangle_v = \langle b(a_1), a_2\rangle_v = \langle a_1, b(a_2)\rangle_v$ ■

Here $b(a_1)$ is the image of a_1 under the action of correspondence b. (Note that if $p_1^*(a_1)$ and b doesn't intersect properly, then the cycle $b(a_1)$ is not defined in a unique way, but it has correctly defined class in the cohomology group with support in $p_2(\text{supp } b \cap p_1^{-1}(\text{supp } a_1))$ that suffice for our purposes.)

The lemma shows for example that the computation of $\langle a_1, a_2\rangle_v$ in case when one of a_i is algebraically equivalent to zero, may be reduced to the computation of link index between a zero cycles on a curve.

Our pairing behaves in a usual way under the change of the base field. Namely let $\eta_{v'}/\eta_v$ be a degree n extension, X_v is a scheme over η_v and $Y_{v'}$ is one over $\eta_{v'}$. Put $X_{v'} = X_v \times_{\eta_v} \eta_{v'}$ and Y_v be $Y_{v'}$ considered as a scheme over η_v. We have the obvious arrows $Z^{\cdot}(X_v) \hookrightarrow Z^{\cdot}(X_{v'})$, $Z^{\cdot}(Y_{v'}) = Z^{\cdot}(Y_v)$ and the following holds (recall that k is separably closed):

Lemma 2.1.4. Let a_i be cycles on X_v and b_i the ones on $Y_{v'}$. Then $\langle a_1, a_2\rangle_v = 1/n \langle a_1, a_2\rangle_{v'}$, $\langle b_1, b_2\rangle_{v'} = \langle b_1, b_2\rangle_v$ ■

In particular by means of first formula we may define the height pairing between cycles on $X_{\bar{\eta}_v}$; this pairing is clearly Galois-invariant.

Let us relate local link pairings with the global height pairing of §1. Assume that we are in global geometric situation of 1.1. As in 2.0 to each closed point on C corresponds the local picture over corresponding local field η_v. Let a_1, a_2 be cycles on X of codimensions d_i s.t. $d_1 + d_2 = N+1$ and $\text{supp} a_1 \cap \text{supp} a_2 = \emptyset$. If a_i belongs to $CH^{d_i}(X)^{\circ}$ then 1.12. shows that a_i is homologous to zero on any X_v. So the link index $\langle a_1, a_2\rangle_v$ is defined, and it is easy to see that the definition 2.1.1.c implies the formula (*) from the introduction:

Lemma 2.1.5. We have $\langle a_1, a_2\rangle = \sum \langle a_1, a_2\rangle_v$ (sum over all closed points of C). ■

Finally let me compare the local pairings from 2.0 and 2.1. Suppose that we are in a situation 2.0. Put $CH^{\cdot}(X_{C_v})^{\circ} := \text{Ker}(CH^{\cdot}(X_{C_v}) \longrightarrow H^{2\cdot}(X_{C_v}, \mathbb{Q}_\ell(\cdot)))$. Let a_{1C_v}, a_{2C_v} be cycles on X_{C_v} as in 2.0; assume that both a_i are homologous to zero on X_v (here $a_i :=$

$a_{iC_v} \cap X_v$). Then 2.1.1c implies.

Lemma 2.1.6. If one of a_{iC_v} also belongs to $CH^{\cdot}(X_{C_v})^{\circ}$, then $(a_{1C_v}, a_{2C_v})_v = \langle a_1, a_2 \rangle_v$. ■

This is local analog of 1.2.1. In particular in this situation $\langle a_1, a_2 \rangle_v \in \mathbb{Q}$ and doesn't depends on the choice of $\ell \neq$ char k.

2.2. Here are some conjectures about local pairings. Put $CH^{\cdot}(X_v)^{\circ} := \mathrm{Ker}\,(CH^{\cdot}(X_v) \longrightarrow H^{2\cdot}(X_{\bar{\eta}_v}, \mathbb{Q}_\ell(\cdot)))$

Conjecture 2.2.1. One has $CH^i(X_v)^{\circ} = \mathrm{Ker}\,(CH^i(X_v) \to H^{2i}(X_v, \mathbb{Q}_\ell(i)))$ i.e. any cycle whose intersection with a generic geometric fiber is homologous to zero is homologous to zero on X_v ■

Lemma 2.2.2. This conjecture is true in following cases:

a. Good reduction case.

b. For cycles algebraically equivalent to zero (in particular the case $i = 1$ and $i = N$).

c. Geometric case.

Proof. a is obvious; c was proved in 1.1.2; b may be reduced using correspondences to the case of zero cycles on curve, where it follows, say, from 2.2.6b. ■

Remark. The conjecture would follow if one knows the information on weights on $H^{\cdot}(X_{\bar{\eta}_v}, \mathbb{Q}_\ell)$ similar to those one has in geometric case. If η_v is a p-adic local field, then the thing we need is the usual conjecture on poles of local L-multiples.

Conjecture 2.2.3. The local link pairing is Q-valued and independent of $\ell \neq$ char k.

Lemma 2.2.4. This conjecture is true:

a. In good reduction case.

b. In case char $s = 0$.

c. When one of the cycles is algebraically equivalent to zero (in particular for the pairing between divisors and zero cycles).

Proof. b is obvious; a and c follow from 2.2.6, for c use correspondences to reduce to 2.2.6b. ■

Suppose that we are in a situation 2.0. Clearly the restriction arrow $CH^{\cdot}(X_{C_v}) \to CH^{\cdot}(X_v)$ maps $CH^{\cdot}(X_{C_v})^{\circ}$ into $CH^{\cdot}(X_v)^{\circ}$.

Conjecture 2.2.5. $CH^{\cdot}(X_C)^{\circ} \twoheadrightarrow CH^{\cdot}(X_v)^{\circ}$. ■

Clearly 2.2.5 implies both 2.2.1 and 2.2.3.

Lemma 2.2.6. This conjecture is true

a. In good reduction case.

b. If X_v is a curve.

Proof. a is obvious; b follows from the well-known fact that the intersection matrix between components of the special fiber is almost negative-definite. ∎

The property, 2.2.6b together with 2.1.6, shows that in case of zero cycles on a curve our link index coincides with Néron's local symbol; using the correspondences one may see that the same is true for the pairing between divisors and zero cycles in arbitrary dimensions (or you may directly verify that $\langle , \rangle_v$ satisfies the conditions of Néron's theorem [12].) For the review of heights on curves together with the study of important examples see [7].

We wish to use the formula 2.1.5. for varieties over number field as the definition of left-hand side, just as Neron did in the case of divisors and zero cycles. The only thing that remains to be defined is the link index for archimedean local fields.

§3. Local indices over $\mathbb{C}$ and $\mathbb{R}$

In this § our base field will be $\mathbb{C}$ or $\mathbb{R}$

3.0. We have the following dictionary (see [2] for notations and details)

Non-archimedean case (η_v is the spectrum of a p-adic local field)	Archimedean case (η_v = Spec $\mathbb{C}$ or Spec $\mathbb{R}$)
Gal $\bar{\eta}_v/\eta_v$-modules	R-mixed Hodge structures
Étale cohomology groups $H^\bullet(X_{\bar{v}}, \mathbb{Q}_\ell(*))$ of the geometric fiber with the Galois action	Ordinary cohomology groups $H^\bullet(X_v(\mathbb{C}), \mathbb{R}(*))$ with Deligne's mixed Hodge structure
Étale cohomology groups $H^\bullet(X_v, \mathbb{Q}_\ell(*))$	Absolute Hodge-Deligne cohomology groups $H^\bullet_{\mathcal{H}}(X_v, \mathbb{R}(*))$
Canonical arrow $H^\bullet(X_v, \mathbb{Q}_\ell(*)) \longrightarrow H^\bullet(X_{\bar{v}}, \mathbb{Q}_\ell(*))$	Canonical arrow $H^\bullet_{\mathcal{H}}(X_v, \mathbb{R}(*)) \longrightarrow H^\bullet_{\mathcal{B}}(X_v, \mathbb{R}(*)) \subset H^\bullet(X_v(\mathbb{C}), \mathbb{R}(*))$

Note that the canonical arrow $H^{2i}(X_v, \mathbb{R}(i)) \longrightarrow H^{2i}(X_v, \mathbb{R}(i))$ is injective; so whenever we have an algebraic cycle homologous to zero as topological cycle it has zero class in $H_{\mathcal{H}}$. This shows that the analog of conjecture 2.2.1 is obviously true in archimedean case (this is one of the sides of the fact that smooth proper varieties have "good reduction" at archimedean places from the Hodge-theoretic point of view). We may translate the definition 2.1.1 a or b using the above dictionary to define the link index in our situation. Namely, this way for any two cycles a_1, a_2 on X_v of codimensions d_1, d_2 s.t. $d_1 + d_2 = N+1$, the supports of a_i doesn't intersect and both a_i are homologous to zero, we get the link index $\langle a_1, a_2 \rangle_v \in \mathbb{R}$. The lemmas 2.1.1 (a $\Longleftrightarrow$ b), 2.1.2. - 2.1.4. remain true, together with their proofs, in our situation.

3.1. Now let me sketch the definition of the analog of intersection index à la Arakelov; since this will not be needed in the main body of the paper I'll omit the details. The similar construction was found by H. Gillet and Ch. Soulé [10]. The role of the model X_{C_v} of X_v over the ring of integers of p-adic field plays now (in our archimedean situation) the Kähler metrics ω on X_v (see [15]). Let us define the $\mathbb{R}$-vector space $\mathcal{Z}^i(X_v, \omega)$ - an analog of the group $\mathcal{Z}^i(X_{C_v})$. In what follows for $c \in H^{\cdot}(X_v, \mathbb{C})$ we denote by $\tilde{c}$ its ω-harmonic representative, and for a cycle $z \in \mathcal{Z}^{\cdot}(X_v) \otimes \mathbb{R}$ let δ_z be its δ-current. Say that a current α is i-Green current if it is $\mathbb{R}(i-1)$-valued of type (i-1, i-1) and for certain (unique) $\alpha_f \in \mathcal{Z}^i(X_v) \otimes \mathbb{R}$ one has $\partial\bar\partial\alpha = \delta_{\alpha_f} - \widetilde{cl\,\alpha_f}$; say that α is regular if it is of C^∞-class off the support of α_f. A (regular) Green current α is trivial if $\alpha = \partial\nu \pm \bar\partial\nu$ for certain (C^∞-class) current ν. Put $\mathcal{Z}^i(X_v, \omega)$ to be the factorspace of i-Green currents modulo trivial ones (you may take all Green currents or regular ones only - this doesn't changes $\mathcal{Z}^i(X_v, \omega)$). For $\alpha \in \mathcal{Z}^i(X_v, \omega)$ let $\alpha_\infty \in H^{i-1,i-1}(X_v, \mathbb{R}(i-1)) (:= H^{i-1,i-1}(X_v) \cap H^{2i-2}_B(X_v, \mathbb{R}(i-1))$ be the (only) class such that for any ω-harmonic 2(N-i+1)-form ν one has $\int_{X_v} \alpha\wedge\nu = \int_{X_v} \alpha_\infty \wedge \nu$. One may see that the arrow $\mathcal{Z}^i(X_v, \omega) \longrightarrow H^{i-1,i-1}(X_v, \mathbb{R}(i-1)) \oplus \mathcal{Z}^i(X_v) \otimes \mathbb{R}$, $\alpha \longmapsto (\alpha_\infty, \alpha_f)$, is isomorphism. This direct sum decomposition of $\mathcal{Z}^i(X_v, \omega)$ is analogous to the decomposition of $\mathcal{Z}^i(X_C)$ into the sum of the group of cycles supported in the special fiber and the group of cycles that intersect the special fiber properly.

Here is an important example of Green current; we'll need it in n° 4.1. For a scheme S put $C^i(S) = \bigoplus_{\substack{\eta \in S \\ \mathrm{Codim}\,\eta = i}} \mathcal{O}^*(\eta)$: an element $\varphi \in C^i(S)$

is a finite number of invertible meromorphic functions φ_η defined at the generic points of codimension i subschemes η (this is $E^{-i-1,i}$-term of Quillen's spectral sequence for $K'_\cdot(S)$); we have the exact sequence $C^{i-1}(S) \xrightarrow{\text{div}} Z^i(S) \longrightarrow CH^i(S) \longrightarrow 0$. Now in our situation for $\varphi \in C^{i-1}(X_v)$ we have Green current $\overline{\text{div}}(\varphi)$ defined by $\int \overline{\text{div}}(\varphi) \wedge \nu = \sum_\eta \int_\eta \log|\varphi_\eta| \cdot \nu|_\eta$; clearly $(\overline{\text{div}}\, \varphi)_f = \text{div}\, \varphi$. This way we get an arrow $\overline{\text{div}} : C^{i-1}(X_v) \longrightarrow \mathcal{Z}^i(X_v, \omega)$.

Now let me define the intersection index. Let $a_1 \in \mathcal{Z}^{d_1}(X_v, \omega)$, $a_2 \in \mathcal{Z}^{d_2}(X_v, \omega)$ be elements such that $d_1 + d_2 = N+1$ and the supports of a_{if} are disjoint. Put $(a_1, a_2)_v = \int_{X_v} (\alpha_1 \wedge \delta_{a_{2f}} + \delta_{a_{1f}} \wedge \widetilde{a_{2\infty}}) \in \mathbb{R}$ (here α_1 is a regular Green current that represents a_1). One may see that $(a_1, a_2)_v$ so defined is independent of the choice of α_1, that $(,)_v$ is symmetric, and in case when both a_{if} are homologous to zero one has $(a_1, a_2)_v = \langle a_{1f}, a_{2f}\rangle_v$. In $N = 1$ case this construction coincides with the original Arakelov one [1] (see also [13]).

§4. Height pairing over number fields

Let us return to the global situation. In this § our base field K will be a finite extension of $\mathbb{Q}$; let $\bar{K}$ be an algebraic closure of K, $\eta = \text{Spec}\, K$, $\bar{\eta} = \text{Spec}\, \bar{K}$. For any place v of K denote by K_v the corresponding local field, and by η_v the spectrum of K_v^{nr} = maximal non-ramified extension of K_v. Define the real number $r(v)$ to be log (number of elements in the residese field of K_v) for non-archimedean v, and $r(v) = 1$ for $K_v = \mathbb{R}$, $r(v) = 2$ for $K_v = \mathbb{C}$.

4.0. Let $X = X_K$ be a N-dimensional smooth projective variety over K; put $CH^i(X)^\circ := \text{Ker}(CH^i(X) \longrightarrow H^{2i}(X_{\bar{\eta}}, \mathbb{Q}_\ell(i)))$, $\overline{CH}^i(X) := CH^i(X)/CH^i(X)^\circ \subset H^{2i}(X_{\bar{\eta}}, \mathbb{Q}_\ell(i))$. We'll assume that for any non-archimedean v the η_v-scheme $X_v = X \times \eta_v$ satisfies the conjectures 2.2.1 and 2.2.3.

Remark 4.0.1. a. If you do not wish to assume this, then change the notations and put $CH^i(X)^\circ := \text{Im}(CH^i(X_{\mathbb{Z}})^\circ \longrightarrow CH^i(X))$; here $X_{\mathbb{Z}}$ is certain regular model of X over the ring of integers of K (we suppose that it exists) and $CH^i(X_{\mathbb{Z}})^o \subset CH^i(X_{\mathbb{Z}})$ is a subgroup of cycles whose intersection with the fiber of $X_{\mathbb{Z}}$ over any prime is homologous to zero. It is easy to see that $CH^i(X)^o$ so defined is independent of choice of particular model $X_{\mathbb{Z}}$. According to conj. 2.2.5. this $CH^i(X)^o$ should coincide with the above one.

b. The same thing happens with the definition of groups $H^i_M(X, \mathbb{Q}(j))_{\mathbb{Z}}$, $i < 2j$ (see [2]); namely we may define them in two ways. The first one: assume the existence of $X_{\mathbb{Z}}$ and take the definition [2] (8.3). The second one: put $H^i_M(X, \mathbb{Q}(j))_{\mathbb{Z}} := \bigcap_{p \text{ prime}} \mathrm{Ker}(H^i_M(X,\mathbb{Q}(j)) \longrightarrow H^1(\mathrm{Gal}\, \bar{\mathbb{Q}}_p/\mathbb{Q}_p^{nr}, H^{i-1}(X_{\bar{\mathbb{Q}}_p}, \mathbb{Q}_\ell(j)))$; here $\ell \neq p$ and the arrow under the bracket is canonical map $H^i_M(X, \mathbb{Q}(j)) \longrightarrow \mathrm{Ker}(H^i(X_{\mathbb{Q}_p^{nr}}, \mathbb{Q}_\ell(j)) \longrightarrow H^i(X_{\bar{\mathbb{Q}}_p}, \mathbb{Q}_\ell(j))) = H^1(\mathrm{Gal}\, \bar{\mathbb{Q}}_p/\mathbb{Q}_p^{nr}, H^{i-1}(X_{\bar{\mathbb{Q}}_p}, \mathbb{Q}_\ell(j)))$. According to [2] conj. 8.3.3 these two definitions should give the same $H_M(X, \mathbb{Q}(j))_{\mathbb{Z}}$. ∎

Let a_1, a_2 be a cycles on X of codimensions d_1, d_2 s.t. $d_1 + d_2 = N+1$, both $cl(a_i) \in CH^{d_i}(X)^o$ and supports of a_i doesn't intersect. For any place v we get the corresponding objects on X_v. This way we get the rational numbers $\langle a_1, a_2\rangle_v$ for non-archimedean v (according to 2.2.1. and 2.2.3, or, if you prefer 4.0.1, according to 2.1.6) and real numbers $\langle a_1, a_2\rangle_v$ for archimedean v. Note, that all but finitely many $\langle a_1, a_2\rangle_v$ are zero. Put $\langle a_1, a_2\rangle := \sum_v r(v)\langle a_1, a_2\rangle_v \in \mathbb{R}$.

<u>Lemma-definition 4.0.2</u>. There exists unique bilinear pairing $\langle , \rangle : CH^{\cdot}(X)^o \times CH^{N+1-\cdot}(X)^o \longrightarrow \mathbb{R}$ s.t. $\langle cl\, a_1, cl\, a_2\rangle = \langle a_1, a_2\rangle$ if supports of a_i doesn't intersect.

<u>Proof</u>. By moving lemma any pair of classes in $CH^{\cdot}$, $CH^{N+1-\cdot}$ may be represented by non-intersecting cycles, so it remains to prove that $\langle a_1, a_2\rangle$ depends only on rational-equivalence class of a_i (the bilinearity is obvious). This reduces to the proof that $\langle a_1, a_2\rangle = 0$ whenever a_1 is rationally equivalent to zero. To see this choose a correspondence b between P^1 and X s.t. $b(0) = a_1$, $b(\infty) = 0$. If the supports of cycles $p_1^*((0) - (\infty))$, b, $p_2^*(a_2)$ intersect by an empty set (here $p_i : P^1 \times X \longrightarrow P^1, X$ are projections), we may apply 2.1.3. to see that $\langle a_1, a_2\rangle = \langle (0) - (\infty), {}^t b(a_2)\rangle$ and we are done by standard arguments with product formula. If not (this means that $\mathrm{supp}\, a_2 \cap Z \neq \emptyset$, where $Z := (\mathrm{supp}\, b \cap p_1^{-1}(0)) \cup (\mathrm{supp}\, b \cap p_1^{-1}(\infty))$ $\supset \mathrm{supp}\, a_1$ is a d_1-codimensional subset in X), then you can move a_2 a little (in the same rational equivalence class) to get a_2' s.t. $\mathrm{supp}\, a_2' \cap Z = \emptyset$. Then, by above, $\langle a_1, a_2'\rangle = 0$ and by the same argu-

ments $\langle a_1, a_2\rangle = \langle a_1, a_2'\rangle$, q.e.d. ■

This is the desired height-pairing. Here are its first properties, that follow directly from the corresponding local properties plus moving lemma.

4.0.3. Let Y be another variety, and $b \in CH^d(X \times Y)$ be a correspondence. Then for $a_1 \in CH^{i_1}(X)^o$, $a_2 \in CH^{i_2}(Y)^o$, $i_1 + i_2 + d = \dim X + \dim Y + 1$, one has $\langle b(a_1), a_2\rangle = \langle a_1, {}^tb(a_2)\rangle$.

4.0.4. The height pairing between divisors and zero cycles coincides with the one of Néron [17], [13].

4.0.5. The pairing $\langle,\rangle$ doesn't depend on the choice of the base field: if you'll consider X as a scheme over $\mathbb{Q}$ this will give the same pairing between cycles.

4.0.6. Let K'/K be an extension of degree n. Then $CH(X_K)^o \otimes \mathbb{Q} \subset CH(X_{K'})^o \otimes \mathbb{Q}$ and for $a_i \in CH(X_K)^o$ one has $\langle a_1, a_2\rangle_K = 1/n\langle a_1, a_2\rangle_{K'}$ By means of this formula you may define the height pairing on $CH(X_{\bar{K}})^o$; this pairing is Galois-invariant.

The following lemma is a particular case of the conjecture 5.6.

Lemma 4.0.7. Let $a \in CH^i(X)^o$ be a cycle such that a is algebraically equivalent to zero and, under some point $K \hookrightarrow \mathbb{C}$, the image of a under the Abel-Jacobi map $CH^i(X_{\mathbb{C}}) \to \mathcal{J}^i(X_{\mathbb{C}})$ is zero (here $\mathcal{J}^i$ is i-th Griffiths Jacobian). Then a lies in the kernel of height pairing.

Proof. Consider a correspondence b between a curve C and X such that $a = b(a_c)$ for certain $a_c \in CH^1(C)^o = J(C)(K)$. Let $A \subset J(C)$ be the abelian subvariety generated by ${}^tb(CH^{N+1-i}(X)^o)$. According to 4.0.3 to prove our lemma it suffice to show that a_c lies in an orthogonal complement to A (under the usual self-duality on $J(C)$); clearly we may consider the situation over $\mathbb{C}$. But there we have the arrow ${}^tb : J^{N+1-i}(X_{\mathbb{C}}) \to J(C_{\mathbb{C}})$ between intermediate Jacobians that commutes with Abel-Jacobi map, and so ${}^tbJ^{N+1-i}(X_{\mathbb{C}}) \supset A_{\mathbb{C}}$. The condition of lemma claims that a_c lies in the orthogonal complement to ${}^tbJ^{N+1-i}(X_{\mathbb{C}})$ and so to A. We are done. ■

Remark 4.0.8. In fact the height pairing between cycles algebraically equivalent to zero may be canonically reduced to the Neron-Tate pairing. Namely for fixed $K \hookrightarrow \mathbb{C}$ consider the Abel-Jacobi images of cycles on $X_{\mathbb{C}}$ algebraically equivalent to zero. They form an abelian subvariety A^i in $J^i(X_{\mathbb{C}})$; it is easy to see that A^i is defined over K. The duality between $J^i(X_{\mathbb{C}})$ and $J^{N+1-i}(X_{\mathbb{C}})$ defines the pairing between A^i and A^{N+1-i}. This pairing gives rise to the Neron-Tate pairing between $A^i(K)$ and $A^{N+1-i}(K)$ that may be identified with the height pairing by means of Abel-Jacobi map.

4.1. In the rest of the § I'll sketch a construction of the Arakelov-type global intersection pairing on A-varieties. The similar construction was found by H. Gillet and Ch. Soulé; I refer to their paper [10] for details. Let (X, ω) be an A-variety [15], i.e. a regular scheme X projective over Spec $\mathbb{Z}$ together with Kähler metrics ω on $X_{\mathbb{R}}$. Let $\mathcal{Z}^i(X_{\mathbb{Z}}, \omega) \subset \mathcal{Z}^i(X_{\mathbb{R}}, \omega) \oplus \mathcal{Z}^i(X_{\mathbb{Z}})$ be the subgroup of elements (a, b) such that $a_f = b_{\mathbb{R}}$. We have the natural map $\overline{\mathrm{div}} : C^{i-1}(X_{\mathbb{Z}}) \longrightarrow \mathcal{Z}^i(X_{\mathbb{Z}}, \omega)$; denote its cokernel by $CH^i(X_{\mathbb{Z}}, \omega)$. There is an obvious exact sequence $C^{i-1}(X_{\mathbb{Z}})^o \xrightarrow{\overline{\mathrm{div}}^o} H^{i-1,i-1}(X_{\mathbb{R}}, \mathbb{R}(i-1)) \longrightarrow CH^i(X_{\mathbb{Z}}, \omega) \xrightarrow{\pi} CH^i(X_{\mathbb{Z}}) \longrightarrow 0$; here $C^{i-1}(X_{\mathbb{Z}})^o := \mathrm{Ker}(\mathrm{div}: C^{i-1}(X_{\mathbb{Z}}) \longrightarrow \mathcal{Z}^i(X_{\mathbb{Z}}))$. Put $CH^i(X_{\mathbb{Z}}, \omega)_\infty := \mathrm{Ker}\ \pi = \mathrm{Coker}\ \overline{\mathrm{div}}^o$. Note that the map $\overline{\mathrm{div}}^o$ coincides with the composition $C^{i-1}(X_{\mathbb{Z}})^o \longrightarrow H_M^{2i-1}(X_{\mathbb{Z}}, \mathbb{Q}(i)) \xrightarrow{r} H_{\mathcal{H}}^{2i-1}(X_{\mathbb{R}}, \mathbb{R}(i)) = H^{i-1,i-1}(X_{\mathbb{R}}, \mathbb{R}(i-1))$, where the first arrow comes from Quillen's spectral sequence and r is regulator map (see [2] n 6). According to [2] conj. 8.4.1.b $\overline{\mathrm{div}}^o(C^{i-1}(X_{\mathbb{Z}})^o) + cl_B(CH^{i-1}(X_{\mathbb{Z}}))$ is a $\mathbb{Z}$-lattice in $H^{i-1,i-1}(X_{\mathbb{R}}, \mathbb{R}(i-1))$, so $CH^i(X_{\mathbb{Z}}, \omega)_\infty$ should be equal to the product of $(\mathcal{I}m\ \overline{\mathrm{div}}^o) \otimes \mathbb{R}/\mathcal{I}m\ \overline{\mathrm{div}}^o$ (= maximal compact subgroup) and $\mathbb{R}$-vector space $\overline{CH}^{i-1}(X_{\mathbb{Q}}) \otimes \mathbb{R}$.

Now let $a_1 \in Z^{d_1}(X_{\mathbb{Z}}, \omega)$; $a_2 \in Z^{d_2}(X_{\mathbb{Z}}, \omega)$ be two cycles such that $d_1 + d_2 = N+1$ and supports of a_{if} doesn't intersect on $X_{\mathbb{Q}}$ (or $X_{\mathbb{R}}$). Put $(a_1, a_2) := (a_1, a_2)_\infty + \sum_p (\log p)(a_1, a_2)_p$ (see 2.0, 3.1.). One may see that this formula defines the pairing $(,) : CH^{d_1}(X_{\mathbb{Z}}, \quad) \otimes CH^{d_2}(X_{\mathbb{Z}}, \quad) \longrightarrow \mathbb{R}$. It has the following properties: ① (,) is symmetric ② the image of $\overline{\mathrm{div}}^o(C^{i-1}(X_{\mathbb{Z}})^o)\cdot\mathbb{R}$ in $CH^i(X_{\mathbb{Z}}, \omega)_\infty$ lies in the kernel of (,) ③ if both $\pi(a_i) \in CH^{d_i}(X_{\mathbb{Z}})^o$ (see 4.0.1.), then $(a_1, a_2) = \langle\pi(a_1), \pi(a_2)\rangle$ ④ Let $a_1 \in CH^d(X_{\mathbb{Z}}, \omega)_\infty$ be the image of $\tilde{a}_1 \in H^{i-1,i-1}(X_{\mathbb{R}}, \mathbb{R}(i-1))$; then $(a_1, a_2) = \int_{a_{2f\mathbb{R}}} \tilde{a}_1$. (Note that 2 and 4 imply that for $i \leqslant \frac{N+1}{2}$ the

$\mathbb{R}$-subspaces in $H^{i-1,i-1}$ generated by Im $\widetilde{\operatorname{div}}°$ and $\widetilde{CH}^{i-1}(X_{\mathbb{Q}})$ doesn't intersect; this fits into above conjecture on $CH^i(X_{\mathbb{Z}}, \omega)_\infty$)

In the next § the conjectures will be formulated for the pairing < , > ; but some of them have sense for (,) also. To find such a variants of them is an exercise for the reader.

§ 5. Some conjectures and problems

Let X be a smooth projective N-dimensional scheme over Q. Assume for X the local conjectures 2.2.1, 2.2.3, and also assume the standard conjectures about analytic continuation and functional equation for $L(H^{\cdot}(X), s)$

Conjecture 5.0 (Swinnerton-Dyer). The groups $CH^{\cdot}(X)°$ are finitely generated and rk $CH^i(X)°$ = order of zero of $L(H^{2i-1}(X), s)$ at $s = i$ ■

The only fact I know supporting this conjecture off the B-Sw-D D case (i.e. $i = 1$ case) is Bloch's calculation for the Jocobian of guartic curve, see [4].

Lemma 5.1. (modulo the conjecture 5.0) The canonical arrow $CH^N(X)° \longrightarrow \mathrm{Alb}(X)(\mathbb{Q})$ is isomorphism up to torsion.

Proof. It is clear (e.g. by norm map considerations) that the arrow is epimorphic up to torsion so it suffice to see that the ranks are the same. According to 5.0. we have rk $CH^N(X)°$

$$= \operatorname{ord}_{s=N} L(H^{2N-1}(X), s) = \operatorname{ord}_{s=1} L(H^1(X), s) =$$

rk $CH^1(X)°$ = rk $\mathrm{Pic}°(X)(\mathbb{Q})$ = rk $\mathrm{Alb}(X)(\mathbb{Q})$, since Pic° is isogeneous to Alb and $H^{2N-1}(X)(N-1) \simeq H^1(X)$ ■

In view of Roitman's theorem on torsion zero cycles (see e.g. [7] lecture 5), the statement of 5.1 is equivalent to the following.

Conjecture 5.2. For any smooth projective $\tilde{X}$ over $\bar{\mathbb{Q}}$ the Chow group of zero cycles of degree zero on $\tilde{X}$ coincides with Alb $(X)(\mathbb{Q})$ ■

Note that according to Mumford's theorem ([16], [7] lecture 1) the situation radically changes if there are parameters in the base field (see the discussion below), so it would be very interesting to try to handle 5.2. in any non-trivial particular case. For example, whether every zero-cycle of degree zero on K3 over $\bar{\mathbb{Q}}$ is rationally equivalent to zero?

Conjecture 5.3. a. (hard Lefschetz). Let $L \in CH^1(X)$ be the class of a hyperplane section. Then for $i \leqslant (N+1)/2$ the arrow $L^{N+1-2i}: CH^i(X)^\circ \otimes \mathbb{Q} \to CH^{N+1-i}(X)^\circ \otimes \mathbb{Q}$ is isomorphism

b. Let Y be smooth N-dimensional affine scheme over $\mathbb{Q}$, and, as in the proper case, put $CH^i(Y)^\circ := \mathrm{Ker}\,(CH^i(Y) \to H_B^{2i}(Y \otimes \mathbb{C}, \mathbb{Z}(i)))$. Then one should have $CH^i(Y)^\circ = 0$ for $i > \frac{\dim Y + 1}{2}$. ■

For $a_1, a_2 \in \mathbb{R}$ say that $a_1 \sim a_2$ if they differ by non-zero rational multiple. Consider the height pairing $\langle , \rangle$ between $\mathbb{R}$-vector spaces $CH^i(X)^\circ \otimes \mathbb{R}$ and $CH^{N+1-i}(X)^\circ \otimes \mathbb{R}$. Denote by $\det \langle , \rangle_i$ its determinant in some rational baseses of $CH^i(X)^\circ \otimes \mathbb{Q}$, $CH^{N+1-i}(X)^\circ \otimes \mathbb{Q}$;

clearly $\det \langle , \rangle_i$ is correctly defined equivalence class of real numbers (we assume rk CH (X) is finite, see 5.0) and $\det \langle , \rangle_i \sim \det \langle , \rangle_{N+1-i}$.

Conjecture 5.4. *) a) The height pairing is non-degenerate.

b) $\det \langle , \rangle_i$, multiplied by the determinant of the period matrix for $H^{ri-1}(X)$ (see [21]), is equivalent to the leading coefficient in Taylor serie expansion of $L(H^{2i-1}(X), s)$ at $s = i$. ■

Conjecture 5.5. (Hodge-index). Assume 5.3 and consider the primitive-cycle decomposition. Then the form $\langle ., L^{N+1-2i} \rangle$ is definite

*) This conjecture has been formulated independently by S.Bloch [6].

of sign $(-1)^i$ on the primitive i-cycles ($i \leq \frac{N+1}{2}$). ■

For i = 1 5.4a and 5.5 is the well-known positivity property of Neron-Tate height, and 5.4b is the weak form of Birch-Swinnerton-Dyer conjecture.

Note that 5.4a together with 4.0.7 imply the following generalisation of 5.1.

Lemma 5.6.*) (modulo 5.4). The Abel-Jacobi map $CH^i(X)^\circ \to \mathcal{J}^i(X_{\mathbb{C}})$ is injective up to torsion on the subgroup of cycles algebraically equivalent to zero. ■

It would be natural to suppose that Abel-Jacobi is injective on the whole $CH^i(X)^\circ \otimes Q$ but I have no definite reason for this.

Now I am going to formulate the motivic versions of 5.0 and 5.4. To do this I need to assume.

Conjecture 5.7. The usual intersection pairing $Ch^i(X)^\circ \otimes CH^j(X)^\circ \longrightarrow CH^{i+j}(X)^\circ$ is zero up to torsion. ■

This is a particular case of the [2], conj. 8.5.1. Note that its geometric analog follows directly from Leray spectral sequence considerations (since H^j of the affine curve are zero for $j>2$). Modulo 5.4 the conjecture 5.7 is equivalent to

Conjecture 5.8. For any three cycles $a_i \in CH^{d_i}(X)^\circ$, $\sum d_i = N+1$, one has $\langle a_1, a_2, a_3 \rangle = 0$.

Clearly 5.7. implies that a correspondence homologous to zero acts in a trivial way on the CH ()° -groups. This shows that we may define $CH^{\cdot}()^\circ$ groups for Grothendieck's motives. More precisely, consider the category $\mathcal{M}$ of motives over $\mathbb{Q}$ with coefficients in $\bar{\mathbb{Q}}$ for the theory of algebraic correspondences modulo homological equivalence. For any i we have $\bar{\mathbb{Q}}$-linear functor $CH^i()^\circ$ on $\mathcal{M}$ with values in finite dimensional $\bar{\mathbb{Q}}$-vector spaces, such that $CH^i(\mathcal{M}(X))^\circ = CH^i(X)^\circ \otimes \bar{\mathbb{Q}}$ where $\mathcal{M}(X)$ is motive of a smooth projective variety X (see [2] for details). Note that now (due to 5.7) the conjecture 5.3.

*)This conjecture has been formulated independently by S.Bloch[5] (1,4)

is implied by the standard conjectures on algebraic cycles. We get also the $\bar{\mathbb{Q}} \otimes \mathbb{R}$ - valued height pairing between $CH^{\cdot}()^{\circ}$ -groups of a motive M and its dual; according to 5.4. this pairing is non-degenerate. Denote by $\det_i\langle,\rangle(M)$ its determinant in the $\bar{\mathbb{Q}}$-baseses of $CH^{\cdot}()^{\circ}$; this is an element of $(\bar{\mathbb{Q}} \otimes \mathbb{R})^*$ correctly defined up to $\bar{\mathbb{Q}}$-multiple. Also for any motive M we have $\bar{\mathbb{Q}} \otimes \mathbb{Q}_\ell$ [Gal] - modules $H^{\cdot}(M,\mathbb{Q}_\ell)$ and $\bar{\mathbb{Q}} \otimes \mathbb{C}$-valued L-functions $L(H^{\cdot}(M), s)$. According to Gross's conjecture, for $i \in \mathbb{Z}$ all the $\mathbb{C}$-valued components of $L(H(M),s)$ (these components are numbered by embeddings $\bar{\mathbb{Q}} \hookrightarrow \mathbb{C}$) should have the same order of zero at i. Denote it by $\operatorname{ord}_i L(H^{\cdot}(M), s)$; so we have $L(H^{\cdot}(M), s-i) = l(H^{\cdot}(M), i).(s-i)^{\operatorname{ord}_i L(H^{\cdot}(M), s)} + O(s-i)^{\operatorname{ord}}$ for some $l(H^{\cdot}(M), i) \in (\bar{\mathbb{Q}} \otimes \mathbb{R})^*$. The following conjecture generalises 5.0 and 5.4:

<u>Conjecture 5.9.</u> We have $\operatorname{ord}_i L(H^{2i-1}(M), s) = \dim_{\bar{\mathbb{Q}}} CH^i(M)^{\circ}$ and $l(H^{2i-1}(M), i) = \det_i\langle,\rangle(M)$ multiplied by the determinent of the period matrix for $H^{2i-1}(M)$ (see [21]) (up to a multiple from $\bar{\mathbb{Q}}^*$). ■

See the beautiful computations of Gross-Zagier [11] in favor of this.

Recall that according to standard conjectures on algebraic cycles the category M is abelian semisimple. This semisimplicity implies that any additive functor on M, e.g. $CH^{\cdot}()^{\circ}$, $H^{\cdot}(\ ,\mathbb{Q}_\ell)$, is exact. On any motive M we have the colevel filtration N. defined by $N_j(M)$ = the sum of components of M that occure in some $\mathcal{M}(Y)(*)$, where Y is projective smooth scheme of dimension $\leqslant j$. It is easy to see that in case $M = \mathcal{M}(X)$ the corresponding filtrations on $CH^{\cdot}(\mathcal{M}(X))^{\circ}$ $CH^{\cdot}(X)^{\circ} \otimes \bar{\mathbb{Q}}$, $H^{\cdot}(X,\mathbb{Q}_\ell)$ are $N_j CH^{\cdot}(X)^{\circ} = \{$ cl a: there exists Y, $\operatorname{supp} a \subset Y \subset X$ s.t. $\operatorname{codim} Y \geqslant \cdot - j$ and a is homologous to zero in Y $\}$ (in particular N_1 = cycles algebraically equivalent to zero), $N_j H^{\cdot}(X,\mathbb{Q}_\ell)$

$$= \{\alpha : \text{there exists } Y \subset X \text{ s.t. } \operatorname{codim} Y \geqslant \tfrac{\cdot - j}{2} \text{ and } \alpha|_{X \setminus Y} = 0\}.$$

Now 5.9. claims that the rank of $Gr^N_j(CH^i(X)^{\circ})$ equals to the order of zero of $L(Gr^N_j H^{2i-1}(X), s)$ at s=i. This sharpened form of Swinnerton-Dyer conjecture was first found by Bloch [5] (1.3). Note that this conjecture for $j = 1$ also implies 5.6. (see [5] (1.4)).

5.10. In the rest of the paper I will try to discuss some general conjectural framework for motives, regulators and the like.

A. Motivic sheaves. One hopes that for any scheme S and an appropriate commutative ring A (say $A = \mathbb{Z}, \mathbb{Z}/n, \mathbb{Q}, \bar{\mathbb{Q}}$) there exists certain abelian A-category $M(S, A)$ of (mixed) motivic (perverse) A-sheaves over S together with corresponding derived category $DM(S,A)$. This categories should resemble very much the categories of mixed ℓ-adic sheaves;

- there should be inner $\otimes$ and Hom
- there should be Tate sheaves $A_{\mathcal{M}}(i)$; the Tate twist $M \mapsto M(i) = M \otimes A_{\mathcal{M}}(i)$ is automorphism of $M(S,A)$
- for any morphism $S_1 \to S_2$ of finite type there should be the corresponding functors f_*, f^*, $f_!$, $f^!$ between $DM(S_i)$
- for $A \supset \mathbb{Q}$ there should be canonical weight filtration $W.$ on the objects of $M(S,A)$ such that any morphism is strictly compatible with $W.$ and any $Gr^W_{\cdot}(M)$ is semisimple.

All this things should behave the same way as in mixed ℓ-adic situation. One should also have realisation functors r on $M(S)$ and $DM(S)$ with values in mixed ℓ-adic sheaves, or Hodge sheaves (if S is an $\mathbb{R}$-scheme, see n° E below)... If $A \supset \mathbb{Q}$ this realisation functors should be exact, faithful on $M(S)$, and should commute with all the above structures. Note that such r induces the morphism between corresponding (absolute) cohomology groups: say $r : H^{\cdot}(S, M(*)) := \mathrm{Ext}^i(\mathbb{Q}_{\mathcal{M}}(0), M(*)) \to \mathrm{Ext}^i(\mathbb{Q}_\ell, r_{\mathbb{Q}_\ell}(M)(*)) = H^i(S, r_{\mathbb{Q}_\ell}(M)(*))$.

If $S = \mathrm{Spec}\, k$, k is a field, and $A \supset \mathbb{Q}$, then the category of semisimple objects in $M(S,A)$ should be equivalent to the category of Grothendieck's A-motives over k constructed by means of algebraic correspondences modulo homological equivalence.

B. Relations with K-theory and relative cycles.. The cohomology groups $H^{\cdot}(S, A_{\mathcal{M}}(*))$ form Poincaré duality theory with supports" in the sense of Bloch-Ogus. The corresponding canonical map $H^{\cdot}_{\mathcal{M}}(S,\mathbb{Q}(*)) \longrightarrow H^{\cdot}(S,\mathbb{Q}_{\mathcal{M}}(*))$ should be isomorphism if S is regular (recall that $H^{\cdot}_{\mathcal{M}}(S,\mathbb{Q}(*))$ is a part of $K_{\cdot}(S) \otimes \mathbb{Q}$ and the canonical map comes from the Chern character). All the canonical maps $H^{\cdot}_{\mathcal{M}} \longrightarrow H^{\cdot}_{?}$ where $H^{\cdot}_{?}$ is ℓ-adic, Hodge-Deligne or some other cohomology, should coincide with $H^{\cdot}_{\mathcal{M}}(S,\mathbb{Q}(*)) \xrightarrow{\sim} H^{\cdot}(S,\mathbb{Q}_{\mathcal{M}}(*)) \xrightarrow{r_?} H^{\cdot}(S,\mathbb{Q}_?(*)) = H^{\cdot}_?(S,\mathbb{Q}(*))$.

More precisely, for S regular there should exist the Atiyah-Hirzebruch type spectral sequence converging to $K_{\cdot}(S)$ with second term

$H(S,\mathbb{Z}_{\mathcal{M}}(*))$; it should degenerate up to factorials at E_2 (one likes to have the same picture for arbitrary singular S, but this possibly forces to replace $\mathcal{M}(S)$ by more clever category of properties I can't imagine). Note that for a smooth varieties over a field S.Landsburg [12] has found a spectral sequence of such a type using relative cycles machine; hopefully it should coincide with the conjectural spectral sequence above. V.Schechtman (report on Landsburg's work, february 1983) and S.Bloch [23], [24] independently have shown that Landsburg's E_2 form Poincaré duality theory with supports. Bloch [23] has shown that landsburg's spectral sequence degenerates at E_2 up to torsion. Note that for $i<0$ one should have $H^i(S,\mathbb{Z}_{\mathcal{M}}(*))=0$ (as Ext^i in a t-category); the corresponding property for Landsburg's E_2 remains unproven (it was conjectured independently by Ch.Soule and the author [3], [18]). See also [25].

C. Application to algebraic cycles. Let us see what B gives being applied to cycles (compare with [7], lecture 1). For regular X one should have $H^{2i}(X,\mathbb{Z}_{\mathcal{M}}(i)) = CH^i(X)$. Let S be regular and $\pi : X_S \to S$ be a smooth projective map. We should have the Leray spectral sequence $E_2^{p,q} = H^p(S, R^q\pi_*\mathbb{Z}_{\mathcal{M}}(i)) \Rightarrow H^{p+q}(X_S, \mathbb{Z}_{\mathcal{M}}(i))$ that degenerates up to torsion. So ony any $H(X_S,\mathbb{Z}_{\mathcal{M}}(i))$ we have canonical filtration $\mathcal{F}^{\cdot}$ s.t. $Gr^p_{\mathcal{F}}H^n(X_S,\mathbb{Q}_{\mathcal{M}}(i)) = H^p(S, R^{n-p}\pi_*\mathbb{Q}_{\mathcal{M}}(i))$. (Note that $CH^{\cdot}()^{\circ}$ in the main body of the paper is $\mathcal{F}^1CH^{\cdot}()$). In particular $CH^i(X_S)$ is controlled by relative motivic cohomology of X with numbers $\leq 2i$.

The group $\mathcal{F}^2CH^i(X_S)$ is something like "nonrepresentable" part of $CH^i(X_S)$: if S is of finite type over $\mathbb{Z}$ or $\mathbb{Q}$, then any element of $\mathcal{F}^2$ is zero at any $\bar{\mathbb{Q}}$-point of S. More precisely, assuming S affine of finite type over $\mathbb{Z}$ we may describe the filtration $\mathcal{F}^{\cdot}$ as follows. Just as in ℓ-adic situation for $M\in\mathcal{M}(S)$ one should have $H^j(S,M) = 0$ for $j>\dim S$. Moreover, for any $j\leq\dim S$ there exists certain $T\subset S$, $\dim T = j$, such that $H^j(S,M)\hookrightarrow H^j(T,M|_T)$. This shows that

$$\mathcal{F}^jCH^{\cdot}(X_S) = \bigcap_{\substack{T\subset S\\ \dim T = j-1}} \mathrm{Ker}\,(CH^{\cdot}(X_S)\to CH^{\cdot}(X_T)).$$

Now assume that S is a spectrum of a field. for any smooth projective Y_S of dimension $i-1$ one has $CH^i(Y_S) = 0$, so in particular any $H^{2i-\ell}(S,R\pi_{Y*}\mathbb{Q}_{\mathcal{M}}(*))$ is zero. For a motive M over S define the level filtration $N^{\cdot}$ on M by formula $N^jM := \bigcap_{Y\,\mathrm{s.t.}\,\dim Y=j} \mathrm{Ker}\,(M\to R\pi_{Y_S*}\mathbb{Q})$.

The above shows that $H^{2i-\ell}(S,R^{\ell}\pi_{X*}\mathbb{Q}_{\mathcal{M}}(i)) = H^{2i-\ell}(S,N^{i-1}R^{\ell}\pi_{X*}\mathbb{Q}_{\mathcal{M}})(i))$.

In particular this groups are non-zero for $\ell = 2i, \dots, i$ only.

D. The structure of $H^\cdot(\ ,A_{\mathcal{M}}(*))$. Consider the complexes $R\Gamma(\ ,A_{\mathcal{M}}(i)) = R\,Hom(A_{\mathcal{M}}(0), A_{\mathcal{M}}(i))$ Zariski locally. These are the complexes $A_{\mathcal{M}}(i)_{Zar}$ in the derived category of sheaves on the big Zariski topology together with canonical commutative multiplication $A_{\mathcal{M}}(i)_{Zar} \overset{L}{\otimes} A_{\mathcal{M}}(j)_{Zar} \longrightarrow A_{\mathcal{M}}(i+j)_{Zar}$. We have $H^\cdot(S, A_{\mathcal{M}}(i)) = H^\cdot(S_{Zar}, A_{\mathcal{M}}(i)_{Zar})$. The following properties of $A_{\mathcal{M}}(i)$ should hold:

(i) $A_{\mathcal{M}}(i)_{Zar} = A \overset{L}{\otimes} \mathbb{Z}_{\mathcal{M}}(i)_{Zar}$; $A_{\mathcal{M}}(i)_{Zar}$ are zero for $i < 0$

(ii) $\mathbb{Z}_{\mathcal{M}}(1)_{Zar} = \mathcal{O}^*[-1]$; for any $i > 0$ the complex $\mathbb{Z}_{\mathcal{M}}(i)_{Zar}$ is acyclic off the degrees $1, \dots, i$

(iii) If k is a field then $H^i(\mathrm{Spec}\ k, \mathbb{Z}_{\mathcal{M}}(i)_{Zar}) = K_i^{Milnor}(k)$; the isomorphism induced by the product $\mathbb{Z}_{\mathcal{M}}(1)_{Zar}^{\otimes i} \longrightarrow \mathbb{Z}_{\mathcal{M}}(i)_{Zar}$. So the sheaf $H^i(\mathbb{Z}_{\mathcal{M}}(i)_{Zar})$ coincide on regular S with the sheaf of Milnor's K-functors (and the whole $\mathbb{Z}_{\mathcal{M}}(i)_{Zar}$ is something like non-abelian left-derived functor of K^{Milnor})

(iv) If k is a field, then there exists a canonical differential bigraded algebra $\tilde{\mathbb{Q}}^\cdot_{\mathcal{M}}(*)$ that represents $\mathbb{Q}_{\mathcal{M}}(*)_{Zar}$ ($\tilde{\mathbb{Q}}^\cdot_{\mathcal{M}}(*)$ is the Sullivan's minimal model of $\mathbb{Q}_{\mathcal{M}}(*)_{Zar}$) such that $\tilde{\mathbb{Q}}^\cdot_{\mathcal{M}}(*) = \Lambda^\cdot(\bigoplus_{i \geq 1} \tilde{\mathbb{Q}}^1(i))$. So $\tilde{\mathbb{Q}}^\cdot_{\mathcal{M}}(*)$ is standard chain algebra of the graded Lie coalgebra $\bigoplus_{i \geq 1} \tilde{\mathbb{Q}}^1(i)$. This conjecture is due to Schechtman. It is closely related to the conjecture that we may compute $H^\cdot(\mathrm{Spec}\ k, \mathbb{Q}_{\mathcal{M}}(i))$ as Ext's not in the whole category $M(\mathrm{Spec}\ k)$, but in the far smaller category $M_{Tate}(\mathrm{Spec}\ k)$ of objects with all irreducible constituents isomorphic to some Tate's modules.

(v) (Étale descent). We may consider the analogous complexes $A_{\mathcal{M}}(i)_{ét}$ on the big étale topology; if $\P : Zar \longrightarrow Ét$ is a canonical morphism, then $A_{\mathcal{M}}(i)_{ét} = \P^* A_{\mathcal{M}}(i)_{Zar}$. Note, that one should have usual étale descent for $A_{\mathcal{M}}(i)_{Zar}$ only in the trivial case $A \supset \mathbb{Q}$. In general one hopes that $A_{\mathcal{M}}(i)_{Zar} = \tau_{\leq i} R\P_* A_{\mathcal{M}}(i)_{ét}$ and also $H^{i+1} R\P_* A_{\mathcal{M}}(i)_{ét} = 0$ (this "Hilbert theorem 90" property was found independently by S. Lichtenbaum [14]; his paper also contains some discussion on this subject).

(vi) (Finite coefficients). Let ℓ be prime to characteristics of our schemes. The theorem of Suslin on K-functor of an algebraically closed fields dictates that one should have $(\mathbb{Z}/\ell^n)_{\mathcal{M}}(i)_{ét} = \mathbb{Z}/\ell^n(i)_{ét}$. So, by (v), one should have $(\mathbb{Z}/\ell^n)_{\mathcal{M}}(i)_{Zar} = \tau_{\leqslant i}R\P_* \mathbb{Z}/\ell^n(i)$. In particular, if k is a field, then $H^j(\mathrm{Spec}\ k, (\mathbb{Z}/\ell^n)_{\mathcal{M}}(i))$ is zero for $j > i$, and coincides with $H^j(\mathrm{Gal}\ \bar{k}/k, \mathbb{Z}/\ell^n(i))$ for $j \leqslant i$. The n°B implies that we should have the corresponding spectral sequence that converges to $K.(\mathrm{Spec}\ k, \mathbb{Z}/\ell^n)$. R. Thomason noticed (letter from 10-5-1984) that such a spectral sequence degenerates at E_2 if $\mu_{\ell^n} \subset k$, and so the usual Atiyah-Hirzebruch spectral sequence for $K/\ell^n{}^{top}$ also should degenerate - this thing is unknown yet. Anyway, it is easy show that $H^{\cdot}_{fine}(S, \mathbb{Z}/\ell^n(i)) := H^{\cdot}(S_{Zar}, R\P_* \mathbb{Z}/\ell^n(i))$ form the Poincaré duality theory with supports, and so we have the characteristic classes with values in this groups; note that $H^{2i}_{fine}(S, \mathbb{Z}/\ell^n(i)) = CH^i(S)/\ell^n$. It would be of interest to construct explicitely the category of "fine étale $\mathbb{Z}/\ell^n$-sheaves" or "fine $\mathbb{Z}/\ell^n$-Galois modules" such that Ext's between $\mathbb{Z}/\ell^n(i)$ give $H^{\cdot}_{fine}$.

E. Hodge sheaves. For a scheme S over $\mathbb{C}$ or $\mathbb{R}$ one should have a category $\mathcal{H}_S$ of Hodge sheaves, analogous to the category of mixed $\mathbb{Q}_\ell$ perverse sheaves. At a moment one definitely knows what lisse Hodge sheaves on smooth varieties are (the pure ones are Griffiths's polarisable variations of Hodge structures, and in the mixed case we have the definition of Steenbrink and Zucker); the attempts to define the whole $\mathcal{H}_S$ as well as to define the functors f_*,.. etc. are quite painful. This contrasts much with the clear étale cohomology situation, where all the constructions are quite simple and natural. Thus the Galois action on étale cohomology groups comes from the Galois symmetries of étale site. It would be marvellous if one may do Hodge theory in a parallel way. Namely, whether there exists the " $\mathbb{R}$-site" $\mathrm{Top}_{\mathbb{R}}(S)$ *) of S together with the action of the group $G_{\mathcal{H}}$ (see [8]) **) such that the $\mathbb{R}$-Hodge sheaves on S would be the $G_{\mathcal{H}}$-sheaves on $\mathrm{Top}_{\mathbb{R}}(S)$? Thus the cohomology of $\mathrm{Top}_{\mathbb{R}}(S)$ with values in a constant sheaf $\mathbb{R}$ should coincide with the usual cohomology of S with real coefficients, and the natural action of $G_{\mathcal{H}}$ on them should give

*) $\mathrm{Top}_{\mathbb{R}}(S)$ should have somewhat a probabilistic nature.

**) $G_{\mathcal{H}}$ is a real proalgebraic group such that the category of real representations of $G_{\mathcal{H}}$ coincides with the category of $\mathbb{R}$-Hodge structures.

the usual Hodge structure. Passing back from $Top_{\mathbb{R}}$ to the usual topology (on the level of cohomology groups this means restoring of the integral structure) breaks the Hodge symmetry.

F. Absolute cohomology with compact supports. Let now S be a scheme over $\mathbb{Z}$. For any $M \in DM(S)$ there should exist "absolute cohomology groups with compact supports" $H^{\cdot}_c(S, M)$ with the following properties

(i) For any $f : S \to S'$ one should have $H^{\cdot}_c(S, M) = H^{\cdot}_c(S', f_!(M))$.

(ii) Let $S = \operatorname{Spec} \mathbb{Z}$. For $M \in \mathcal{M}(S, \mathbb{Z})$ one should have an exact sequence $\cdots \to H^1_c(S, M) \to H^1(S, M) \xrightarrow{r} H^1_{\mathcal{H}}(M_{\mathcal{H}} \otimes \mathbb{R}) \to H^2_c(S, M)$. where r is a regulator map. So for M of weight ≥ 0 all $H^{\cdot}_c(S, M)$ are zero; for M of weight -1 the only non-zero H_c may be the discret group $H^1_c(S, M)$; for M of weight ≤ -2 the only non-zero H_c is $H^2_c(S, M)$, this group is compact for $M \neq \mathbb{Z}_{\mathcal{M}}(1)$ and $H^2_c(S, \mathbb{Z}_{\mathcal{M}}(1)) = \mathbb{R}$. The Euler-Poincaré volumes of this H^2_c should be related with the values of L-functions, and the height pairing should coincide with $H^1_c(S, M) \otimes H^1_c(S, M^\circ) \longrightarrow H^2_c(S, \mathbb{Z}_{\mathcal{M}}(1))$.

References

1. S. Arakelov Vancouver ICM report 1974.
2. A. Beilinson Notes on absolute Hodge cohomology, *Contemporary mathematics, vol 55, pp 35-68*
3. A. Beilinson Higher regulators and values of L-functions (in Russian) in Modern problems in mathematics VINITI series, vol. 24, 1984, 181-238.
4. S. Bloch Algebraic cycles and values of L-functions, preprint, 1983.
5. S. Bloch Algebraic cycles and values of L-functions II, preprint, 1984.
6. S. Bloch Height pairings for algebraic cycles, preprint 1984.
7. S. Bloch Lectures on algebraic cycles, Duke university math. series IV, 1980.
8. P. Deligne Structures de Hodge mixtes reeles, preprint IHES,1982.
9. P. Deligne La conjecture de Weil II, Publ. math. IHES, 52(1980).
10. H. Gillet, Ch. Soule Intersection sur les varietes d'Arakelov, CRAS t. 299, ser. 1, 12, 1984, 563-566.
11. B. Gross, D. Zagier Paper on heights of Heegner points, to appear.
12. S. Landsburg Relative cycles and algebraic K-theory, preprint 1983.
13. S. Lang Fundamentals of Diophantine geometry, Springer 1983.
14. S. Lichtenbaum Values of zeta-functions at non-negative integers, preprint 1983.

15. Yu.I. Manin New dimensions in geometry, Arbeitstagung, Bonn, 1984, in Lecture note 1000.

16. D. Mumford Rational equivalence of o-cycles on surfaces, J.Math. Kyoto Univ. 9, 195-204, 1968.

17. S.Lang Fundamentals of diophantine geometry, Springer-Verlag 1983

18. Ch. Soulé Operations en K-theorie algebrique, preprint, 1983.

19. Ch. Soulé Regulateurs, Sem. Bourbaki, exp. 644, Fevrier 1985.

20 H.P.F. Swinnerton-Dyer.The conjectures of Birch-Swinnerton-Dyer and of Tate, in Proceedings of a Conference on Local Fields, Springer-Verlag, Berlin, 1967.

21. P.Deligne Valeurs de fonctions L et periodes d'integrales Proc. Symp. Pure math, Vol 33, p. 2, 1979, pp. 313-346.

22. Seminaire sur le pinceaux arithmetiques: la conjecture de Mordell, Asterisque 127, 1985

23. S.Bloch Algebraic cycles and higher K-theory, Advances in Math., to appear.

24. S.Bloch Algebraic cycles and the Beilinson conjectures, Contemporary math., Vol.58, part 1, 186, pp. 65-79

25. A.Beilinson, R.Macpherson, V.Schechtman Notes on motivic cohomology, to appear in Duke. math. journ.

Corrections to [2]

p. 36 ligne 5. Add "This exact sequence corresponds to the exact sequence

$$0 \longrightarrow H^1(Gal\,\bar{K}/K, H^{i-1}_{ét}(X\otimes\bar{K},\mathcal{F})) \longrightarrow H^i_{ét}(X,\mathcal{F}) \longrightarrow H^0(Gal\,\bar{K}/K, H^i(X\otimes\bar{K},\mathcal{F})) \longrightarrow 0$$

in étale situation with cohomological dimension of K equals to one."

p. 36 ligne 6. Omit "in contrast to the étale situation"

p. 38 ligne 4. Replace "ought to be known to specialists" by" where first done by J.Carlson see his paper "Extensions of mixed Hodge Structures" in Journees de geometrie algebrique d'Angers, juillet 1979, Sijthoff & Noordhoff 1980); unfortunately I was not acquainted with this very nice paper when I wrote the text."

p. 67. Replace the part of n° 8.5.2. from words "One may reformulate..." up to the end of n° 8.5.2. by "Recently P.Deligne (letter to Soulé from 20-1-85) has found a very nice reformulation of the above conjecture for the values of L in the half-plane of convergence. Here it is. Let M be a motive of weight $\leqslant -2$ that does not contains $\mathbb{Q}(1)$. Then L(M,S) converges at S=0, L(M,0)$\neq$0. The $\mathbb{R}$-vector space $H_{DR}(M\otimes\mathbb{R})/F^0H_{DR}(M\otimes\mathbb{R})=(H_{DR}(M)/F^0H_{DR})\otimes\mathbb{R}$ has natural $\mathbb{Q}$ structure, hence the volume form defined up to $\sim$. It contains $\mathbb{Q}$-vector subspace $H^+_B(M)$, and the regulator maps $H_{\mathcal{M}}(M)$ to the corresponding quotient. The conjecture claims that $H_{\mathcal{M}}(M)+H^+_B(M)$ is a $\mathbb{Q}$ - lattice and $L(M,0)\sim$ $\mathrm{vol}\left(\frac{H^+_B\backslash H_{DR}(M)/F^0H_{DR}}{H_{\mathcal{M}}(M)}\right)$. The equivalence of this and the previous formulations may be easily seen by using the functional equation together with computations in Deligne's paper [5]"

p. 68 ligne 3. Add "The computations for modular curves where transmitted by D.Ramakrishnan to Shimura curves (this volume). He also considered the case of K_1 of Hilbert-Blumenthal surfaces (preprint 1985)."

On the derived category of perverse sheaves

by A.A. Beilinson

Let $D = D^b(X, \mathbb{Q}_\ell)$ be the usual derived category of $\mathbb{Q}_\ell$-sheaves on a certain scheme X, and $M = M(X) \subset D$ be the category of perverse sheaves for middle perversity. Now consider the derived category $D^b(M)$ of an abelian category M; we have the natural exact functor $D^b(M) \to D$. The aim of this note is to show that this functor is an equivalence of categories. The same result result holds for M = the category of algebraic holonomic $\mathcal{D}$-modules and D = the derived category of complexes of $\mathcal{D}$-modules with holonomic cohomology.

One may look at this from two complementary points of view. First we see that Yoneda-type Ext's in M are computable by easy topological means (since they coincide with Ext's in D). Secondly, the niche D where M dwells, may be recovered from M (note, that a priori D is quite transcendental with respect to M); this may be of use in a future motivic sheaf theory.

I would like to thank P. Deligne for useful discussions.

§1. Notation and statement of the main theorem

1.1. Fix a base field k; in what follows the base schemes will be separated of finite type over k. For a scheme X denote $D(X)^{(i)}$, $i = 1,\dots,5$, the following triangulated categories:.

(i) $D(X)^{(1)} := D^b_c(X,R)$ = the derived category of complexes of étale R-sheaves having bounded constructible cohomology; here R is a finite ring of characteristic prime to char k.

(ii) $D(X)^{(2)} := D^b_c(X, \bar{\mathbb{Q}}_\ell)$. Here $\ell \neq$ char k and k is assumed to be algebraically closed; see e.g. [1] (2.2.18).

(iii) $D(X)^{(3)} := D^b_c(X(\mathbb{C}), R)$ = the derived category of complexes of R-sheaves on the classical topology of X, having algebraically constructible cohomology; here $k = \mathbb{C}$ and R is any field, see e.g. [1] (2.2.1.).

(iv) $D(X)^{(4)} = D^b_m(X, \bar{\mathbb{Q}}_\ell)$ = the derived category of mixed sheaves, see e.g. [1] (5.1.5.).

(v) $D(X)^{(5)} = D_{Hol}(X)$ = the derived category of complexes of $\mathcal{D}$-modules having bounded holonomic cohomology, see e.g. [3] §4 (here char k= 0).

Each of these triangulated categories $D(X)^{(i)}$ has a canonical filtered counterpart - the f-category $DF(X)^{(i)}$ over $D(X)^{(i)}$ (for f-categories see the appendix). In the cases i = 1,3 this is the derived category of complexes of sheaves with finite decreasing filtration such that each graded quotient belongs to $D(X)^{(i)}$; in the other cases $D(X)^{(i)}$ is the corresponding $\mathbb{Q}_\ell$ - or $\mathcal{D}$-module analogue.

There are various standard functors between $D(X)^{(i)}$ such as $\otimes$, Hom, the direct and inverse image functors $f_!$, f_*, $f^!$, f^* (more precisely, in case i = 1 $\otimes$ and Hom may take values in unbounded complexes); all these functors have a canonical f-lifting to $DF(X)^{(i)}$. We will consider $D(X)^{(i)}$ as t-categories with a t-structure defined by the middle perversity for i = 1,...,4 and with the obvious t-structure for i = 5. The hearts $M(X)^{(i)}$ of this t-structures are categories of constructible perverse sheaves in the cases i = 1,...,4 and the category of holonomic modules $M_{Hol}(X)$ in the case i = 5.

1.2. Assume that for any scheme X over k we are given a strictly full t-subcategory D(X) in $D(X)^{(i)}$ above (i = 1,...,5 is fixed) closed under $\otimes$, Hom and f_*, $f_!$, f^*, $f_!$ (i.e. for any morphism $f : X \to Y$ of schemes one should have $f_*, f_! : D(X) \to D(Y)$; $f^*, f_! : D(Y) \to D(X)$).

Examples. Clearly, we may take $D(X) \equiv D(X)^{(i)}$ or $D(X) \equiv 0$. In the case i = 1 this are the only possibilities. In the cases i = 2,3 we may take for D(X) the subcategories generated by quasiunipotent local systems (according to Kashiwara and Gabber, see [5]), or by local systems having geometric origin ([1] (6.2.4.)). In the case i = 5 we may take $D(X) = D_{RS}(X)$ = the subcategory generated by lisse holonomic modules having regular singularities at ∞ (see [3] §4), or, more generally, $D(X) = D_{RS\Lambda}(X)$ ([3] (4.8)). In what follows assume that D(X) is not identically zero.

Remark 1.2.1. a). In case $k = \mathbb{C}$ we have the canonical t-exact functor $DR : D(X)^{(5)} = D_{Hol}(X) \longrightarrow D(X)^{(3)} = D^b_c(X(\mathbb{C}), \mathbb{C})$, whose restriction on $D_{RS}(X) \subset D_{Hol}(X)$ is an equivalence of categories commuting with any standard functor (see [3] §5).

b) clearly D (spec k) contains all the Tate modules R(j) (cases i = 1,3,5), or $\bar{\mathbb{Q}}_\ell(j)$ (cases i = 2,4); hence the functors Φ_f, Ψ_f,

Ξ_f (see [2]) preserve D(X).

1.3. Let $M(X) \subset D(X)$ be the heart of D(X). Clearly, D(X) coincides with the full subcategory in $D(X)^{(i)}$ of complexes having all t-cohomology in M(X). Let $DF(X) \subset DF(X)^{(i)}$ be the full subcategory of objects having each a graded quotient in D(X). Clearly, DF(X) is an f-category over D(X). It defines a canonical t-exact functor $real_X : D^b(M(X)) \to D(X)$ that induces the identity functor between hearts M(X) (see appendix; in holonomic case real is obvious functor). Now we may formulate

Main theorem 1.3. This functor is an equivalence of categories.

Remarks. a) The corresponding statement for the category of sheaves lisse along a fixed stratification is usually false.

b) I don't know whether the analogous fact remains true for perverse sheaves of other perversities different from the middle one, say, for ordinary constructible $\mathbb{Q}_\ell$-sheaves. Also I am ignorant of the analytic cases, both constructible and holonomic.

1.4. Note that the main theorem just claims that Yoneda-type Ext's between the objects of M(X) (i.e. Ext's computed in $D^b(M(X))$ coincide with usual Ext's computed in D(X). Namely, the following simple general lemma holds (proof is similar to [1] (3.1.16)).

Lemma 1.4. Let $F : D_1 \to D_2$ be a t-exact functor between t-categories. D_i with hearts $C_i \subset D_i$. Assume that $F/_{C_1} : C_1 \to C_2$ is an equivalence of categories, and $D_2 = D_2{}^b$. Then the following statements are equivalent:

(i) F is an equivalence of categories;

(ii) For any $M,N \in \mathrm{Ob}\, C_1$ and $i > 0$ the map $\mathrm{Hom}^i_{D_1}(M,N) \longrightarrow \mathrm{Hom}^i_{D_2}(F(M), F(N))$ is an isomorphism;

(iii) Assume that $D_1 = D^b(C_1)$. For any $M,N \in \mathrm{Ob}\, C_1$, $i > 0$ and $x \in \mathrm{Hom}^i_{D_2}(F(M), F(N))$ there exists an injection $N \hookrightarrow N'$ in C_1 such that the image of x in $\mathrm{Hom}^i_{D_2}(F(M), F(N'))$ is zero. □

Clearly 1.3. falls into this situation, so it suffices to prove for $F = real_X$ either 1.4. (ii) or 1.4 (iii). For $M,N \in \mathcal{M}(X)$ put $\mathrm{Ext}^i_{M(X)}(M, N) := \mathrm{Hom}^i_{D^b(M(X))}(M, N)$, $\mathrm{Ext}^i_{D(X)}(M,N) := \mathrm{Hom}^i_{D(X)}(M,N)$. So to prove 1.3. we have to show that these Ext's coincide.

§2. Proofs

The proof of the theorem 1.3 is divided into two steps: first we show that it is valid at the generic point of X (lemma 2.1.1.), and then by means of glueing (see [2]) we use this to reduce the problem to lower dimensions.

2.1. Let $\eta \in X$ be a generic point; $D(\eta) = 2\text{-}\varinjlim_{\eta \in U} D(U)$ the 2-limit of t-categories $D(U)$, U runs the Zariski open sets containing η. Clearly $D(\eta)$ is t-category with the heart $\mathcal{M}(\eta) = 2\text{-}\lim \mathcal{M}(U)$; we also have our t-exact functor $\text{real}_\eta : D^b(\mathcal{M}(\eta)) = 2\text{-}\lim D^b(\mathcal{M}(U)) \longrightarrow D(\eta)$.

Lemma 2.1.1. $\text{real}_\eta : D^b(\mathcal{M}(\eta)) \longrightarrow D(\eta)$ is an equivalence of categories.

Proof. First notice that in case $D(X) = D(X)^{(1)}$ (see 1.1.) the lemma is trivial since here $\mathcal{M}(\eta)$ = finite Galois R-modules, $D(\eta)$ = derived category of complexes of arbitrary Galois R-modules with finite bounded cohomology groups. Therefore we assume that we are in one of the situations (ii)-(v) of 1.1; in particular the coefficient ring is a field. Assume also that X is reduced; this changes nothing.

According to 1.4. (iii) it suffices to show the following. Let $U \subset X$ be a Zariski open set, $\eta \in U$, and M_U, N_U are in $\mathcal{M}(U)$. Then for some open set $V \subset U$, $\eta \in V$, there exists O_V in $\mathcal{M}(V)$ and an injection $N_V := N_U|_V \hookrightarrow O_V$ such that for any $i > 0$ the induced arrow $\text{Ext}^i_{D(U)}(M_U, N_U) \longrightarrow \text{Ext}^i_{D(V)}(M_V, O_V)$ is zero.

The proof will be carried by induction in $\dim X$; clearly 2.1.1. holds for X of dimension zero, so assume that we have 2.1.1. for any Y of dimension less then X.

Shrinking U if necessary we may assume that M_U, N_U are lisse, that U is irreducible and there exists a smooth affine $\pi : U \longrightarrow Z$ with 1-dimensional fibers such that Z is regular and $L^q = R^q\pi_*\underline{\text{Hom}}(M_U, N_U) = R^q\pi_*(M_U^* \otimes N_U)$ are lisse sheaves on Z. Clearly $L^q = 0$ unless $q = 0, 1$, so the Leray spectral sequence $E_2^{pq} = H^p(Z, L^q) \Rightarrow \text{Ext}^{p+q}_{D(U)}(M_U, N_U)$ becomes degenerate at E_3.

Remark. Certainly, L^q are usual lisse constructible sheaves in constructible situation; in $\mathcal{D}$-module situation they are lisse holonomic modules placed in degree $\dim Z$ in $D(Z)$.

We will need the following lemma (here for an open $Y \subset Z$ we put $\pi_Y : U_Y := \pi^{-1}(Y) \longrightarrow Y$):

Lemma 2.1.2. a) There exists an open $Y \subset Z$, a lisse P_{U_Y} in $\mathcal{M}(U_Y)$ and an injection $N_{U_Y} \hookrightarrow P_{U_Y}$ such that the corresponding arrow $R^1\pi_{Y*}\text{Hom}(M_{U_Y}, N_{U_Y}) \longrightarrow R^1\pi_{Y*}\text{Hom}(M_{U_Y}, P_{U_Y})$ is zero.

b) There exists an open $Y' \subset Z$, a lisse $Q_{U_{Y'}}$ in $M(U_{Y'})$ and an injection $N_{U_{Y'}} \hookrightarrow Q_{U_{Y'}}$ such that the corresponding arrows $H^p(Z, R^0\pi_*Hom(M_U, N_U)) \longrightarrow H^p(Y', R^0\pi_{Y'*}(Hom(M_{U_{Y'}}, Q_{U_{Y'}}))$ are zero for $p > 0$.

Lemma 2.1.2. $\Longrightarrow$ Lemma 2.1.1: first choose $N_{U_Y} \hookrightarrow P_{U_Y}$ as in 2.1.2. a). Then the Leray spectral sequence shows that the image of $Ext^i_{D(U)}(M_U, N_U)$ in $Ext^i_{D(U_Y)}(M_{U_Y}, P_{U_Y})$ is contained in the image of $H^i(Y, R^0\pi_*Hom(M_{U_Y}, P_{U_Y}))$. Now apply 2.1.2. b) to Z replaced by Y, and sheaves M_{U_Y} and P_{U_Y}. We get $Y' \subset Y$ and $P_{U_{Y'}} \hookrightarrow Q_{U_{Y'}}$ such that $H^i(Y, R^0\pi_*Hom(M_{U_Y}, P_{U_Y})) \longrightarrow H^i(Y', R^0\pi_*Hom(M_{U_{Y'}}, Q_{U_{Y'}}))$ is zero for $i > 0$. This shows that for the composite map $N_{U_{Y'}} \hookrightarrow Q_{U_{Y'}}$ on $V = U_{Y'}$, all the arrows $Ext^i_{D(U)}(M_U, N_U) \to Ext^i_{D(V)}(M_V, O_V)$ are zero for $i > 0$, Q.E.D. $\square$

Proof of 2.1.2. a) First notice that we may easily construct such P along each closed fiber. Namely, consider the canonical element $\alpha \in H^0(Z, L^{1*} \otimes L^1) = H^0(Z, R^1\pi_*Hom(\pi^*L^1 \otimes M, N))$. If this α came from the global extension $\tilde{\alpha} \in Ext^1(\pi^*L^1 \otimes M, N)$ we are done: just take $Y = Z$ and define P_{U_Y} from the extension $0 \to N \to P_{U_Y} \to \pi^*L^1 \otimes M \to 0$ of class $\tilde{\alpha}$. If not, consider the obstruction to existense of $\tilde{\alpha}$: the Leray spectral sequence defines the exact sequence

$$Ext^1(\pi^*L^1 \otimes M,N) \to H^0(Z,R^1\pi_*Hom(\pi^*L^1 \otimes M,N)) \xrightarrow{\partial} H^2(Z,R^0\pi_*Hom(\pi^*L^1 \otimes M,N))$$

$$H^0(Z,L^{1*} \otimes L^1) \qquad\qquad H^2(Z, L^{1*} \otimes L^0)$$

so the obstruction is $\partial(\alpha)$. To kill this obstruction we replace L^1 by a certain extension. To construct this extension we will use the inductive hypothesis applied to L^{0*} and L^{1*}. They say that there exist an open set $Y \subset Z$, a lisse sheaf K_Y on Y and an injective arrow $\varphi: L^{1*}{}_Y \hookrightarrow K_Y$ such that the induced arrow $H^2(Z,L^{1*} \otimes L^0) \xrightarrow{\varphi} H^2(Y, K_Y \otimes L^0_Y)$ is zero. In particular $\varphi(\partial\alpha)$ is zero. Now consider the element $\varphi(\alpha) \in H^0(Y, K_Y \otimes L^1_Y) = H^0(Y, R^1\pi_{Y*}Hom(\pi^*K_Y \otimes M_Y, N_Y))$. This element comes from certain global $\widetilde{\varphi(\alpha)} \in Ext^1(\pi^*K_Y \otimes M_{U_Y}, N_{U_Y})$, since the corresponding obstruction is $\partial\varphi(\alpha) = \varphi(\partial\alpha) = 0$. Now define P_{U_Y} from the extension $0 \to N_{U_Y} \to P_{U_Y} \to \pi_*K_Y \otimes M_{U_Y} \to 0$ of class $\widetilde{\varphi(\alpha)}$. It satisfies all the needed properties, since fiberwise it came from the class of α . □

Proof of 2.1.2. b. Apply the inductive hypothesis to Z, a constant sheaf and L^o. We get $Y' \subset Z$, a lisse $Q_{Y'}$ on Y' and an injection $L^o_{Y'} \hookrightarrow Q_{Y'}$ such that the corresponding arrow $H^i(Z, L^o) \longrightarrow H^i(Y', Q_{Y'})$ is zero for $i > 0$. Define $O_{Y'}$ by means of the cocartesian square

$$\begin{array}{ccc} \pi^* Q_{Y'} \otimes M_{U_{Y'}} & \longrightarrow & O_{U_{Y'}} \\ \uparrow & & \uparrow \\ \pi^* L^o_{Y'} \otimes M_{U_{Y'}} & \longrightarrow & N_{U_{Y'}} \end{array}$$

where $\pi^* L^o_{Y'} \otimes M_{U_{Y'}} \longrightarrow N_{U_{Y'}}$ is the canonical arrow. The obvious commutative diagram

$$\begin{array}{ccc} R^o\pi_*Hom(M_{U_{Y'}}, N_{U_{Y'}}) & \longrightarrow & R^o\pi_*Hom(M_{U_{Y'}}, O_{U_{Y'}}) \\ \| & & \uparrow \\ L^o_{Y'} & \longrightarrow & Q_{Y'} \end{array}$$

shows that our $N_{U_{Y'}} \hookrightarrow O_{U_{Y'}}$ is what we need. □

So 2.1.1. is proven, and we may pass to the

2.2. Proof of Theorem 1.3. We will also use the induction in dim X; so assume that we have 1.3. for any variety of dimension less then dim X.

First note that the statement of 1.3. is Zaris ki local: let us consider 1.3. in the form 1.4. (iii); assume that M,N are in $\mathcal{M}(X)$ and we have found an affine Zariski covering $\{U_i\}$ of X together with injections $N_{U_i} \hookrightarrow N'_{U_i}$ such that all the maps $Ext^j_{D(U_i)}(M_{U_i}, N_{U_i}) \longrightarrow Ext^j_{D(U_i)}(M_{U_i}, N'_{U_i})$ are zero for any $j > 0$; then the injection $N \hookrightarrow N' := \bigoplus_i j_{i*}N'_{U_i}$ kills all $Ext^j_{D(X)}(M, N)$ (note that $j_{i*}N'_{U_i} \in Ob\, \mathcal{M}(X)$ since $j_i: U_i \hookrightarrow X$ is affine).

So we may assume X to be affine. Let us prove 1.3. in the form 1.4. (ii): we have to show that for any $M,N \in \mathcal{M}(X)$ one has $Ext^i_{\mathcal{M}(X)}(M,N) = Ext^i_{D(X)}(M,N)$.

2.2.1. First assume that the dimensions of supports of M,N are less than dim X. Then for certain $f \in \mathcal{O}(X)$ this supports lie in $Y = f^{-1}(0)$, dim Y < dim X. We have $Ext^i_{D(X)}(M,N) = Ext^i_{D(Y)}(M,N) =$

$= \mathrm{Ext}^i_{M(Y)}(M,N)$ (the first equality follows from adjunction of i_* and i^*, and the second one is the inductive hypothesis). So it remains to prove that the embedding $M(Y) \hookrightarrow M(X)$ induces the isomorphism $I : \mathrm{Ext}^i_{M(Y)}(M,N) \xrightarrow{\sim} \mathrm{Ext}^i_{M(X)}(M,N)$. Let us construct the inverse map to I. Consider the vanishing cycles functor $\Phi_f : M(X) \to M(Y)$ (see 1.2.1. b)). Since Φ_f is exact, it defines the map $\Phi_{f_*} : \mathrm{Ext}^i_{M(X)}(P, Q) \to \mathrm{Ext}^i_{M(Y)}(\Phi_f(P), \Phi_f(Q))$. But $\Phi_f |_{M(Y)}$ is identity functor, so for $M,N \in M(Y)$ we get the arrow $\Phi_{f_*} : \mathrm{Ext}^i_{M(X)}(M,N) \to \mathrm{Ext}^i_{M(Y)}(M,N)$ left-inverse to I. It remains to show that it is also right-inverse, i.e. that $I\Phi_{f_*} = \mathrm{id}_{\mathrm{Ext}^i_{M(X)}}$.

At this point it is convenient to use Yoneda's construction of $\mathrm{Ext}^i_{M(X)}$. Let me recall it briefly. Namely, let $E^i_{M(X)}(M,N)$ be the category of acyclic complexes in M of type $0 \to N \to C^1 \to \ldots \to C^i \to M \to 0$, the morphisms in $E^i_{M(X)}(M,N)$ being morphisms of complexes that induce identity maps on the ends M,N. Then $\mathrm{Ext}^i_{M(X)}(M,N)$ is just the set of connected components of $E^i_{M(X)}(M,N)$, i.e. Ext^i is the set of equivalence classes of objects of E^i, where two objects are equivalent if you may connect them by a sequence of morphisms.

So for $M,N \in M(Y)$ let $C^\cdot = (N \to C^1 \to \ldots \to C^i \to M)$ be an object in $E^i_{M(X)}(M,N)$. We have to show that $C^\cdot$ is equivalent to $\Phi_f(C^\cdot) = (N \to \Phi_f(C^1) \to \ldots \to \Phi_f(C^i) \to M)$. Here is the sequence of morphisms in $E^i_{M(X)}(M,N)$ that connects them (it comes from [2]n°3(*)):

$$C^\cdot \to C^\cdot \oplus \Xi_f(C^\cdot_U) \to C^\cdot \oplus \Xi_f(C^\cdot_U)/j_!(C^\cdot_U) \leftarrow \Phi_f(C^\cdot).$$

2.2.2. We will use the following remark. Let $j : U \hookrightarrow X$ be an affine embedding. Then the functors j^*, $j_!$ and j_* between $M(U)$ and $M(X)$ are exact and pairs $(j_!, j^*)$, (j^*, j_*) are adjoint. This implies that for any $A \in M(U)$, $B \in M(X)$ one has

$$\mathrm{Ext}^\cdot_{M(X)}(j_!A,B) = \mathrm{Ext}^\cdot_{M(U)}(A,B_U), \quad \mathrm{Ext}^\cdot_{M(X)}(B,j_*A) = \mathrm{Ext}^\cdot_{M(U)}(B_U, A).$$

2.2.3. Consider now the case when $\dim \operatorname{supp} N < \dim X$. Choose an open affine $j : U \hookrightarrow X$ such that $\dim(X-U) < \dim X$ and $X-U \supset \mathrm{supp}\, N$. The exact sequence of Ext's shows that it suffices to consider the case when M is irreducible. We may also assume that $\mathrm{supp}\, M \cap U \neq \emptyset$ (otherwise see 2.2.1). Then the canonical map $j_! j^* M \to M$ is epimorphism; let K be its kernel. We have $\mathrm{supp}\, K \subset X-U$, Hence $\mathrm{Ext}^{\cdot}_{M(X)}(K,N) = \mathrm{Ext}^{\cdot}_{D(X)}(K,N)$ by 2.2.1. Also $\mathrm{Ext}^{\cdot}_{M(X)}(j_! M_U, N) = \mathrm{Ext}^{\cdot}_{D(X)}(j_! M_U, N)$ since by 2.2.2. both parts are zero. Now the exact sequence of Ext's shows that $\mathrm{Ext}^{\cdot}_{M(X)}(M,N) = \mathrm{Ext}_{D(X)}(M,N)$.

2.2.4. Now let we show that $\mathrm{Ext}^{\cdot}_{M(X)} = \mathrm{Ext}^{\cdot}_{D(X)}$ for an irreducible N supported at a generic point ; then the exact sequence of Ext's plus 2.2.3. prove the result for general N. For any open affine $j_U : U \hookrightarrow X$, $\eta \in U$, we have canonical injection $N \hookrightarrow j_*(N_U)$; put $L^U = j_*(N_U)/N$, clearly $\mathrm{supp}\, L^U \subset \bar{U}-U$. Consider the morphism between the long exact sequences of $\mathrm{Ext}^{\cdot}_{M(X)}$ and $\mathrm{Ext}^{\cdot}_{D(U)}$ for $0 \to N \to j_*(N_U) \to L^U - 0$. By 2.2.3. this morphism is an isomorphism on $\mathrm{Ext}^{\cdot}(M, L^U)$, hence $\mathrm{Ker}^{\cdot}(M,N) := \mathrm{Ker}\,(\mathrm{Ext}^{\cdot}_{M(X)}(M,N) \to \mathrm{Ext}^{\cdot}_{D(U)}(M,N)) = \mathrm{Ker}\,(\mathrm{Ext}^{\cdot}_{M(X)}(M, j_*N_U) \to \mathrm{Ext}^{\cdot}_{D(X)}(M, j_*N_U))$, the same with Coker. According to 2.2.2., one has $\mathrm{Ext}^{\cdot}(M, j_*N_U) = \mathrm{Ext}^{\cdot}(M_U, N_U)$ for both $\mathrm{Ext}_{M(X)}$, $\mathrm{Ext}_{D(X)}$, therefore $\mathrm{Ker}^{\cdot}(M,N) = \mathrm{Ker}^{\cdot}(M_U, N_U)$. Shrinking U to η we get $\mathrm{Ker}^{\cdot}(M,N) = \mathrm{Ker}^{\cdot}(M_\eta, N_\eta) = 0$ by 2.1.; the same with Coker. This proves Theorem 1.3. □

§3. Direct images as derived functors

3.1. Recall that all the standard functors between categories $D(X)$ come naturally together with their f-liftings. So we may use the construction from A 7. In particular, A 7.1 implies

Lemma 3.1. Let F be a t-exact standard functor between $D(X)$, say, F is Verdier's duality, or F is the nearby cycles functor, or F is the vanishing cycles functor; let $F_M = HF|_M$ be corresponding exact functor between hearts $M(X)$ and $\tilde{F}_M$ the functor between $D^b(M(X))$ induced by F_M. Then $\tilde{F}_M = F$ via the identification established by the main theorem. □

3.2. Now let us pass to the direct image functors. Let $f : X \to Y$ be a morphism of schemes. It determines, under the identification of the main theorem, the exact functors $f_*, f_! : D^b(M(X)) \to D^b(M(Y))$. Assume that f is affine (for a non-affine case, see n° 3.4). Then $f_!$ is left t-exact, and f_* is right t-exact; let $f_{!M} = Hf_!|_{M(X)}$,

$f_{*M} : \mathcal{M}(X) \to \mathcal{M}(Y)$ be corresponding functors between $\mathcal{M}$'s. According to A 7. we have natural morphisms $Rf_{!M} \to f_!$, $f_* \to Lf_{*M}$.

Theorem 3.2. This morphisms are isomorphisms.

Proof. It suffices to treat the case of f_*, the case of $f_!$ follows by duality. The theorem would follow if we show that for any $M \in \mathcal{M}(X)$ there exists $N \twoheadrightarrow M$ such that $H^i f_*(N) = 0$ for any $i < 0$ (hence for any $i \neq 0$). This fact is a particular case of the following more general lemma, valid for arbitrary f (not necessary affine; in full generality this will be needed later; for the aimes of 3.2. you may assume that $U_i = X$)

Lemma 3.3. Let $f : X \to Y$ be any morphism of schemes, $r_\ell : U_\ell \hookrightarrow X$ a finite set of affine open embeddings, and $M \in \operatorname{Ob}\mathcal{M}(X)$. Then there exists $N \twoheadrightarrow M$ such that the perverse sheaves $H^i f|_{U_\ell *}(N_{U_\ell})$ on Y are zero for any U_ℓ and $i < 0$.

Proof. 3.3.1. First assume that X is quasiprojective. Our N will be of the form $j_!(M_V)$ for a certain affine open embedding $j : V \hookrightarrow X$. Consider a diagram $\begin{array}{ccc} X & \xrightarrow{k_X} & \bar{X} \\ f\downarrow & & \downarrow\bar{f} \\ Y & \xrightarrow{k_Y} & \bar{Y} \end{array}$, where k_X is an affine open embedding and $\bar{X}$ is projective. Then $k_\ell := k_X r_\ell : U_\ell \hookrightarrow \bar{X}$ are also affine, therefore $\bar{M} := k_{X*}M$, $\bar{M}_\ell := k_{\ell *}(M_{U_\ell})$ belong to $\mathcal{M}(\bar{X})$. Find a hyperplane section $H \subset \bar{X}$ such that the following conditions hold (here $V := \bar{X} \setminus H \xrightarrow{j} \bar{X}$).

$$\begin{array}{ccc} & & \\ \uparrow k_{\ell V} & & \uparrow k_\ell \\ V_\ell := V \cap U_\ell & \xrightarrow{j_\ell} & U_\ell \end{array}$$

a_H. The canonical arrow $j_!(\bar{M}_V) = j_! j^* \bar{M} \to \bar{M}$ is surjective

b_H. The canonical arrows $j_!(\bar{M}_{\ell V}) = j_! k_{\ell V*}(M_{V_\ell}) \to k_{\ell *} j_{\ell !}(M_{V_\ell})$ are isomorphisms.

Suppose we have found such an H; then set $N = k_X^* j_! \bar{M}_V = j_{X!} M_{V_X}$ (here $j_X : V_X := V \cap X \hookrightarrow X$); the condition a_H shows that the canonical arrow $N \to M$ is surjective, and condition b_H implies that $H^i(f|_{U_\ell *}(N_{U_\ell})) = H^i(\bar{f}_* k_{\ell *} j_{\ell !} M_{U_\ell})_Y = H^i(\bar{f}_* j_!(\bar{M}_\ell)_V)_Y = H^i((\bar{f} \circ j)_!(\bar{M}_\ell)_V)_Y$ are zero for $i < 0$ (since $\bar{f} \circ j$ is affine). Thus 3.3. in our case is proved.

To find H let us rewrite the conditions a_H and b_H as follows (here $H \xrightarrow{i_H} \bar{X}$):

$$\begin{array}{ccc} & & \\ k_{\ell H}\uparrow & & \uparrow k_\ell \\ H_{U_\ell} & \xrightarrow{i_{\ell H}} & U_\ell \end{array}$$

a'$_H$. The objects $H^a i_H^*(\bar{M}) \in \mathcal{M}(H)$ are zero for $a \neq -1$

b'$_H$. The canonical arrows $i_H^* k_{\ell *}(M_{U_\ell}) \longrightarrow k_{\ell H *} i_{\ell H}^*(M_{U_\ell})$ are isomorphisms.

(To see that b <=> b' use the octahedron of the commutative diagram

$$k_{\ell *}M_{U_\ell} \nearrow i_{H*}i_H^* k_{\ell *}(M_{U_\ell}) \downarrow \searrow i_{H*} k_{\ell H*} i_{\ell H}^*(M_{U_\ell}) \ .)$$

Now let us consider all the hyperplane sections H simultaneously: they are parametrized by a projective space $\mathcal{P}$, and the above pictures for different H's are the fibers of the global picture over the scheme $\mathcal{P}$. For example we have canonical closed hyperplane section subspace $i : H_{\mathcal{P}} \longrightarrow \bar{X}_{\mathcal{P}} := \bar{X} \times \mathcal{P}$, the object $M_{\mathcal{P}} := P_X^*(M)[\dim \mathcal{P}]$ in $\mathcal{M}(X_{\mathcal{P}})$ and so on.

The statements a'$_H$ and b'$_H$ have obvious analogues $a_{\mathcal{P}}$, $b_{\mathcal{P}}$ in this situation, which are clearly true, since the projection $H_{\mathcal{P}} \longrightarrow \bar{X}$ is smooth. So a_H and b_H for a particular H would follow if we know that the objects from a_H and b_H coincide with the restriction of objects from $a_{\mathcal{P}}$, $b_{\mathcal{P}}$ to the fiber over the closed point "H" of $\mathcal{P}$. But this is true over some Zariski open subset of $\mathcal{P}$ (see [4] th. 1.9. for the constructible case, the holonomic case is quite easy), so we are done (in fact, if the base field is finite, this Zariski open may have no points over our field, so first one has to pass to a certain finite extension, and then use the trace).

3.3.2. To get Lemma 3.3. without quasiprojectivity assumption on X, one proceeds as follows. First, you may assume that for a certain open $r : U \hookrightarrow X$ such that U is affine one has $M = r_! M_U$ (to see this choose an affine open covering $\{U_\nu\}$ of X; then the canonical arrow $\oplus r_{\nu !} M_{U_\nu} \longrightarrow M$ is surjective, so if $N_\nu \longrightarrow r_{\nu !} M_{U_\nu}$ satisfy 3.3, then $N = \oplus N_\nu \longrightarrow M$ also fits 3.3.). Now take $U \overset{\tilde{r}}{\nearrow} \tilde{X} \downarrow \P \underset{r}{\searrow} X$, where $\P$ is proper, $\tilde{X}$ is quasiprojective and $\tilde{r}$ is an open embedding. Apply 3.3.1. to $f \circ \P : \tilde{X} \longrightarrow Y$, $\tilde{M} = \tilde{r}_! M_U$ and $\tilde{U}_\ell = \P^{-1}(U_\ell)$. We get an affine $\tilde{j} : \tilde{V} \hookrightarrow \tilde{X}$ such that $\tilde{N} = \tilde{j}_! \tilde{M}_V = (\tilde{r} \circ j)_!(M_V) \longrightarrow M$ (here $j : V := \tilde{V} \cap U \hookrightarrow U$) is surjective and $H^i(f \circ \P)_*(\tilde{N}_{\tilde{U}_\ell})$ are zero for $i < 0$. Consider $N := (r \circ j)_!(M_V) \longrightarrow M$: clearly this arrow is surjective (it coincides with $\tilde{N} \longrightarrow \tilde{M}$ on U and $r_!$ is exact) and $f_*(N_{U_\ell}) = (f \P)_*(\tilde{N}_{\tilde{U}_\ell})$ (since $\P_* \tilde{N} = \P_! \tilde{N} = N$) are acyclic in negative degrees. So we are done. □

3.4. If $f : X \longrightarrow Y$ is an arbitrary morphism of schemes, then f_*

may be neither left, nor right t-exact. To recover f_* (or $f_!$) from f_*^o one may proceed as follows. Fix some open covering $\{U_i\}$, $\dot{\bigcup} U_i = X$, such that every U_i is affine. We get the functor $f^o_{\{U_i\}*} : M(X) \to M(Y)^\Delta$ (= the category of cosimplicial objects in $M(Y)$) by the formula $f^o_{\{U_i\}*}(M)^n = \bigoplus_{(U_{i_0},\dots,U_{i_n})} f^o_*(M_{U_{i_0}\cap\dots\cap U_{i_n}})$ with obvious face and degeneracy maps (this is f^o_* applied to the Cech complex of M). Clearly, $f^o_{\{U_i\}*}$ is right exact and 3.3. implies that $f_* = \mathrm{Tot}\, L f^o_{\{U_i\}*}$, where $\mathrm{Tot} : D^b(M(Y)^\Delta) \to D^b(M(Y))$ is passing to the total complex functor. One may say this in a more invariant way, without fixing $\{U_i\}$. Consider the category $Sh_{X,Y}$ of $M(Y)$-valued sheaves on X_{Zar}; we have the global section functor $\Gamma : Sh_{X,Y} \to M(Y)$ and its right-derived functor $R\Gamma : D^b(Sh_{X,Y}) \to D^b(M(Y))$. For $M \in M(X)$ let $f^\sim_*(M)$ be the sheaf that corresponds to the presheaf $U \longmapsto f^o_*(M_U)$; clearly $f^\sim_* : M(X) \to Sh_{X,Y}$ is a right-exact functor of finite cohomology dimension, and 3.3. shows that $f_* = R\Gamma \cdot Lf^\sim_*$.

Appendix. Filtered categories and realization functor

What follows is a variation on theme [1] (3.1). We deal with the following problem: given a t-category D with the heart $\mathcal{C}$; construct the t-exact functor real : $D^b(\mathcal{C}) \to D$ that induces the identity functor on $\mathcal{C}$. To do this one needs some extra structure on D: namely one should fix a filtered category over D. Here are convenient definitions.

Definition A 1. a) A filtered triangulated category, or f-category for short, is a triangulated category DF together with two strictly full triangulated subcategories DF (≤ 0) and DF (≥ 0), an exact automorphism $s : DF \to DF$ (called "shift of filtration"), and morphism of functors $\alpha : Id_{DF} \to s$. The following axioms should hold (here $DF(\leq n) := s^n DF(\leq 0)$, $DF(\geq n) := s^n DF(\geq 0)$):

(i) $DF(\geq 1) \subset DF(\geq 0)$, $DF(\leq 1) \supset DF(\leq 0)$, $\bigcup_{n\in\mathbb{Z}} DF(\leq n) = \bigcup_{n\in\mathbb{Z}} DF(\geq n) = DF$

(ii) For any X in DF one has $\alpha_X = s(\alpha_{s^{-1}X})$.

(iii) For any X in $DF(\geq 1)$, Y in $DF(\leq 0)$ one has $Hom(X,Y) = 0$ and α induces isomorphisms $Hom(Y,X) = Hom(Y, s^{-1}X) = Hom(sY,X)$.

(iv) For any X in DF there exists a distinguished triangle $A \to X \to B$ with A in $D(\geq 1)$ and B in $D(\leq 0)$.

b) An f-functor between f-categories is an exact functor that commutes with s and α and conserves $DF(\leqslant 0)$, $DF(\geqslant 0)$.

c) Let D be a triangulated category. An f-category over D is an f-category DF together with equivalence of triangulated categories $i : D \longrightarrow DF(\leqslant 0) \cap DF(\geqslant 0)$. If $\Phi : D_1 \to D_2$ is an exact functor between triangulated categories, and DF_i is an f-category over D_i $(i = 1,2)$ then an f-lifting of Φ is an f-functor $\Phi F :$ $DF_1 \to DF_2$ such that $i \circ \Phi = \Phi F \circ i$.

Example A 2. Let A be ab abelian category, $D^{\bullet}A$ its derived category, and $DF(A)$ the filtered derived category of complexes with finite decreasing filtration. Then $D^{\cdot}F(A)$ is an f-category over $D^{\cdot}(A)$: take $D^{\cdot}F(A)(\geqslant n) = \{(C^{\cdot}, F) : gr^i_F(C^{\cdot}) = 0 \text{ for } i < n\}$, $D^{\cdot}F(A)(\leqslant n) =$ $= \{(C^{\cdot}, F) : gr^i_F(C^{\cdot}) = 0 \text{ for } i > n\}$; $s(C^{\cdot}, F) = (C^{\cdot}, F^{\cdot -1})$; $\alpha_{(C^{\cdot},F)} = id_{C^{\cdot}} : (C^{\cdot}, F) \longrightarrow (C^{\cdot}, F^{\cdot -1})$; $i(C^{\cdot}) = (C^{\cdot}, Tr)$, where Tr is the trivial filtration : $gr^i_{Tr} = 0$ for $i \neq 0$. If $T : A_1 \to A_2$ is a left exact functor between abelian categories, and $RT : D^{+}(A_1) \longrightarrow D^{+}(A_2)$, $RFT : D^{+}F(A_1) \to D^{+}F(A)$ are corresponding right derived functors, then RFT is an f-lifting of RT; same for left derived functors.

Note that we have canonical exact functors $\sigma_{\geqslant n} : DF(A) \to DF(A)(\geqslant n)$, $\sigma_{\leqslant n} : DF(A) \longrightarrow DF(A)(\leqslant n)$ defined by formulas $\sigma_{\geqslant n}(C^{\cdot}, F) =$ $F^n(C^{\cdot})$ with the induced filtration, $\sigma_{\leqslant n}(C^{\cdot}, F) = C^{\cdot}/F^{n+1}(C^{\cdot})$ with the induced filtration, and also the forgetting of filtration functor $\omega : DF(A) \to D(A)$. One may define them for arbitrary f-categories as follows.

Proposition A 3. Let DF, D be as in A 1.c. Then

(i) The inclusion of $DF(\geqslant n)$ in DF has *right* adjoint $\sigma_{\geqslant n} : DF \to DF(\geqslant n)$, and the inclusion of $DF(\leqslant n)$ in DF has *left* adjoint $\sigma_{\leqslant n} : DF \longrightarrow DF(\leqslant n)$. The functors $\sigma_{\leqslant n}$, $\sigma_{\geqslant n}$ are exact; they preserve subcategories $DF(\leqslant a)$, $DF(\geqslant a)$; there is the unique isomorphism $\sigma_{\leqslant a}\sigma_{\geqslant b} \xrightarrow{\sim} \sigma_{\geqslant b}\sigma_{\leqslant a}$ such that the diagram

$$\begin{array}{ccccccc} & & & X & & & \\ & \nearrow & & & & \searrow & \\ \sigma_{\geqslant b}X & \to & \sigma_{\leqslant a}\sigma_{\geqslant b}X & \xrightarrow{\sim} & \sigma_{\geqslant b}\sigma_{\leqslant a}X & \to & \sigma_{\leqslant a}X \end{array}$$

commutes. Put $gr^n_F := i^{-1}s^{-n}\sigma_{\leqslant n}\sigma_{\geqslant n} : DF \to D$.

(ii) For any X in DF there exists unique morphism $d \in Hom^1(\sigma_{\leqslant 0}X, \sigma_{\geqslant 1}X)$ such that the triangle $\sigma_{\geqslant 1}X \to X \to \sigma_{\leqslant 0}X \xrightarrow{d} \dots$

is distinguished; this is the only, up to unique isomorphism, triangle (A,X,B) with A in $DF(\geq 1)$ and B in $DF(\leq 0)$.

(iii) There exists the only, up to unique isomorphism, exact functor $\omega : DF \to D$ such that

1). $\omega|_{DF(\leq 0)} : DF(\leq 0) \to D$ is right adjoint to $i : D \to DF(\leq 0)$

2) $\omega|_{DF(\geq 0)} : DF(\geq 0) \to D$ is left adjoint to $i : D \to DF(\geq 0)$

3) For any X in DF the arrow $\omega(\alpha_X) : \omega(X) \to \omega(sX)$ is an isomorphism.

4) For any A in $DF(\leq 0)$, B in $DF(\geq 0)$ we have $\omega : Hom(A,B) \xrightarrow{\sim} Hom(\omega A, \omega B)$.

In fact, ω is uniquely determined by properties 1,3 or 2,3. □

One may see that all the standard constructions in usual filtered derived categories may be carried over in the f-category framework. Exercise: do this for [1] (3.1.2.6).

Definition A 4. Let D, DF be as in 1.5.1., and assume that we are given t-structures on D and DF. Say that they are compatible if $i : D \to DF$ is t-exact, and $s(DF^{\leq 0}) = DF^{\leq -1}$

Proposition A 5. a) Given a t-structure on D, there exists unique t-structure on DF compatible with it, namely $DF^{\leq 0} = \{X : gr_F^i(X) \in D^{\leq i}\}$, $DF^{\geq 0} = \{X : gr_F^i(X) \in D^{\geq i}\}$

b) Assume we are given compatible t-structures on D, DF with hearts $\mathcal{C}$, $\mathcal{C}F$ respectively; let $H : D \to \mathcal{C}$ be corresponding cohomology functor. Define the cohomology functor $H_F : DF \to C^b(\mathcal{C})$ by the formula $H_F(X)^i = H^i gr_F^i(X)$, the differential $H_F(X)^i \to H_F(X)^{i+1}$ comes from the distinguished triangle $\omega(\sigma_{\leq i+1}\sigma_{\geq i+1} \to \sigma_{\leq i+1}\sigma_{\geq i} \to \sigma_{\leq i}\sigma_{\geq i})(X)$ in D. Then $H_F|_{\mathcal{C}F} : \mathcal{C}F \to C^b(\mathcal{C})$ is an equivalence of categories, and, via this equivalence, H_F becomes a canonical cohomology functor of our t-structure on DF. □

A 6. Let $\widetilde{real} : C^b(\mathcal{C}) \to D$ be the composition $C^b(\mathcal{C}) \xrightarrow{\alpha} \mathcal{C}F \subset DF \xrightarrow{\omega} D$, where $\alpha = H_F|_{\mathcal{C}F}^{-1}$. One may see that $H \circ \widetilde{real} : C^b(\mathcal{C}) \to$ is a usual cohomology functor, and $\widetilde{real}$ factors (uniquely) through the t-exact functor $real : D^b(\mathcal{C}) \to D$, $real|_{\mathcal{C}} = id_{\mathcal{C}}$. This is the functor we looked for.

A 7. This construction has the following functorial properties. Let D_i, $i = 1,2$, be triangulated categories, and DF_i some f-categories over D_i. Let $\Phi : D_1 \to D_2$ be an exact functor, and $\Phi F : DF_1 \to DF_2$ an f-lifting of Φ. Assume that we are given compatible t-structures on D_i, DF_i.

If Φ is t-exact, then it induces an exact functor $\Phi_{\mathcal{C}} : \mathcal{C}_1 \longrightarrow \mathcal{C}_2$, hence the t-exact functor $D^b(\mathcal{C}_1) \xrightarrow{D\Phi_{\mathcal{C}}} D^b(\mathcal{C}_2)$. Since, by 5.a, ΦF is also t-exact, we get

Lemma A 7.1. In this situation real $D\Phi_{\mathcal{C}} = \Phi$ real. □

More generally, suppose that Φ is left t-exact. Consider the left exact functor $\Phi_{\mathcal{C}} = H\Phi|_{\mathcal{C}_1} : \mathcal{C}_1 \to \mathcal{C}_2$ and the corresponding graded functor $\Phi^{\cdot}_{\mathcal{C}} : C^b(\mathcal{C}_1) \to C^b(\mathcal{C}_2)$. Clearly, ΦF is also left t-exact, and $\Phi^{\cdot}_{\mathcal{C}} = H_F \circ \Phi F \circ \alpha$, so for any $C^{\cdot}$ in $C^b(\mathcal{C}_1)$ we have the canonical arrow $\alpha\Phi^{\cdot}_{\mathcal{C}}(C^{\cdot}) = \tau_{\leq 0}\Phi F\alpha(C^{\cdot}) \to \Phi F\alpha(C^{\cdot})$. Applying ω we get the morphism $\widetilde{\text{real}}\,\Phi^{\cdot}_{\mathcal{C}} \to \Phi\,\widetilde{\text{real}}$ of D_2-valued functors on $C^b(\mathcal{C}_1)$, or, passing to the derived functor $R\Phi^{\cdot}_{\mathcal{C}} : D^b(\mathcal{C}_1) \to \varinjlim D^b(\mathcal{C}_2)$, the morphism real$\circ R\Phi^{\cdot}_{\mathcal{C}} \to \Phi\circ$real. This construction has an obvious analogue for the right t-exact functors.

References

1. A. Beilinson, J. Bernstein, P. Deligne. Faisceaux pervers, Astérisque 100, 1982.
2. A. Beilinson. How to glue perverse sheaves, this volume
3. J. Bernstein. Algebraic theory of D-modules, to appear in Astérisque.
4. P. Deligne. Theoremes de finitude en cohomologie l-adique, dans SGA 4½, Lect. notes in math. 569 (1977).
5. Le D.-T. Faisceaux constructibles quasiunipotents, Sem. Bourbaki exp. 581 (1981).

Complement to §3.

Here we will see how a variant of 3.3.1. gives a very concrete chain complexes that compute $\int$ (:= direct image to a point) of perverse sheaves on projective variety.

These chain complexes will be constructed by means of a pair of transversal flags (F, F') on $\mathbb{P}^n$. Clearly, such pairs are the same as systems of coordinate axes; they are parametrized by GL(n+1)/ diagonal matrices. We put $F = (F_n \subset \ldots \subset F_0 = \mathbb{P}^n)$, $F' = (F'_n \subset \ldots \subset F'_0 = \mathbb{P}^n)$ $i_k : F_k \hookrightarrow \mathbb{P}^n$, $i'_k : F'_k \hookrightarrow \mathbb{P}^n$.

Let M be a perverse sheaf on $\mathbb{P}^n$. Say that (F,F') is in a generic position with respect to M if for any pair (a,b) of indices (we may assume $0\leqslant a,\ b\leqslant n,\ a+b\leqslant n$) the following holds:

(i) The objects $M^a_b := i_{a*}i^!_a i'_{b*}i'^*_b M[a-b]$, $\tilde{M}^a_b := i'_{b*}i'^*_b i_{a*}i^!_a M[a-b]$ of $D(\mathbb{P}^n)$ (supported on $F_a\cap F'_b$) are perverse sheaves

(ii) The natural arrow $\tilde{M}^a_b \to M^a_b$ is isomorphism.

One may see, as in 3.3.1, that a generic (F,F') satisfies this conditions. Assume that our (F,F') is of such kind. Then for any (a,b) we have canonical commentative diagram of perverse sheaves, supported on $F_a\cap F'_b$, with exact rows and columns:

$$\begin{array}{ccccccccc}
 & & 0 & & 0 & & 0 & & \\
 & & \uparrow & & \uparrow & & \uparrow & & \\
0 & \to & M^a_b & \to & j_*j^*M^a_b & \to & M^{a+1}_b & \to & 0\\
 & & \uparrow & & \uparrow\beta & & \uparrow & & \\
0 & \to & j'_!j'^*M^a_b & \xrightarrow{\alpha'} & \mathcal{J}^a_b(M) & \xrightarrow{\beta'} & j'_!j'^*M^{a+1}_b & \to & 0\\
 & & \uparrow & & \uparrow\alpha & & \uparrow & & \\
0 & \to & M^a_{b+1} & \to & j_*j^*M^a_{b+1} & \to & M^{a+1}_{b+1} & \to & 0\\
 & & \uparrow & & \uparrow & & \uparrow & & \\
 & & 0 & & 0 & & 0 & &
\end{array}$$

where $j:\ \mathbb{P}^n\setminus H\hookrightarrow\mathbb{P}^n$, $j':\ \mathbb{P}^n\setminus H'\hookrightarrow P^n$, H, H' are hyperplanes such that $H\cap F_a\cap F'_b = F_{a+1}\cap F'_b$, $H'\cap F_a\cap F'_b = F_a\cap F'_{b+1}$, and $\mathcal{J}^a_b(M) := j'_!\ j'^*\ j_*\ j^*\ M = j_*\ j^*\ j'_!\ j'^*\ M^a_b$ (the property (ii) guarantees that these are the same objects). So the sheaves $\mathcal{J}^a_b(M)$ form a bicomplex with differentials $d=\alpha\beta$, $d'=\alpha'\beta'$. Clearly it is d-acyclic if $b\neq 0$, and d'-acyclic if $a\neq 0$. Let $\mathcal{J}^{\cdot}(M)$ be the corresponding total complex ($\mathcal{J}^n = \bigoplus_{a-b=n}\mathcal{J}^a_b$). The diagram also shows that it is canonically quasiisomorphic to M (i.e. one has $H^i\ \mathcal{J}^{\cdot}(M)=0$ for $i\neq 0$, $H^0\mathcal{J}^{\cdot}(M)=M$): the quasiisomorphism is $M=M^0_0 \to (\ j_*j^*M^0_0 \to j_*j^*M^1_0 \to \ldots)\xleftarrow{\beta}$ $\mathcal{J}^{\cdot}(M)$ (you may also use the 0's column to get the same quis). Now note that for $i\neq 0$ and any (a,b) $H^i\int\ \mathcal{J}^a_b(M)=0$ (see 3.3.1.). So $\int^0\mathcal{J}^{\cdot}(M)$ is chain complex that represents $\int M$.

How to glue perverse sheaves

A.A. Beilinson

The aim of this note is to give short self-contained account forthe vanishing cycle constructions of perverse sheaves, e.g. for needs of [1]. It differs somewhat from alternative approaches offered by Macpherson-Villonen [6] and Verdier [8],[9], that justifies, possibly, its publication.

The text follows closely the notes written down by S.I. Gel'fand in 1982. I am much obliged to him. I am also much obliged to Y.Bernstein: the construction of the functor Ψ in n°2 below was found in a joint work with him in spring 1981.

We will consider the algebraic situation only; for notations see [1] §1.

1. The monodromy Jordan block. In this preliminary we will fix the notation concerning the standard local systems on $\mathbb{A}^1 \setminus \{0\}$ with unipotent monodromy of single Jordan block.

1.1. Let us start with the classical topology situation, so in this n° $k=\mathbb{C}$. For an integer i as usually $\mathbb{Z}(i) := (2\pi\sqrt{-1})^i \mathbb{Z} = \mathbb{Z}(1)^{\otimes i}$, $\mathbb{Z}(1) = \pi_1((\mathbb{A}^1 \setminus \{0\})(\mathbb{C}), 1)$. Consider the group algebra $\mathbb{Z}[\mathbb{Z}(1)]$; for $l \in \mathbb{Z}(1)$ let $\tilde{l}$ denote l as an (invertible) element in $\mathbb{Z}[\mathbb{Z}(1)]$. Let $I = \mathbb{Z}[\mathbb{Z}(1)](\tilde{t}-1)$ be the angmentation ideal, here t is a generator of $\mathbb{Z}(1)$. Put $A^\circ :=$ I-adic completion of $\mathbb{Z}[\mathbb{Z}(1)] = \mathbb{Z}[[\tilde{t}-1]]$, $A^i = (\tilde{t}-1)^i A^\circ = (A^1)^i$ $(i \geq 0)$. The graded ring $\mathrm{Gr} A^\circ = \bigoplus_{i \geq 0} A^i/A^{i+1}$ is canonically isomorphic to the polynomial ring $\mathbb{Z} \oplus \mathbb{Z}(1) \oplus \mathbb{Z}(2) + \ldots$ Put $A^{i*} = \{x \in A^i : A^i = xA^\circ\}$ $= \{x \in A^i :\ x \bmod A^{i+1}$ generates $A^i/A^{i+1} = \mathbb{Z}(i)\}$; one has $A^{i*} \cdot A^{j*} = A^{i+j*}$, so $\bigcup_{i \geq 0} A^{i*}$ is a multiplicative system. Let $A \supset A^\circ$ be the corresponding localisation of A°; one has $A = A^\circ_{(\tilde{t}-1)} = \mathbb{Z}((\tilde{t}-1))$. The ring A has a natural $\mathbb{Z}$-filtration $A^i = (A^{1*})^i A^\circ = (\tilde{t}-1)^i A^\circ$; for $i > 0$ this A^i conincide with the ones above; one has $\mathrm{Gr} A = \bigoplus_{i \in \mathbb{Z}} A^i/A^{i+1} = \oplus \mathbb{Z}(i)$ is a

Laurent polynomial ring.

Define $\mathbb{Z}$-bilinear pairing $\langle,\rangle : A\times A \to \mathbb{Z}(-1)$ by the formula $(\langle f,g\rangle, t) = \mathrm{Res}_{\tilde{t}=1}(f\, g^{-} d\log \tilde{t})$; here $g \mapsto g^{-}$ is the canonical involution of the ring $A: \tilde{1}^{-} := (\tilde{1})^{-1} = \widetilde{(-1)}$. This pairing is skew symmetric ($\langle f,g\rangle = -\langle g,f\rangle$), $\mathbb{Z}(1)$ - invariant ($\langle 1\, f,\ 1\, g\rangle = \langle f,g\rangle$); compatible with filtration and non-degenerate: ($A^{i} = (A^{-i})^{\perp}$ and $A^{a,b} := A^{a}/_{A}b \xrightarrow{\sim} \mathrm{Hom}\,(A^{-b}/_{A}{-a},\ \mathbb{Z}(-1))$; the pairing induced on $\mathrm{Gr}A$ is $\langle S_{i}\ ,\ S_{-i-i}\rangle = (-1)^{i} s_{i}\cdot s_{i-1}$, $s_{j} \in \mathbb{Z}(j)$.

Let I be the local system of invertible A-modules on $(\mathbb{A}^{1}\setminus\{0\})(\mathbb{C})$ such that the fiber of I over $1 \in \mathbb{A}^{1}(\mathbb{C})$ is A, and monodromy action of $1 \in \pi_{1}((\mathbb{A}^{1}\setminus\{0\})(\mathbb{C}), 1)$ = is the multiplication by $\tilde{1}$. We have the canonical filtration I^{i} on I, such that $I^{i}{}_{1} = A^{i}$, $I^{i}/I^{i+1} = \mathbb{Z}(i)$, together with the non-degenerate skew-symmetric pairing $\langle,\rangle : I\times I \to \mathbb{Z}(-1)$ compatible with the filtrations, that coincides on I_{1} with the above $\langle,\rangle$. Put $I^{a,b} = I^{a}/I^{b}$. We may consider $I = \varinjlim\!\!\!\!\varprojlim\ I^{a,b}$ as filtered A-object in the category $\varinjlim\!\!\!\!\varprojlim$ (local systems on $(\mathbb{A}^{1}\setminus\{0\})(\mathbb{C})$ (for definition of $\varinjlim\!\!\!\!\varprojlim$ see appendix, n°3).

1.2. These definitions have an obvious étale version: just replace $\mathbb{Z}(i)$ above by $\mathbb{Z}/\ell^{n}(i)$ and repeat 1.1. word-for-word. We get the ring $A_{ét} = \varinjlim\!\!\!\!\varprojlim\ A^{a,b}$ in $\varinjlim\!\!\!\!\varprojlim$ (sheaves on $\mathrm{spec}\,k_{ét}$) and $A_{ét}$- object $I_{ét} = \varinjlim\!\!\!\!\varprojlim\ I^{a,b}{}_{ét}$ in $\varinjlim\!\!\!\!\varprojlim$ (sheaves on $\mathbb{A}^{1}\setminus\{0\}_{ét}$). In the same way we get $\mathbb{Q}_{\ell}$- and mixed variants.

1.3. The holonomic counterpart of above is as follows. Put $A_{hol} = k((s))$, $A^{i}{}_{hol} = s^{i}k[[s]]$; define the pairing $\langle,\rangle: A_{hol} \times A_{hol} \to k$ by the formula $\langle f(s), g(s)\rangle = \mathrm{Res}_{s=0}\, f(s)\, g(-s)\, ds$. This pairing has the same properties as the one above (invariance: $\langle sf,g\rangle + \langle f,sg\rangle = 0$). For intgers $a \geqslant b$ let $I^{a,b}{}_{hol}$ be a D-module on $(\mathbb{A}^{1}\setminus\{0\})_{k}$ such that $I^{a,b}{}_{hol} \otimes \mathcal{O}$ as an $\mathcal{O}$- module, and $\nabla\alpha = s\alpha \frac{dx}{x}$ for $A^{a,b}{}_{hol}$ (here x is the parameter on $\mathbb{A}^{1}\setminus\{0\} = G_{m}$). Put $I_{hol} = \varinjlim\!\!\!\!\varprojlim\ I^{a,b}{}_{hol}$. This is filtered A_{hol} - object in $\varinjlim\!\!\!\!\varprojlim\ (M_{hol}(G_{m}))$. Clearly, $\mathrm{Gr}\, I = \oplus\ \mathcal{O}_{G_{m}}\cdot s^{i}$ (recall that the Tate twist is the identity functor in holonomic situation).

1.4. For any $\pi \in A^{1*}$ we have an isomorphism $\sigma_\pi : I^{a,b} \to I^{a+n,\, b+n}{}_{(-n)}$

$\sigma_\pi(x) = \pi^n x \otimes \bar{\pi}^{-n}$, where $\bar{\pi} = \pi \bmod A^2$. In holonomic or in $\mathbb{Q}_\ell$- situation we have the canonical choice σ of σ_π ; in holonomic case put **$\sigma = \sigma_s$ = multiplication by s^n, in $\mathbb{Q}_\ell$-case put** $\sigma = \sigma_{\log\, t}$, where t is a generator of $\mathbb{Q}_\ell(1)$.

In what follows I consider $I^{a,b}$ as ordinary sheaves on $A^1 \setminus \{0\}$, so $\underline{I}^{a,b}$ lives in $M(A^1 \setminus \{0\})[-1] \subset D(A^1 \setminus \{0\})$.

2. <u>The unipotent nearby cycles functor Ψ^{un}</u>. Let X be a scheme, and $f \in \mathcal{O}(X)$ a fixed function. Put $Y := f^{-1}(0) \overset{i}{\hookrightarrow} X \overset{j}{\hookleftarrow} U := X - Y$, $\underline{I}^{a,b}{}_f := f|_U^* (\underline{I}^{a,b})$ (so, in holonimic case, $\underline{I}_f$ is D-module generated by f^s). For any M in M(V) consider $M \otimes \underline{I}_f = \varprojlim M \otimes \underline{I}^{a,b}{}_f$ in $\varprojlim M(V)$. The ring A acts on $M \otimes \underline{I}_f$ via $\underline{I}_f$, and the pairing < , > defines the canonical isomorphism $*(M \otimes \underline{I}_f) = *(M) \otimes \underline{I}_f(1)$ compatible with the A-action (here $*$ is Verdier duality).

<u>Key-lemma 2,1.</u> The canonical arrow $\alpha : j_!(M \otimes \underline{I}_f) \to j_*(M \otimes \underline{I}_f)$ in $\varprojlim M(X)$ is an isomorphism.

<u>Proof.</u> The lemma would follow if for $\pi \in A^{1*}$ we find a certain $N \geq 0$ and a compatible system of morphisms $\beta^{a,b} : j_*(M \otimes \underline{I}^{a,b}{}_f) \to j_!(M \otimes \underline{I}^{a,b}{}_f)$ such that $\alpha^{ab} \circ \beta^{ab} = \pi^N = \beta^{ab} \alpha^{ab}$ (here α^{ab} are (a,b) - components of α), then $\alpha^{-1} = \pi^{-N} \varprojlim \beta^{ab}$. To do this it suffices to show that all $\operatorname{Ker} \alpha^{ab}$, $\operatorname{Coker} \alpha^{ab}$ are annihilated by some π^m independent of (a,b): just take $N = 2m$ and $\beta^{ab} = \beta_! \beta_*$ in the commutative diagram

$$\begin{array}{ccccc}
j_!(M \otimes \underline{I}^{ab}) & & & & j_* M \otimes \underline{I}^{ab} \\
\downarrow \pi^m & \searrow & & \swarrow \beta_* & \downarrow \pi^m \\
 & & \operatorname{Coim} \alpha^{ab} = \operatorname{Im} \alpha^{ab} & & \\
 & \swarrow \beta_! & & \searrow & \\
j_!(M \otimes \underline{I}^{ab}) & & & & j_* M \otimes \underline{I}^{ab}
\end{array}$$

By duality we may consider Coker's only. In holonomic case the desired fact follows from the lemma on b-functions (see [2] (3.8)), since $M \otimes \underline{I}_f = M \cdot f^s((s))$: take $m = \sum l(u)$, u runs a finite set of generators of M and l(u) is the number of integral roots of b-function of u. In the constructible case one should use the finitness theorem for the usual nearby cycles functor $R\Psi$ (see [5]). Note that for any $\mathcal{F}$ in D(X) we have a distinguished triangle $i_* j_* \mathcal{F} \to R\Psi^{un}(\mathcal{F}) \xrightarrow{1-t} R\Psi^{un}(\mathcal{F}) \to ..$ in D(Y), where $R\Psi^{un}$ is the part of $R\Psi$ on which monodromy acts in a unipotent way, and t is a generator of the monodromy group $\mathbb{Z}_\ell(1)$. Therefore Cone $(j_!(M \otimes I^{ab}_f) \to j_*(M \otimes I^{ab}_f)) = i^* j_*(M \otimes I^{ab}_f) = \mathrm{Cone}\,(R\Psi^{un}(M) \otimes I^{ab}_1 \xrightarrow{1-t} R\Psi^{un}(M) \otimes I^{ab}_1)[-1]$, where t acts on $R\Psi^{un}(M) \otimes I^{ab}_1 = R\Psi^{un}(M) \otimes A^{a,b}$ via $t \otimes \tilde{t}$. Since $A^{a,b}$ is a $\mathbb{Z}_\ell(1)$ - module with one generator, the power of 1-t that annulates $R\Psi^{un}(M)$, annulates the cone also, and we are done by finitness theorm for $R\Psi$ (see [5]). □

2.2.*) Put $\Pi_f(M) := j_!(M \otimes I_f) = j_*(M \otimes I_f)$; clearly $\Pi_f : M(U) \to \varprojlim M(X)$ is an exact functor (since so are $j_!$ and j_*), and < , > defines the canonical isomorphism $*\Pi_f(M) = \Pi_f(*M)(1)$. On Π_f, there are two admissible filtrations $\Pi^{\cdot}_!(M) = j_!(M \otimes I^a_f)$, $\Pi^{\cdot}_*(M) = j_*(M \otimes I^{\cdot}_f)$; one has $\Pi^{\cdot}_! \subset \Pi^{\cdot}_*$, $Gr^{\cdot}_{\Pi_!}(M) = j_! M(\cdot)$, $Gr^{\cdot}_{\Pi_*}(M) = j_* M(\cdot)$. By 2.1. for $a \leq b$ any $\Pi^{ab}_{!*}(M) := \Pi^a_*(M)/\Pi^b_!(M)$ belongs to $M(X) \subset \varprojlim M(X)$. Clearly $\Pi^{ab}_{!*} : M(U) \to M(X)$ are exact functors, $\Pi = \varprojlim \Pi^{ab}_{!*}$; one has $*\Pi^{ab}_{!*}(M) = \Pi^{-b-a}_{!*}(*M)(1)$; and (see 1.4) we have isomorphisms $\sigma_\pi : \Pi^{ab}_{!*} \xrightarrow{\sim} \Pi^{a+n,b+n}_{!*}(-n)$.

2.3. We will need the following particular $\Pi^{a,b}_{!*}$ - functors. The first is the functor $\Psi^{un}_f := \Pi^{0,0}_{!*}$ or simply Ψ_f, for short, of unipotent nearby cycles, and its relatives $\Psi_f^{(i)} := \Pi^{i,i}_{!*} \xrightarrow[\sigma_\pi]{\sim} \Psi_f(i)$. These take values in $M(Y) \subset M(X)$, and we have $\Psi_f^{(i)} * = * \Psi_f^{(-i)}(1)$. The second is $\Xi_f := \Pi^{01}_{!*}$ = the maximal extension functor, and the corresponding $\Xi_f^i : \Pi^{i,i+1}_{!*} \xrightarrow[\sigma_\pi]{\sim} \Xi_f(i)$.

We have the canonical exact sequences

$$0 \to j_!(M)(a) \xrightarrow{\alpha_-} \Xi^a_f(M) \xrightarrow{\beta_-} \Psi^{(a)}_f(M) \to 0$$

$$0 \to \Psi^{(a+1)}_f(M) \xrightarrow{\beta_+} \Xi^a_f(M) \xrightarrow{\alpha_+} j_*(M)(a) \to 0$$

*) The following constructions are quite paralled to the Lax - Phillips scheme in the scattering theory: the multiplication by s is time translation, $\Pi^{\circ}_!$, Π/Π°_* are in - and out-spaces, etc.

that are interchanged by duality. Here $\alpha_+ \circ \alpha_- = \alpha$ is the canonical morphism $j_! \to j_*$, and $\beta = \beta_- \beta_+ : \Psi_f^{(1)} \to \Psi_f^{(0)}$ is the canonical arrow $\Pi_{!*}^{1,1} \to \Pi_{!*}^{0,0}$; under the isomorphism $\sigma_\pi : \Psi_f^{(0)} \to \Psi_f^{(1)}(-1)$ the arrow becomes the multiplication by π.

3. <u>Vanishing cycles and glueing functor.</u> Let M_X be a perverse sheaf on X, and M_U be its restriction to U. Consider the following complex

$$(*) \qquad j_!(M_U) \xrightarrow{(\alpha_-,\gamma_-)} \Xi_f(M_U) \oplus M_X \xrightarrow{(\alpha_+,-\gamma_+)} j_*(M_U)$$

where $\alpha_\pm, \gamma_\pm$ are the (only) arrows that coincide with $\mathrm{id}\, M_U$ on U. Put $\Phi_f(M_X) := \mathrm{Ker}\,(\alpha_+, -\gamma_+) / \mathrm{Im}\,(\alpha_-, \gamma_-)$. Clearly $\Phi_f(M_X)$ is supported on Y and $\Phi_f : M(X) \to M(Y)$ is exact functor (since α_- is injective and α_+ is surjective). We have canonical arrows $\Psi_f^{(1)}(M_U) \xrightarrow{u} \Phi_f(M_X) \xrightarrow{v} \Psi_f(M_U)$ given by formulas $u(\psi) = (\beta_+(\psi), 0)$, $v(\xi, m) = \beta_-(\xi)$; clearly $vu = \beta_-\beta_+ = \beta$

Define a vanishing cycles data for f, f-data for short, to be a quadruple (M_U, M_Y, U, V), with M_U in $M(U)$, M_Y in $M(Y)$ and $\Psi_f^{(1)}(M_U) \xrightarrow{u} M_Y \xrightarrow{v} \Psi_f(M_U)$ such that $vu = \beta$. These f-datas form an abelian category $M_f(U,Y)$ in an obvious way. Put $F_f(M_X) := (M_U, \Phi_f(M_X), u, v)$; clearly this defines exact functor $F_f : M(X) \to M_f(U,Y)$.
Conversly, let (M_U, M_Y, u, v) be an f-data. Consider the complex

$$(**) \qquad \Psi^{(1)}_f(M_U) \xrightarrow{(\beta_+,u)} \Xi_f(M_U) \oplus M_Y \xrightarrow{(\beta_-,-v)} \Psi_f(M_U)$$

Put $G_f(M_U, M, u, v) = \mathrm{Ker}\,(\beta_-, -v) / \mathrm{Im}\,(\beta_+, u)$: this defines exact functor $G_f : M_f(U,Y) \to M(X)$ (G_f is exact since β_+ is mono and β_- is epi.). Call G_f the glue ing functor.

<u>Proposition.</u> 3.1. The functors $M(X) \underset{G_f}{\overset{F_f}{\rightleftarrows}} M_f(U,Y)$ are mutually inverse equivalences of categories.

<u>Proof.</u> For M_X in $M(X)$ consider $(*)$ as a diad (for diads see appendix A.3.); this way we may identify $M(X)$ with the category of diads of type $(*)$ having the property that both $\gamma_+|_U$ and $\gamma_-|_U$ are isomorphisms. Same way, for (M_U, M_Y, u, v) in $M_f(U,Y)$ take the diad

(**); this way we identify $\mathcal{M}_f(U,Y)$ with the category of diads of type (**) having the property that M_Y is supported on Y. After this identification was done, we see that F_f and G_f are just the reflexion functors; since $r \circ r = Id$, we are done. □

Here is the simplest case of the above proposition:

<u>Corollary</u> 3.2. Let X be a small disk round 0 in a complex line. Then the category M (X) of perverse sheaves on X (with singularities at 0 only) is equivalent to the category $\mathcal{C}$ of diagrams $V_o \underset{u}{\overset{v}{\rightleftarrows}} V_1$ of vector spaces such that both operators $Id_{V_o} - uv$ and $Id_{V_1} - vu$ are invertible.

<u>Proof</u> For a vector space V and φ in End V let $(V,\varphi)^o \subset V$ be the maximal subspace on which φ acts in a nilpotent way. Consider the category $\mathcal{C}'$ of diagrams $(V_o', V_1', \varphi, u, v)$, where V_o', V_1' are vector spaces, $\varphi \in Aut\, V_1'$, and $(V_1', \varphi)^o \underset{v}{\overset{u}{\rightleftarrows}} V_o'$ are such that $vu = Id_{V_1} - \varphi$

The category $\mathcal{C}$ is equivalent to $\mathcal{C}'$ via the functor $(V \underset{u}{\overset{v}{\rightleftarrows}} V_1) \mapsto ((V_o, uv)^o, V_1, Id_{V_1} - vu, u, v)$. The category $M_f(X \setminus \{0\}, \{0\})$, where $f: X \hookrightarrow \mathbb{A}^1$, is equivalent to $\mathcal{C}'$, since $M(\{0\}) = \{$vector spaces$\}$, $M(U) = \{$vector spaces with an automorphism (monodromy)$\}$, and, under this identification, $\Psi_f((V,\varphi)) = (V,\varphi)^o$. Now apply 2.3. □

<u>Remark.</u> The end of the proof of 2.1. in fact shows that $\Psi \circ \Psi^{un}$ as defined in 2.3 coincides with standard $R\Psi^{un}[-1]$; same is true for Φ. One may recover all $R\Psi(M)$ applying Ψ^{un} to $M \otimes f^*(?)$, where ? runs through the irreducible local systems on a punctured disk. For example, if M is RS holonomic, then the component of $R\Psi$ (DR (M)) that corresponds to the eigenvalue $\alpha \in \mathbb{C}^*$ of the monodromy is just $DR\Psi^{un}(M \cdot f^a)$, where $\exp(2\pi i a) = \alpha$ (since Ψ^{un}, obviously, commutes with DR). This fact was also found by Malgrange [7] and Kashiwara [5]. See also [3] where the total nearby cycles functor for arbitrary holonomic modules (not necessary RS) was introduced.

Appendix.

Here some linear algebra consructions, needed in the main body of the paper, are presented. Below A will be an exact category in the sense of Quillen; as usually $\rightarrowtail$, $\twoheadrightarrow$ denote admissible monomorphism, resp. epimorphism. If $\mathcal{C}$ is any category, then $\mathcal{C}^\circ$ is its dual.

A 1. Monads. A monad $\mathcal{P}$ is a complex of type $P_- \overset{\alpha_-}{\rightarrowtail} P \overset{\alpha_+}{\twoheadrightarrow} P_+$. Denote by $H(\mathcal{P}) := \operatorname{Ker}\alpha_+ / \operatorname{Im}\alpha_- \in \operatorname{Ob}A$ the cohomology of $\mathcal{P}$. The category of monads $A^\sim$ is an exact category: the exact sequences in $A^\sim$ are the ones which are componentwise exact; it is easy to see that $H: A^\sim \to A$ is an exact functor. An exact functor between exact categories induces the one between the categories of monads; these functors commute with H. Also one has $A^{\sim\circ} = A^{\circ\sim}$.

Often it is convenient to represent monads by somewhat another types of diagrams. Namely, let $A_1^\sim$ be the category of objects together with 3-step admissible filtration $\mathcal{P}_1 = (P_{-1} \overset{\gamma_{-1}}{\rightarrowtail} P_\circ \overset{\gamma_\circ}{\rightarrowtail} P_1)$; and $A_2^\sim$ be the category of short exact sequences $\mathcal{P}_2 = (L_- \overset{(\delta_-, \varepsilon_-)}{\rightarrowtail} A \oplus B \overset{(\delta_+, \varepsilon_+)}{\twoheadrightarrow} L+)$ such that $\delta-$ is admissible monomorphism, $\delta+$ is admissible epimorphism. These are exact categories same way as $A^\sim$ was.

Lemma. The categories $A^\sim$, $A_1^\sim$ and $A_2^\sim$ are canonically equivalent.

Here are the corresponding functors: the functor $A^\sim \to A_1^\sim$ maps $\mathcal{P}$ above to $P_- \rightarrowtail \operatorname{Ker}\alpha_+ \rightarrowtail P$; the one $A_1^\sim \to A_2^\sim$ maps $\mathcal{P}_1$ above to $P_\circ \overset{(\gamma_\circ, \varepsilon_-)}{\rightarrowtail} P_1 \oplus P_\circ/P_{-1} \overset{(\delta_+, -\gamma_\circ)}{\twoheadrightarrow} P_1/P_{-1}$ where ε_-, δ_+ are natural projections; finally $A_2^\sim \to A^\sim$ maps $\mathcal{P}_2$ above to $\operatorname{Ker}\varepsilon_- \rightarrowtail A \twoheadrightarrow A/\delta_-(L_-)$, where $\alpha_- = \delta_-|_{\operatorname{Ker}\varepsilon_-}$ and α_+ is natural projection. I leave the proof of lemma to the reader; note that B in $A_2^\sim$-avatar is $H(\mathcal{P})$. □

A.2. Diads and reflexion functor. A diad in A is a commutative diagram Q of type

$$\begin{array}{ccccc} & & A & & \\ & \overset{\alpha_-}{\nearrow} & & \overset{\alpha_+}{\searrow} & \\ C_- & & & & C_+ \\ & \underset{\beta_-}{\searrow} & & \underset{\beta_+}{\nearrow} & \\ & & B & & \end{array}$$

Clearly diads with component wise exact sequences form an exact category $A^{\#}$; one has $A^{\#\circ} = A^{\circ\#}$ and exact functors between A's induce the ones between $A^{\#}$'s.

As in the case of monads, we may represent diads by another diagrams. Let $A_1^{\#}$ be the category of monads of type $Q_1 = (C_- \rightarrowtail A \oplus B \twoheadrightarrow C_+)$ such that the corresponding arrow $C_- \to A$ is admissibe monomorphism, and the one $A \to C_+$ is admissible epimorphism. Let $A_2^{\#}$ be the category of short exact sequences $Q_2 = (D_- \xrightarrow{(\gamma_-, \delta_-^1, \delta_-^2)} A \oplus B^1 \oplus B^2 \xrightarrow{(\gamma_+, \delta_+^1, \delta_+^2)} D_+)$ such that both $D_- \xrightarrow{(\gamma_-, \delta_-^i)} A \oplus B^i$ are admissible monomorphisms, and $A \oplus B^i \xrightarrow{(\gamma_+, \delta_+^i)} D_+$ are admissible monomorphisms. This are exact categories same way as $A^{\#}$ is.

<u>Lemma.</u> The categories $A^{\#}$, $A_1^{\#}$ and $A_2^{\#}$ are canonically equivalent.

The corresponding functors are: the one $A^{\#} \to A_1^{\#}$ maps Q above to $C_- \xrightarrow{(\alpha_-, -\beta_-)} A \oplus B \xrightarrow{(\alpha_+, \beta_+)} C_+$ and the one $A_1^{\#} \to A_2^{\#}$ maps Q_1 above to its $A_2^{\sim}$ - avatar (so $B^1 = B$, $B^2 = H(Q_2)$). The first functor is obviously equivalence of categories; as for the second one this follows from the lemma A1. □

Note that in the diagram of type $A_2^{\#}$ one may interchange the objects B^i. This defines the important automorphism r of $A_2^{\#}$, and thus of $A^{\#}$, with $r \circ r = Id_{A^{\#}}$; call it the reflexion functor. Clearly it transforms a diad Q above into $r(Q) = \mathrm{Ker}\,\alpha_+ \begin{smallmatrix} \nearrow A \searrow \\ \searrow H(Q) \nearrow \end{smallmatrix} \mathrm{Coker}\,\alpha_-$, where $H(Q) := H(C_- \longrightarrow A \oplus B \longrightarrow C_+)$ is homology of the corresponding monad.

A 3. <u>Generalities on lim.</u> In what follows I will make no difference between an ordered set S and the usual category it defines; so for a category $\mathcal{C}$ an S-object in $\mathcal{C}$ is a functor $S \to \mathcal{C}$ i.e. the set of objects C_i, $i \in S$, and arrows $C_i \to C_j$ defined for $i \leqslant j$ with usual compatibilities. Let $\mathbb{Z}$ denotes the set of integers with standard order, and $\Pi = \{ (i,j) : i \leqslant j \} \subset \mathbb{Z} \times \mathbb{Z}$ with order induced from $\mathbb{Z} \times \mathbb{Z}$

Let A^{Π} be the category of Π-objects in A; this is exact category in a usual way; a short sequence $X_{ij} \to Y_{ij} \to Z_{ij}$ is exact iff any corresponding (ij)'s-sequence in A is exact. Say that an object

X_{ij} of A is admissible if for any $i \leqslant j \leqslant k$ the corresponding sequence $X_{ij} \to X_{ik} \to X_{jk}$ is short exact. Let $A^{\sqcap}_{a} \subset A^{\sqcap}$ be the full subcategory of admissible objects. In any short exact sequence if two objects are admissible then the third is, so $A^{\sqcap}_{a}$ is exact category in an obvious way. Clearly A imbeds in $A^{\sqcap}_{a}$ as full exact subcategory of objects X_{ij} with $X_{i,-1}=0$, $X_{1,j}=0$.

Let φ be any order-preserving map $\mathbb{Z} \to \mathbb{Z}$ such that $\lim\limits_{i \to \pm\infty} \varphi(i) = \pm\infty$. Then for any $\sqcap$-object X_{ij} we have $\sqcap$-object $\tilde{\varphi}(X_{ij}) := X_{\varphi(i)\varphi(j)}$; clearly

$X \mapsto \tilde{\varphi}(X)$ is exact functor that preserves $A^{\sqcap}_{a}$ and $\widetilde{id}$ is identity functor. If $\varphi \leqslant \psi$ i.e. $\varphi(i) \leqslant \psi(i)$ for any i, then we have an obvious morphism of functors $\tilde{\varphi} \to \tilde{\psi}$.

Define the category $\underleftrightarrow{\lim}\, A$ to be the localisation of $A^{\sqcap}_{a}$ with respect to morphisms $\tilde{\varphi}(X) \to \tilde{\psi}(X)$, $X \in A^{\sqcap}_{a}$, and $\varphi \leqslant \psi$ are as above. The natural functor $\underleftrightarrow{\lim}$: $A^{\sqcap}_{a} \to \underleftrightarrow{\lim} A$ is surjective on (isomorphism classes of) objects. A morphism $\underleftrightarrow{\lim}\, X_{ij} \to \underleftrightarrow{\lim}\, Y_{ij}$ is given by a pair (α, f_{α}), where $\alpha : \mathbb{Z} \to \mathbb{Z}$ is order-preserving function with $\lim\limits_{i \to -\infty} \alpha(i) = -\infty$, and $f_{\alpha} : X_{ij} \to Y_{\alpha(i)\alpha(j)}$ is compatible system of morphisms; two pairs (α, f_{α}) and (β, f_{β}) give the same morphism if f, g give the same maps $X_{ij} \to Y_{\max(\alpha(i),\beta(i)), \max(\alpha(j),\beta(j))}$.

Say that a short sequence in $\underleftrightarrow{\lim}\, A$ is exact if it is isomorphic to $\underleftrightarrow{\lim}$ of certain exact sequence in $A^{\sqcap}_{a}$. The routine verification of Quillen's axioms shows that this way $\underleftrightarrow{\lim}\, A$ becomes exact category. The functor $\underleftrightarrow{\lim}$ is exact ; it defines the faithful exact embedding $A \hookrightarrow \underleftrightarrow{\lim}\, A$. Any exact functor between A's induces the one between their $\underleftrightarrow{\lim}$'s; we also have $\lim (A^{\circ}) = (\lim A)^{\circ}$.

<u>Remarks.</u> a. We have $Q\, \underleftrightarrow{\lim}\, A = \underrightarrow{\lim}\, QA$ (where Q is Quillen's Q construction)

b. If $A \neq 0$, then $\lim A$ is <u>not</u> abelian.

c. One also can take lim's along any ordered sets with any finite subset having an upper and lower bounds.

References

1. A.Beilinson, On the derived category of perverse sheaves, this volume.

2. J.Bernstein. Algebraic theory of D-modules, to appear.

3. P.Deligne. Letter to Malgrange from 20-12-83.

4. P.Deligne. Théorèmes de finitude en cohomologie l-adique, Lect. Notes Math. 569, 233-261 (1977)

5. M.Kashiwara. Vanishing cycles for holonomic systems, Lect.Notes Math 1016, 134-142 (1983)

6. R.Macpherson, L.Villonen. Elementary construction of perverse sheaves, Inventiones math. 84

7. B.Malgrange. Polynomes de Bernstein-Sato et cohomologie évanescente. Astérisque 101-102, 243-267 (1983)

8. J.-L.Verdier. Extension of a perverse sheaf over a closed subspace. Astérisque 130, 210-217 (1985)

9. J.-L. Verdier. Prolongement des faisceaux pervers monodromiques Asterisque 130, 218-236 (1985)

Sheaves of the Virasoro and Neveu-Schwarz algebras.

A.A. Beilinson, Yu.I. Manin, V.V. Schechtman.

Introduction.

The quantum string theory ties together two rather disparate branches of algebra: representation theory of certain infinite-dimensional Lie (super) algebras, and theory of algebraic curves (complex Riemann surfaces) and their moduli spaces. One may expect that a mathematical explanation of this connection should be roughly similar to the classical Borel-Weil construction, in the sense that the representations of the relevant Lie algebras may be realized in sections or cohomology of certain natural sheaves on appropriate moduli spaces.

Our note is a step in this direction. We construct certain sheaves of Lie algebras on algebraic curves (or 1|1 - dimensional super manifolds with contact structure) which are essentially sheaves of Virasoro and Neveu-Schwarz type. They are however globalized in the sense that a central charge is asscociated with each point of a Riemann surface and each closed oriented curve on it.

The starting point of our construction is the sheaf of formal pseudodifferential operators well known in the theory of two-dimensional integrable systems (like KdV and KP) and its superextension introduced in [MR].

This note consists of two closely parallel parts devoted to the bosonic and to the fermionic strings respectively.

A few words about some related work.

Probably, it was A.A. Kirillov who first noticed [K] that Schrödinger operators form coadjoint representation of the Virasoro algebra. A result closely connected with our theorem I.2.2 is contained in [Kh].

We understand also that a part of our results was known to A.O. Radul [R] and B.L. Feigin.

We are especially grateful to A.O. Radul for very stimulating discussions that initiated this work.

One of us (Yu.I.M.) would like to thank heartily T.M. Kirillova for assistance which allowed him to work upon this paper in a hospital bed.

Since this note was written some of us have managed to prove a variant of Conjecture I.2.4, see [B Sch] .

PART I. Virasoro.

§1. Pseudodifferential operators.

1. ΨDO. Let A be an associative ring, ∂: A→A a derivation. The ring of differential operators $\underline{D}$ generated by ∂ consists of formal polynomials $\sum_{i=0}^{n} a_i \partial^i$, $a_i \in A$, with the multiplication rule provided by the Leibniz formula

$$\left(\sum_i a_i \partial^i\right) \circ \left(\sum_j b_j \partial^j\right) = \sum_{i,j,k} \binom{i}{k} a_i \partial^k(b_j) \partial^{i+j-k} \qquad (1)$$

The ring of (formal) pseudodifferential operators $\underline{E}$ consists of formal series $\sum_{i=-\infty}^{n} a_i \partial^i$, $a_i \in A$ with the same multiplication rule (1) where $\binom{i}{k}$ is defined for all integer i,k, by the formulae $\binom{i}{k} = 0$ for i<k; 1 for i=k; (k+1) ... i / (i-k) ! for i>k. For example, $\partial \circ a = a\partial + \partial a$ and

$$\partial^{-1} \circ a = a\partial^{-1} - (\partial a)\partial^{-2} + (\partial^2 a)\partial^{-3} - \dots$$

We note the following properties of $\underline{E}$.

a). Set $\underline{E}_n = \left\{\sum_{i \le n} a_i \partial^i\right\}$. Then $\dots \subset \underline{E}_{-1} \subset \underline{E}_0 \subset \underline{E}_1 \subset \underline{E}_2 \dots$ We have $\underline{E} = \underline{D} \oplus \underline{E}_{-1}$, $\underline{E}_i \circ \underline{E}_j \subset \underline{E}_{i+j}$, $[\underline{E}_i, \underline{E}_j] \subset \underline{E}_{i+j-1}$.

b). Let $\partial' = a\partial$, $a \in A$ invertible, and denote by $\underline{E}'$ the corresponding ring generated by ∂'. Then we can canonically identify $\underline{E}'$

with $\underline{E}$ in such a way that $\partial'^i \mapsto (a\partial) \circ \ldots \circ (a\partial)$ (i times) for $i > 0, \partial'^i \mapsto (\partial^{-1} \circ a^{-1}) \circ \ldots \circ (\partial^{-1} \circ a^{-1})$ ($|i|$ times) for $i < 0$. Clearly, $\underline{E}'_i = \underline{E}_i$ and $\underline{D}' = \underline{D}$.

This simple remark allows us to associate a sheaf of formal ΨDO's with an arbitrary one-dimensional distribution on a differentiable or complex manifold, using for A a ring of local functions and for ∂ any convenient local vector field generating the distribution.

From this point on we assume A commutative.

c) Put $\Omega = \underline{E}_{-1}/\underline{E}_{-2}$. This is a free A-module of rank 1. If $\partial z = 1$ for a certain $z \in A$, it is convenient to write $\partial = \partial_z$ and to denote $dz = \partial_z^{-1}$ mod $\underline{E}_{-2}$ so that $\Omega = Adz$. The map $d: A \to \Omega$, $da = dz \cdot \partial_z a$ enjoys the usual formal properties of differential.

Generally, we shall write $\nu_\partial = \partial^{-1}$ mod E_2 instead of d_z.

d). We have $\underline{E}_i/\underline{E}_{i-1} \simeq \Omega^{-i} = \Omega^{\otimes(-i)}$ for all i. The corresponding symbol map

$$\sigma_i: \underline{E}_i \longrightarrow \Omega^{-i}, \quad \sigma_i\Big(\sum_{j\le i} a_j \partial^j\Big) = a_i (\nu_\partial)^{-i}$$

is clearly independent on the choice of ∂ in $\{a\partial \mid a$ invertible $\}$.

2. <u>Twisting.</u> Let $\underline{L}, \underline{M}$ be two A-modules, $\underline{L}^* = \mathrm{Hom}(\underline{L}, A)$. We set $\underline{E}_{\underline{L}\to\underline{M}} = \underline{M} \otimes_A \underline{E} \otimes_A \underline{L}^*$ (note that $\underline{E}$ is an A-bimodule since $A \subset \underline{E}$ as a subring). The subspace $\underline{D}_{\underline{L}\to\underline{M}} = \underline{M} \otimes_A \underline{D} \otimes_A \underline{L}^*$ can be realized as the space of differential operators $L \to M$: for $m \in \underline{M}$, $F \in \underline{D}$, $\lambda \in L^*$, $l \in L$ put $(m \otimes F \otimes \lambda)(l) = F(\lambda(l))\, m$ (note that $\underline{E}$ acts upon A via $(\sum a_i \partial^i) f = \sum a_i \partial^i f$). Of course, whole $\underline{E}_{\underline{L}\to\underline{M}}$ does not act upon $\underline{L}$ in a natural way but the formal properties of this construction are the same as if $\underline{E}_{\underline{L}\to\underline{M}}$ were morphisms from $\underline{L}$ to $\underline{M}$ in a category. In particular, there is a natural multiplication $\underline{E}_{\underline{M}\to\underline{N}} \times \underline{E}_{\underline{L}\to\underline{M}} \to \underline{E}_{\underline{L}\to\underline{N}}$ and $\underline{E}_{\underline{L}} = \underline{E}_{\underline{L}\to\underline{L}}$ is a ring. A symbol map $\sigma_i : \underline{E}_{i,\,\underline{L}\to\underline{M}} \to \underline{M} \otimes \Omega^{-i} \otimes \underline{L}^*$ is naturally defined as in 1d).

3. Conjugation. There is a unique isomorphism of additive groups $t: \underline{E} \to \underline{E}_{\Omega} = \Omega \underset{A}{\otimes} \underline{E} \underset{A}{\otimes} \Omega^{-1}$, $F \mapsto F^t$, with the following properties:

a). $f^t = v_\partial \otimes f \otimes v_\partial^{-1}$ for $f \in A$, where $v_\partial = \partial^{-1} \mod \underline{E}_{-2}$.

b). $\partial^t = -v_\partial \otimes \partial \otimes v_\partial^{-1}$.

c). $(F \circ G)^t = G^t \circ F^t$.

d). t is continuous in ∂^{-1} - adic topology.

A key property is again the independence of t on the choice of ∂ in $\{a\partial \mid a \text{ invertible}\}$. It is given by the formula

$$\left(\sum a_i \partial^i\right)^t = v_\partial \otimes \left(\sum (-\partial)^i \circ a\right) \otimes v_\partial^{-1}. \tag{2}$$

For example, if $\partial' = a\partial$, we have:

in $\underline{E}'$: $(\partial')^t = -v_{\partial'} \otimes \partial' \otimes v_{\partial'}^{-1} = -v_\partial a^{-1} \otimes a \partial \otimes a v_\partial = -v_\partial \otimes (\partial \circ a) \otimes v_\partial^{-1}$

in $\underline{E}$: $(a\partial)^t = \partial^t \circ a^t = -v_\partial \otimes (\partial \circ a) \otimes v_\partial^{-1}$,

and the results agree.

Hence t extends to a sheafification of $\underline{E}$, as in 1 b).

Similarly, we can define a map $t: \underline{E}_{\underline{L} \to \underline{M}} \to \underline{E}_{\underline{M}^t \to \underline{L}^t}$, where $\underline{L}^t = \underline{L}^* \underset{A}{\otimes} \Omega$, by the formula

$$(m \otimes F \otimes \lambda)^t = \lambda \otimes F^t \otimes m \in \underline{L}^* \otimes \Omega \otimes \underline{E} \otimes \Omega^{-1} \otimes \underline{M}$$

(we must assume here that $\underline{L}, \underline{M}$ are free finite rank A-modules).

If $\underline{L} = \underline{M}^t$ (and hence $\underline{M} = \underline{L}^t$), then t acts upon $\underline{E}_{\underline{L} \to \underline{M}}$. In particular, t acts upon $\Omega^\alpha \otimes \underline{E} \otimes \Omega^{\alpha - 1}$. For each i we can define a map

$$\sigma_i : \Omega^\alpha \otimes \underline{E}_i \otimes \Omega^{\alpha-1} \to \Omega^\alpha \otimes \underline{E}_i / \underline{E}_{i-1} \otimes \Omega^{\alpha-1} = \Omega^{2\alpha-1-i}$$

Clearly, $\sigma_i(F^t) = (-1)^i \sigma_i(F)$. Hence we can define i- self-conjugate operators by

$$\underline{E}_i^{+} = \{ F \in \Omega^\alpha \otimes \underline{E}_i \otimes \Omega^{\alpha-1} \mid F^t = (-1)^i F \} \tag{3}$$

4. Non commutative residue. Set $\underline{H} = \Omega / dA$ (the de Rham cohomology of (A, ∂)) where $da = v_\partial^{-1} \partial a$. The map

Res: $\underline{E} \to \underline{H}$, $\mathrm{Res}\,(\sum a_i \partial^i) = a_{-1} v_\partial \bmod dA$

is well defined i.e. independent on the choice of ∂. Its main property is

Res $[F,G] = 0$ for $F, G \in \underline{E}$.

5. The Virasoro algebra. For $A = \mathbb{C}[t, t^{-1}]$, $\partial = \frac{d}{dt}$ the classical residue $\mathrm{res}: \Omega \to \mathbb{C}$, $\mathrm{res}\,(\sum c_i t^i dt) = c_{-1}$ defines the isomorphism $\Omega / dA = H = \mathbb{C}$.

The Virasoro algebra V is a central extension of the Lie algebra A d/dt by $\mathbb{C}z_o$ with the commutation rule

$$\left[f \frac{d}{dt}, g \frac{d}{dt}\right]_{z_o} = \left(f \frac{dg}{dt} - g \frac{df}{dt}\right) \frac{d}{dt} + \frac{1}{12} \mathrm{res}\, g \frac{d^3 f}{dt^3} dt \cdot z_o \qquad (4)$$

The element z_o or its proper value in a representation space of V is called the central charge.

We shall see in the next section that a natural generalization of (4) arises automatically in our setting (A, ∂), with H replacing $\mathbb{C}z_o$.

§2. Virasoro sheaves.

1. Princip construction. In the above notation (3) we set

$$\tilde{V}_n = [\Omega^{n/2} \otimes_A \underline{E}_n / \underline{E}_{n-3} \otimes_A \Omega^{n/2-1}]^+$$

The following lemma describes this space in down-to-earth terms. We choose $z \in A$ with $\partial = \partial_z$.

Lemma. V_n consists of expressions

$$F_{a,c} = (dz)^{\frac{n}{2}} \otimes \{ a \partial^n + \frac{n}{2} \partial a \cdot \partial^{n-1} + c \partial^{n-2} \bmod \partial^{n-3} \} (dz)^{\frac{n}{2}-1} \qquad (5)$$

where a, $c \in A$ are arbitrary. In particular, there is an exact sequence independent on the choice of ∂

$$0 \to \Omega \to \tilde{V}_n \xrightarrow{\sigma_n} \mathcal{T} \to 0, \qquad (6)$$

where $\mathcal{T} = \Omega^{-1} = A\partial$, $\sigma_n(F_{a,c}) = a\partial$, $\mathrm{Ker}\, \sigma_n = \{F_{0,c}\}$

is identified with Ω via $\sigma_{n-2} : F_{o,c} \mapsto c dz$. □

The proof is a direct computation using (2) and (3).

In order to make sense of $(dz)^{\frac{n}{2}}$ in (5) we have either to assume $n \equiv 0$ (2), or to define certain fractional powers of Ω formally. E.g. on Riemann surface we shall be able to use odd n after a choice of a theta characteristics.

We now set $V_n = \tilde{V}_n / d$ where $dA \subset \Omega \subset \tilde{V}_n$. We get from (6):

$$0 \to \underline{H} \to V_n \xrightarrow{\sigma_n} T \to 0$$

2. Theorem. V_n is endowed with a natural structure of a Lie algebra, $\underline{H}$ belongs to its center and σ_n is a homomorphism of Lie algebras. This extension corresponds to the cocycle

$$c\,(b\,\partial_z, a\,\partial_z) = \frac{1}{12}\, n\,(n-1)(n-2)\, a\,\partial_z^3\, b\,\partial_z \mod dA \qquad (8)$$

which is a direct generalization of (4) (central charge $n(n-1)(n-2)$ (cf. [Kh]).

Sketch of proof. There is a natural action of the Lie algebra $T = A\partial \subset \underline{E}$ on all natural A-modules ("Lie derivative"). Namely, T acts upon E via adjoint representation: $[X,F] = X \circ F - F \circ X$ for $X \in T$, $F \in \underline{E}$. Since $[T, \underline{E}_i] \subset \underline{E}_i$, this induces an action on $\Omega^i = \underline{E}_{-i} / \underline{E}_{-i-1}$. Finally, the Leibniz rule defines an action on tensor products of these modules. In particular, we obtain from (1)

$$\mathrm{Lie}_{\partial_z} (dz^m) = [b\,\partial_z, dz^m] = (b\,\partial_z a + m a\,\partial_z b)\,(dz)^m,$$

and, after a straight forward calculation

$$[\,b\,\partial_z, (dz)^{n/2} \otimes \{\, a\partial_z^n + \tfrac{n}{2}\,\partial_z a\,\partial^{n-1} + c\,\partial_z^{n-2} \mathrm{mod}\,\partial_z^{n-3}\,\} \otimes (dz)^{\frac{n}{2}-1}\,] =$$

$$= (dz)^{\frac{n}{2}} \otimes \{\,(b\,\partial_z a - a\,\partial_z b)\,\partial_z^n + \tfrac{n}{2}\,\partial_z\,(b\,\partial_z a - a\,\partial_z b)\,\partial_z^{n-1} +$$

$$+ \frac{n(n-1)(n-2)}{12}\, a\,\partial_z^3\, b\,\partial_z^{n-2} + \partial_z(bc)\,\partial_z^{n-2} \mod \partial_z^{n-3}\,\} \otimes (dz)^{\frac{n}{2}-1}$$

Therefore V_n is a T - module, σ_n is a morphism of T-modules and the action of T maps $\Omega \subset \tilde{V}_n$ into $dA \subset \tilde{V}_n$ (in our coordinate notation, Ω corresponds to $a=0$). Hence the induced action on $V_n = \tilde{V}_n / dA$ is trivial on H and (7) becomes a central extension of T. □

3. Virasoro sheaves. Let now $\pi : X \to S$ be an algebraic (or analytic) family of smooth compact complex curves. $T = TX/S$ the sheaf of vertical vector fields. We can apply all previous constructions to the structure sheaf $\underline{O}_X$ and T instead of A and ∂. In this way we get algebraic or analytic sheaves on X which we denote by the same letters $\underline{E}, \underline{D}, \underline{E}_{\underline{L} \to \underline{M}}$ ($\underline{L}$, $\underline{M}$ being locally free $\underline{O}_X$-sheaves), $\underline{H}$.

A warning on $\underline{H}$ is in order here. Take S= { point }. If we consider X with its natural Hausdorff topology, then the analytic de Rham sheaf vanishes since locally any holomorphic 1-form is a complete differential! This is a nuisance since in (7) V_n becomes just T. Therefore it is reasonable to consider X in the Zariski topology where open sets are just X minus several points. On such $V \neq X$ both $\underline{H}_{an}(V) = \mathcal{O}_{X, an}(V) / d\, \mathcal{O}_X(V)$ and a similar algebraic sheaf $\underline{H}_{alg}$ are nontrivial and can be considered as spaces of periods and residues.

Hence H embodies central charges associated with points and 1-homology classes of X, and we get our Virasoro - Lie sheaves on X/S:

$$0 \to \underline{H} \to V_n \to TX/S \to 0 \tag{9}$$

Assume now that $S = M_g$, the moduli space of curves of genus $g \geq 1$ or its covering like the Teichmuller space. Taking the direct images of (9) we get an exact sequence

$$0 \to R^1\pi_* (\underline{H}) \to R^1\pi_* V_n \to R^1\pi_* (TX/S) \to 0.$$

The map $\Omega X/S \to \underline{H}$ induces and isomorphism

$$\mathcal{O}_S \simeq R^1\pi_* \Omega X/S \simeq R^1\pi_* \underline{H}.$$

On the other hand, the exact sequence $0 \to TX/S \to TX \to \pi^* TS \to 0$ induces the Kodaira - Spencer isomorphism (at least at smooth points of S)

$$TS = R^{\circ}\pi_{*}(\pi^{*}TS) \xrightarrow{\sim} R^{1}\pi_{*}TX/S.$$

Hence we finally get a canonical extension of the tangent sheaf of the moduli space by its structure sheaf

$$0 \to \mathcal{O}_S \to R^1\pi_* \quad V_n \xrightarrow{\sigma} TS \to 0 \tag{10}$$

4. Conjecture. The sheaf $R^1\pi_* V_n$ in (10) can be canonically identified with the sheaf of differential operators of order ≤ 1 on the Mumford sheaf

$$\lambda_{n/2} = \det R\pi_*(\Omega^{1-n/2} \ X/S) = \det R\pi_*(\Omega^{n/2} \ X/S)$$

in such a way that σ becomes the symbol map and $f \in \mathcal{O}_S$ acts as a multiplication by $c_{n/2} f$ for some constant $c_{n/2}$. □

As was mentioned in the Introduction, a version of this Conjecture is now proved [B Sch]. It differs from the above statement, however, since a different construction of the Virasoro sheaves is used, whose relation to (9) is not clear.

The explicitation of this proof gives a complete system of differential equations satisfied by the Mumford forms and therefore by the Polyakov string measure: cf [BM], [B Sch] .

PART II. Neveu - Schwarz.

§1. Pseudodifferential operators.

1. Ψ DO. We now plunge into superalgebra where all additive groups are Z_2 - graded and signs and oddity effects abound. We used conventions of [M1] and denote by $\tilde{a} \in Z_2$ the parity of an object a.

Let A be a Z_2 -graded ring, $D: A \to A$ an odd derivation:

$$\widetilde{Da} = \tilde{a}+1,\ D(a+b) = Da+Db,\ D(ab) = Da\cdot b + (-1)^{\tilde{a}} aDb.$$

The ring of differential operators $\underline{D}^s$ generated by D consists of formal polynomials $\sum_{i=0}^{n} a_i D^i$, $a_i \in A$, with the multiplication rule furnished by super-Leibniz formula:

$$(\sum a_i D^i)\circ(\sum b_j D^j) = \sum_{i,j,k} \left[{i \atop k}\right] a_i (D^k b_j) D^{i+j-k} \qquad (1)'$$

where the superbinomial coefficients are defined, as in [MR]:

$$\left[{i \atop k}\right] = \begin{cases} & \text{for } k>i \text{ and for } (i,k) = (0,1) \bmod 2, \\ \binom{[i/2]}{[k/2]} & \text{for } k \leq i, \quad (i,k) \neq (0,1) \bmod 2. \end{cases}$$

(Here $[i/2]$ is the integer part of $i/2$). Of course, $\underline{D}^s$ is Z_2-graded by $(a_i D^i)^\sim = \tilde{a}_i + i \bmod 2$.

The ring of ψDO's $\underline{E}^s$ consists of formal series $\sum_{i=-\infty}^{n} a_i D^i$, $a_i \in A$, with the multiplication rule (1)'.

In order to understand the relation of $\underline{E}^s$ to $\underline{E}$ of Part I, note that $\partial = D^2$ is a derivation of the even subring $A_0 \subset A$. If we construct E from $(A_0, \partial = D^2)$, then we can embed $\underline{E}$ into $\underline{E}^s$ as $\{\sum a_{2i} D^{2i} | a_{2i} \in A_0\}$. Therefore $\underline{E}^s$ is an extension of $\underline{E}$ by $\sqrt{\partial}$ and odd superfunctions. This justifies the following definitions.

a). Set $\underline{E}^s_{n/2} = \{\sum_{i\leq n} a_i D^i\}$. Then $\ldots \subset \underline{E}^s_{-1/2} \subset \underline{E}^s_0 \subset \underline{E}^s_{1/2}$.

We have $E^s = \underline{D}^s \oplus \underline{E}^s_{-1/2}$, $\underline{E}^s_i \circ \underline{E}^s_j \subset \underline{E}^s_{i+j}$. (Note however, that $[\underline{E}^s_i, \underline{E}^s_j] \not\subset \underline{E}^s_{i+j-1/2}$ in general, since e.g. $[D,D] = 2D^2$, where $[a,b] = ab-(-1)^{\tilde{a}\tilde{b}} ba$ is the super commutator).

b) Let $D' = aD$, $a \in A$ an invertible even element, and let $\underline{E}'^s$ be the corresponding ring. Then we can identify $\underline{E}^s$ and $\underline{E}'^s$ like in § I.1. Again, this will allow us to pass from rings to sheaves.

From this point on we assume A supercommutative, i.e. $[a,b]=0$ for all $a,b \in A$.

c). Set $\omega = \underline{E}^s_{-1/2} / \underline{E}^s_{-1}$. This is a free A-module of rank 0|1 generated by $D^{-1} \bmod \underline{E}^s_{-1}$. If for certain $z \in A_0$, $\zeta \in A_1$ we have $Dz = \zeta$, $D\zeta = 1$, we denote $Z = (z, \zeta)$ and $D = D_Z$. When $Z = (z, \zeta)$ is an admissible local coordinate system on a SUSY -curve (cf. [M2]) we have $D_Z =$

$\partial_\zeta + \zeta\partial_Z$. We write then $dZ = D_Z^{-1} \bmod \underline{E}^s_{-1}$. The map $\delta: A \to \omega$, $\delta a = dZ \cdot D_Z a$ is well defined, i.e. remains the same when D_Z is replaced by $D_{Z'} = aD_Z$, a even invertible. Note however that ω is not a module of 1-differentials but that of Berezin volume forms.

In general we shall write $\nu_D = D^{-1} \bmod \underline{E}^s_{-1}$, $\delta a = \nu_D \cdot Da \in \omega$.

d) We have the symbol maps

$$\sigma_i : \underline{E}_i \to \underline{E}_i/\underline{E}_{i-1/2} \simeq \omega^{\otimes 2i} = \omega^{2i},\ i \in \tfrac{1}{2}\mathbb{Z},$$

$$\sigma_i\Big(\sum_{j\le i} a_j D^j\Big) = a_i \nu_D^{-i}.$$

2. Twisting. Repeat everything from I.1.2 for Z_2 -graded A-modules $\underline{L},\underline{M}$ in order to define $\underline{E}^s_{\underline{L}\to\underline{M}}$ etc.

3. Conjugation. There is a unique isomorphism of additive groups

$$t: E^s \to E^s_\omega = \omega \underset{A}{\otimes} \underline{E}^s \underset{A}{\otimes} \omega^{-1}, \quad F \mapsto F^t$$

with the following properties:

a) $f^t = \nu_D \otimes f \otimes \nu_D^{-1}$ for $f \in A$.

b) $D^t = -\nu_D \otimes D \otimes \nu_D^{-1}$.

c) $(F \circ G)^t = (-1)^{\tilde F \tilde G} G^t \circ F^t$ (note the sign!).

Of course, t is independent on the choice of D and is given by the formula

$$\Big(\sum a_i D^i\Big)^t = \nu_D \otimes \Big(\sum (-1)^{\frac{i(i+1)}{2} + \tilde a_i \cdot i} D^i \circ a_i\Big) \otimes \nu_D^{-1}. \quad (2)'$$

Putting $\underline{L}^t = \underline{L}^* \otimes \omega$ we can define $t: \underline{E}_{\underline{L}\to\underline{M}} \to \underline{E}_{\underline{M}^t \to \underline{L}^t}$ by the formula

$$(m \otimes F \otimes \lambda)^t = (-1)^{\tilde\lambda\tilde F + \tilde\lambda\tilde m + \tilde m\tilde F} \lambda \otimes F^t \otimes m.$$

Again, t acts upon E^s a $^{-1}$, and the symbol map

$$\sigma_i : \omega^\alpha \otimes \underline{E}^s_i \otimes \omega^{\alpha-1} \to \omega^\alpha \otimes \underline{E}_i/\underline{E}_{i-1/2} \otimes \omega^{\alpha-1} \in \omega^{2\alpha-1-2i}$$

verifies

$$\sigma_i(F^t) = \varepsilon_F\, \sigma_i(F),\ i \in \tfrac{1}{2}\mathbb{Z}, \quad \varepsilon_F = (-1)^{\tilde F(2i+1) + i(2i+3)}$$

Hence we can define i-selfconjugate operators by

$$\underline{E}^s_i{}^+ = \{ F \in \omega^{\alpha} \otimes \underline{E}^s_i \otimes \omega^{\alpha-1} \mid F^t = \varepsilon_F F \} . \qquad (3)'$$

4. Noncommutative residue. Set $\underline{H}^s = \omega / \delta A$. The map

$$\mathrm{Res}: \underline{E}^s \longrightarrow \underline{H}^s, \quad \mathrm{Res}\,(\Sigma\, a_i D^i) = a_{-1} \sqrt{}_D \mod \delta A$$

is well defined and verifies

$$\mathrm{Res}\,[F,G] = 0, \quad F,G \in \underline{E}^s$$

(see [MR], Lemma 1,5).

5. The Neveu - Schwarz superalgebra and the contact Lie algebra.

The Neveu - Schwarz Lie superalgebra NS over $\mathbb{C}$ is traditionally defined by even generators e_i, $i \in \mathbb{Z}$, odd generators f_α, $\alpha \in \frac{1}{2} + \mathbb{Z}$, and the central generator Z_0, with the commutator rules

$$[e_i, e_j] = (j-i)\, e_{i+j} + \delta_{i,-j} \frac{j^3-j}{12} Z_0 ,$$

$$[e_i, f_\alpha] = \left(\alpha - \frac{i}{2} \right) f_{\alpha+i} ,$$

$$[f_\alpha , f_\beta] = 2\, e_{\alpha+\beta} + \frac{1}{3} \delta_{\alpha,-\beta} \left(\alpha^2 - \frac{1}{4}\right) Z_0 .$$

For arbitrary supercommutative $\mathbb{C}$-algebra we put $NS_\Lambda = \Lambda \otimes_{\mathbb{C}} NS$.

The (central extension of the) contact Lie algebra K_Λ can be defined as $\Lambda_0 Z_0 \oplus A_0$, where Λ_0, A_0 are even parts of Λ and $A = \Lambda[t, t^{-1}, \tau]$ respectively, $\tilde{t}=0$, $\tilde{\tau}=1$, with the commutation rule

$$\{a,b\} = aD^2b - bD^2a + \frac{1}{2} DaDb - \frac{1}{12} \rho\,(aD^5b)\, Z_0 .$$

Here $a,b \in A_0$, $D = \frac{\partial}{\partial\tau} + \tau\frac{\partial}{\partial t}$; $\rho: A_1/DA_0 \to \Lambda_0$, $\rho\,(t^{-1}\tau) = 1$, is the isomorphism of Λ_0-modules, changing parity.

These two objects are related by the following isomorphisms, functorial in Λ :

$$(NS_\Lambda)_0 \simeq K_\Lambda : \quad ae_i \mapsto at^{i+1}, \quad \lambda f_\alpha \mapsto 2\lambda t^{\alpha+\frac{1}{2}}\tau , \quad z_0 \mapsto z_0$$

where $a \in \Lambda_0$, $\lambda \in \Lambda_1$.

Thus one can say that the contact Lie algebra (in dimension 1|1) is the universal even part of the Neveu-Schwarz superalgebra.

In the next section we shall construct the sheaf version of K rather than that of NS, but since our construction will be valid over an arbitrary base supermanifold, no information on NS will be lost.

§2. Neveu-Schwarz sheaves.

1. Principla construction. We choose $n \equiv 1 \bmod 2$ and put

$$\tilde{K}_n = \text{even part of } [\omega^{\frac{n-1}{2}} \otimes \underline{E}^s_{\frac{n}{2}} \;/\; \underline{E}^s_{\frac{n-1}{2}} \otimes \omega^{\frac{n-3}{2}}]^+ .$$

As in part I, we have the following Lemma (which we write in terms of $D_Z = D$ and dZ only to make it look more alike the bosonic version).

Lemma. $\tilde{K}_n$ consists of expressions

$$F_{a,c} = (dZ)^{\frac{n-1}{2}} \otimes \{ aD^n + \tfrac{1}{2} Da \cdot D^{n-1} + \tfrac{n-1}{4} D^2 a \cdot D^{n-2} + cD^{n-3} \bmod D^{n-4} \} \otimes$$

$$\otimes (dZ)^{\frac{n-3}{2}} \qquad (5)'$$

where $a \in A_0$, $c \in A_1$. In particular, there is an exact sequence, independent of the choice of D in $\{aD \mid a \in A_0^*\}$:

$$0 \to \omega_0 \to \tilde{K}_n \xrightarrow{\sigma_{n/2}} T_0 \to 0 \qquad (6)'$$

where T_0 = even part of $T = \omega^{-2}$, $\sigma_{n/2}(F_{a,c}) = aD^2$,

ω_0 = even part ω, $\operatorname{Ker} \sigma_{n/2} = \{F_{0,c} \mid c \in A_1\}$ is identified with ω_0 via $\sigma_{\frac{n-3}{2}}$: $F_{0,c} \mapsto cdZ$. □

The proof is a lenthy computation using (2)' and (3)'. We only hope having got all signs straight.

We now set $K_n = \tilde{K}_n / \delta A_0$, where $\delta A_0 \subset \omega_0 \subset \tilde{K}_n$,

$\delta a = dZ \cdot D_Z a$. We get from (6)'

$$0 \to \underline{H}_0{}^s \to K_n \to T_0 \to 0.$$

A minor surprise is that à priori T_0 is not a Lie algebra. However, consider the 1|1 - contact Lie algebra in its usual definition:

$$T = \{X \in (AD+AD^2)_o \mid [X, gD] \in A_o D \text{ for all } g \in A_o\} .$$

One easily sees that

$$T = \{ bD^2 + \tfrac{1}{2} Db \cdot D \mid b \in A_o \}$$

and the symbol map $T \to T_o$: $bD^2 + \frac{1}{2} Db \cdot D \longmapsto bD^2$ is a well defined isomorphism. Set $X_b = bD^2 + \frac{1}{2} Db \cdot b$.

Then

$$[X_f, X_g] = X_{\{f,g\}} , \ \{f,g\} = \tfrac{1}{2} Df \cdot Dg + fD^2g - gD^2f \qquad (6)''$$

which explains our definition of $K_\wedge$ in II.1.5.

Identifying T_o with T by means of symbol map, we get finally from (6)' an exact sequence

$$0 \to H^s{}_o \to K_n \xrightarrow{\sigma} T \to 0. \qquad (7)'$$
$$K_n = NS_{n,o}$$

2. Theorem. K_n is endowed with a natural structure of a Lie algebra, $H_o{}^s$ lies in its center and σ is a homomorphism of Lie algebras. This extension corresponds to the cocycle

$$c\,(X_b, X_a) = \frac{(n-1)\ (n-3)}{16} \quad aD^5b\ dZ \bmod \delta A. \qquad (8)'$$

Sketch of proof. As in part I, the Lie algebra T acts upon $\underline{E}^s$ via adjoint representation since $T \subset \underline{E}^s$. One checks directly that $[T, E^s{}_n] \subset E^s{}_n$. Hence T acts upon ω^i and upon $\tilde{K}_n$ by super-Leibniz rule. In particular,

$$\text{Lie}_{X_f} \ (g(dZ)^m) = [X_f, g(dZ)^m] = (\tfrac{1}{2} Df \cdot Dg - fD^2g + \tfrac{m}{2} gD^2f)(dZ)^m$$

and after a heart-breaking computation we get

$$[X_b, (dZ)^{\frac{n-1}{2}} \otimes \{ aD^n + \tfrac{1}{2} Da \cdot D^{n-1} + \tfrac{n-1}{4} D^2aD^{n-2} + cD^{n-3} \bmod D^{n-4}\} \otimes$$
$$\otimes (dZ)^{\frac{n-3}{2}} =$$
$$= (dZ)^{\frac{n-1}{2}} \otimes \{ \{b,a\} D^n + \tfrac{1}{2} D \{b,a\} D^{n-1} + \tfrac{n-1}{4} D^2 \{b,a\} D^{n-2} +$$

$$+ \frac{(n-1)(n-3)}{16} aD^5 b \cdot D^{n-3} + D\,(Dc \cdot b - \tfrac{1}{2} c \cdot Db)\, D^{n-3} \bmod D^{n-4}\} \otimes$$

$$\otimes (dZ)^{\frac{n-3}{2}}.$$

Therefore $\widetilde{K}_n$ is a T-module, σ is a morphism of T-modules and commutation with T maps $\omega_\circ \subset \widetilde{K}_n$ into $\delta A_\circ \subset \omega_\circ$. Hence the induced action on $K_n = \widetilde{K}_n / \delta A_\circ$ is trivial on $\underline{H}_\circ^s$, and (7)' becomes a central extension of T. □

3. Sheaves. Let $\pi : X \to S$ be a family of $1|1$ - dimensional complex compact manifolds parametrized by a complex supermanifold S. Such families arising in superstring theory are endowed with an additional set of data which is called in [M2] a (relative) SUSY - structure. Mathematically it is simply a relative contact structure, which can be described by a maximally non-integrable $0|1$ - dimensional distribution $T^1 \subset TX/S$. Using local generators D of T^1, we construct on such X the sheaves, denoted by $\underline{E}^s$, $\underline{D}^s$, $\underline{E}^s_{\underline{L} \to \underline{M}}$, $\underline{H}^s$, etc.

The appropriately modified arguments from I.2.3 are applicable here and we leave to the reader a pleasant exercise to extend to this case the Conjecture I.2.4 (cf. [M2]).

REFERENCES

[BM] Beilinson A.A., Manin Yu.I. The Mumford form and the Polyakov measure in string theory. - Comm. Math. Phys., 1986, 107:3, p. 359-376.

[B Sch] Beilinson A.A., Schechtman V.V. (to appear)

[K] Kirillov A.A. On the orbits of the circle diffeomorphism group and local Lie superalgebras. - Funkc. Analiz e ego pril, 1981, 15:2, pp. 75-76.

[Kh] Khovanova T.G. The Gelfand-Dikii and the Virasoro Lie algebras. - Fankc. Analiz i ego. pril., 1986, 20:4, pp. 89-90

[M1] Manin Yu.I. Gauge fields and complex geometry. Nauka, 1985; Springer Verlag (to appear).

[M2] Manin Yu.I. a). Critical dimensions in string theories and the dualizing sheaf of the moduli space. - Funkc. Analiz i ego pril., 1986, 20:3, pp. 88-89.

b) Quantum strings and algebraic curves (Berkeley ICM talk, 1986).

[MR] Manin Yu. I., Radul A.O. A supersymmetric extension of the Kadomtsev - Petviashvili hierarchy. - Comm. Math. Phys., 1985, 98, pp. 65-77.

[R] Radul A.O. The Schwarz derivative and the Bott cocycle for Lie superalgebras of string theories. In: Numerical methods for differential equations, Moscow, MGU 1986, pp. 53-67.

Additive K-theory

B.L. Feigin

B.L. Tsygan

Contents

Introduction

0.1. The best known among the various definitions of algebraic K-theory is due to D. Quillen. Recall his construction. Let A be a ring and $GL(A)$ be the group of infinite invertible matrices $E + m$ where E is the identity matrix and the number of non-zero entries in m is finite. Let $BGL(A)$ be the classifying space of $GL(A)$; consider the "quillenization" $BGL(A) \longrightarrow BGL(A)^+$. This means that the induced map $H_\cdot(BGL(A); \mathbb{Z}) \longrightarrow H_\cdot(BGL(A)^+; \mathbb{Z})$ is an isomorphism and $\pi_1(BGL(A)^+) = GL(A)/[GL(A), GL(A)]$. The algebraic K-functor $K_i(A)$ is defined as the i-th homotopy group $\pi_i(BGL(A)^+)$, $i > 0$ (cf. [21]).

There is a homomorphism $GL(A) \times GL(A) \longrightarrow GL(A)$. It induces a multiplication $H_\cdot(BGL(A); \mathbb{Z}) \otimes H_\cdot(BGL(A); \mathbb{Z}) \longrightarrow H_\cdot(BGL(A); \mathbb{Z})$. Moreover, it induces an H-space structure on $BGL(A)^+$. So the space $H_\cdot(BGL(A); \mathbb{Z}) \otimes_{\mathbb{Z}} \mathbb{Q}$ becomes a Hopf algebra. By Hopf's theorem, it is freely generated by the graded space of its primitive elements $\mathrm{Prim}_\cdot(H_\cdot(BGL(A)) \otimes_{\mathbb{Z}} \mathbb{Q}))$; on the other hand, the well known Cartan-Milnor-Moore theorem [46] shows that $\mathrm{Prim}_i(H_\cdot(BGL(A); \mathbb{Z}) \otimes_{\mathbb{Z}} \mathbb{Q}) \cong \pi_i(BGL(A)^+) \otimes_{\mathbb{Z}} \mathbb{Q} = K_i(A) \otimes_{\mathbb{Z}} \mathbb{Q}$. So the spaces $K_i(A) \otimes_{\mathbb{Z}} \mathbb{Q}$ may be defined as the spaces of generators in the homology ring of the discrete group $GL(A)$ with values in the trivial one-dimensional module $\mathbb{Q}$. So K-theory is closely related to the problem of computing the homology groups of $GL(A)$.

This problem is extremely complicated. Even in the easiest case $A = \mathbb{Z}$ the full structure of the ring $H_\cdot(GL(A); \mathbb{Z})$ is unknown. Therefore, one very often studies not the ring $H_\cdot(GL(A); \mathbb{Z})$ but rather its image under a map into some group. For example, let V be a subspace of $H^\cdot(GL(A); \mathbb{R})$. There exists a pairing $V \otimes H_\cdot(GL(A); \mathbb{Z}) \to \mathbb{R}$ which defines a map $H_\cdot(GL(A); \mathbb{Z}) \to V^*$. The most frequently used method for constructing V is the following.

Let A be an algebra over $\mathbb{R}$. Then $GL(A)$ may be regarded in various cases as an infinite-dimensional Lie group. The continuous cohomology groups $H_c^\cdot(GL(A); \mathbb{R})$ map naturally into the usual $H^\cdot(GL(A); \mathbb{R})$. In other words there is a map

$$H_\cdot(GL(A); \mathbb{Z}) \longrightarrow H_c^\cdot(GL(A); \mathbb{R})^*.$$

The continuous cohomology groups of a Lie group may be determined using the Van Est theorem (cf. [61]). It asserts that there exists a spectral sequence $E_2^{ij} = H_c^i(G; R) \otimes H_{top}^i(G) \Rightarrow H_c^{i+j}(g; R)$. Here g is the Lie algeabra of G; in our case, the Lie algebra of $GL(A)$ is $gl(A)$. It consists of the infinite matrices over A having finitely many non-zero entries. This shows that the question about the Lie algebra homology of $gl(A)$ may be closely related to algebraic K-theory.

Let A be an algebra over a field k of characteristic zero. Then the homology space $H_\cdot(gl(A); k)$ is a Hopf algebra. The space of its primitive elements of degree i is denoted by $K_i^+(A)$. We call this the i-th additive K-functor of A.

Certainly the above program cannot be applied directly in the general case. There are three cases when the attaching of a Lie algebra to a group in order to study the interplay of corresponding homology theories may be done in the most direct way.

In the first case the ring under consideration contains a two-sided nilpotent ideal. More precisely, let $B = A \oplus I$ be an A-bimodule direct sum where I is a two-sided nilpotent ideal in B. Following standard notation we put $K_\cdot(A \oplus I, I) = \ker(K_\cdot(A \oplus I) \to K_\cdot(A))$; if $A_{\mathbb{Q}} = A \otimes_{\mathbb{Z}} \mathbb{Q}$, $I_{\mathbb{Q}} = I \otimes_{\mathbb{Z}} \mathbb{Q}$, we write $K_\cdot^+(A_{\mathbb{Q}} \oplus I_{\mathbb{Q}}, I_{\mathbb{Q}}) = \ker(K_\cdot^+(A_{\mathbb{Q}} \oplus I_{\mathbb{Q}}) \to K^+(A_{\mathbb{Q}}))$. We construct the maps

$$K_i^+(A \oplus I, I) \otimes_{\mathbb{Z}} \mathbb{Q} \xrightarrow{\iota} K_i(A \oplus I_{\mathbb{Q}}, I_{\mathbb{Q}}) \xrightarrow{\rho} K_i^+(A_{\mathbb{Q}} \oplus I_{\mathbb{Q}}, I_{\mathbb{Q}})$$

and show that ι is an isomorphism. If $A = \mathbb{Z}$, then ρ is an isomorphism; in the general case it is an isomorphism when $i = 2$ and an epimorphism when either $i = 3$ or $I^2 = 0$. This enables one to generalize some well known results on the deformations of algebraic K-theory (see 7.5). The corresponding question has been put by A.A. Beilinson in [2]. If $\mathfrak{a}$ is a commutative Q-algebra without unit and $\mathfrak{a}^n=0$ for some n, we establish the isomorphisms

$$K_2(\mathfrak{a}) \simeq \Omega^1_{\mathfrak{a}} / d\Omega^0_{\mathfrak{a}} \quad \text{(S. Bloch, [4]);}$$

$$F^2_\gamma K_i(\mathfrak{a})/F^3_\gamma K_i(\mathfrak{a}) \simeq D_{i-2}(\mathfrak{a},\mathfrak{a})$$

where $D.$ is the Quillen-Harrison homology (cf. [48]); furthermore, we show the connection between $K.(\mathfrak{a})$ and the de Rham complex of $\mathfrak{a}$ (or rather with the derived de Rham complex in sense of L. Illusie).

Note that our construction does not use the Cartan-Milnor-Moore theorem. We need only a realization of the space $BGL(\mathbb{Q} \oplus I_{\mathbb{Q}})$ provided by Sullivan theory. Using technique of operads which is represented in the paper by V.A. Hinich and V.V. Schechtman [25] in these proceedings, one can obtain a generalization of the map ρ which need not the assertion of zero characteristic.

The second case was investigated by D. Burghelea in [7]. Here A is not a (topological) ring but a reduced simplicial ring with a certain supplementary condition. The algebraic K-theory of such a ring (for instance when A is the group ring of Kan's simplicial loop group of a simply connected simplicial set) has very rich topological applications (F. Waldhausen, [63]). But in this case there is a very good correspondence between the Lie algebras and the "Lie groups" given by Quillen rational homotopy theory. Burghelea has shown that it is possible to pass from the group homology to the Lie algebra homology. Moreover, the K-functors of A tensored by $\mathbb{Q}$ turn out to be (almost) isomorphic to their additive analogues, i.e., the primitive part of $H.(gl(A); \mathbb{Q})$ (see A.9.4).

In the third case, A is a Banach algebra. Then application of the Van Est spectral sequence seems to be reasonable. In the first place, $GL(A)$ is then a Banach-Lie group; in the second place, the second tensor factor in the second term of the spectral sequence is of great interest. In fact, the primitive part of $H^{top}(GL(A), \mathbb{R})$, i.e. the homotopy groups $\pi_i(GL(A)) \otimes \mathbb{R}$, are nothing other than the topological K-functors $K_i^{top}(A) \otimes \mathbb{R}$. For $A = C(X)$ they give the usual theory of X.

The interplay of $K_i(A)$ with $K_i^{top}(A)$ and $K_i^+(A)$ was investigated by M. Karoubi (his approach does not directly involve the Lie algebra homology; see [33], [34], [35], [15]. Karoubi constructed regulator maps from $K_i(A)$ to an object which involves $K_\cdot^{top}$ and $K_\cdot^+$. This generalizes the well known Borel construction and is closely related to the constructions of S. Bloch, P. Deligne, and A. Beilinson (cf. [2], [54]). Karoubi's construction might be carried out using the Van Est spectral sequence; for instance, the maps of transgression in it are the Chern characters in topological K-theory.

In the general case we cannot apply the Lie theory directly. However, we shall see in Chapter 7 that there are maps from multiplicative K-theory to additive K-theory (we follow Karoubi's papers [31], [32], [35]; see also [19]). It is perhaps of greater importance that the properties of the K^+-functors are parallel to those of Quillen's K-theory. Since the computations of additive K-functors are far easier, we may hope that the development of this subject will give analogies and conjectures for algebraic K-theory.

0.2. There exists another definition of the K-functors due to S. Gersten and R.Swan (cf. [21], [56]). Let A be a ring. Consider a free simplicial resolution of A, i.e., a simplicial ring R together with an epimorphic weak equivalence $R \longrightarrow A$ such that R_n are free for $n \geqslant 0$ and the degeneracy maps preserve the set of generators. Apply the functor GL to this resolution. We obtain a simplicial group GL(R).

Now regard GL(R) as a simplicial set. Then $\pi_{i-1}(GL(R)) = K_i(\mathbb{Z} \longrightarrow A)$ if $i \geqslant 2$. Here $K_\cdot(\mathbb{Z} \longrightarrow A)$ denotes the homotopy groups of the homotopy fibre of the map $BGL(\mathbb{Z})^+ \longrightarrow BGL(A)^+$. If $i = 0$, then $\pi_0(GL(R)) \simeq St(A)$.*) In this definition the algebraic K-functors arise as the derived functors of GL.

The additive K-functor has an analogous definition. Namely, apply the functor "quotient by commutant" to the resolution R. Suppose that A is an algebra over a field k of characteristic zero. We obtain a simplicial vector space; the homology groups of the corresponding complex turn out to be equal to the relative additive K-functors of the homomorphism $k \longrightarrow A$. In other words, the K^+-functors are the higher derived functors for the functor $A \longrightarrow A/[A, A]$.

The construction of Gersten-Swan uses the following fact. Let R be a free associative algebra over k. Then the kernel K of the map $GL(R) \longrightarrow GL(\mathbb{Z})$ is acyclic (i.e., $H_i(K; \mathbb{Z}) = 0$, $i > 0$). This is established by Swan. If R is a k-algebra, then the group FL(R) may be regarded as a Lie group over k. For instance, if $k = \mathbb{C}$, then every element may be included in a one-parameter subgroup. This means that

*) More precisely, there is an exact sequence $1 \longrightarrow St(A) \longrightarrow \pi_0 GL(R) \longrightarrow K_1\mathbb{Z} \longrightarrow 1$

there is a Lie algebra corresponding to GL(R). It is well known that GL(R) is generated by the invertible matrices over k and the elementary matrices. This shows that the Lie algebra corresponding to GL(R) is isomorphic to $\mathfrak{gl}(k) + [\mathfrak{gl}(R), \mathfrak{gl}(R)]$. The map $H.([\mathfrak{gl}(R), \mathfrak{gl}(R)]) \to H.(\mathfrak{sl}(k))$ happens to be an isomorphism. This is the analogue of the theorem about the K-groups of a free ring. Note that this result was stated earlier by W.C. Hsiang and R.E. Staffeldt in [27].

Now we give another definition of K^+ which is closer to that of Gersten and Swan. Apply the functor Lie GL to the resolution R of A. We obtain a simplicial vector space Lie GL(R). It can be shown that $H_i(\text{Lie GL}(R)) \simeq K_{i+1}(k \to A)$, $i > 0$, and $H_o(\text{Lie GL}(R))$ is the Lie-Steinberg algebra, i.e., the universal central extension of $[\mathfrak{gl}(A), \mathfrak{gl}(A)]$.*) This algebra was investigated by S. Bloch, C. Kassel and J.L. Loday, cf. [3], [38]. So we may say that the additive K-functors are the higher derived functors for Lie GL. We see that the Gersten-Swan definition enables one to define the additive K-theory directly using the Lie theory (because of the fact that the Lie theory works rather well for general linear groups over free algebras).

0.3. The additive K-functors were introduced independently in [12], [13]. A. Connes calls them the cyclic homology:

$$HC_i(A) = K^+_{i+1}(A).$$

In his works they arise in the trace computations. We shall explain this by the following example.

Let M be a smooth compact manifold without boundary, E be a vector bundle over M and $D : \Gamma(E) \to \Gamma(E)$ a skew self-adjoint elliptic differential operator (for example, the Dirac operator). We suppose that we have fixed riemannian metrics on M and E. The space $\Gamma(E)$ is the direct sum $\Gamma(E) = \Gamma(E)^+ \oplus \Gamma(E)^-$; $\Gamma(E)^+$ is generated by the eigenvectors corresponding to positive eigenvalues, and $\Gamma(E)^-$ is generated by the eigenvectors corresponding to the nonpositive eigenvalues. Let P^+ be the orthogonal projection onto $\Gamma(E)^+$. In the algebra $\text{End}(\Gamma(E)^+)$ consider the ideal L^n ($n = \dim M$) of operators B such that $\text{tr}\,|B|^n < \infty$. Let a, b be elements of $C^\infty(M)$. The endomorphisms of $\Gamma(E)$ given by multiplication by a and b will also be denoted by the letters a and b. The operators $P^+a\,P^+$ and $P^+b\,P^+$ preserve $\Gamma(E)^+$. It may be verified that

$$(P^+a\,P^+b\,P^+ - P^+\,ab\,P^+) \in L^n.$$

So the formula $a \overset{\varphi}{\mapsto} P^+a\,P^+$ gives an "almost representation" of

*) More precisely, there is an exact sequence $0 \to \mathfrak{st}(A) \to H_o(\mathfrak{gl}(R)) \to K_1^+k \to 0$.

A in $\Gamma(E)^+$. It is convenient to pass to the "almost representation" of the Lie algebra. Multiplying φ by the matrix algebra $M_\infty(\mathbb{C})$, we obtain a map

$$\tilde{\varphi} : \mathfrak{gl}(A) \simeq M_\infty \otimes A \longrightarrow M_\infty \otimes \mathrm{End}(\Gamma(E)^+) \xrightarrow{\sim} \mathfrak{gl}(\mathrm{End}\,\Gamma(E)^+).$$

Set

$$\Theta(X_1, X_2) = \tilde{\varphi}([X_1, X_2]) - [\tilde{\varphi}(X_1), \tilde{\varphi}(X_2)].$$

This is a curvature form on the Lie algebra $\mathfrak{gl}(A)$ with values in $\mathfrak{gl}(L^n) \subset (\mathrm{End}\,\Gamma(E)^+)$. There is an invariant polynomial on the Lie algebra $\mathfrak{gl}(L^n)$: $P(X) = \mathrm{tr}(X^n)$. Substituting Θ in place of X in the polarization of P, we obtain a 2n-dimensional closed cochain in the standard cohomology complex of the Lie algebra $\mathfrak{gl}(A)$.

Let $f_1, \ldots, f_{2n} \in A$ and $f_i\, E_{11}$ be elements of $\mathfrak{gl}(A)$; $(f_i\, E_{11})_{kl} = \delta^1_k\, \delta^1_l\, f_i$. It was shown by Connes that

$$\mathrm{tr}\, \Theta^n (f_1 E_{11}, \ldots, f_{2n} E_{11}) = \varkappa \cdot \int_M f_1\, df_2 \ldots df_{2n}.$$

Here d is the de Rham differential and $\varkappa$ does not depend on $f_1, \ldots, f_{2n}$.

(Note that for $n = 1$ this construction was used by J. Tate to define the residue of a differential form on a complex algebraic curve, cf. [57]).

More generally, let A be a C*-algebra and $\varphi : A \longrightarrow \mathrm{End}\, V$ be an almost representation. We may proceed as above and define a class in $H^{2n}(\mathfrak{gl}(A); k)$. This gives a pairing between Kasparov's homological K_1 and the additive K-functor K^+_{2n} (cf. [12]).

The construction discussed above associates to the space $\Gamma(E)^+$ a number which is an analogue of $\dim \Gamma(E)^+$. This number is connected with the η-invariant of the operator D (cf. [66]).

A. Connes uses an analogous construction to formulate a generalization of the Atiyah-Singer index theorem.

Now we shall say in few words why the classical homological algebra of the ring A arises in this approach. Consider an $(n+1)$-linear form ψ on A satisfying the following equations:

a) $\psi(a_1, \ldots, a_n, a_o) = (-1)^n\, \psi(a_o, \ldots, a_n)$;

b) $\sum_{i=0}^{n-1} (-1)^{i+1}\, \psi(a_o, \ldots, a_i\, a_{i+1}, \ldots, a_n) + \psi(a_1, \ldots, a_n a_o) = 0$

There are two important examples. In the first place, let τ be a trace on A; then for even n the form $\psi(a_0,\dots,a_n) = \tau(a_0 \dots a_n)$ satisfies the equations a) and b). In the second place, let $A = C^\infty(M)$ where M is a smooth manifold. Let c be an n-dimensional cycle on M. Then the form

$$\psi(a_0,\dots,a_n) = \int_c da_0 \dots da_{n-1} a_n$$

also satisfies the above equations.

So, in the first place, these equations show how one can generalize the notion of trace on an algebra; in the second place, they show how to describe the cycles (or rather the closed de Rham currents) in purely algebraic way.

On the other hand, the equation b) means that ψ is a cocycle in the standard complex for computing Hochschild cohomology of A with values in the bimodule A* (cf. [10]).

The basic object of our work is the following:

let

$$C_{n+1}^{cycl}(A) = \overset{n+1}{\otimes} A/\mathrm{im}(1-\tau), \quad \tau(a_0 \otimes \dots \otimes a_n) =$$

$$= (-1)^n a_1 \otimes \dots \otimes a_n \otimes a_0;$$

set

$$\delta(a_0 \otimes \dots \otimes a_n) = \sum_{i=0}^{n-1} (-1)^i a_0 \otimes \dots \otimes a_i a_{i+1} \otimes \dots \otimes a_n +$$

$$+ (-1)^n a_1 \otimes \dots \quad a_n a_0;$$

we show in 1.2 that δ turns $C^{cycl}(A)$ to be a homological complex. We prove in Chapter 4 that the homology of this complex is isomorphic to the additive K-functors of A (it was shown independently by J.L. Loday and D. Quillen in [42]).

The above definition is correct when the base field is of characteristic zero; in the general case there is another definition (see 1.2).

0.4. Let A be the algebra of C^∞-functions on a smooth manifold M. The Lie algebra $\mathfrak{gl}(A)$ is often called the current algebra over M with values in $\mathfrak{gl}(\mathbb{R})$. It is regarded as a topological Lie algebra with the C^∞-topology. The problem of computing the continuous Lie algebra cohomology of $\mathfrak{gl}(A)$ is closely related to the analogous question for the Lie algebra of smooth vector fields. We maintain that this analogy is rather essential. In particular, the methods of computations are very similar: in both cases the invariant theory of the Lie algebra $\mathfrak{gl}_n$ is used (see [20], [42], [58]). We now state the answer for the Lie algebra cohomology of $H_c^{\bullet}(\mathfrak{gl}(A); \mathbb{R})$ (see [12]). Speaking somewhat imprecisely, the algebra $H_c^{\bullet}(\mathfrak{gl}(A); \mathbb{R})$ is freely generated by the graded space $\bigoplus_{i>0} L_i$;

$$L_i = (\Omega_M^{i-1}/d\Omega_M^{i-2})^* \oplus H_{i-3}(M, \mathbb{R}) \oplus H_{i-5}(M, \mathbb{R}) \oplus \dots$$

the first summand in the right hand side is the space of closed de Rham currents.

Note that the L_i are dual to the cohomology of the trucated de Rham complexes introduced in [2] (cf. also [54]) for constructing characteristic classes in algebraic K-theory.

0.5. The main technical tool needed for computing $H_.(\mathfrak{gl}(A); k)$ is invariant theory. We now say a few words about it. Let $C(\mathfrak{gl}_n(A))$ be the standard complex of the Lie algebra $\mathfrak{gl}_n(A)$ with values in the trivial one-dimensional module k. Suppose that A contains a unit. Then $\mathfrak{gl}_n(k) \subset \mathfrak{gl}_n(A)$.

The complex $C_.(\mathfrak{gl}_n(A))$ has the same homology as its subcomplex consisting of chains invariant with respect to $\mathfrak{gl}_n(k)$. Denote the latter subcomplex by $C_.(\mathfrak{gl}_n(A))^{inv}$. The invariant theory enable one to introduce a complex $\mathcal{K}_.(A, n)$ together with an epimorphism $\mathcal{K}_.(A, n) \longrightarrow C(\mathfrak{gl}_n(A))^{inv}$. The complex $\mathcal{K}_.(A, n)$ is isomorphic to $C_.(\mathfrak{gl}(A))^{inv}$ as a graded vector space; the differential in it depends linearly on n. This allows one to define a complex $\mathcal{K}_.(A, n)$ for an arbitrary n belonging to the ground field k of characteristic zero; $C_.(\mathfrak{gl}_n(A))^{inv}$ is a quotient of $\mathcal{K}_.(A, n)$.

Furthermore, there is a map

$$\bar{\theta}(n) : \mathcal{K}_.(A, n) \longrightarrow C_.(\mathfrak{gl}(A))^{inv}$$

which depends polynomially on n. For $n \notin \mathbb{Z}$, $\bar{\theta}(n)$ is an isomorphism. At the singular points $(n \in \mathbb{Z})$ the family $\bar{\theta}(n)$ defines a filtration of $\mathcal{K}_.(A, n)$ and $C(\mathfrak{gl}(A))^{inv}$. More precisely, let $n_0 \in \mathbb{Z}$; then there is a filtration of $C_.(\mathfrak{gl}(A))^{inv}$ by the subcomplexes

$$\operatorname{im}\bar{\theta}(n_0) \subseteq \operatorname{im}\bar{\theta}(n_0) + \operatorname{im}\frac{\partial\bar{\theta}}{\partial n}(n_0) \subseteq \dots$$

All these filtrations $(n_0 \in \mathbb{Z})$ define a filtration of $C_.(\mathfrak{gl}(A))^{inv}$ indexed by the partially ordered set of Young diagrams (see Chapter 4). The subcomplex corresponding to the infinite Young diagram having n cells in each column is isomorphic to $C_.(\mathfrak{gl}_n(A))^{inv}$. Speaking informally, the situation is the same as if there existed a Lie algebra $\mathfrak{gl}_{\P}(A)$ for an arbitrary Young diagram $\P$. At least the homology groups $H_{\cdot}(\mathfrak{gl}_{\eta}(A))$ are well defined. It would be interesting to find an analogy of this phenomenon in algebraic K-theory. The idea that the parameter n need not be an integer is used once more in Chapter 5, where we construct the algebra of operations in additive K-theory. Note that the algebra constructed there is rather similar to one in complex cobordism theory. Here, as in the work by Buchschtaber, there is a relation with the formal diffeomorphism group of the line. We hope to discuss this in another work.

0.6. The contents of the paper are as follows. In Chapter 1 we introduce the additive K-functors and study their properties. In particular, we construct a spectral sequence connecting them to the Hochschild homology. We also introduce the Bott periodicity operator. We define the cohomology of A to be, roughly speaking, the additive K-functor $K_0 \oplus K_1$ with the inverted Bott operator (as in topological K-theory). In 1.4, following Connes, we compute the K^+-functors of the coordinate rings of affine smooth algebraic varieties. Note that the definitions in Chapter 1 make sense for an arbitrary scalar ring k.

In Chapter 2 we compute the K^+-functors of free algebras. As a consequence, we find that the K^+-functors are the higher derived functors of $K_1^+(A) = A/[A, A]$. We also introduce the relative additive K-functors.

In Chapter 3 we prove Waldhausen type theorems for additive K-theory. In particular, we define additive analogues of the functors Nil. (cf. [62]) and compute the K^+-functors of amalgamated products of algebras. The K^+-functors may be characterized as the functors on homotopy category of simplicial k-algebras, enjoying the following properties:

a) $K_1^+(A) \cong A/[A, A]$;

b) $K_{\cdot}^+(k[t]) \cong K_{\cdot}^+(k)$;

c) there exists a functorial exact sequence

$$\ldots \longrightarrow K_i^+(A) \oplus K_i^+(B) \longrightarrow K_i^+(A \underset{k}{*} B) \longrightarrow K_{i-1}^+(k) \longrightarrow \ldots$$

The basic facts in Chapter 3 are proved for a scalar ring k of characteristic zero (i.e., $\mathbb{Q} \subseteq k$). The general case is contained in Appendix (A8).

In Chapter 4 (4.1-4.3) we compute the Lie algebra homology of $\mathfrak{gl}(A)$ (char $k = 0$) and prove that its primitive part coincides with the K^+-functors introduced in Chapter 1. This result was obtained independently by Loday and Quillen, cf. [42]. We also calculate the stable homology of orthogonal and symplectic Lie algebras over an arbitrary k-algebra with involution (4.7). Furthermore, we compute the homology of $\mathfrak{gl}(A)$ with values in modules of indecomposable tensors over the adjoint representation, i.e., in the interior and symmetric powers of $\mathfrak{gl}(A)$, etc. This is contained in 4.8. In Subsections 4.4-4.6 we study the various filtrations on $H.(\mathfrak{gl}(A))$, in particular, the filtration by the images of the maps $H(\mathfrak{gl}_n(A)) \to H(\mathfrak{gl}(A))$.

Let A be an algebra without unit. Then the problem of computing $H.(\mathfrak{gl}(A))$ may be reduced to a similar problem for $H.(\mathfrak{gl}(\tilde{A}); M)$, where $\tilde{A}$ is the algebra obtained from A by adjoining a unit and M is a module over $\mathfrak{gl}(k)$. We calculate $H.(\mathfrak{gl}(A); M)$ in 4.8; in the case $A = x\,k[x]$ we apply this information to write down the stable Macdonald-Kac identity.

In Chapter 5 we construct the operations in additive K-theory. These are, first, the multiplication (introduced by Loday and Quillen in [43]) and the comultiplication (introduced by Connes in [12]). Let A and B be two algebras. Then there exist homomorphisms: $K_i^+(A) \otimes K_j^+(B) \to K_{i+j}^+(A \otimes B)$, $K_{i+j-1}^+(A \otimes B) \to K_i^+(A) \otimes K_j^+(B)$. We obtain these two operations as a result of a single construction. The operations appear in the following exact sequence:

$$\ldots \to K_i^+(A \otimes B) \to \bigoplus_{\alpha+\beta=i+1} K_\alpha^+(A) \otimes K_\beta^+(B) \to$$

$$\to \bigoplus_{\alpha+\beta=i-1} K_\alpha^+(A) \otimes K_\beta^+(B) \to K_{i-1}^+(A \otimes B) \to \ldots$$

It is interesting that in algebraic K-theory multiplication exists and comultiplication probably does not.

We interpret the differential $B : H_i(A, A) \to H_{i+1}(A, A)$ introduced in 1.2 in terms of the comultiplication (note that if $A = k[X]$ for a smooth affine algebraic variety X, then $H_i(A, A) = \Omega_X^i$ and $B : \Omega_X^i \to \Omega_X^{i+1}$ is the de Rham differential).

If A is commutative, then $K_\cdot^+(A)$ is a skew commutative graded algebra. If A is a Hopf algebra, then $K_\cdot^+(A)$ is a coalgebra.

The second half of Chapter 5 is devoted to the operations in $H(\mathfrak{gl}(A); k)$ for a commutative ring A. These operations are endomorphisms of the functor $A \mapsto H.(\mathfrak{gl}(A); k)$. We describe a large subalgebra of the algebra of all operations. We shall discuss this material in more detail elsewhere. At the end of Chapter 5 we construct the Grothedieck and Adams operations in K^+.

Chapter 6 is devoted to the K^+-functors of commutative noethrian algebras over a field k of characteristic zero. Let A be such an algebra, and let $R \longrightarrow A$ be a free resolution of A in the category of differential graded algebras. This means that $R = \bigoplus_{i=0} R_i$ is a free skew-commutative algebra endowed with a derivation $\partial : R_i \longrightarrow R_{i-1}$ and an epimorphism $R \longrightarrow A$ such that $\partial^2 = 0$, $H.(R) \xrightarrow{\sim} A$. The algebra A is the ring of functions on a superspace V. Consider the algebra Ω_V of differential forms on V. The differential ∂ induces the differentials $\Omega_V^i \xrightarrow{\partial} \Omega_V^i$ for all i. The homology of the complex Ω_V^1 is called the Quillen-Harrison homology of A with values in A; the complex $\Omega_V^1 \otimes_R A$ is the cotangent complex of the homomorphism $k \longrightarrow A$. In general, the homology groups of Ω_V^i are by definition the higher derived functors of $A \longrightarrow \Omega_A^i$ on the category of commutative algebras. The de Rham differential $d : \Omega_V^i \longrightarrow \Omega_V^{i+1}$ is a homomorphism of the complexes. We show that the relative K^+-functors of the arrow $k \longrightarrow A$ is the homology of direct sum of the complexes $\Omega_V^i / d\Omega_V^{i-1}$; this is the spectral decomposition of the Adams operations. Speaking informally, when restricted to the category of commutative algebras the K^+-functors become the higher derived functor of $A \mapsto \Omega_A^i / d\Omega_A^{i-1}$. The homology groups of $\Omega_V^i / d\Omega_V^{i-1}$ form a reasonable cohomology theory. Similar objects were considered by Illusie in [28]. These homology groups map to the cohomology groups of the quotient of the de Rham complex of some smooth variety containing Spec A by the n-th term of the Hodge filtration. The latter cohomology groups are crystalline cohomology with values in the sheaf $\mathcal{O} / \mathcal{J}^n$; the higher derived functors of $A \mapsto \Omega_A^i / d\Omega_A^{i-1}$ may also be interpreted in terms of sheaf cohomology groups on a certain Grothendieck topology associated to Spec A.

In Chapter 6 we also show that the cohomology groups introduce in 1.5 coincide with the crystalline cohomology of Spec A with values in the structure sheaf (if $k = \mathbb{C}$, this is the usual singular cohomology).

In Subsection 6.6 we provide an application of the results of Chapter 6 to rational homotopy theory. For a simply connected topological space X let $A'(X)$ be its Sullivan model (cf. [5]). We show that the relative K^+-functors of $k \longrightarrow A'(X)$ are isomorphic to the reduced cohomology of the space $ES^1 \times_{S^1} X^{S^1}$ with coefficients in k;

here X^{S^1} is the free loop space of X and ES^1 is the total space of the universal S^1-bundle.

On the other hand, Burghelea, Goodwillie, Dwyer and Jones showed that rational homology of $ES^1 \times_{S^1} X^{S^1}$ is isomorphic to Waldhausen rational algebraic K-functor of X. Thus, Theorem 6.6.1 enables one to compute $K.(X) \times \mathbb{Q}$ for some examples. For instance, if X is formal, then

$$K.(X) \otimes \mathbb{Q} \xrightarrow{\sim} K.(\mathbb{Z}) \otimes \mathbb{Q} \oplus K^{\cdot}_{+}(\mathbb{Q} \longrightarrow H^{\cdot}(X, \mathbb{Q})).$$

Note that algebraic geometry of spectrum of superalgebra $H^{\cdot}(X; \mathbb{Q})$ may carry important topological information.

In Chapter 7, following [31], [32] and [19], we define the Chern character from algebraic K-theory to additive K-theory. We discuss two constructions: the first uses Quillen's definition; the second uses the definition by Gersten-Swan.

In Subsection 7.5 we define a regulator map from relative algebraic K-theory of a two sided nilpotent ideal to relative additive K-theory and study its properties. As a consequence of the general construction we obtain that algebraic K-functors of an artinian commutative local algebra over a number field are isomorphic to the additive K-functors (see 7.5 for more explicit statement). So, because of Theorem 6.1.1., the graded quotients of the γ-filtration are isomorphic to homology of the higher de Rham complexes in sence of Illusie.

In the Appendix we discuss another approach to defining the additive K-functors, which was due to Connes (cf. [14]). One defines a sequence of functors from the category of algebras to the category of k-modules $F_i(A) = \overset{i+1}{\otimes} A$, $i \geq 0$. Let $\mathcal{K}$ be the category whose objects are indexed by the nonnegative integers and whose morphisms from i to j are the natural transformations $F_i \longrightarrow F_j$. For example, the multiplication $A \otimes A \longrightarrow A$ determines a morphism $F_2 \longrightarrow F_1$. Then an algebra A clearly determines a functor $A^{\natural}$ from $\mathcal{K}$ to the category $\mathfrak{A}$ of k-modules. For an arbitrary small subcategory $\Lambda \subseteq \mathcal{K}$, the category of functors from Λ to $\mathcal{K}$ is abelian. Denote it by $\mathfrak{A}^{\Lambda}$. Now we may obtain the homological functors of A by considering the higher derived functors $\mathbb{L}.\varinjlim (A^{\natural})$. Connes shows that for a suitably defined $\Lambda \subset \mathcal{K}$ these derived functors are isomorphic to $K^{+}_{\cdot+1}(A)$. It is also possible to define the bifunctors $\mathrm{Ext}^{\cdot}_{\mathfrak{A}^{\Lambda}}(A^{\natural}, B^{\natural})$. They are related to Kasparov KK-bifunctors.

This idea for defining the (co-)homology groups depending on A seems to be very promising. For example, it appears in the recent Hopf algebra cohomology theory of Drinfeld; a similar approach was also used by Beilinson in his absolute Hodge cohomology theory.

We formulate Connes' results in a slightly more general form. Namely, we attach K^+-groups to a ring A together with an action of a cyclic group G by automorphisms of A. If G is infinite, then our construction gives usual Hochschild cohomology. If it is finite, then the K^+-functors introduced in A3 (denote them by $K^+_\cdot(G;A)$) are more interesting. For example, let A be commutative and $X = \operatorname{Spec} A$; consider the "fixed point homotopy cosimplicial scheme" $\underleftarrow{\operatorname{holim}}_G X$:

$$X \rightleftarrows \operatorname{Hom}(G, X) \;\substack{\leftarrow\\ \rightarrow\\ \rightarrow\\ \rightarrow\\ \leftarrow}\; \operatorname{Hom}(G \times G, X) \ldots$$

Then

$$K^+_\cdot(G; A) \simeq K^+_\cdot(k[\underleftarrow{\operatorname{holim}}_G X])$$

if $|G|$ is invertible in k; the right hand side is the K^+-functor of a simplicial ring. We also may construct the cohomology groups as in 1.5; then for $\operatorname{char} k = 0$ we shall have $H^\cdot(G; A) \simeq H_{cris}((\operatorname{Spec} A)^G)$.

0.7. Now a few words about notation.

All the algebras are defined over an arbitrary commutative ring k; we suppose for simplicity that they are projective k-modules. In Chapters 2, 3, 5, 6 k is a field of characteristic zero. All the algebras are assumed to contain the unit (unless the contrary is explicitly stated).

Let $\mathfrak{A}$ be a category. We regard an object of $\mathfrak{A}$ automatically as a constant (co-)simplicial object; if $\mathfrak{A}$ is abelian, we also regard its objects as complexes contained in degree zero; a simplicial object of $\mathfrak{A}$ is also regarded as a complex (with differential $d_0 - d_1 + \ldots$). Furthermore, the category of rings is assumed to be embedded in the obvious way in the category of graded or differential graded rings.

In this paper, we have attempted to describe some results obtained in recent years concerning new homology theory. Our main sources are the papers [12], [13], [19], [35], [44], [58], etc. Some facts and constructions are published for the first time.

We are greatly indebted to the participants at Yu.I. Manin's seminar on algebraic K-theory and especially to A.A. Beilinson, Yu.I. Manin, V.S. Retakh, V.V. Schechtman, M. Wodzicki, with whom we discussed frequently the additive analogues of the algebraic K-functors. We also have the pleasure of thanking I.M. Gelfand, Yu.L. Daletskii, D.B. Fuks, A.K. Tolpygo and A.M. Vershik.

Chapter 1. Additive K-functors

1.1. Hochschild homology. We recall here some facts and definitions from [9]. Let A be an algebra over k, A^o the opposite algebra, M a bimodule over A. Then M is a left $A \otimes A^o$ -module. The Hochschild homology groups of A with coefficients in M are by definition the groups $\mathrm{Tor}_i^{A\otimes A^o}(A, M)$. We denote them by $H_i(A, M)$, $i > 0$. These homology groups may be computed by means of the standard bar-resolution of the bimodule A:

$$\dots \longrightarrow \mathcal{B}_n(A) \xrightarrow{b} \mathcal{B}_{n-1}(A) \xrightarrow{b} \dots \xrightarrow{b} \mathcal{B}_o(A) \xrightarrow{\varepsilon} A \longrightarrow 0, \qquad (1.1.1)$$

$$\mathcal{B}_n(A) = \overset{n+2}{\otimes} A; \quad b(a_{-1} \otimes a_o \otimes \dots \otimes a_n) =$$

$$= \sum_{i=0}^{n} (-1)^i a_{-1} \otimes \dots \otimes a_{i-1} a_i \otimes \dots \otimes a_n;$$

$\varepsilon(a_{-1} \otimes a_o) = a_{-1} a_o$. The operator s of left tensoring by the unit is a contracting homotopy, whence $\mathcal{B}.(A)$ actually is a resolution. Let $C.(A, M) = \mathcal{B}.(A) \underset{A\otimes A^o}{\otimes} M$.

$$C_n(A, M) \cong \overset{n}{\otimes} A \otimes M; \quad \delta(a_o \otimes \dots \otimes a_{n-1} \otimes m) = \qquad (1.1.2)$$

$$a_1 \otimes \dots \otimes a_{n-1} \otimes m a_o + \sum_{i=1}^{n-1} (-1)^i a_o \otimes \dots \otimes a_{i-1} a_i \otimes \dots$$

$$\dots \otimes a_{n-1} \otimes m + (-1)^n a_o \otimes \dots \otimes a_{n-2} \otimes a_{n-1} m$$

Let $\mathcal{B}.'(A)$ be the quotient complex of $\mathcal{B}.(A)$ by the subcomplex consisting of linear combinations of degenerate elements $a_{-1} \otimes \dots \otimes a_n$, i.e. such that $a_i = 1$ for some $i \in [0, n-1]$. We call $\mathcal{B}.'(A)$ the normalized bar-resolution of A; put also $C.'(A,M) = \mathcal{B}.'(A) \underset{A\otimes A^o}{\otimes} M$. Thus $C_n'(A,M) = \overset{n}{\otimes} (A/k) \otimes M$ and the differential is given by (1.1.2). It is well known that the projections $\mathcal{B}.(A) \longrightarrow \mathcal{B}.'(A)$, $C.(A,M) \longrightarrow C'(A,M)$ are quasi-isomorphisms.

Fix an $N \in \mathbb{N}$. Let $M_N(A)$ be the algebra of $N \times N$ -matrices over A; we denote by $M_N(M)$ the bimodule over $M_N(A)$; as a k-module, it is equal to the module of $N \times N$ -matrices with entries in M; if $\alpha = (a_{ij})_{1\leqslant i,\, j\leqslant N}$, $\mu = (m_{ij})_{1\leqslant i,\, j\leqslant N}$, then we define the bimodule structure by the following formulas:

$$(\alpha\cdot\mu)_{ik} = \sum_{j=1}^{N} a_{ij}m_{jk}; \qquad (\mu\cdot\alpha)_{ik} = \sum_{j=1}^{N} m_{ij}a_{jk}$$

Define a map

$$tr_N\colon C.(M_N(A);\, M_N(M)) \longrightarrow C.(A;M);$$

if $\alpha^{(k)} = (a_{ij}^{(k)})$, $\mu = (m_{ij})$, then

$$tr_N\,(\alpha^{(o)}\otimes \dots \otimes \alpha^{(n)}\otimes \mu) =$$

$$= \sum a^{(o)}_{i_o i_1}\otimes \dots \otimes a^{(n)}_{i_n i_{n+1}} \otimes m_{i_{n+1} i_o} \quad ;$$

on the other hand, we have the obvious inclusion

$$\iota_N\colon C.(A;\, M) \hookrightarrow C.(M_N(A);\, M_N(M)),$$

$$\iota_N(a_o\otimes \dots \otimes a_n \otimes m) = (E_{11}\ a_o)\otimes \dots \otimes(E_{11}\ a_n)\otimes(E_{11}\ m)$$

where E_{11} is the matrix such that $(E_{11})_{ij} = \delta^1_i\ \delta^1_j$.

<u>Proposition 1.1.1</u>. The maps ι_N, tr_N are the homomorphisms of complexes; they induce homology isomorphisms which are inverse to each other.

<u>1.2. Definition of the additive K-functors</u>. Consider the operators t_n, τ_n on $C_n(A,A)$:

$$t_n(a_o\otimes \dots \otimes a_n) = a_1 \otimes \dots \otimes a_n \otimes a_o; \quad \tau_n = (-1)^n\, t_n \qquad (1.2.1)$$

Sometimes we shall omit the index n.

Put also $N_n = 1 + \tau_n + \tau_n^2 + \dots + \tau_n^n$. Let δ be the differential in $C.(A,A)$ and b be the differential in $\mathcal{B}.(A)$:

$$b(a_o\otimes \dots \otimes a_n) = \sum_{i=0}^{n-1} (-1)^i a_o\otimes\dots\otimes a_i a_{i+1}\otimes\dots\otimes a_n.$$

<u>Lemma 1.2.1</u>. (cf. [12], [58]). The diagram

$$\begin{array}{ccccccccc}
 & & \vdots & & \downarrow\delta & & \downarrow b & & \vdots \\
\dots & \xrightarrow{N} & \overset{n}{\otimes}A & \xrightarrow{1-\tau} & \overset{n}{\otimes}A & \xrightarrow{N} & \overset{n}{\otimes}A & \xrightarrow{1-\tau} & \overset{n}{\otimes}A \\
 & & \downarrow b & & \downarrow\delta & & \downarrow b & & \downarrow\delta \\
 & & \vdots & & \vdots & & \vdots & & \vdots \\
 & & \downarrow b & & \downarrow\delta & & \downarrow b & & \downarrow\delta \\
\dots & \xrightarrow{N} & \overset{2}{\otimes}A & \xrightarrow{1-\tau} & \overset{2}{\otimes}A & \xrightarrow{N} & \overset{2}{\otimes}A & \xrightarrow{1-\tau} & \overset{2}{\otimes}A \\
 & & \downarrow b & & \downarrow\delta & & \downarrow b & & \downarrow\delta \\
\dots & \xrightarrow{N} & A & \xrightarrow{1-\tau} & A & \xrightarrow{N} & A & \xrightarrow{1-\tau} & A
\end{array} \qquad (1.2.2)$$

is a double complex.

Proof. Let $d_i(a_0 \otimes \ldots \otimes a_n) = a_0 \otimes \ldots \otimes a_{i-1}a_i \otimes \ldots \otimes a_n$, $i = 1, \ldots, n$; $d_0(a_0 \otimes \ldots \otimes a_n) = a_1 \otimes \ldots \otimes a_n a_0$. It is easy to see that

$$b = d_1 - d_2 + \ldots + (-1)^{n-1} d_n; \quad \delta = d_0 - b; \quad d_0 t = t d_1; \ldots ; d_{n-1} t = t d_n; \quad d_n t = d_0$$

Therefore

$$\delta(1 - \tau) = \sum_{i=0}^{n} (-1)^i d_i \, (1 - (-1)^n t) = \sum_{i=0}^{n} (-1)^i d_i - \sum_{i=1}^{n} (-1)^{i+n-1} t d_i - d_0 = (1 - \tau) \sum_{i=1}^{n} (-1)^i d_i = -(1 - \tau) b;$$

$$-b\,N = \sum_{i=1}^{n} (-1)^i t^{-i} d_0 t^i \sum_{j=0}^{n} (-1)^{nj} t^j = \sum_{i=1}^{n} \sum_{j=0}^{n} (-1)^{i+nj} t^{-i} d_0 t^{i+j} =$$

$$= \sum_{i=1}^{n} \sum_{j=0}^{n} (-1)^{(n-1)i+jn} t^{-i} d_0 t^j = N d_0 N = \sum_{k=1}^{n} \sum_{j=0}^{n} (-1)^{kn+j-k} t^{k-j} d_0 t^j =$$

$$= N\delta$$

Denote the double complex (1.2.2) by $L_{..}(A)$. More precisely, $L_{ij}(A) = \overset{j+1}{\otimes} A$, $i \geq 0$, $j \geq 0$. The associated simple complex will be denoted by $L_.(A)$.

Definition 1.2.2. $K^+_{n+1}(A) = H_n(L_.(A))$, $n \geq 0$.

Example 1.2.3. $K_1(A) = A/[A, A]$.

If k is of characteristic zero (i.e. $\mathbb{Q} \subseteq k$), then the horizontal homology of $L_{..}(A)$ vanish for $i > 0$. Put $C^{cycl}_{n+1}(A) = C_n(A, A)/\mathrm{Im}(1-\tau)$; $C^{cycl}_.(A)$ is a complex with respect to δ and the first spectral sequence of $L_{..}(A)$ shows that $H_.(C^{cycl}_.(A)) \xrightarrow{\sim} K^+_.(A)$.

The second spectral sequence corresponding to $L_{..}(A)$ gives the following.

Theorem 1.2.4. There exists a spectral sequence

$$E_{ij} = H_i(A,A) \otimes K^+_{j+1}(k) \Rightarrow K^+_{i+j+1}(A), \quad i, j \geq 0,$$

where $K^+_{i+1}(k) = 0$ if $i \equiv 1 \pmod 2$ and $K^+_{i+1}(k) \simeq k$ if $i \equiv 0 \pmod 2$

It follows from the fact that b (1.1.1) is acyclic.

Corollary 1.2.5. $K^+_.(A) \simeq K^+_.(M_n(A))$, $n > 0$.

Proof. This follows from Proposition 1.1.1.

Remark 1.2.6. There arises a differential $B : H_n(A, A) \to H_{n+1}(A, A)$ in the third term of the spectral sequence. It may be given at the chain level: for $a \in \overset{n+1}{\otimes} A$ $Ba = -(1 - \tau)sNa = (-1)^{n+1}(1-\tau)[Na \otimes 1]$ $\in \overset{n+2}{\otimes} A$.

Note that the shift of $L..(A)$ two times to the right gives the maps

$$S : K^+_{n+1}(A) \to K^+_{n-1}(A)$$

(the Bott homomorphism).

Theorem 1.2.7. There is an exact sequence

$$\ldots \to H_n(A,A) \to K^+_{n+1}(A) \xrightarrow{S} K^+_{n-1}(A) \to H_{n-1}(A,A) \to \ldots$$

Proof. Consider a subbicomplex $\underset{i<2}{\oplus} L_i \subset L..$. This subcomplex is quasi-isomorphic to $C.(A,A)$ because the differential b is acyclic. The exact sequence of the pair is the desired one.

Now let char $k = 0$. Put $D.(A) = (\otimes^* A, b)$. The spaces $\ker(1-\tau)$ form a subcomplex in $D.(A)$.

Proposition 1.2.8. $K^+_*(A) \cong H.(\ker(1 - \tau))$.

Proof. $C./\text{im}(1-\tau) \cong C./\ker N \cong \text{Im } N = \ker(1-\tau)$.

It is convenient in many cases to change the double complex $L..(A)$ by the bicomplex introduced by Connes in [13]:

$$
\begin{array}{ccccccc}
A & \xrightarrow{B} & \overset{2}{\otimes}A & \xrightarrow{B} & \ldots & \xrightarrow{B_{n+1}} & \overset{n+1}{\otimes} A \\
& & \downarrow{\delta} & & & & \downarrow{\delta} \\
A & \xrightarrow{B} & \ldots & \xrightarrow{B} & \overset{n}{\otimes}A & & \\
& & & & \downarrow{\delta} & & \\
& & & & \vdots & & \\
& & & & \downarrow{\delta} & & \\
& & & & A & &
\end{array}
$$

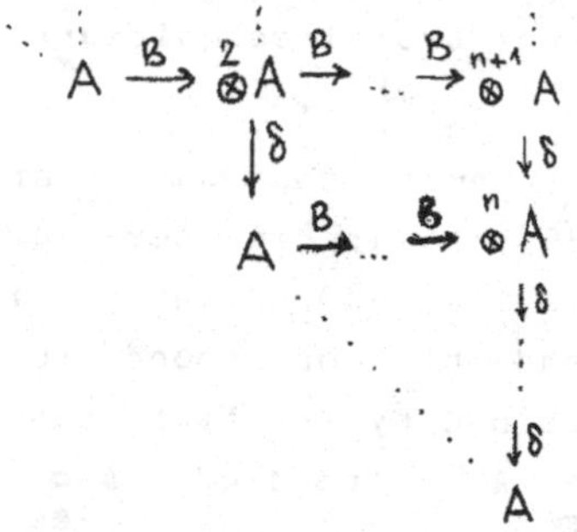

More precisely, let $F_{ij}(A) = \overset{j-i+1}{\otimes} A$, $j \geqslant i \geqslant 0$; $F_{ij}(A) = 0$, $0 \leqslant j < i$. Let $s(a_0 \otimes \ldots \otimes a_n) = 1 \otimes a_0 \otimes \ldots \otimes a_n$. We have

$$B = -(1-\tau)sN; \quad B^2 = (1-\tau)sN(1-\tau)sN = 0; \quad \delta B + B\delta =$$

$$= -\delta(1-\tau)sN - (1-\tau)sN\delta = (1-\tau)bsN + (1-\tau)sbN =$$

$$= (1-\tau)(bs + sb)N = (1-\tau)N = 0;$$

therefore $F..(A)$ is really a bicomplex. The corresponding simple complex is denoted by $F.(A)$.

Proposition 1.2.9. The complexes $L.(A)$ and $F.(A)$ are quasi-isomorphic.

Proof. Let $\alpha \in F_{i,j-1}(A)$; $\varphi(\alpha) \in L_{2i,j-i-1}(A) + L_{2i-1,j-i}(A)$: $\varphi(\alpha) = \alpha - sN\alpha$. We claim that φ gives a quasi-isomorphism $F.(A) \longrightarrow L.(A)$. The differential of $\varphi(\alpha)$ in $L.(A)$ is equal to

$(\delta\alpha + N\alpha) - (bsN\alpha + (1-\tau)sN\alpha) = \delta\alpha + N\alpha - N\alpha + sbN\alpha + (1-\tau)sN\alpha =$

$= \delta\alpha - sN\delta\alpha + (1-\tau)\delta N\alpha = \varphi(\delta\alpha + B\alpha)$;

so φ is a homomorphism of complexes. To show that it is a quasi-isomorphism consider the filtrations

$$\mathrm{filt}^n L..(A) = \bigoplus_{0\leqslant i\leqslant 2n} L_{i.}; \quad \mathrm{filt}^n F..(A) = \bigoplus_{0\leqslant i\leqslant n} F_{i.} .$$

Since b is acyclic we see that φ induces a quasi-isomorphism of the graded complexes. Therefore φ is a quasi-isomorphism.

Now we may replace $F..(A)$ by the normalized bicomplex $F'_{ij}(A) = \overset{j-i}{\otimes}(A/k) \otimes A$; δ is the same;

$$B(a_o \otimes \ldots \otimes a_n) = (-1)^{n+1}N(a_o \otimes \ldots \otimes a_n) \otimes 1 .$$

This double complex is more convenient for computations in many cases. Note that the spectral sequences for $F..$, $F'..$ are in fact the same as for $L..$ but they converge two times quicker: $E^{2r}_{2i,j}$ for $L..$ is isomorphic to $E^{r}_{i,i+j}$ for $F..$, $F'..$.

Remark 1.2.10. One has to remember that our notation is distinct from this in the papers [13], [35], [42]; in these works another formulas for the differential in $C.(A,A)$ and for the cyclic shift t are used; b corresponds to our δ, and b' corresponds to our b. The distinction in definitions is explained by the fact that we work with the complex $C.(A,A) = \mathcal{B}.(A) \underset{A\otimes A^o}{\otimes} A$ instead of $A \underset{A\otimes A^o}{\otimes} \mathcal{B}.(A)$.

1.3. We prove here that the operator $B: H_i(A,A) \longrightarrow H_{i+1}(A,A)$ is a derivation with respect to exterior multiplication in Hochschild homology. The following chain maps are defined in [9]:

$$\top : C.(A,A) \otimes C.(B,B) \rightarrow C.(A \otimes B, A \otimes B);$$

$$\perp : C.(A \otimes B, A \otimes B) \longrightarrow C.(A,A) \otimes C.(B,B);$$

$$(a_o \otimes \ldots \otimes a_{n-1} \otimes a) \top (b_o \otimes \ldots \otimes b_{m-1} \otimes b) = \sum (-1)^{\mathrm{sgn}\sigma} c_{\sigma^{-1}(0)} \otimes \ldots$$

$$\dots \otimes c_{\sigma^{-1}(n+m-1)} \otimes (a \otimes b)$$

where $(c_0, c_1, \dots, c_{n+m-1}) = (a_0 \otimes 1, \dots, a_{n-1} \otimes 1, 1 \otimes b_0, \dots, 1 \otimes b_{m-1})$ and sum is taken over the set of the permutations $\sigma \in \Sigma_{n+m}$ such that $\sigma(0) < \dots < \sigma(n-1)$; $\sigma(n) < \dots < \sigma(n+m-1)$;

$$\perp((a_0 \otimes b_0) \otimes \dots \otimes (a_{n-1} \otimes b_{n-1}) \otimes (a \otimes b) =$$

$$= \sum_{i=-1}^{n-1} (a_0 \otimes a_1 \otimes \dots \otimes a_i \otimes a_{i+1}\, a_{i+2} \dots a_{n-1}\, a) \otimes$$

$$\otimes (b_{i+1} \otimes b_{i+2} \otimes \dots \otimes b_{n-1} \otimes b\, b_0\, b_1 \dots b_i)$$

It has been shown in [9] that the maps $\top$ and $\perp$ induce the homology isomorphisms inverse to each other. We have a map

$$H.(A,A) \otimes H.(B,B) \to H.(C.(A,A) \otimes C.(B,B)) \to H.(A\otimes B,\ A\otimes B);$$

we denote it by $\top$ and call an exterior multiplication in Hochschild homology.

<u>Proposition 1.3.1</u>. Let $\alpha \in H_n(A,A)$ and $\beta \in H_m(B,B)$. Then

$$B(\alpha \top \beta) = B\alpha \top \beta + (-1)^n \alpha \top B\beta.$$

<u>Proof</u>. Compute an element $\perp B(\alpha\top\beta) = \perp((-1)^{n+m}(N_{n+m}(\alpha\top\beta) \otimes (1\otimes 1)))$. We may restrict ourselves with such permutations σ in the formula for $\top$ that either $\sigma(i) = i + \sigma(0)$ for all $i \in [0, n-1]$ or $\sigma(n+j) = \sigma(n) + j$ for all $j \in [0, m-1]$. Indeed, the chains $\perp N(c_{\sigma^{-1}0} \otimes \dots)$ are degenerate for all other σ.

Therefore

$$\perp B(\alpha\top\beta) = (-1)^{n+m+1}[\sum_{i=0}^{n} (a_{n-i} \otimes \dots \otimes a_{n-1} \otimes a \otimes a_0 \otimes \dots \otimes a_{n-1-i} \otimes 1)$$

$$\otimes (b_0 \otimes \dots \otimes b_{m-1} \otimes b)(-1)^{m+n(n-i)} + \sum_{j=1}^{m-1} (-1)^{j(m-j)+m-j} (a_0 \otimes \dots$$

$$\dots \otimes a_{n-1} \otimes a) \otimes (b_{m-j} \otimes \dots \otimes b_{m-1} \otimes b \otimes b_0 \otimes \dots \otimes b_{m-j-1} \otimes 1) +$$

$$+ (-1)^m (a_0 \otimes \dots \otimes a_{n-1} \otimes a) \otimes (b \otimes b_0 \otimes \dots \otimes b_{m-1} \otimes 1) +$$

$$+ (a_0 \otimes \dots \otimes a) \otimes (b_0 \otimes \dots \otimes b \otimes 1)] + (\text{degenerate chains}) =$$

$$= \sum_{i=0}^{n} (-1)^{1+n+ni} (t^i \alpha \otimes 1) \otimes \beta + \sum_{j=0}^{m} (-1)^{n+1+mj} \otimes t^{-1-j} \beta \quad \text{Q.E.D.}$$

1.4. Additive K-functors of the smooth rings. Let k be a field, char $(k) = 0$.

Theorem 1.4.1. (Connes, [12]). Let X be an affine nonsingular algebraic variety over k. Then

$$K^{+}_{n+1}(k[X]) \xrightarrow{\sim} \Omega^{n}_{X}/d\Omega^{n-1}_{X} \oplus H^{n-2}_{DR}(X) \oplus H^{n-4}_{DR}(X) \oplus \dots$$

where $0 \longrightarrow \Omega^{0}_{X} \dots \xrightarrow{d} \Omega^{i}_{X} \xrightarrow{d} \dots$ is the de Rham complex of the algebraic forms and $H_{DR}(X)$ is its cohomology.

Proof. (we follow [50]). Consider a map $\Omega^{n}_{X} \longrightarrow H_{n}(k[X], k[X])$: $df_{0} \dots df_{n-1} f_{n} \mapsto (f_{0} \otimes 1) \top \dots \top (f_{n-1} \otimes 1) \top f = \sum_{\sigma \in \Sigma_n} \operatorname{sgn}\sigma \times f_{\sigma^{-1}0} \otimes \dots \otimes f_{\sigma^{-1}(n-1)} \otimes f$ (for any commutative algebra a multiplication $\top : H.(A,A) \otimes H.(A,A) \longrightarrow H.(A\otimes A, A\otimes A) \longrightarrow H.(A,A)$ is defined). It is well known from [26] that this map is bijective. Actually, an ideal $\ker(k[X] \otimes k[X] \longrightarrow k[X])$ is locally generated by a regular sequence and therefore

$$H.(k[X], k[X]) = \operatorname{Tor}^{k[X]\otimes k[X]}(k[X], k[X])$$

may be computed with use the Koszul complex. Therefore we may replace $H.$ by $\Omega^{\cdot}_{X}$ in the spectral sequence of Theorem 1.2.4; we see from Proposition 1.3.1. that the differential in the second term coincides with one in the de Rham complex. Therefore

$$E^{3}_{ij} = \begin{cases} \Omega^{j}_{X}/d\Omega^{j-1}_{X}, & i = 0; \\ H^{j}_{DR}(X), i & \text{even}, \ i > 0; \\ 0, \ i & \text{odd}. \end{cases}$$

We must show that the spectral sequence degenerates. Put $\mu(a_{0} \otimes \dots \otimes a_{n}) = \frac{(-1)^{n}}{n!} da_{0} \dots da_{n-1} a_{n} \in \Omega^{n}_{X}$. It may be verified by direct computation that $\mu B = d\mu$; $\mu\delta = 0$. So we obtain a map from $F..(A)$ to the bicomplex

$$
\begin{array}{ccccccc}
\vdots & & \vdots & & & & \vdots \\
\Omega^0_X & \xrightarrow{d} & \Omega^1_X & \cdots & & \xrightarrow{d} & \Omega^n_X \\
 & & \downarrow 0 & & & & \downarrow 0 \\
 & & \Omega^0_X & \xrightarrow{d} & \cdots & \xrightarrow{d} & \Omega^{n-1}_X \\
 & & & & & & \downarrow 0 \\
 & & & & \Omega^0_X & \xrightarrow{d} & \Omega^1_X \\
 & & & & & & \downarrow 0 \\
 & & & & & & \Omega^0_X
\end{array}
$$

This shows that $F..(A)$ is quasi-isomorphic to the term of the spectral sequence. The theorem is proved.

<u>Corollary 1.4.2</u>. Let V be a finite dimensional vector space. Then the map of the complex $C^{cycl}_{.+1}(k[V])$ into the complex

$$\Omega^n_V/d\Omega^{n-1} \xrightarrow{0} \Omega^{n-1}_V/d\Omega^{n-2}_V \xrightarrow{0} \cdots \xrightarrow{0} \Omega^0_V$$

which takes $a_0 \otimes \ldots \otimes a_n$ to $\frac{(-1)}{n!} da_0 \ldots da_{n-1}\, a_n (\mathrm{mod}\ d\Omega^{n-1}_V)$ gives a quasi-isomorphism

$$K^+_{n+1}(k[V]) \xrightarrow{\sim} K^+_{n+1}(k) \oplus \Omega^n_V/d\Omega^{n-1}_V$$

1.5. In this subsection we define the cohomology of an algebra A using the analogy given by Theorem 1.4.1. We show in Chapter 6 that for a commutative algebra over $\mathbb{C}$ this definition gives usual singular cohomology of the space of maximal ideals (regarded as a closed subspace of $\mathbb{C}^N$).

Extend the double complex $L..(A)$ to the right:

$$\mathcal{L}_{ij}(A) = \overset{j+1}{\otimes} A, \quad i \in \mathbb{Z}; \qquad j \geqslant 0;$$

the differentials are the same as in $L..$; $\mathcal{L}_n(A) = \bigoplus_{i \in \mathbb{Z};\, j \geqslant 0;\, i+j=n} \mathcal{L}_{ij}(A)$. There is a decreasing filtration on $\mathcal{L}.(A)$:

$$\mathrm{filt}^n \mathcal{L}.(A) = \bigoplus_{i \leqslant -2n} \mathcal{L}_{i.}(A).$$

Let $\hat{\mathcal{L}}.(A) = \varprojlim \mathcal{L}.(A)/\mathrm{filt}^n \mathcal{L}.(A)$. Put

$$H^{odd}(A) = H_1(\hat{\mathcal{L}}.(A)); \quad H^{ev}(A) = H_0(\hat{\mathcal{L}}.(A));$$

$$H^{\cdot}(A) = H^{odd}(A) \oplus H^{ev}(A); \; H^{2n+1}(A) = \mathrm{filt}^n H^{odd}(A)/\mathrm{filt}^{n+1} H^{odd}(A);$$

$$H^{2n}(A) = \mathrm{filt}^n H^{ev}(A)/\mathrm{filt}^{n+1} H^{ev}(A), \quad n \geqslant 0.$$

The cohomology $H^{\cdot}(A)$ is connected with the additive K-functors by an exact sequence

$$0 \to \varprojlim K^{+}_{2n+1}(A) \to H^{ev}(A) \to R^{1}\varprojlim K^{+}_{2n+2}(A) \to 0;$$

$$0 \to \varprojlim K^{+}_{2n+2}(A) \to H^{odd}(A) \to R^{1}\varprojlim K^{+}_{2n+1}(A) \to 0$$

In case when A is a coordinate ring of a nonsingular affine variety over a field of characteristic zero we see from Theorem 1.4.1. that the inverse system $\xrightarrow{S} K^{+}_{n} \xrightarrow{S} K^{+}_{n-2} \xrightarrow{S} \dots$ stabilizes and that

Proposition 1.5.1. There are the functorial isomorphism on the category of nonsingular affine varieties over a field of characteristic zero:

$$H^{\cdot}(k[X]) \xrightarrow{\sim} \operatorname{gr} H^{\cdot}(k[X]) \to H^{\cdot}_{DR}(X).$$

A generalisation will be given in Chapter 6.

Chapter 2. Derived functors and relative additive K-functors

The aim of the present Chapter is to express the additive K-functors in terms of the derived functors in sense of [47] on the category of associative algebras. In Subsection 2.1. we recall briefly the notions of a resolution and a derived functor on a non-abelian category. In 2.2. we reformulate the main statements and results of Chapter 1 in the case when A is a differential graded algebra (see in more detail in [7]). In 2.3 we compute the additive K-functors of the free algebras. According to this we define in 2.4 the relative K^{+}-functors in terms of the derived functors and discuss their connection to the basic objects of Chapter 1. In this Chapter a differential graded algebra R has the grading $R = \bigoplus_{i=-\infty}^{\infty} R_i$ and the differential $\partial : R_i \to R_{i-1}$, $i \in \mathbb{Z}$.

Let $V = \bigoplus_{n=0}^{\infty} V_n$ be a graded space and $k\langle V\rangle$ the tensor algebra of V over k; for an arbitrary graded algebra R put $R\langle V\rangle = R *_k k\langle V\rangle$. Let $R^{(1)} \to R^{(2)}$ be a morphism of differential graded algebras. We say that $R^{(2)}$ is free over $R^{(1)}$ if there exist V and $\alpha : R^{(1)}\langle V\rangle \xrightarrow{\sim} R^{(2)}$ such that the diagram

$$\begin{array}{ccc} R^{(1)} & \longrightarrow & R^{(2)} \\ & \searrow & \uparrow \wr\, \alpha \\ & & R^{(1)}\langle V\rangle \end{array}$$

is commutative.

For homogeneous elements r_1, r_2 we put as usually $[r_1, r_2] = r_1 r_2 - (-1)^{\deg r_1 \deg r_2} r_2 r_1$, and denote by $[R,R]$ the linear space spanned by all such elements.

2.1. <u>A resolution of an algebra</u>. Let $A \xrightarrow{f} B$ be a homomorphism of k-algebras.

<u>Lemma 2.1.1.</u> (cf. [47]). There exist a differential graded algebra $R = \overset{\infty}{\underset{i=0}{\oplus}} R_i$ and the homomorphisms of the differential graded algebras $A \xrightarrow{\iota} R \xrightarrow{\pi} B$ such that:

') π is an epimorphism; the diagram

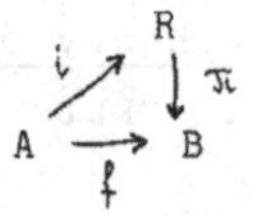

is conmutative;

2) π is a quasi-isomorphism;

3) the graded algebra R is free over A.

<u>Definition</u>. We call a graded algebra satisfying the conditions 1)-3) a free resolution of B over A.

<u>Proof of the Lemma</u>. First we construct a commutative diagram of algebras

$$\begin{array}{ccc} & R^{(o)} & \\ \iota_o \nearrow & \downarrow \pi_o & \\ A \xrightarrow{f} & B & \end{array}$$

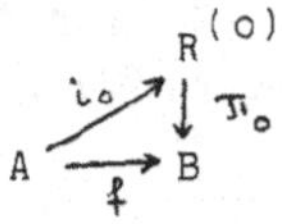

such that $R^{(o)}$ is free over A and π_o is onto. Let $R^{(o)} = A\langle V_o \rangle$, $I_o = \ker \pi_o$. Let $R^{(1)}$ be the following differential graded algebra: $R^{(1)} = A\langle V_o \oplus V_1 \rangle$, $V_1 = \ker \pi_o$; if ∂_1 is the differential in $R^{(1)}$, r an element of I_o and v_r the corresponding element of V_1, then $\partial_1(v_r) = r$. Put $I_1 = R^{(1)}_1 \cap \ker \partial_1$; construct $R^{(2)} = A\langle V_o \oplus V_1 \oplus V_2 \rangle$; $V_2 \cong I_1$; if $r \in I_1$ and v_r corresponds to it in V_2, then $\partial_2(v_r) = r$; $\partial_2 (V_o + V_1) = \partial_1$. Iterating this construction we obtain a sequence

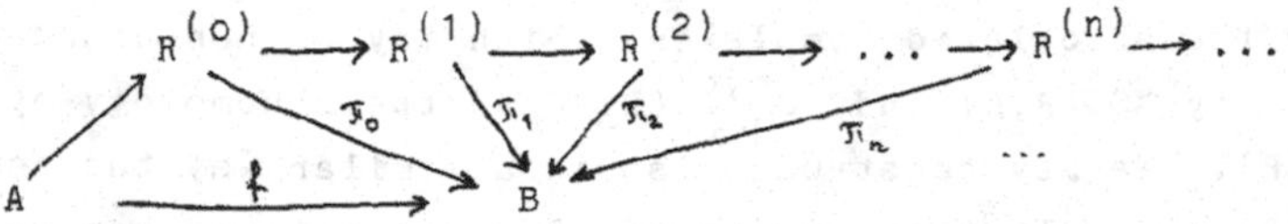

Put $R = \lim\limits_{\rightarrow} R^{(n)}$. It is easily seen that this differential graded algebra satisfies the conditions 1)-3).

Let A be an algebra and $\mathcal{C}_A$ the category whose objects are arrows $A \to B$ and whose morphisms are commutative diagrams

$$\begin{array}{ccc} A & \longrightarrow & B \\ & \searrow & \downarrow \\ & & B' \end{array}$$

Suppose that F is a functor from $\mathcal{C}_A$ to an abelian category which may be extended in a natural way to be a functor F from the category of the differential graded algebras over A to the category of complexes. Then it is natural to regard the homology of F(R) where R is a free resolution of B over A as the derived functors of the functor F. (cf. [47]). The interpretation of the K^+-functors in these terms is discussed in 2.4.

<u>2.2. Hochschild homology and additive K-theory of the differential graded algebras</u>. Let R be a graded algebra. Put $C_{nm}(R, R) = (\overset{n+1}{\otimes} R)_m$; $\delta : C_{nm}(R, R) \to C_{n-1,m}(R, R)$;

$$\delta(r_0 \otimes \ldots \otimes r_n) = (-1)^{\deg r_0(\deg r_1 + \ldots + \deg r_n)} r_1 \otimes \ldots \otimes r_n r_0 + $$
$$+ \sum_{i=0}^{n-1} (-1)^{i-1} r_0 \otimes \ldots \otimes r_i r_{i+1} \otimes \ldots \otimes r_n$$

for the homogeneous elements $r_0, \ldots, r_n \in R$.

The homology groups of the differential δ are naturally graded; denote them by $\tilde{H}.(R, R)$: $\tilde{H}_n(R, R)_m$ is the homology group in C_{nm}, $n \geq 0$; $m \in \mathbb{Z}$.

Furthermore, put for the homogeneous r_i

$$\tau(r_0 \otimes \ldots \otimes r_n) = (-1)^{\deg r_0 (\deg r_1 + \ldots + \deg r_n) + n} r_1 \otimes \ldots \otimes r_n \otimes r_0;$$

then $\operatorname{im}(1 - \tau)$ forms a subcomplex in $C..(R, R)$ with respect to ∂; put

$$C^{cycl}_{\cdot,\cdot+1}(R) = C..(R,R)/\operatorname{im}(1 - \tau); \ \tilde{K}^+(R). = H.(C^{cycl}(R)).$$

Now let R be a differential graded algebra. Then all tensor powers of R are complexes; the differentials will be denoted by the same letter ∂. It is clear that δ and $(-1)^n\partial$ turn $C..(R,R)$ into a bicomplex and that $\operatorname{im}(1 - \tau)$ is a subbicomplex. So $C^{cycl}_{\cdot}(R)$ is a bicomplex; denote the associated complexes (which may be non-bounded in the both sides) by $C.(R,R)$ and $C^{cycl}_{\cdot}(R)$ and their homology by $H.(R,R)$ and $K^+(R)$. We may construct also in a similar way the complexes $L.(R)$, $F.(R)$ etc. If the grading of R is positive then we obtain usual homology complexes.

Note also that $\tilde{K}^+_\cdot(R)$, $\tilde{H}.(R, R)$ are the complexes with respect to differentials induced by ∂. The following obvious statements hold.

Lemma 2.2.1. A quasi-isomorphism $R^{(1)} \longrightarrow R^{(2)}$ induces the isomorphisms (if either $R^{(i)}_n = 0$ for $n \gg 0$ or $R^{(i)}_n = 0$ for $n \ll 0$):

$$H.(R^{(1)}, R^{(1)}) \xrightarrow{\sim} H.(R^{(2)}, R^{(2)}); \quad K^+_\cdot(R^{(1)}) \xrightarrow{\sim} K^+_\cdot(R^{(2)}).$$

Lemma 2.2.2. Let $R = \bigoplus_{i\geq 0} R_i$. The spectral sequences of the bicomplexes $C..(R,R)$ and $C^{cycl}_\cdot(R)$ are as follows:

$$'E^2_{ij} = H_i(\tilde{H}_j(R,R)) \Rightarrow H_{i+j}(R,R);$$

$$''E_{ij} = \tilde{H}_i(H(R),H(R))_j \Rightarrow H_{i+j}(R,R);$$

$$'E^2_{ij} = H_i(\tilde{K}^+_\cdot(R)) \Rightarrow K^+_{i+j}(R);$$

$$''E^2_{ij} = \tilde{K}^+_i(H(R)) \Rightarrow K^+_{i+j}(R).$$

Here $H(R)$ is the graded homology ring of R.

Lemma 2.2.3. Let $R = \bigoplus_{i\geq 0} R_i$. There are the spectral sequences

$$E^2_{ij} = K^+_i(k) \otimes \widetilde{H}^+_j(R,R) \Rightarrow \tilde{K}^+_{i+j}(R); \quad E^2_{ij} = K^+_i(k) \otimes H_j(R,R) \Rightarrow K^+_{i+j}(R).$$

Lemma 2.2.4. There is an exact sequence

$$\ldots \to H_{i-1}(R,R) \to K^+_i(R) \xrightarrow{S} K^+_{i-2}(R) \to H_{i-2}(R,R) \to \ldots$$

and similarly for $\tilde{H}.,\tilde{K}$.

We leave the proofs to the reader.

2.3. Homology of free algebras. Let R be a graded algebra and V a graded k-module: $V = \bigoplus_{i=0}^{\infty} V_i$, $R = \bigoplus_{i=0}^{\infty} R_i$

Lemma 2.3.1. (cf. also [44], [49]). Let k be an arbitrary commutative ring. Then

$$\tilde{K}^+_{i+1}(R\langle V\rangle) \simeq \tilde{K}^+_{i+1}(R) \oplus \left(\bigoplus_{m\geq 1} H_i\left(\mathbb{Z}/m\mathbb{Z}\,;\ \overset{m}{\otimes}(V \otimes R)\right)\right)$$

Proof. We want to use Proposition 2.2.3. In order to do this compute the Hochschild homology $\tilde{H}.(R\langle V\rangle, R\langle V\rangle)$.

Construct a complex $\mathcal{P}.$ of the graded $R\langle V\rangle$-bimodules:

$$\ldots \xrightarrow{b'} R\langle V\rangle \otimes R \otimes R \otimes R\langle V\rangle \xrightarrow{b'} \begin{matrix} R\langle V\rangle \otimes R \otimes R\langle V\rangle \\ \oplus \\ R\langle V\rangle \otimes V \otimes R\langle V\rangle \end{matrix} \xrightarrow{b' \,,\ \alpha} R\langle V\rangle \otimes R\langle V\rangle$$

$$b'(r_{-1}\otimes\ldots\otimes r_{n-1}\otimes r_n)=\sum_{i=0}^{n}(-1)^i r_{-1}\otimes\ldots\otimes r_{i-1}r_i\otimes\ldots\otimes r_n,\ r_{-1},r_n\in R\langle V\rangle,$$

$r_i\in R$, $0\leqslant i<n$; $\alpha(r_{-1}\otimes v\otimes r_o)=r_{-1}v\otimes r_o-r_{-1}\otimes v\,r_o$. It is easy to see that $\mathcal{P}_\cdot$ is a resolution of the graded bimodule R. It is clear also that $\tilde{H}_\cdot(R,R)=\mathrm{Tor}_\cdot^{R\otimes R^o}(R,R)$ for an arbitrary graded algebra R; $\mathrm{Tor}^{R\otimes R^o}(R,N)$ means the derived functors of the functor $N\mapsto N\otimes_{R\otimes R^o}R=N/[R,N]=N/\langle r\cdot n-(-1)^{\deg r\deg n}n\cdot r\rangle$ on the category of graded bimodules. So

$$H_\cdot(\mathcal{P}_\cdot\otimes_{R\langle V\rangle\otimes R\langle V\rangle^o}R\langle V\rangle)\simeq\tilde{H}_\cdot(R\langle V\rangle,R\langle V\rangle).$$

Let $R\langle V\rangle_+$ be the ideal in $R\langle V\rangle$ generated by $\bigoplus_{m\geqslant 0}V_m$. It is a free R-bimodule; thus

$$\tilde{H}_i(R,R\langle V\rangle_+)=0,\ i>0;\quad \tilde{H}_i(R,R)\xrightarrow{\sim}\tilde{H}_i(R,R\langle V\rangle),\quad i>0.$$

This implies straightforward that

$$\tilde{H}_i(R\langle V\rangle,R\langle V\rangle)\xrightarrow{\sim}H_i(\mathcal{P}_\cdot\otimes_{R\langle V\rangle\otimes R\langle V\rangle^o}R\langle V\rangle)\xrightarrow{\sim}\tilde{H}_i(R,R),\quad i>1;$$

$$\tilde{H}_i(R\langle V\rangle,R\langle V\rangle)\xrightarrow{\sim}H_i(\mathcal{P}_\cdot\otimes_{R\langle V\rangle\otimes R\langle V\rangle^o}R\langle V\rangle)=\tilde{H}_i(R,R)\oplus H_i(\tilde{\mathcal{C}}_\cdot),\ i=0,1,$$

here $\tilde{\mathcal{C}}_\cdot$ is the following complex:

$$0\longrightarrow V\otimes R\langle V\rangle\xrightarrow{\tilde{\alpha}}R\langle V\rangle_+\otimes_{R\otimes R^o}R\longrightarrow 0;$$

$$\tilde{\alpha}(v\otimes x)=v\,x-(-1)^{\deg v\deg x}x\cdot v.$$

Furthermore, consider a commutative diagram

$$\begin{array}{ccc}\mathcal{C}_1 & \xrightarrow[\varphi_1]{\sim} & \bigoplus_{m\geqslant 1}(\overset{m}{\otimes}(V\otimes R))\\ \tilde{\alpha}\downarrow & & \downarrow 1-\theta\\ \mathcal{C}_o & \xrightarrow[\varphi_o]{\sim} & \bigoplus_{m\geqslant 1}(\overset{m}{\otimes}(V\otimes R))\end{array}$$

in which

$$\varphi_o(v_1r_1\ldots v_{m-1}r_{m-1}v_mr_m)=\varphi_1(v_1\otimes(r_1v_2\ldots r_{m-1}v_mr_m)=$$

$$=((v_1\otimes r_1)\otimes\ldots\otimes(v_m\otimes r_m));\quad\text{for}\quad w_i=v_i\otimes r_i$$

$$\theta(w_1\otimes\ldots\otimes w_m)=(-1)^{\deg w_1(\deg w_2+\ldots+\deg w_m)}w_2\otimes\ldots\otimes w_m\otimes w_1.$$

It is easy to see that the boundary operator $B:\tilde{H}_o(R\langle V\rangle,R\langle V\rangle)\to H_1(R\langle V\rangle,R\langle V\rangle)$

in the second term of the spectral sequence from the Proposition 2.2.3 sends the summand $\tilde{H}_0(R,R)$ to the summand $\tilde{H}_1(R,R)$ and the summand $H_0(\tilde{\mathcal{C}}.)$ to the summand $H_1(\tilde{\mathcal{C}}.)$; $\varphi_1 B\varphi_0^{-1}$ acts on the summand $\overset{m}{\otimes}(V\otimes R)$ as $1+\theta+\ldots+\theta^{m-1}$. Indeed, if we put $p_i = \deg(v_i r_i)$, then

$$B(v_1r_1v_2r_2\ldots v_mr_m) = (v_1r_1\ldots v_mr_m)\otimes 1 = -\delta((v_1r_1)\otimes(v_2r_2\otimes\ldots)\otimes 1)+$$

$$+(-1)^{p_1(p_2+\ldots+p_m)}v_2r_2\ldots v_mr_m\otimes v_1r_1 + v_1r_1\otimes(v_2r_2\ldots v_mr_m) =$$

$$= -\delta((v_1r_1)\otimes(v_2r_2\ldots v_mr_m)\otimes 1) - (-1)^{p_1(p_2+\ldots+p_m)}\delta((v_2r_2)\otimes(v_3r_3)\ldots x(v_mr_m)$$

$$\otimes(v_1r_1)) + v_1r_1\otimes(v_2r_2\ldots v_mr_m) + (-1)^{p_1(p_2+\ldots+p_m)}v_2r_2\otimes$$

$$x(v_3r_3\ldots v_mr_mv_1r_1) = \ldots \equiv \sum_{k=1}^{m} \pm v_kr_k\otimes(v_{k+1}r_{k+1}\ldots v_{k-1}r_{k-1}) \pmod{\operatorname{im}(\delta)}.$$

So for the map $\varphi_1 B\varphi_0^{-1}: \tilde{\mathcal{C}}_0/\operatorname{im}\tilde{\alpha} \longrightarrow \ker\tilde{\alpha}$

$$\ker(\varphi_1 B\varphi_0^{-1}) = \ker(1+\theta+\ldots+\theta^{m-1}/\operatorname{im}(1-\theta)$$

and similarly for the cokernel. This proves the Lemma.

Corollary 2.3.2. Let char $k = 0$. Then $\tilde{K}_i^+(R) \xrightarrow{\sim} \tilde{K}_i^+(R\langle V\rangle)$, $i \geqslant 2$.

2.4. The relative additive K-functors. Let $f: A \longrightarrow B$ be a homomorphism of algebras and R a free resolution of B over A.

The space $[R,R] + \operatorname{im}(i)$ is a subcomplex in R. Put $K_i^+(A\longrightarrow B) = H_i(R/[R,R]+\operatorname{im}(i))$.

Theorem 2.4.1. 1) The spaces $K_i(A\longrightarrow B)$ do not depend on a resolution; they are the covariant functors on the category $\mathcal{C}_A$.

2) $K_{i+1}^+(A\longrightarrow 0) \xrightarrow{\sim} K_{i+1}^+(A)$.

3) Let $A\longrightarrow B\longrightarrow C$ be a sequence of homomorphisms of algebras. Then there is an exact sequence

$$\ldots\longrightarrow K_i^+(A\longrightarrow B)\longrightarrow K_i^+(A\longrightarrow C)\longrightarrow K_i^+(B\longrightarrow C)\longrightarrow K_{i-1}^+(A\longrightarrow B)\ldots\longrightarrow$$

$$\longrightarrow K_0^+(A\longrightarrow B)\longrightarrow 0.$$

Proof. The complex $C^{cycl}_{\cdot}(R)$ is mapped onto

$$C^{cycl}_{1\cdot}(R)/\operatorname{im}(\delta: C^{cycl}_{2\cdot}(R)\longrightarrow C^{cycl}_{1\cdot}(R)) \cong (R/[R,R])[-1].$$

It follows from Lemma 2.3.1. and from the third spectral sequence of 2.2.1 that the horizontal sequence in the following diagram is a distinguished triangle in the derived category:

$$C^{cycl}(A) \longrightarrow C^{cycl}(R) \longrightarrow (R/([R,R] + \mathrm{im}(i)))\,[-1]$$
$$\downarrow\wr$$
$$C^{cycl}(B)$$

Corollary 2.2.2. shows that the vertical map is a quasi-isomorphism. Thus, in the derived category the complex $(R/([R,R] + \mathrm{im}(i)))\,[-1]$ is isomorphic to the cone of $C^{cycl}(A) \longrightarrow C^{cycl}(B)$; this proves the statements 1) and 3) (in combination with the octahedra axiom).

It remains to prove 2). Consider the differential graded algebra $R(k)$ which is the tensor algebra of the graded space $V = k\,t$, $\deg t = 1$, with the differential ∂: $\partial t = 1$. Put $R(A) = A *_k R(k)$. The algebra $R(A)$ is a free resolution of the algebra $B = 0$ over A. Actually, $H(R(A))$ is a ring with unit; on the other hand $1 = \partial t = 0$ in $H(R(A))$. Therefore $H(R(A)) = 0$.

Clearly, the map of complexes

$$R(A)/([R(A),\, R(A)] + A) \longrightarrow C^{cycl}(A)$$

which sends $ta_0ta_1 \ldots ta_n$ to $(a_0 \otimes \ldots \otimes a_n) (\mathrm{mod}\ \mathrm{im}(1 - \tau))$ is an isomorphism. This ends the proof of Theorem 2.4.1.

<u>Remark 2.4.2</u>. Consider a homomorphism $f : A \longrightarrow B$. Let $F'(B)$ be as in the end of 1.2. One may take the quotient by the subcomplex

$$\Big(\bigoplus_{j-i\geq 2} F_{ij}\Big) + \Big(\bigoplus_i F_{i,i+1} \cap \mathrm{im}\,\delta\Big).$$

We obtain a complex

$$\ldots \longrightarrow ((B/k) \otimes B)/\mathrm{im}(\delta) \longrightarrow B \longrightarrow ((B/k) \otimes B)/\mathrm{im}(\delta) \longrightarrow B \longrightarrow 0;$$

it may be projected itself onto the complex $\Omega.(A \longrightarrow B)$:

$$\ldots \longrightarrow (\bar{B} \otimes_{A\otimes A^{\circ}} B)/\mathrm{im}(\delta) \xrightarrow{\partial_1} \bar{B} \otimes_{A\otimes A^{\circ}} A \xrightarrow{\partial_0} (\bar{B} \otimes_{A\otimes A^{\circ}} B)/\mathrm{im}(\delta) \longrightarrow \bar{B} \otimes_{A\otimes A^{\circ}} A \longrightarrow 0\,;$$

$\Omega_i(A \longrightarrow B) = \bar{B} \otimes_{A\otimes A^{\circ}} A$, $i \equiv 0 \ (\mathrm{mod}\ 2)$; $\Omega_i(A \longrightarrow B) = (\bar{B} \otimes_{A\otimes A^{\circ}} B)/\mathrm{im}\,\delta$, $i \equiv 1 \ (\mathrm{mod}\ 2)$, where $\bar{B} = B/\mathrm{im}\, f$; $\partial_0(b) = b \otimes 1$; $\partial_1(b_0 \otimes b_1) = b_1 b_0 - b_0 b_1$. If $B = A\langle V\rangle$ then $\Omega.(A \longrightarrow B) \xrightarrow{\sim} \bigoplus_{m=1} C.(\mathbb{Z}/m\mathbb{Z},\ \otimes^m (V \otimes A))$; the proof of 2.3.1 shows that (for an arbitrary scalar ring) the triangle

$$F'(A) \longrightarrow F'(B) \longrightarrow \Omega.(A \longrightarrow B)$$

is distinguished. Now define the bicomplex $\Omega.(A \longrightarrow R)$ where R is a free resolution of B over A. When $\mathrm{char}\ k = 0$ it is quasi-isomorphic to $R/([R,R] + A)$; in the case $B = 0$, $R = R(A)$ (the end of the proof of Theorem 2.4.1) we obtain the exact sequence of complexes

$$0 \longrightarrow D.(A)[1] \longrightarrow \Omega.(A \longrightarrow R(A)) \longrightarrow L.(A)[-1] \longrightarrow 0.$$

When k is arbitrary the relative K^+-functors may be defined as the derived functors of the functor $(A \longrightarrow B) \mapsto \Omega.(A \longrightarrow B)$ from the category $\mathcal{C}^A$ to the category of complexes (cf. A 8).

For an algebra B put $\bar{C}_n^{cycl}(B) = \overset{n}{\otimes}(B/k)$. More generally, let $A \longrightarrow B$ be a monomorphism of k-algebras. Denote

$$C_n^{cycl}(B/A,A) = (((B/A) \otimes_A \dots \otimes_A (B/A)) \otimes_{A\otimes A^\circ} A)/\mathrm{im}(1-\tau);$$

$$\tau(b_o \otimes \dots \otimes b_{n-1}) = (-1)^{n-1} b_1 \otimes \dots \otimes b_{n-1} \otimes b_o; \quad \delta(b_o \otimes \dots \otimes b_{n-1}) =$$

$$= b_1 \otimes \dots \otimes b_{n-1} b_o + \sum_{i=1}^{n-1} (-1)^i b_o \otimes \dots \otimes b_{i-1} b_i \otimes \dots \otimes b_{n-1}$$

<u>Proposition 2.4.3</u>. Assume that B/A is a free bimodule over A. Then

$$K^+_{\cdot-1}(A \longrightarrow B) \xrightarrow{\sim} H.(C^{cycl}(B/A,A)).$$

In particular,

$$K^+_{\cdot-1}(k \longrightarrow B) \xrightarrow{\sim} H.(\bar{C}^{cycl}(B)).$$

(This object is the reduced cyclic homology from [32], [35], [50] etc.).

<u>Proof of 2.4.3</u>. Consider the map

$$C^{cycl}(B/A, A) \longleftarrow C^{cycl}(R/A, A).$$

Our assertion implies that this map is a quasi-isomorphism. Now, we may construct the complex $F^!_\cdot(R/A, A)$ in the obvious way as in 1.2: $F^!_{ij}(R/A, A) = (\underbrace{(R/A) \otimes_A \dots \otimes_A (R/A)}_{(j-i+1)\text{ times}}) \otimes_{A\otimes A^\circ} A$, etc.; it is quasi-isomorphic to $C^{cycl}(R/A, A)[1]$. It remains to verify that the projection $F^!_\cdot(R/A, A) \longrightarrow \Omega.(A \longrightarrow R)$ is a quasi-isomorphism. We leave this to the reader.

<u>Remark 2.4.4</u>. It is clear that $K_i(A \longrightarrow B) \xrightarrow{\sim} H_{i-1}([R, R])$, $i > 1$. There is a filtration

$$[R, R] \supset [R, [R, R]] \supset \dots$$

It would be interesting to compare the induced filtration on the K^+-functors with ones constructed in 1.4 and 4.4-4.6.

Chapter 3. Generalized free products.

3.1. Consider a diagram of rings and ring homomorphisms

$$\begin{array}{ccc} C & \overset{j_1}{\underset{i_1}{\leftrightarrows}} & A \\ {\scriptstyle i_2}\downarrow\uparrow{\scriptstyle j_2} & & \\ B & & \end{array} \qquad (3.1.1)$$

such that $j_1 \circ i_1 = j_2 \circ i_2 = 1_C$. Put $\bar{A} = \ker j_1$, $\bar{B} = \ker j_2$. It has been shown in Waldhausen's paper [62] that for the Quillen K-functors

$$K_{i+1}(C) \oplus K_{i+1}(A \underset{C}{*} B) \xrightarrow{\sim} K_{i+1}(A) \oplus K_{i+1}(B) \oplus \mathrm{Nil}_i(C;\ A,\ B)$$

if A and B are free C-modules with respect to the right and left multiplications. Here $\mathrm{Nil}.(C;\ A,\ B)$ are some functors depending only on the C-bimodule structure of A and B.

Furthermore, let A be a ring, $\alpha \in \mathrm{Aut}\ A$; define the Laurent and polynomial extensions of A corresponding to α : $A_\alpha[t] = \bigoplus_{i\geq 0} A\cdot t^i$; $A_\alpha[t,t^{-1}] = \bigoplus_{i\in Z} A\cdot t^i$; $(a\cdot t^i)\times(b\cdot t^j) = a\,\alpha^i(b)\cdot t^{i+j}$. Waldhausen has shown that there exist the functors $\mathrm{Nil}.(A;\ M)$ depending on a ring A and an A-bimodule M such that

$$K_{i+1}(A_\alpha[t]) \to K_{i+1}(A) \oplus \mathrm{Nil}_i(A;\ A_\alpha);$$

moreover, let M be free A-module (right and left), then

$$K_{i+1}(T_A(M)) \xrightarrow{\sim} K_{i+1}(A) \oplus \mathrm{Nil}_i(A;\ M);$$

at last, there exists an exact sequence

$$0 \to K_{i+1}(A)_\alpha \to K_{i+1}(A_\alpha[t,t^{-1}]) \to K_i(A)^\alpha \oplus \mathrm{Nil}_i(A,\ A_\alpha) \oplus$$

$$\oplus\ \mathrm{Nil}_i(A,\ A_{\alpha^{-1}}) \to 0.$$

(For a bimodule M and for $\alpha, \beta \in \mathrm{Aut}\ A$ we denote by ${}_\alpha M_\beta$ a bimodule which is equal to M as an abelian group and such that $a_0 m a_1$ in ${}_\alpha M_\beta$ is equal to $\alpha(a_0)m\beta(a_1)$ in M; $M_\alpha := {}_1M_\alpha$; $K.(A)_\alpha$, $K.(A)_\alpha$ are respectively the kernel and cokernel of $1 - \alpha_*$ in $K.$).

Our aim is to formulate and prove the additive analogues of Waldhausen's results. First we define the additive functors Nil. .

Let C be a k-algebra and $P.$ a C-bimodule complex. Consider a complex

$$D^{(k)}(C,P.) = \underbrace{(P.\underset{C}{\otimes} P.\underset{C}{\otimes} \dots \underset{C}{\otimes} P.)}_{(k\ \text{times})} \underset{C\otimes C^{\circ}}{\otimes} C, \quad k \geqslant 1.$$

There is an isomorphism t_k of the complex $D^{(k)}$:

$$t_k(p_o \otimes \dots \otimes p_{k-1}) = (-1)^{\deg p_o(\deg p_1 + \dots + \deg p_{k-1})} p_1 \otimes \dots \otimes p_k \otimes p_o.$$

We write here for simplicity

$$p_o \otimes \dots \otimes p_{k-1} = (p_o \underset{C}{\otimes} \dots \underset{C}{\otimes} p_{k-1}) \underset{C\otimes C^{\circ}}{\otimes} 1.$$

Lemma 3.1.1. Let P. and Q. be homotopically equivalent. Then $D^{(k)}(C,P.)$ and $D^{(k)}(C,Q.)$ are homotopically equivalent complexes of $\mathbb{Z}/k\mathbb{Z}$-modules.

Proof. If suffices to show that a map homotopic to zero induces a map homotopic to zero. Let $f : P. \to Q.$, ∂_P and ∂_Q be differentials in P. and Q.; let s be a homotopy:

$$s : P_i \to Q_{i+1}, \ i \geqslant 0; \quad s\,\partial_P + \partial_Q\, s = f.$$

Put

$$\delta(p_o \otimes \dots \otimes p_{k-1}) = \sum_{i=0}^{k-1} (-1)^{\deg p_o + \dots + \deg p_{i-1}} p_o \otimes \dots \otimes sp_i \otimes \dots \otimes p_{k-1};$$

$$\tilde{f}(p_o \otimes \dots \otimes p_{k-1}) = \sum_{i=0}^{k-1} p_o \otimes \dots \otimes fp_i \otimes \dots \otimes p_{k-1}$$

It is clear that $\tilde{f} : D^{(k)}(C,P.) \to D^{(k)}(C,Q.)$ are correctly defined and commute with t_k; moreover, δ is a homotopy for f.

Let now M. be a complex of C-bimodules. Put

$$\mathcal{N}^{(k)}_{\bullet}(C,M.) = C.(\mathbb{Z}/k\mathbb{Z};\ D^{(k)}(C,P.))$$

where $k > 0$ and P. is a projective bimodule resolution of M. . It follows from Lemma 3.1.1. that the homotopy type of $\mathcal{N}^{(k)}$ does not depend on a choise of P. . We write

$$\mathcal{N}il^{+}(C,M.) = \bigoplus_{k=1} \mathcal{N}^{(k)}(C,M); \ Nil^{+}_{i}(C,M) = H_i(\mathcal{N}il^{+}_{\cdot}(C,M.)) =$$

$$= \bigoplus_{k=1} \mathbb{H}_i(\mathbb{Z}/k\mathbb{Z};\ D^{(k)}(C,P.)).$$

3.2. Consider now the diagram (3.1.1.) with A, B, C being k-algebras.

Theorem 3.2.1. Suppose that $Tor^C_i(\bar{A}, \bar{A}) = Tor^C_i(\bar{A}, \bar{B}) = Tor^C_i(\bar{B}, \bar{A}) = Tor^C_i(\bar{B}, \bar{B}) = 0$, $i > 0$. Then

$$K^{+}_{i+1}(\bar{A} \underset{C}{*} \bar{B}) \oplus K^{+}_{i+1}(C) \xrightarrow{\sim} K^{+}_{i+1}(A) \oplus K^{+}_{i+1}(B) \oplus Nil^{+}_{i}(C, A \underset{C}{\otimes} B).$$

Theorem 3.2.2. There is an exact sequence for $i \geqslant 0$:

$$0 \to K^+_{i+1}(A)_\alpha \to K^+_{i+1}(A_\alpha[t,t^{-1}]) \to K^+_i(A)^\alpha \oplus Nil^+_i(A,A_\alpha) \oplus Nil^+_i(A,A_{\alpha^{-1}}) \to 0$$

Theorem 3.2.3. Let M be an A-bimodule and $Tor^A_i(M,M) = 0$, $i > 0$. Then

$$K^+_{i+1}(T_A(M)) \xrightarrow{\sim} K^+_{i+1}(A) \oplus Nil^+_i(A,M), \quad i \geqslant 0.$$

In particular,

$$K^+_{i+1}(A_\alpha[t]) \to K^+_{i+1}(A) \oplus Nil^+_i(A,A_\alpha).$$

Theorem 3.2.4. $D^{(k)}(A,A_\alpha) \xrightarrow{\sim} C.(A,A_\alpha)$ in the derived category of the category of $k[\mathbb{Z}]$-modules (the action of $\mathbb{Z}$ in the left hand side is induced by t_k and in the right hand side by α^{k-1}).

Proof of Theorem 3.2.4. We may choose such a free resolution $P. \xrightarrow{\varepsilon} A$ that for any $\alpha \in Aut\ A$ there exists an automorphism $\tilde{\alpha}$ of $P.$ such that $\varepsilon\cdot\tilde{\alpha} = \alpha\cdot\varepsilon$, $\tilde{\alpha}\cdot\tilde{\beta} = \widetilde{\alpha\beta}$, $\tilde{\alpha}(a\cdot p\cdot b) = \alpha(a)\tilde{\alpha}(p)\alpha(b)$; for example it may be the standard bar-resolution $\mathcal{B}.(A)$; $\tilde{\alpha}(a_{-1}\otimes\ldots\otimes a_{n-1} \otimes a_n) = \alpha(a_{-1}) \otimes\ldots\otimes \alpha(a_{n-1}) \otimes \alpha(a_n)$. For all $\beta, \gamma \in Aut\ A$ $_\beta(P.)_\gamma$ is a free resolution of $_\beta A_\gamma$.

Consider the isomorphism

$$1 \otimes \tilde{\alpha} \otimes\ldots\otimes \tilde{\alpha}^{k-1} : (P.)_\alpha \otimes_A \ldots \otimes_A (P.)_\alpha \xrightarrow{\sim} (P.)_\alpha \otimes_A \ldots \otimes_{A_{\alpha^{k-1}}} (P.)_{\alpha^k}$$

The right hand side is a free resolution of A_{α^k}. We obtain an isomorphism

$$\Phi : [(P.)_\alpha \otimes_A \ldots \otimes_A (P.)_\alpha] \underset{A\otimes A^o}{\otimes} A \xrightarrow{\sim} ((P.)_\alpha \otimes_A \ldots {}_A\underset{\alpha^{k-1}}{\otimes} (P.)_{\alpha^k}) \underset{A\otimes A^o}{\otimes} A$$

such that

$$(\Phi t_k \Phi^{-1})(p_0 \otimes\ldots\otimes p_{k-1}) = (-1)^{\deg p_0(\deg p_1+\ldots+\deg p_{k-1})}\tilde{\alpha}^{-1}(p_1)\otimes\ldots$$
$$\ldots\otimes\tilde{\alpha}^{-1}(p_{k-1})\otimes\tilde{\alpha}^{k-1}(p_0)$$

Note that there exist two chain bimodule maps

$$\varphi_0, \varphi_1 : (P.)_\alpha \underset{A}{\otimes} \ldots \underset{A^{\alpha^{k-1}}}{\otimes} (P.)_{\alpha^k} \longrightarrow (P.)_{\alpha^k};$$

$$\varphi_0(p_0 \otimes\ldots\otimes p_{k-1}) = \varepsilon(p_0)\ldots\varepsilon(p_{k-2})p_{k-1}; \quad \varphi_1(p_0 \otimes\ldots\otimes p_{k-1}) =$$
$$= p_0\,\varepsilon(p_1)\ \ldots\ \varepsilon(p_{k-1}).$$

We obtain a commutative diagram

$$\begin{array}{ccc}
[(P.)_\alpha \otimes_A \cdots \otimes_{A_{\alpha^{k-1}}} (P.)_{\alpha^k}] \otimes_{A\otimes A^\circ} A & \xrightarrow{\varphi_1 \otimes_{A\otimes A^\circ} 1_A} & (P.)_{\alpha^k} \otimes_{A\otimes A^\circ} A \\
\downarrow{\scriptstyle \varphi t_k \varphi^{-1}} & & \downarrow{\scriptstyle \tilde{\alpha}^{k-1}\otimes \alpha^{k-1}} \\
[(P.)_\alpha \otimes_A \cdots \otimes_{A_{\alpha^{k-1}}} (P.)_{\alpha^k}] \otimes_{A\otimes A^\circ} A & \xrightarrow{\varphi_0 \otimes_{A\otimes A^\circ} 1_A} & (P.)_{\alpha^k} \otimes_{A\otimes A^\circ} A
\end{array}$$

Here the horizontal maps are homotopic (since they are induced by the chain maps of the free resolutions). Q.E.D.

We see that $H.(D^{(k)}(A,(P.)_\alpha)$ is isomorphic to $H.(A,A_{\alpha^k})$; t_k acts under this isomorphism as $\alpha_*^{k-1} = \alpha_*^{-1}$ (we know that α_*^k is the identity automorphism; cf. A 3.5).

The proofs of the rest of the theorems of 3.2 are given below for the case char k = 0. We may suppose that $\mathcal{N}^{(k)}(C,M) = D^{(k)}(C,P.)/\mathrm{Im}(1-t_k)$. The necessary changes for the general case contain in p.A8.

Proof of Theorem 3.2.3. let P. be a free bimodule resolution of M. Then $T_A(P.)$ is a free resolution of $T_A(M)$ over A and the statement follows from Theorem 2.4.1.

Proof of Theorem 3.2.1. Let R^A be a free resolution of A over C and R^B a free resolution of B over C (cf. 2.1.):

$$\begin{array}{ccc}
 & & R^A \\
 & \overset{\tilde{i}^A}{\nearrow} & \downarrow{\scriptstyle \P^A} \\
C & \xrightarrow[i^A]{} & A
\end{array}
\qquad\qquad
\begin{array}{ccc}
 & & R^B \\
 & \overset{\tilde{i}^B}{\nearrow} & \downarrow{\scriptstyle \P^B} \\
C & \xrightarrow[i^B]{} & B
\end{array}$$

We may choose such R^A and R^B that $\tilde{i}^A$, $\tilde{i}^B$ split, $\P^A : R^A \xrightarrow{\sim} A$, $\P^B : R^B \xrightarrow{\sim} B$. By the hypothesis of Theorem 3.2.1 $R^A \underset{C}{*} R^B$ is a free resolution of $A \underset{C}{*} B$. We have

$$R^A \underset{C}{*} R^B = C \oplus R^A \oplus R^B \oplus (R^A \underset{C}{\otimes} R^B) \oplus (R^B \underset{C}{\otimes} R^A) \oplus (R^A \underset{C}{\otimes} R^B \underset{C}{\otimes} R^A) \oplus \cdots$$

It is clear that the map

$$\bigoplus_{k=1}^{\infty} \mathcal{N}^{(k)}(C; R^A \underset{C}{\otimes} R^B) \longrightarrow R^A * R^B/(C + R^A + R^B + [R^A \underset{C}{*} R^B, R^A \underset{C}{*} R^B])$$

which takes $(p_o \otimes q_o) \otimes \cdots \otimes (p_{k-1} \otimes q_{k=1})$ to $p_o q_o \cdots p_{k-1} q_{k-1}$ (mod C+...) is an isomorphism. Again by our hypothesis $R^A \otimes_C R^B$ is a free C-bimodule resolution of $A \otimes_C B$. Therefore

$$K_i^+(C \to A \underset{C}{*} B) \xrightarrow{\sim} K_i^+(C \to A) \oplus K_i^+(C \to B) \oplus \mathrm{Nil}_i^+(C, A \underset{C}{\otimes} B)$$

The proof is given now by Theorem 2.4.1.

3.3. An another example where the functors $Nil^+_\cdot$ arise is connected with the deformations of K^+-functors. Let M be a bimodule over A. There is an algebra structure on $A \oplus M$: $(a_1, m_1)(a_2 m_2) = (a_1 a_2,\ a_1 m_2 + m_1 a_2)$.

<u>Proposition 3.3.1.</u> $K^+_\cdot(A \oplus M) \xrightarrow{\sim} K^+_\cdot(A) \oplus Nil^+_\cdot(A,\ M[-1])$.

<u>Corollary 3.3.2.</u> $K^+_{n+1}(A \oplus A_\alpha) \xrightarrow{\sim} K^+_{n+1}(A) \oplus [\bigoplus_{i=0} H_{n-i}(A, A\ )/Im(1-(-1)^i \alpha_*)]$

if char $k = 0$.

In particular

$$K^+_{n+1}(A[\varepsilon]) \xrightarrow{\sim} K^+_{n+1}(A) \oplus H_n(A,A) \oplus H_{n-2}(A,A) + \dots \qquad (\varepsilon^2 = 0)$$

<u>Proof of Proposition 3.3.1.</u> First we prove the following elementary statement.

<u>Lemma 3.2.</u> Let M, N be A-bimodules and P^M, P^N their bimodule resolutions such that for any $i, j \geq 0$ $P^M_i = A \otimes_k V_i$, $P^N_j = A \otimes_k W_j$ where W_j, V_i are some right A-modules. Then

$$Tor^{A\otimes A^\circ}(M,\ N) \simeq H_\cdot(P^M \underset{A\otimes A^\circ}{\otimes} P^N).$$

<u>Proof.</u> If suffices to show that $Tor^{A\otimes A^\circ}_\ell(P^M_i,\ P^N_j) = 0$, $i,j \geq 0$, $\ell > 0$. If $\mathcal{P}^{(j)}_\cdot$ is a projective resolution of the bimodule P^N_j then

$$Tor^{A\otimes A^\circ}_\ell(P^M_i,\ P^N_j) = H_\ell(P^M_i \otimes_{A\otimes A^\circ} \mathcal{P}^{(j)}_\cdot) = H_\ell(V_i \otimes_A \mathcal{P}^{(j)}_\cdot) = Tor^A_\ell(V_i, P^N_j) = 0,\quad \ell > 0.$$

Now in zero characteristic

$$C^{cycl}_\cdot(A \oplus M) \to C^{cycl}_\cdot(A) \oplus (\bigoplus_{k=1} \mathcal{N}^{(k)}_\cdot(A; \mathcal{B}_\cdot(A) \underset{A}{\otimes} M[-1])) \xleftarrow{\sim}$$

$$\xleftarrow{\sim} C^{cycl}(A) \oplus (\bigoplus_{k=1} \mathcal{N}^{(k)}_\cdot(A;\ M[-1])) \quad \text{(by Lemma 3.3.2);}$$

this gives the proof; cf. A8 for the general case.

Chapter 4. Lie algebra homology

4.1. Let B be an associative algebra (possibly without unit). Connect with B a Lie algebra $\mathcal{L}ie\,(B)$; the latter is equal to B as a linear space and the bracket in it is $ab-ba$. A question about computation of the Lie algebra homology of $\mathcal{L}ie(B)$ arises naturally. In this chapter we express these in terms of the associative algebra homology of B in some cases. In the Chapter 4, 5, 6 the scalar field k is supposed to be a field of characteristic zero.

Recall that the Lie algebra homology with values in the trivial module are the higher derived functors of the quotient by commutant. It means that they may be computed as follows. Take a Lie algebra g and construct it's free resolution, i.e. a free differential graded Lie algebra $R = \bigoplus_{i>0} R_i$, $\partial : R_i \to R_{i-1}$ together with an epimorphism $\pi : R \to g$ such that the sequence

$$\dots \to R_i \xrightarrow{\partial} R_{i-1} \xrightarrow{\partial} \dots \xrightarrow{\partial} R_o \xrightarrow{\pi} g \to 0$$

is exact. Then ∂ induces a differential on the graded space $R/[R,R]$. Its homology group in dimension i is equal to $H_{i+1}(g, k) = \mathrm{Tor}^{U(g)}_{i+1}(k, k)$; the usual definition of these groups using the standard complex will be given below; we write for simplicity $H.(g)$ for $H.(g, k)$.

We see that there is a certain analogy between Lie algebra homology and additive K-theory (in view of the results of Chapter 2).

Let A be an algebra with the unit over k. Put $B = M_\infty(A)$. This is the algebra of the infinite matrices over A such that almost all their entries are equal to zero.

Theorem 4.1.1. $H.(\mathrm{Lie}(B))$ is a free skew-commutative graded coalgebra with the space of primitive elements naturally isomorphic to $K^+_.(B)$.

It is very probably that there exist another associative algebras (except of $M_\infty(A)$) enjoying such a property.

4.2. Here we formulate a slightly more precise form of the Theorem 4.1.1.

Let $M_n(A)$ be $n\times n$ matrix algebra over A. Put $\mathfrak{gl}_n(A) = \mathrm{Lie}\,(M_n(A))$. Consider the obvious embeddings $j_{n,n+1} : \mathfrak{gl}_n(A) \to \mathfrak{gl}_{n+1}(A)$, $j_{n,n+1} : (a_{ij})_{1\leqslant i,j\leqslant n} \mapsto (a'_{ij})_{1\leqslant i,j\leqslant n+1}$: $a'_{ij} = a_{ij}$ if $i,j < n$ and $a'_{ij} = 0$ if $i = n+1$ or $j = n+1$. Let $\mathfrak{gl}(A) = \varinjlim \mathfrak{gl}_n(A)$. Clearly $\mathfrak{gl}(A) = \mathrm{Lie}\,(M_\infty(A))$.

Let $a = (a_{ij})_{1\leqslant i,j\leqslant n}$ and $b = (b_{ij})_{1\leqslant i,j\leqslant n}$ be in $\mathfrak{gl}_n(A)$. We define the direct sum $a\oplus b = c$ in $\mathfrak{gl}_{2n}(A)$ to be a matrix $(c_{k\ell})_{1\leqslant k,\ell\leqslant 2n}$ such that $c_{2i,2j} = b_{ij}$; $c_{2i-1,2j-1} = a_{ij}$; $c_{2i,2j-1} = c_{2i-1,2j} = 0$. This operation defines a homomorphism $\mathfrak{gl}_n(A) \oplus \mathfrak{gl}_n(A) \to \mathfrak{gl}_{2n}(A)$. Passing to the direct limits we obtain a homomorphism $\oplus : \mathfrak{gl}(A) \oplus \mathfrak{gl}(A) \to \mathfrak{gl}(A)$. The map $\oplus$ defines a multiplication $H.(\mathfrak{gl}(A) \otimes H.(\mathfrak{gl}(A)) \to H.(\mathfrak{gl}(A))$. This one is associative and skew commutative. Really, the matrices $a\oplus b$ and $b\oplus a$, $a\oplus(b\oplus c)$ and $(a\oplus b)\oplus c$ are adjoint to each other by means of the permutation matrices from $GL_{2n}(Z)$ and $GL_{3n}(Z)$ respectively. There is the comultiplication in the Lie algebra homology (cf. [10]). It is easily seen that it together with the multiplication defined above turns $H.(\mathfrak{gl}(A))$ to a Hopf algebra. By the well known Hopf theorem (cf. [46]) $H.(\mathfrak{gl}(A))$ is isomorphic to the free skew commutative algebra generated by the graded space $\mathrm{Prim}.(H.(\mathfrak{gl}(A))$ of primitive elements.

Recall that the homology of an arbitrary Lie algebra $\mathfrak{g}$ with values in the trivial module are defined as ones of the standard complex $C.(\mathfrak{g}): C_k(\mathfrak{g}) = \Lambda^k \mathfrak{g}$; the differential $d : \Lambda^k \mathfrak{g} \to \Lambda^{k-1} \mathfrak{g}$ acts as follows:

$$d(\ell_0 \wedge \ldots \wedge \ell_{k-1}) = \sum_{0 \leq i,j \leq k-1} (-1)^{i+j-1} [\ell_i, \ell_j] \wedge \ldots \wedge \hat{\ell}_i \wedge \ldots \hat{\ell}_j \wedge \ldots \ldots \wedge \ell_{k-1} \tag{4.2.1}$$

Here $\ell_i \in \mathfrak{g}$; the cap $\wedge$ means that ℓ_i, ℓ_j do not enter the corresponding summand.

Now we construct a map T from the standard complex of the Lie algebra $\mathfrak{gl}_n(A)$ to the complex computing the additive K-functors of A. Let $m_0, m_1, \ldots, m_{k-1}$ be matrices with entries in k and $a_0, \ldots, a_{k-1} \in A$. It suffices to define T on the elements $(m_0 \otimes a_0) \wedge \ldots \ldots \wedge (m_{k-1} \otimes a_{k-1})$ in $C_k(\mathfrak{gl}_n(A))$. Put

$$T((m_0 \otimes a_0) \wedge \ldots \wedge (m_{k-1} \otimes a_{k-1})) = \sum_{\sigma} (\mathrm{sgn}\,\sigma)\, \mathrm{tr}(m_{\sigma 0} \ldots m_{\sigma(k-1)}) \otimes \otimes (a_{\sigma 0} \otimes \ldots \otimes a_{\sigma(k-1)}) \tag{4.2.1}$$

The sum is taken over all permutations on k elements. We obtain a map $C_k(\mathfrak{gl}_n(A)) \to \overset{k}{\otimes} A$. It is clear that $\mathrm{Im}(T) \subset \ker(1 - \tau)$, $\tau(a_0 \otimes \ldots \otimes a_{k-1}) = (-1)^{k-1} a_1 \otimes \ldots \otimes a_k \otimes a_0$. Furthermore, $T(d\omega) = b(T\omega)$ where $b : \overset{k}{\otimes} A \to \overset{k}{\otimes} A$ acts as in (1.2.2). We have seen that the complex $C^{cycl}_{\cdot}(A)$ is embedded into $D.(A) = (\overset{\cdot}{\otimes} A, b)$ (Proposition 1.2.8); if we denote the image of this embedding again by $C^{cycl}(A)$ we obtain a map

$$C.(\mathfrak{gl}_n(A)) \to C^{cycl}(A)$$

and hence a map

$$\mu : H.(\mathfrak{gl}_n(A)) \to K^+_\cdot(A).$$

heorem 4.2.1. 1) The restriction of μ to the primitive elements gi-es the isomorphisms

$$\mathrm{Prim}_i H.(\mathfrak{gl}(A)) \to K^+_i(A), \quad i > 0.$$

) The map $(j_{n,n+1})_*: H_i(\mathfrak{gl}_n(A)) — H_i(\mathfrak{gl}_{n+1}(A))$ is bijective for $\leqslant$ n. In particular

$$H_i(\mathfrak{gl}_n(A)) \xrightarrow{\sim} H_i(\mathfrak{gl}(A)), \quad i \leqslant n.$$

It is clear that Theorem 4.2.1. implies Theorem 4.1.1. ecause $K^+_\cdot(A) \simeq K^+_\cdot(M_n(A))$ (cf. Corollary 1.2.5).

.3. Proof of Theorem 4.2.1. The Lie algebra $\mathfrak{gl}_n(A)$ contains the ubalgebra $\mathfrak{gl}_n(k)$ (since $k \subsetneq A$). The standard complex $C.(\mathfrak{gl}_n(A))$ ay be decomposed into a direct sum of the subcomplexes K_ρ (indexed y the set of the finite dimensional irreducible representations of $\mathfrak{l}_n(k)$) such that the action of $\mathfrak{gl}_n(k)$ on K_ρ is an isotypical re-resentation of type ρ. Furthermore, all subcomplexes K_ρ corres-onding to non-trivial ρ are acyclic. This follows from the well known fact that a Lie algebra acts trivially on its homology. So $C.(\mathfrak{gl}_n(A))$ is quasi-isomorphic to its invariant subcomplex $C.(\mathfrak{gl}_n(A))^{inv}$.

Now we shall describe the structure of $C.(\mathfrak{gl}_n(A))^{inv}$. The com-plex $C.(\mathfrak{gl}_n(A))^{inv}$ is the sum of subspaces $((\mathfrak{gl}_n(k) \otimes a_o) \wedge \ldots$ $\ldots \wedge (\mathfrak{gl}_n(k) \otimes a_{k-1}))^{inv}$, $a_o,\ldots,a_{k-1} \in A$. We may apply invariant theo-ry to study these subspaces (cf. [64]). We shall formulate a state-ment of invariant theory in the form we need here.

Let V be an n-dimensional vector space over k. Identify $\mathfrak{gl}_n(k)$ with the space of all linear maps $V \to V$. The space V turns in such a way into a $\mathfrak{gl}_n(k)$-module. The adjoint representation of $\mathfrak{gl}_n(k)$ is identified naturally with the tensor product $V^* \otimes V$. Fur-thermore, we have $\otimes^p \mathfrak{gl}_n(k) \simeq \otimes^p (V^* \otimes V) \simeq \mathrm{End}(\otimes^p V)$. So every invariant element of $\otimes^p \mathfrak{gl}_n(k)$ determines an endomorphism of the $\mathfrak{gl}_n(k)$-module $\otimes^p V$.

Proposition 4.3.1. 1) The space of endomorphisms of module $\overset{p}{\otimes} V$ is linearly generated by the permutations of tensor factors. I.e., there exists an epimorphism

$$\varphi : k[\Sigma_p] \longrightarrow \mathrm{End}_{\mathfrak{gl}_n(k)} (\overset{p}{\otimes} V) \simeq (\overset{p}{\otimes}\mathfrak{gl}_n(k))^{inv};$$

$$\varphi(\sigma)(v_o \otimes \ldots \otimes v_{p-1}) = v_{\sigma^{-1}o} \otimes \ldots \otimes v_{\sigma^{-1}(p-1)}.$$

2) The kernel of φ is the two-sided ideal in the group algebra $k[\Sigma_p]$. If π is an irreducible representation of Σ_p, then $\ker\varphi$ annihilates π iff the Young diagram corresponding to π has more than n cells in the first column. In particular, φ is an isomorphism iff $n = \dim V \geqslant p$.

The proof contains in [64].

Reformulate this proposition. We shall describe all multilinear invariant functions depending on p matrices, i.e., all invariant linear maps $\otimes^p \mathfrak{gl}_n(k) \to k$. Let σ be a permutation. Represent it as a product of cycles: $\sigma = \sigma_1^{-1} \ldots \sigma_r^{-1}$ where σ_i is a cyclic permutation

$j^{(i)} \to j^{(i)} \to \ldots \to j^{(i)}_{\alpha_i-1} \to j^{(i)}_o$, $\alpha_i = \mathrm{ord}\,\sigma_i$, $\alpha_1 + \ldots + \alpha_r = p$.

Construct a multilinear function T_σ:

$$T_\sigma(m_o,\ldots,m_{p-1}) = \prod_{i=1}^{r} \mathrm{tr}(m_{j_o^{(i)}} \ldots m_{j^{(i)}_{\alpha_i-1}}) \tag{4.3.1}$$

<u>Proposition 4.3.2</u>. The functions T_σ linearly generate the space of invariant forms. They are linearly independent iff $p \leqslant n$. The natural pairing between $(\overset{p}{\otimes}\mathfrak{gl}_n(k))^{inv}$ and the space of invariant p-forms on $\mathfrak{gl}_n(k)$ is nondegenerate.

Note that Theorem 4.2.1 follows imediately from Proposition 4.3.2. Actually, define a map

$$S^\cdot(T) : C.(\mathfrak{gl}_n(A)) \longrightarrow S^\cdot(C^{cycl}_\cdot(A)).$$

Here $S^\cdot(C^{cycl}_\cdot(A))$ is the symmetric algebra of the complex $C^{cycl}(A)$, i.e., the sum $k \oplus C^{cycl}_\cdot(A) \oplus S^2 C^{cycl}_\cdot(A) \oplus \ldots$; the symmetric powers are as usually in the graded case. The homomorphism $S^\cdot(T)$ is determined by the following conditions:

a) the component $C.(\mathfrak{gl}(A)) \longrightarrow S^o(C^{cycl}_\cdot(A)) = k$ is the augmentation; the component $C.(\mathfrak{gl}(A)) \longrightarrow S^1(C^{cycl}_\cdot(A)) = C^{cycl}_\cdot(A)$ is equal to T;

b) $S^\cdot(T)$ is a coalgebra homomorphism.

(The coalgebra structures on $S^\cdot(C^{cycl}(A))$ and $C.(\mathfrak{gl}(A))$ are given by $u \mapsto u \otimes 1 + 1 \otimes u$, if $u \in C_1(\mathfrak{gl}(A)) = \mathfrak{gl}(A)$ or

$\in S^1(C^{cycl}_{\cdot}(A)) = C^{cycl}_{\cdot}(A)$; they are compatible with multiplication. Uniqueness of $S^{\cdot}(T)$ is dual to a trivial fact that a ring homomorphism from the symmetric algebra is determined by its restriction to the space of generators. For instance, in our case the component $S^2(T) : C.(\mathfrak{gl}(A)) \longrightarrow S^2(C^{cycl}(A))$ acts as follows: $(m_o \otimes a_o) \otimes (m_1 \otimes a_1) \mapsto$ $\to tr(m_o)\ tr(m_1) \otimes a_o a_1 \in S^1 \cdot S^1)$.

Proposition 4.3.2. shows that the map $S^{\cdot}(T)|C.(\mathfrak{gl}(A))^{inv}$ is bijective. This ends the proof of Theorem 4.2.1.

4.4. <u>Filtrations on the standard complex $C.(\mathfrak{gl}(k))^{inv}$</u>.

We need a slightly more precise form of the basic statement on invariants. Proposition 4.3.1. associates to a permutation δ an invariant from $\overset{p}{\otimes}\mathfrak{gl}_n(k)$. Proposition 4.3.2 associates to δ an invariant linear functional on $\overset{p}{\otimes}\mathfrak{gl}_n(k)$. So we obtain a pairing between the elements of Σ_p or equivalently a bilinear form on $k[\Sigma_p]$. Denote its value on $\delta, \tau \in \Sigma_p$ by $\langle\delta,\tau\rangle$.

<u>Proposition 4.4.1</u>. $\langle\delta,\tau\rangle = tr(\varphi(\delta\cdot\tau))$.

The trace in the right hand side is one of the operator in $\overset{p}{\otimes} V)$.

<u>Proof</u>. The invariant $\varphi(\tau)$ as an element of $\overset{p}{\otimes}\mathfrak{gl}_n(k)$ is equal to the sum

$$\varphi(\tau) = \sum E_{i_o, i_{\tau 0}} \otimes \dots \otimes E_{i_{p-1}, i_{\tau(p-1)}}$$

over the set of all $(i_o, \dots, i_{p-1})$ such that $1 \leqslant i_k \leqslant n = \dim V$. Let ν be as above (before the formula (4.3.1)).

Then

$$\langle\delta,\nu\rangle = \sum \prod_{i=1}^{r} tr(E_{j_o^{(i)}, \tau j_1^{(i)}} E_{j_1^{(i)}, \tau j_1^{(i)}} \dots E_{j^{(i)}_{\alpha_{i-1}}, j_{\tau\alpha_i}}) =$$

$$= n^{\#\{l : \delta^{-1} l = \nu l\}}$$

But the latter number is easily seen to be equal to the trace of $\varphi(\delta\tau)$. In fact, both are equal to $n^{\varepsilon(\delta\tau)}$ where $\varepsilon(\delta\tau)$ is the number of cycles in decomposition of $\delta\tau$. This provides the proof of Proposition 4.4.1. together with the formula

$$\langle\delta,\tau\rangle = (\dim V)^{\varepsilon(\delta\tau)}.$$

We see that the pairing $\langle,\rangle$ depends polynomially on n. Therefore n may be replaced by an arbitrary parameter belonging to k. The form $\langle,\rangle$ determines a map $k[\Sigma_p] \longrightarrow k[\Sigma_p]^*$. There is the standard identification $k[\Sigma_p] \overset{\sim}{\to} k[\Sigma_p]^*$: $\delta \mapsto f_\delta \in k[\Sigma_p]^*$ $f_\delta(\tau) = 0$ if

$\tau\sigma \neq 1$; $f_\sigma(\tau) = 1$ if $\tau\sigma = 1$.

Using this we obtain a map

$$\theta(n) : k[\Sigma_p] \to k[\Sigma_p];$$

$$\theta(n)(\sigma) = \sum_{\nu\in\Sigma_p} tr(\sigma\nu)\,\nu^{-1} \qquad (4.4.1)$$

It depends polynomially on n. For $n \in \mathbb{N}$ we have $\ker\theta(n) = \ker\varphi$ (Proposition 4.3.1), where corresponds to $\dim V = n$. Furthermore, $\det\theta(n) \neq 0$, $n \geqslant p$.

Now recall a property of matrices depending on a parameter. Let A be a matrix depending polynomially on λ; $A(\lambda) \in \mathrm{End}\, W$. Choose $\lambda_0 \in k$. Construct a sequence of matrices $B_0 = A(\lambda_0)$, $B_1 = \frac{\partial A}{\partial\lambda}(\lambda_0)$, $B_2 = \frac{\partial^2 A}{\partial\lambda^2}(\lambda_0)$, ...

Consider a substitution $\lambda \mapsto g(\lambda)$, $g(\lambda_0) = \lambda_0$, $g'(\lambda_0) \neq 0$. It is clear that $\frac{\partial^i}{\partial\lambda^i}(A(g(\lambda)))\big|_{\lambda=\lambda_0}$ is equal to the sum of $B_i(\lambda_0)\, g'(\lambda_0)^i$ and a linear combination of $B_{i-1}(\lambda_0), B_{i-2}(\lambda_0), \ldots, B_0(\lambda_0)$. We see that the following sequence of maps (up to multiplication by a nonzero scalar) does not depend on a substitution $A(.) \mapsto A(g(.))$:

$$B_0 : W \to W$$

$$B_1 : \ker B_0 \to W/\mathrm{im}\, B_0 \qquad (4.4.2)$$

$$B_i : \ker B_0 \cap \ldots \cap \ker B_{i-1} \to W/(\mathrm{im}\, B_0 + \ldots + \mathrm{im}\, B_{i-1})$$

Here B_i denotes the restriction of B_i to a corresponding subspace. We obtain a pair of filtrations on W:

$$\ker B_0 \supset (\ker B_0 \cap \ker B_1) \supset \ldots; \qquad (4.4.3)$$

$$\mathrm{im}\, B_0 \subset \mathrm{im}\, B_0 + \mathrm{im}\, B_1 \subset \ldots \qquad (4.4.4)$$

The operators B_i map the graded quotients of these filtrations to each other.

The first filtration will be denoted by $F^{\cdot}(\lambda_0)$; $\ker B_0 = F^1(\lambda_0) \supset F^2(\lambda_0) \supset \ldots$ Analogously for the second filtration: $\mathrm{Im}\, B_0 = F_1(\lambda_0) \subset F_2(\lambda_0)$ ...

We want now to find these filtrations for the family $\Theta(n)$, $n \in k$. Consider the direct product $\Sigma_p \times \Sigma_p$ together with an action on Σ_p : $(\sigma_1, \sigma_2)\nu = \sigma_1 \nu \sigma_2^{-1}$, $\nu \in \Sigma_p$, $(\sigma_1, \sigma_2) \in \Sigma_p \times \Sigma_p$. Let

$$k[\Sigma_p] = \oplus L_{\P}$$

be the decomposition of $k[\Sigma_p]$ to a direct sum of the irreducible representations of $\Sigma_p \times \Sigma_p$. Every component is an indecomposable two-sided ideal in $k[\Sigma_p]$, and ¶ denotes a corresponding Young diagram. The map $\Theta(n)$ commutes with the action of $\Sigma_p \times \Sigma_p$ (this follows from Proposition 4.4.1). So $\Theta(n)$ preserves the decomposition $\oplus L_{\P}$ and multiplies a component $L_{\P}$ by a number $K(\P)$. We have seen that

$$\Theta(n)\sigma = \sigma \sum_{\nu} \chi(\nu)\nu^{-1} \tag{4.4.5}$$

where   is the character of the representation of Σ_p in $\overset{p}{\otimes} V$. But it is well known (see [45]) that the character χ is equal to $\sum_{\P} d_{\P} \chi_{\P}$ where $\chi_{\P}$ is the character corresponding to the Young diagram ¶ and $d_{\P}$ is the dimension of the irreducible representation of $\mathfrak{gl}_n(k)$ with the highest weight associated to ¶. So $\frac{1}{p!} K(\P) = d_{\P}$.

Now we shall write down a well known formula for $a_{\P}$ in a form convenient for our purposes. Let the Young diagram ¶ be situated in the first quadrant:

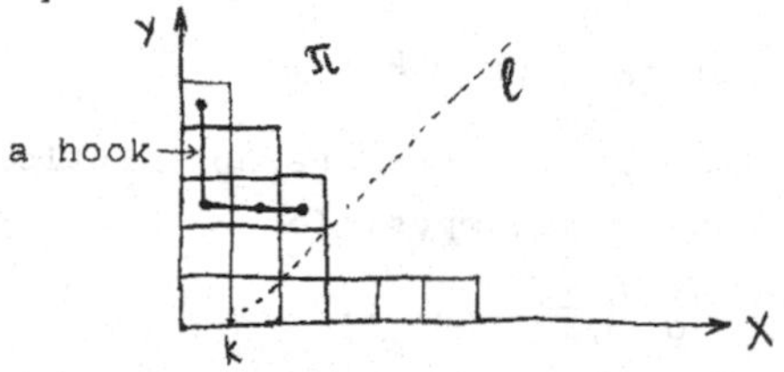

Fig. 4.4.1.

Choose an integral point k of the axis OX and draw a straight line ℓ through it with the slope 45°. Let $a_{\P}(k)$ be the number of cells of ¶ which are intersected by ℓ. These numbers are the so called Frobenius coordinates. Furthermore, for a cell (i,j) in the diagram ¶ let h_{ij} be the number of the cells in the hook with the vertex (i,j), where a hook is the set of the cells (i_1, j) and (i, j_1), $i_1 \geqslant i$, $j_1 \geqslant j$. The following formula is valid:

$$d_{\P} = \frac{K(\P)}{p!} = \frac{\prod_{k=-\infty}^{\infty} (n-k)^{a_{\P}(k)}}{\prod h_{ij}} \tag{4.4.6}$$

The product in the denominator is taken over the set of all cells in ¶. This formula implies the following

Corollary 4.4.2. For $n \in \mathbb{Z}$

$$F_{i+1}(n) = \bigoplus_{\P : a_\P(n) \leqslant i} L_\P, \quad i \geqslant 0;$$
$$F^i(n) = \bigoplus_{\P : a_\P(n) \geqslant i} L_\P, \quad i \geqslant 0. \tag{4.4.7}$$

In other words, the spaces $F_i(n)$, $F^i(n)$ and their sums and intersections with each other define the filtrations on $k[\Sigma_p]$ which is indexed by the set of all Young diagrams:

$$F_\P = \bigcap_{n=-\infty}^{\infty} F_{a_\P(n)+1}(n)$$
$$F^\P = \sum_{n=-\infty}^{\infty} F^{a_\P(n)+1}(n) \tag{4.4.8}$$

Certainly, in this case

$$F_\P = \bigoplus_{\P' \leqslant \P} L_{\P'}; \quad F^\P = \bigoplus_{\P \not\leqslant \P'} L_{\P'}$$

Note that the spaces $F_\P$, $F^\P$ may be defined for an infinite $\P$.

Now we return to the examination of the complex $C.(\mathfrak{gl}_n(A))^{inv}$. This one consists of the elements

$$s = \sum (E_{i_0, i_{\sigma 0}} \otimes a_0) \wedge \ldots \wedge (E_{i_{p-1}, i_{\sigma(p-1)}} \otimes a_{p-1}) \tag{4.4.9}$$

(see Proposition 4.3.1). Let $p \leqslant n$, i.e. there are no relations between these elements for all σ. Take a decomposition

$\sigma = (\alpha_0^{(1)} \to \alpha_1^{(1)} \to \ldots \to \alpha_{k-1}^{(1)})(\alpha_0^{(2)} \to \ldots)(\ldots$

We shall denote s from (4.4.9) by a symbol

$(a_{\alpha_0^{(1)}} | a_{\alpha_1^{(1)}} | \ldots | a_{\alpha_{k-1}^{(1)}}) \; (a_{\alpha_0^{(2)}} | \ldots) \; \ldots$

Write down a formula (4.2.1) for ds. The latter happens to be a sum of the two summands:

$$ds = (nd' + d'')s;$$

$$d'(a_0 | a_1 \ldots | a_{k-1}) = (a_0 a_1 | a_2 | \ldots | a_{k-1}) - (a_0 | a_1 a_2 | \ldots | a_{k-1}) + \ldots$$
$$\pm (a_0 | a_1 | \ldots | a_{k-2} a_{k-1}) - (a_1 | \ldots | a_k a_0);$$

for $\omega_i = (a_{\alpha_0^{(i)}} \ldots a_{\alpha_{k-1}^{(i)}})$

$$d'(\omega_o\wedge\ldots\wedge\omega_{r-1}) = d'\omega_o\wedge\ldots\wedge\omega_{r-1} + (-1)^{\deg\omega_o}\omega_o\wedge d'\omega_1 \ \ldots \pm \ldots$$

We see that "the principal part" of d is nothing else that the differential in $S^\cdot(C^{cycl}_\cdot(A))$ (up to a sign). The component d'' is more complicated. d', d'' do not depend on n. It is clear for d' and may be easily verifyed for d''. For instance,

$$d''(a_o|a_1|a_2|a_3) = d''(\sum(E_{i_oi_1}\otimes a_o)\wedge(E_{i_1i_2}\otimes a_1)\wedge(E_{i_2i_3}\otimes a_2)$$

$$\otimes(E_{i_3i_o}\otimes a_3)) = -\sum(E_{i_3i_2}\otimes a_2a_o)\wedge(E_{i_2i_3}\otimes a_1)\wedge(E_{i_1i_1}\otimes a_2)$$

$$+\sum(E_{i_oi_3}\otimes a_oa_2)\wedge(E_{i_1i_1}\otimes a_1)\wedge(E_{i_3i_o}\otimes a_3) + \ldots =$$

$$= -(a_2a_o|a_1)\wedge(a_2) - (a_oa_2|a_3)\wedge(a_1) + \ldots ;$$

$$d''(a_o|a_1|a_2) = d''(\sum(E_{i_oi_1}\otimes a_o)\wedge(E_{i_1i_2}\otimes a_1)\wedge(E_{i_2i_o}\otimes a_2)) =$$

$$= -\sum(E_{i_1i_1}\otimes a_1a_o)\wedge(E_{i_oi_o}\otimes a_2) + \sum(E_{i_oi_o}\otimes a_o)\wedge(E_{i_1i_1}\otimes a_2a_1) -$$

$$-\sum(E_{i_oi_o}\otimes a_oa_2)\wedge(E_{i_1i_1}\otimes a_1) = -(a_1a_o)\wedge(a_2) + (a_o)\wedge(a_2a_1) -$$

$$-(a_oa_2)\wedge(a_1);$$

$$d''((a_o|a_1)\wedge(a_2)) = (a_2a_o|a_1) - (a_2a_1|a_o) + (a_1a_2|a_o) - (a_oa_2|a_1)$$

<u>Proposition 4.4.3</u>. There is a complex $\mathcal{K}_\cdot(A,n)$ for any $n \geqslant 1$ such that:

1) the underlying graded vector space of $\mathcal{K}_\cdot(A,n)$ is equal to this of $S^\cdot(C^{cycl}_\cdot(A))$ for any n;

2) the differential in $\mathcal{K}_\cdot(A,n)$ is equal to $d(n) = nd' + d''$ where d' is the differential in $S^\cdot(C^{cycl}_\cdot(A))$ and d'' does not depend on n;

3) there are the maps of $\mathcal{K}_\cdot(A,n)$ onto $C_\cdot(\mathfrak{gl}_n(A))^{inv}$ which give the homology isomorphisms in dimensions $p \leqslant n$.

All these statements are straightforward consequences of the above discussion.

Consider the composition

$$\overline{\theta}(n) : \mathcal{K}_\cdot(A,n) \to C_\cdot(\mathfrak{gl}_n(A))^{inv} \to S^\cdot(C^{cycl}_\cdot(A)).$$

It is easily seen that $\overline{\theta}(n)$ depends polynomially on n. So we may consider n as a parameter belonging to k. Now fix an $n_o \in \mathbb{N}$. We may apply the general scheme (4.4.2)-(4.4.4) and receive an increasing filtration $F.(n_o)$ on $S^{\cdot}(C^{cycl}(A))$:

$$F_i(n_o) = \operatorname{Im} \bar{\theta}(n_o) + \operatorname{Im} \frac{\partial\bar{\theta}}{\partial n}(n_o) + \ldots + \operatorname{Im} \frac{\partial^i\bar{\theta}}{\partial n^i}(n_o).$$

It is easily seen that all $F_i(n_o)$ are the subcomplexes in $S^{\cdot}(C^{cycl}(A))$:

$$d' \frac{\partial^i\bar{\theta}(n)}{\partial n^i} = \frac{\partial^i}{\partial n^i}(\bar{\theta}(n)\cdot d(n)) = \frac{\partial^i\bar{\theta}}{\partial n^i}(n)\, d(n) + i \frac{\partial^{i-1}\bar{\theta}}{\partial n^{i-1}}(n)\, d'$$

for all n.

Theorem 4.4.4. The complex $S^{\cdot}(C_{\bullet}^{cycl}(A))$ admits an increasing filtration indexed by the partially ordered set of all (finite and infinite) Young diagrams. The subcomplex $F_{\P}$ corresponding to a diagram $\P$ is equal to

$$F_{\P} = \bigcap_{n=-\infty}^{\infty} F_{a_{\P}(n)+1}(n),$$

where $a_{\P}(n)$ are the Frobenius coordinates of $\P$.

Let $\P$ be a diagram with m cells. Then

$F_{\P}(S^{\cdot}(C_{\bullet}^{cycl}(A))_r) = 0$, $r > m$. Put $V_{\P} = F_{\P}(S^{\cdot}(C_{\bullet}^{cycl}(A))_m)$. Note that $S^{\cdot}(C^{cycl}(A))_m = \oplus V_{\P}$ where the sum is taken over all the diagrams with m entries. Furthermore, $F_{\P} = \bigoplus_{\P' \subseteq \P} V_{\P'}$. It is clear that $dV_{\P} \subset \bigoplus_{\P' \subset \P} V_{\P'}$.

The Young diagrams form a partially ordered set: $\P_1$ is less then $\P_2$ if $\P_1 \subset \P_2$. Consider the corresponding oriented graph (Young's graph). Every indecomposable arrow of it (i.e. a pair $\P \supset \P'$, card $\P$ = 1 + card $\P'$) gives a component in the differential d. It would be interesting to study the structure of this diagram of maps. The next subsection contains the direct description of the structure of $V_{\P}$. In the same time we shall give an another construction for the filtration $F_{\P}$.

4.5. Let B be an associative algebra without unit. Let $d_i : \overset{p+1}{\otimes} B \longrightarrow \overset{p}{\otimes} B$, $d_i(b_o \otimes \ldots \otimes b_p) = b_o \otimes \ldots \otimes b_{i-1}b_i \otimes \ldots \otimes b_p$, $1 \leqslant i \leqslant p$. Then $d_i d_j = d_{j-1} d_i$, $i < j$; $d_1 - d_2 + \ldots + (-1)^{p-1} d_p :=$ $:= \partial$; $\partial^2 = 0$.

$$\begin{array}{c}\xrightarrow{d_1}\\ \rightarrow\\ \xrightarrow[d_3]{}\end{array} B \otimes B \otimes B \begin{array}{c}\xrightarrow{d_1}\\ \xrightarrow[d_2]{}\end{array} B \otimes B \xrightarrow{d_1} B \qquad (4.5.1)$$

Denote by Δ_s the category whose objects are numerated by the nonnegative integers (we write $[p]$ for the object corresponding to p) such that $\mathrm{Mor}\,\Delta_s$ is generated by the morphisms

$$d_i : [p] \longrightarrow [p-1], \quad 1 \leqslant i \leqslant p$$

with the only relations

$$d_i d_j = d_{j-1} d_i, \quad i < j.$$

Call a functor from Δ_s to a category $\mathfrak{A}$ a Δ_s-object of $\mathfrak{A}$. So we may relate to B a Δ_s-space. We call it $\mathcal{J}(B)$ and write $\mathcal{J}_p(B) = \overset{p+1}{\otimes} B$.

Now let $B_n = M_n(k)$. The Lie algebra $\mathfrak{gl}_n(k)$ acts on the tensor powers of B. Clearly, this action commutes with the maps d_i. Our first aim is to describe the Δ_s-space $\mathcal{J}(B_n)^{inv}$.

For this we construct another Δ_s-space S. Put $S_p = k[\Sigma_{p+1}]$;

$$d_i\sigma = 0 \quad \text{if} \quad \sigma i \neq i+1;$$

for $i<j$: $(d_i\sigma)(j) = \sigma(j+1)$ if $\sigma(j+1) \leqslant i$; $(d_i\sigma)(j) = \sigma(j+1) - 1$,
$\sigma(j+1) > i+1$; (4.5.2)

for $i>j$: $(d_i\sigma)(j) = \sigma j$ if $\sigma j \leqslant i$; $(d_i\sigma)(j) = \sigma j - 1$; $\sigma j > i+1$.

<u>Proposition 4.5.1</u>. 1) There exists a monomorphism of Δ_s-spaces

$$\lambda^{(n)} : S \longrightarrow \mathcal{J}(B_n)^{inv}.$$

2) The kernel of $\lambda_{p-1} : S_{p-1} \to \mathcal{J}(B_n)^{inv}_{p-1}$ $(p \geqslant 1)$ is the two-sided ideal which is the direct sum of all the two sided ideals of $k[\Sigma_p]$ such that the corresponding Young diagrams have more then n cells in the first column.

<u>Proof</u>. This follows directly from the invariant theory. Indeed, we have seen in the Proposition 4.3.2 that every permutation from Σ_p defines a linear form on $(\overset{p}{\otimes} B)^{inv}$; so we obtain a pairing $(\overset{p}{\otimes} B_n)^{inv} \otimes k[\Sigma_p] \to k$ and hence a map $(\overset{p}{\otimes} B_n)^{inv} \to k[\Sigma_p]^*$. Identify $k[\Sigma_p]$ with its dual in the standard way (see 4.4, just before the formula

(4.4.1)). So we have defined the maps $(\overset{p}{\otimes} B_n)^{inv} \longrightarrow k[\Sigma_p]$, i.e. $\mathcal{J}_{p-1}(B_n) \to S_{p-1}$. Its commuting with the operators d_i may be shown by direct computation. The statement 2) is obvious (see Proposition 4.3.2).

The quotient of S by the kernel of $\lambda^{(n)}$ we denote by $S(n)$; $n = 0,1,2,\ldots,\infty$; $S(\infty) = S$.

Fix on the space $\mathcal{J}_{p-1}(B_n) = \overset{p}{\otimes} B_n$ the action of the group Σ_p by permutation of the factors. There is a similar action on S_p; $\sigma(x) = \sigma x \sigma^{-1}$, $\sigma \in \Sigma_p$, $x \in k[\Sigma_p]$. The map $\lambda^{(n)}_{p-1}$ commutes with the action of Σ_p.

Now take the differential $\partial = d_1 - d_2 + \ldots \pm d_p$ on $\mathcal{J}(B_n)$.

Proposition 4.5.2. Let A be an algebra. The Δ_s-space $\mathcal{J}(M_n(A))^{inv}$ is isomorphic to $\mathcal{J}(A) \boxtimes S(n)$.

Here the sign $\boxtimes$ means the tensor product of functors; this is the functor whose values on the objects and morphisms of Δ_s are equal to the tensor products of those for $\mathcal{J}(A)$, S.

The proposition follows directly from the above discussion. Indeed,

$$\overset{p}{\otimes} M_n(A) = \overset{p}{\otimes} (M_n(k) \otimes A) = (\overset{p}{\otimes} M_n(k)) \otimes (\overset{p}{\otimes} A);$$

$$(\overset{p}{\otimes} M_n(A))^{inv} = (\overset{p}{\otimes} M_n(k))^{inv} \otimes (\overset{p}{\otimes} A) = S_{p-1} \otimes \mathcal{J}_{p-1}(A).$$

Corollary 4.5.3. The subcomplex of $\mathfrak{gl}_n(k)$-invariants in the standard complex of the Lie algebra $\mathfrak{gl}_n(A)$ is isomorphic to the complex

$$\to (S_{p-1}(n) \otimes (\overset{p}{\otimes} A))^{\Sigma_p} \xrightarrow{\partial} (S_{p-2}(n) \otimes (\overset{p-1}{\otimes} A))^{\Sigma_{p-1}} \longrightarrow \ldots$$

where $\partial(\sigma \otimes a_0 \otimes \ldots \otimes a_{p-1}) = \sum_{i=1}^{p-1} (-1)^{i-1} \partial_i(\sigma) \otimes (a_0 \otimes \ldots \otimes a_{i-1}a_i \otimes \ldots$

$\ldots \otimes a_{p-1})$

and ∂_i are given by (4.5.1); the action of Σ_p on the chain groups is given by

$$\sigma(x \otimes a_0 \otimes \ldots \otimes a_{p-1}) = (\sigma x \sigma^{-1} \otimes a_{\sigma^{-1}0} \otimes \ldots \otimes a_{\sigma^{-1}(p-1)}) \operatorname{sgn}\sigma$$

Proof. For an arbitrary algebra B the homology of $\mathcal{J}(B)^{\Sigma_p}$ with respect to ∂ is nothing but the Lie algebra homology of Lie (B). Now our statement follows from Propositions 4.5.1, 4.5.2.

Remark 4.5.4. This corollary shows that the Lie algebra homology of

$gl_n(A)$ are obtained as the homology of the invariant subcomplex in the tensor product of $\mathcal{J}(A)$ with the standard Δ_s-space with some supplementary properties. Formulate them briefly. Consider an algebra B and a Δ_s-space $\mathcal{J}(B)$. There is an action of Σ_p on $\mathcal{J}_{p-1}(B)$ for every p. There are some relations between the operators d_i and the operators ν, $\nu \in \Sigma_p$, which are universal for all the algebras B. All these relations may be easily described. For instance, let $\nu_o(b_o \otimes b_1 \otimes b_2) = b_1 \otimes b_2 \otimes b_o$ and $\nu_1(b_o \otimes b_1) = b_1 \otimes b_o$; then $d_1 \nu_o = \nu_1 d_2$. We call a Δ_s-space a Σ-space, if all such conditions are valid; then we may replace S by an arbitrary Σ-space and obtain new functors on the category of algebras. Moreover, the computations in the category of Σ-spaces show that we may replace the complex of invariants by the bicomplex whose vertical complex computes the group cohomology of Σ_p (or rather the Tate cohomology). This gives the new definition for $H.(\mathcal{L}ie(B))$ for an algebra B over an arbitary scalar ring but may be useful also in characteristic zero. We suppose to discuss these questions in a separate work (cf. also A5).

Now let ¶ be a Young diagram (finite or infinite). As in 4.4, put

$$F_{¶}k[\Sigma_p] = \bigoplus_{¶' \subset ¶} L_{¶'} ;$$

$$F^{¶}k[\Sigma_p] = \bigoplus_{¶' \not\supset ¶} L_{¶'} .$$

<u>Lemma 4.5.5</u>. $F_{¶}S$, $F^{¶}S$ are the Δ_s-subspaces in S.

<u>Proof</u>. It suffices to show that $d_i L_{¶} \subset \bigoplus_{¶' \subset ¶} L_{¶'}$.

But it is easily seen that the action of d_i on the element σ may be rewrited as

$$d_i \sigma = pr_p^{p-1}(\tau_{i+1}^{-1} \sigma \tau_i)$$

where $\tau_i = (\overleftarrow{i \to i+1 \to \ldots \to p})$; $pr_p^{p-1} : k[\Sigma_p] \to k[\Sigma_{p-1}]$; $\sigma \mapsto \sigma$ if $\sigma(p-1) = p-1$ and $\sigma \mapsto 0$ if $\sigma(p-1) \neq p-1$. But it is well known that such a projection sends $L_{¶}$ to $\bigoplus_{¶' \subset ¶} L_{¶'}$; since $L_{¶}$ are two-sided ideals, we obtain the proof.

Put $H.(gl_{¶}(A)) = H.((F_{¶}S \otimes J(A))^{\Sigma.} ; \partial)$;

$H.(gl^{¶}(A)) = H.((F^{¶}S \otimes J(A))^{\Sigma.} ; \partial)$.

<u>So we put in correpondence to a Young diagram a sequence of the</u>

<u>homological functors on the category of algebras</u>. $H_\cdot(\mathfrak{gl}^{¶}(A)) = H_\cdot(\mathfrak{gl}_n(A))$ if ¶ is an infinite diagram which contains n cells in every column; $H_\cdot(\mathfrak{gl}_{¶}(A)) = H_\cdot(\mathfrak{gl}_n(A))$ if ¶ is a diagram which consists of one column with height $n+1$.

4.6. Here we discuss the filtrations on the additive K-functors induced by those constructed above.

We have constructed an increasing filtration $F_{¶}$ on the invariant subcomplex of the standard complex of the Lie algebra $\mathfrak{gl}(A)$, i.e. i.e. on $S^{\cdot}(C_{\cdot}^{cycl}(A))$. On one hand, the complex $C_{\cdot}^{cycl}(A)$ is embedded into $S^{\cdot}(C_{\cdot}^{cycl}(A))$. On the other hand, there is the projection $S^{\cdot}(C^{cycl}(A)) \to C^{cycl}(A)$. So we may intersect $F_{¶}$ with $C_{\cdot}^{cycl}(A)$ or take the image of $F_{¶}$ under the projection. So we obtain two increasing filtrations $F_{¶}C^{cycl}(A)$, $f_{¶}C^{cycl}(A)$ indexed by the partially ordered set of the Young diagrams. Put also

$$\begin{aligned} f_n C_{\cdot}^{cycl}(A) &= \sum_{h(¶)\leqslant n} f_{¶}C^{cycl}(A); \\ F_n C_{\cdot}^{cycl}(A) &= \sum_{h(¶)\leqslant n} F_{¶}C^{cycl}(A). \end{aligned} \qquad (4.6.1)$$

Here $h(¶)$ denotes the height of the diagram ¶, i.e. the number of cells in the first column. The filtration $F_\cdot$ has been defined by Loday and Quillen in [43]. The results of this Chapter imply

$$\begin{aligned} F_n C_{\cdot}^{cycl}(A) &= i_*(C_\cdot(\mathfrak{gl}_n(A)) \cap C^{cycl}(A)); \\ f_n C^{cycl}(A) &= \operatorname{im}(C_\cdot(\mathfrak{gl}_n(A)) \to C^{cycl}(A)). \end{aligned} \qquad (4.6.2)$$

The filtrations $F_{¶}$, $f_{¶}$ may be defined by means of the analogous spaces attached to "the infinite simal neighbourhoods" of $\mathfrak{gl}_n$.

<u>Proposition 4.6.1</u>. $f_{n+1}C_{2n}^{cycl}(A) = C_{2n}^{cycl}(A)$;

$$f_{n+1}C_{2n+1}^{cycl}(A) = C_{2n+1}^{cycl}(A).$$

<u>Proof</u>. Let $U_p \subset \Sigma_p$ be the set of all cyclic permutations. The group Σ_p acts transitively on U_p by conjugations. Consider the following commutative diagram:

$$\begin{array}{ccc} (k[U_p] \otimes (\overset{p}{\otimes} A))^{\Sigma_p} & \xrightarrow{\sim} & C^{cycl}_p(A) \\ \downarrow & & \downarrow \\ (k[\Sigma_p] \otimes (\overset{p}{\otimes} A))^{\Sigma_p} & \xrightarrow{\sim} & C_p(\mathfrak{gl}(A))^{inv} \\ \uparrow & & \uparrow \\ (\bigoplus_{h(\P) \leq n} L_\P \otimes (\overset{p}{\otimes} A))^{\Sigma_p} & \xrightarrow{\sim} & C_p(\mathfrak{gl}_n(A))^{inv} \end{array} \qquad (4.6.3)$$

Here $L_\P$ means as above the two-sided ideal in $k[\Sigma_p]$ corresponding to $\P$ and the invariants are taken with respect to the action

$$\sigma(\nu \otimes a_o \otimes \ldots \otimes a_{p-1}) = (\sigma \nu \sigma^{-1} \otimes a_{\sigma^{-1}o} \otimes \ldots \otimes a_{\sigma^{-1}(p-1)}) \operatorname{sgn}\sigma \qquad (4.6.4)$$

Let $P_p : k[\Sigma_p] \to k[\Sigma_p]$ be the map defined by

$$P_p(\sigma) = \sigma, \quad \sigma \in U_p; \quad P_p(\sigma) = 0, \quad \sigma \in U_p.$$

The horizontal isomorphisms reduce the problem about f. to the study of the map P_p. More precisely, we must show that

$$P_{2p}(\bigoplus_{h(\P) < p+1} L_\P) = k[U_{2p}];$$
$$(4.6.5)$$
$$P_{2p+1}(\bigoplus_{h(\P) \leq p+1} L_\P) = k[U_{2p+1}].$$

Consider a standard scalar product on $k[\Sigma_p]$: $\langle \sigma_1, \sigma_2 \rangle = 0$ for $\sigma_1 \sigma_2 \neq 1$ and $\langle \sigma_1, \sigma_1^{-1} \rangle = 1$. Consider the adjoint operators P_p^*. Clearly, (4.6.5) may be rewritten as follows: P_p^* of the right hand side does not intersect with the orthogonal complement to the left hand side. We obtain that our statement is equivalent to the following:

$$k[U_{2p}] \cap (\bigoplus_{h(\P) > p+1} L_\P) = 0;$$

$$k[U_{2p+1}] \cap (\bigoplus_{h(\P) > p+1} L_\P) = 0.$$

But this is well known. Indeed, this follows from the Amitsur-Levitskiy theorem (see [51]). Make it more clear. The Amitsur-Levitskiy theorem describes the minimal degree identity in the matrix algebra. Namely, it claims that for any (2p) matrices $m_1, \ldots, m_{2p}$ of order $p \times p$ the following equality holds:

$$\sum \operatorname{sgn}(\sigma)\, m_{\sigma 1}\, m_{\sigma 2} \ldots m_{\sigma(2p)} = 0. \qquad (4.6.6)$$

Moreover, there are no identities on $k < 2p$ matrices of order $p \times p$.

Now (4.6.6) may be rewritten as a trace identity. Multiply it by a matrix m_o and compute the trace:

$$(\mathrm{sgn}\,\sigma)\, \mathrm{tr}(m_o m_{\sigma 1} m_{\sigma 2} \dots m_{\sigma(2p)}) = 0 \qquad (4.6.7)$$

Moreover, there are no trace identities for $k \leqslant 2p$ matrices $p \times p$.

(Note that (4.6.7) admits a cohomological interpretation: the restriction to $\mathfrak{gl}_p(k)$ of the cochain representing the $(2p+1)$th primitive class of $H^{\cdot}(\mathfrak{gl}(k))$ is equal to zero).

Now let

$$\sum_{\sigma} r(\sigma)\, \mathrm{tr}(m_{\sigma o} \dots m_{\sigma(k-1)}) = 0 \qquad (4.6.8)$$

be an identity in $M_p(k)$. Note that $\mathrm{tr}(m_{\sigma o} m_{\sigma 1} \dots) = \mathrm{tr}(m_{\sigma' o} m_{\sigma' 1} \dots)$ where $\sigma' = \sigma\tau$ and τ is the standard cycle: $\tau i = i+1$, $i \leqslant p-2$; $\tau(p-1) = 0$. Construct a function f on U_p : $f(u) = \sum r(\sigma)$ where the sum is taken over all the permutations σ satisfying $\sigma^{-1}\tau\sigma = u$. It is easily seen that the identity is defined by the function f, and the invariant theory (see 4.3) shows that f continued by zero to the whole $k[\Sigma_p]$ lies in $\bigoplus_{h(\P)>n} L_{\P}$. So the Amitsur-Levitskiy theorem gives the proof of our statement.

The Proposition 4.6.1 shows that the filtration $f.K_{\cdot}^{+}$ stabilizes: $f_{[i/2+1]}K_i^+ = K_i^+$. In other words, every primitive class in $H_i(\mathfrak{gl}(A))$ is a projection of some element of $H_i(\mathfrak{gl}_{[i/2+1]}(A))$. An analogous question may be asked about the filtration $F.K_{\cdot}^{+}$ We are interested for which k a primitive class in $K_{\cdot}^{+}(A)$ may lie in the image of $H_i(\mathfrak{gl}_k(A))$. Loday and Quillen conjectured in [43] that $F_p(K_{2p}^+(A)) = F_p(K_{2p+1}^+(A)) = 0$, i.e., that a primitive class in dimension $2p$ cannot lie in the image of $H.(\mathfrak{gl}_p(A))$. But in the general case this is wrong. In order to show this, return to the proof of Proposition 4.6.1. Similar computations show that the statement

$$F_q(K_p^+(A)) = 0 \quad \text{for all } A \qquad (4.6.9)$$

is equivalent to the following:

$$k[U_p] \cap \Big(\bigoplus_{h(\P)\leqslant q} L_{\P} = 0 \qquad (4.6.10)$$

or equivalently to the assertion that the map

$$P_p^* : \bigoplus_{h(\P)>q+1} L_\P \longrightarrow k[U_p] \qquad (4.6.11)$$

s onto.

Actually, the diagram (4.6.1) shows that

$$F_q C_p^{cycl} \simeq (((k[U_p] \cap (\bigoplus_{h(\P)\leq q} L_\P)) \otimes (\overset{p}{\otimes} A))^{\Sigma_p};$$

n the other hand, take $A = k \oplus V$ where $\dim V = \infty$ and $V^2 = 0$; hen we see that for linearly independent $v_o, \ldots, v_{p-1}$ there is a diect summand in $K_p^+(A)$ equal to

$$(k[U_p] \cap (\bigoplus_{h(\P)\leq q} L_\P) \otimes (v_o \otimes \ldots \otimes v_{p-1}))^{\Sigma_p} \simeq k[U_p] \cap (\bigoplus_{h(\P)\leq q} L_\P).$$

We want to estimate for which p, q the dimension of the right and side of (4.6.11) is less than that of the left hand side. It is ell known ([45]) that the dimension of irreducible representation f Σ_p corresponding to $\P$ is equal to

$$D_\P = \frac{p!}{\prod h_{ij}}$$

here the product is taken over the set of all cells and h_{ij} denotes he length of a corresponding hook (see above).

or any r

$$\sum_{h(\P)=r} D_\P^2 \leqslant \sum_{h(\P)=r} \left(\frac{p!}{r!(\prod_{i>1} h_{ij})}\right)^2 = \sum_{h(\P)=r} \left(\frac{p!}{r!(p-r)!}\right)^2 \left(\frac{(p-r)!}{\prod_{i>1} h_{ij}}\right)^2 =$$

$$= \left(\frac{p!}{r!(p-r)!}\right)^2 \cdot \sum D_{\P'}^2$$

here the last sum involves all the Young diagrams $\P'$ with $(p-r)$ ells. So

$$\sum_{h(\P)=r} D_\P^2 \leqslant \frac{p!^2}{r!^2(p-r)!} ;$$

aking the sum over all $r > q$ we obtain

$$\sum_{h(\P)>q} D_\P^2 \leqslant p! \sum_{r=q+1}^{p} \frac{p!}{r!(p-r)!} \frac{1}{r!} < \frac{p!}{(q+1)!} 2^p;$$

n the other hand,

$$\dim k[U_p] = (p-1)! ;$$

therefore we have

$$(F_q K_p^+(A) = 0 \quad \text{for all } A) \Longrightarrow (q+1)! \leqslant p \cdot 2^p .$$

This shows that for any $p \in \mathbb{N}$ there exists such N that $F_N(K_{pN}^+(A)) \neq \neq 0$ where $A = k \oplus V$, $V^2 = 0$, $\dim V = \infty$.

However we show in the Chapter 6 that the conjecture of Loday and Quillen holds if A is a quotient of a polynomial ring by an ideal which is locally generated by a regular sequence.

It may be shown also that for all A, p $F_{p2}(K_p^+(A)) = 0$.

It seems to be very interesting to define the filtrations analogous to $f_\P$, $F_\P$ in Quillen K-theory.

4.7. Here we compute the homology of the stable orthogonal and symplectic Lie algebras over an algebra A with an involution.

Let A be an algebra with an involution $*$; let $\varepsilon = \pm 1$. We denote by ${}_\varepsilon o(n,A)$ the subalgebra in $\mathfrak{gl}(2n,A)$ consisting of the matrices

$$\begin{pmatrix} \alpha & \beta \\ \gamma & \delta \end{pmatrix}$$

where $\alpha, \beta, \gamma, \delta \in M_n(A)$, $\alpha + (\delta^\tau)^* = \varepsilon\beta + (\beta^\tau)^* = \varepsilon\gamma + (\gamma^\tau)^* = = 0$. Here τ means the transposition and the operator $*$ is applied to a matrix elementwise. In other words, ${}_\varepsilon o(n)$ is the subalgebra consisting of the elements fixed by a second order automorphism p_ε :

$$p_\varepsilon \begin{pmatrix} \alpha & \beta \\ \gamma & \delta \end{pmatrix} = \begin{pmatrix} (-\delta^\tau)^* & \varepsilon(\beta^\tau)^* \\ \varepsilon(\gamma^\tau)^* & -(\alpha^\tau)^* \end{pmatrix}$$

The Lie algebra ${}_\varepsilon o(n,A)$ is naturally embedded into $\mathfrak{gl}(2n,A)$. Define a homomorphism ${}_\varepsilon o(n,A) \oplus {}_\varepsilon o(n,A) \longrightarrow {}_\varepsilon o(2n,A)$ as in 4.2; the Lie algebra homology $H.({}_\varepsilon o(n,A))$ is turned into a skew commutative cocommutative Hopf algebra; so it is freely generated by the primitive elements graded space which we denote by $L^+_\cdot(A)$.

Now note that the automorphism p_ε acts on the standard complex $C.(\mathfrak{gl}_n(A))$ and whence on $C.(\mathfrak{gl}_n(A))^{inv}$; it means that p_ε acts on $C^{cycl}_\cdot(A)$. Describe this action.

Define an automorphism ρ of the Hochschild complex $C.(A,A)$ by the formula

$$\rho(a_0 \otimes \ldots \otimes a_n) = (-1)^{(n^2+n)/2}\, a_{n-1}^* \otimes \ldots \otimes a_0^* \otimes a_n^*$$

it is easy to see that $\rho\delta = \delta\rho$). Let $(a_0 \otimes \ldots \otimes a_n) = (-1)^n \times a_1 \otimes \ldots \otimes a_n \otimes a_0$; then $\rho\tau\rho^{-1} = \tau^{-1}$. We obtain a representation of the dihedral group on every $C_n(A,A)$. It is clear that ρ preserves Im$(1-\tau)$, i.e., it acts on $C^{cycl}_{\cdot}(A)$.

Now let $D.(A)$ be as in 1.2. Then we define the action of ρ on $D.(A)$ as follows:

$$\rho(a_0 \otimes \ldots \otimes a_{n-1}) = -(-1)^{n(n+1)/2}\, a^*_{n-1} \otimes \ldots \otimes a^*_0.$$

Clearly $\rho b = b\rho$.

Consider the double complex

$$\ldots \xrightarrow{N} D.(A) \xrightarrow{1-\tau} C_{\cdot-1}(A,A) \xrightarrow{N} D.(A) \xrightarrow{1-\tau} C_{\cdot-1}(A,A) \longrightarrow 0$$

from 1.2. Define an automorphism $\varkappa$ of this bicomplex: $\varkappa = \rho$ on the first $C_{\cdot-1}(A)$ and on the first $D.(A)$; $\varkappa = -\rho$ on the second $C_{\cdot-1}(A,A)$ and on the second $D.(A)$, etc. The subcomplexes in $C_{\cdot-1}(A)$, $D.(A)$, $C^{cycl}(A)$ which are multiplied by ± 1 by the automorphism ρ we define by $C_{\cdot-1}(A,A)^{\pm}$, $D.(A)^{\pm}$, $C^{cycl}_{\cdot}(A)^{\pm}$. We may obtain the following double complex:

$$\ldots \longrightarrow C_{\cdot-1}(A,A)^+ \longrightarrow D.(A) \longrightarrow C_{\cdot-1}(A,A) \longrightarrow D.(A)^+ \longrightarrow C_{\cdot-1}(A,A)^+ \longrightarrow \ldots$$

Theorem 4.7.1. 1) The homology groups of the complex $C^{cycl}_{\cdot}(A)^+$ are equal to ${}_{\varepsilon}L^+(A)$, $\varepsilon = \pm 1$, where ρ acts on C^{cycl} via (4.7.1).
2) There exist a spectral sequence with

$$E^2_{ij} = \begin{cases} H_i(A,A)^+ & \text{if } i \equiv 0 \pmod 4; \\ H_i(A,A)^- & \text{if } i \equiv 2 \pmod 4; \\ 0 & \text{if } i \equiv 1 \pmod 2) \end{cases}$$

converging to ${}_{\varepsilon}L^+_{i+j+1}(A)$.

Here $H_i(A,A)^{\pm}$ mean the subspaces in $H_i(A,A)$ on which ρ acts by multiplication by ± 1.

The proof of the Theorem 4.7.1 is the same as for the Lie algebra $\mathfrak{gl}$; we must use the invariant theory for the groups $o_n(k)$, $sp_n(k)$ (see [64]).

Note that the maps $H.({}_{\varepsilon}o(A)) \longrightarrow H.(\mathfrak{gl}(A))$ induced by the embedding ${}_{\varepsilon}o(A) \longrightarrow \mathfrak{gl}(A)$ are injective. This means that "the additive Witt groups" are equal to zero.

4.8. In this subsection we give a computation of the Lie algebra homology of $\mathfrak{gl}(A)$ with values in some non-trivial modules.

Let A be a k-algebra with an augmentation $\varepsilon : A \longrightarrow k$. The map ε induces the homomorphism $\mathfrak{gl}_n(A) \to \mathfrak{gl}_n(k)$. Every representation ρ_n of the Lie algebra $\mathfrak{gl}_n(k)$ induces a representation $\varepsilon\cdot\rho_n$ of $\mathfrak{gl}_n(A)$. Here we compute $H.(\mathfrak{gl}_n(A), \varepsilon\rho_n)$ "in the stable dimensions".

An irreducible finite-dimensional representation of $\mathfrak{gl}_n(k)$ is defined by it's highest weight. This is the n-tuple of numbers $a_1 \leqslant \dots \leqslant a_n$, $a_i - a_j \in \mathbb{Z}$; $E_{ii}v = a_i v$ where v is the highest-vector of ρ_n. The identity matrix in $\mathfrak{gl}_n(k)$ multiplies $C.(\mathfrak{gl}_n(A), \varepsilon\rho_n)$ by $\sum_{i=1}^{n} a_i$. Therefore the homology groups $H.(\mathfrak{gl}_n(A); \varepsilon\rho_n)$ are trivial if $\sum a_i \neq 0$.

Without loss of generality we may suppose that ρ is contained in the decomposition of $\overset{k}{\otimes}(V\otimes V^*)$ for some k. We call ρ_n small if $2k \leqslant n$. So all a_i are integer.

Let us divide the set of numbers $a_1, \dots, a_n$ into two parts. Changing the numeration we may suppose that the first part is $a_1 \leqslant a_2 \leqslant \dots \leqslant a_s < 0$ and the second is $a_{s+1} \geqslant a_{s+2} \geqslant \dots \geqslant a_n > 0$. Construct the Young diagrams $¶_1(\rho_n), ¶_2(\rho_n)$: $¶_1$ consists of rows with lengths $a_1, \dots, a_s$; $¶_2$ consists of rows with lengths $a_{s+1}, \dots, a_n$. We shall characterize ρ_n by this pair of diagrams and write $\rho_n = \rho_n(¶_1, ¶_2)$. For a diagram ¶ let $|¶|$ be the number of its cells. Note that $|¶_1| = |¶_2| = k$.

<u>Lemma 4.8.1</u>. $H_i(\mathfrak{gl}_n(A), \varepsilon\rho_n(¶_1,¶_2)) \xrightarrow{\sim} H_i(\mathfrak{gl}_{n+1}(A), \varepsilon\rho_{n+1}(¶_1,¶_2))$ if $i + |\cdot¶_1| < n-1$.

<u>Proof</u>. Recall that the Lie algebra homology of an arbitrary $\mathfrak{g}$ with values in a module M (which is equal to $\mathrm{Tor}^{U(\mathfrak{g})}(k,M)$) is computed by means of the standard complex $C.(\mathfrak{g}, M)$: $C_n(\mathfrak{g}, M) = (\Lambda^n\mathfrak{g}) \otimes M$;

$$d((l_0 \wedge l_1 \wedge \dots \wedge l_{n-1}) \otimes m) = \sum_{i<j} (-1)^{i+j-1}[l_1, l_j] \wedge \dots \wedge \dots \wedge \hat{l}_i \wedge \dots \wedge \hat{l}_j \wedge \dots + \sum_{i=0}^{n-1} (-1)^{i-1}(l_0 \wedge \dots \hat{l}_i \wedge \dots l_{n-1}) \otimes l_i m \qquad (4.8.1)$$

(cf. [10]).

Consider the complex $C.(\mathfrak{gl}_n(A), \varepsilon\rho_n(¶_1,¶_2))^{\mathrm{inv}}$.

The invariant theory implies that in the dimensions $i < |¶_1| + n-1$ this complex depends only on $¶_1$ and $¶_2$ by not on n;

$$C_i(\mathfrak{gl}_n(A), \varepsilon\rho_n(\P_1,\P_2))^{inv} \xrightarrow{\sim} C_i(\mathfrak{gl}_{n+1}(A), \varepsilon\rho_{n+1}(\P_1,\P_2))^{inv}.$$

But the invariant subcomplex is quasi-isomorphic to the standard complex whence the Lemma.

The homology under the conditions of Lemma 4.8.1. we shall call $H.(\mathfrak{gl}(A);\ \varepsilon\rho(\P_1,\P_2))$.

Remark. The Lie algebra $\mathfrak{gl}(A)$ is the direct limit of the algebras $\mathfrak{gl}_n(A)$; the modules $\varepsilon\rho_n(\P_1,\P_2)$ form a direct system. We are computing the homology of this direct limit. The morphism $\mathfrak{gl}(A) \oplus \mathfrak{gl}(A) \to \mathfrak{gl}(A)$ gives the following pairing:

$$H.(\mathfrak{gl}(A),k) \otimes H.(\mathfrak{gl}(A), \varepsilon\rho(\P_1,\P_2)) \to H.(\mathfrak{gl}(A), \varepsilon\rho(\P_1,\P_2)).$$

We shall need the following tensor functor (in sense of Macdonald, cf. [45]) on the category of vector spaces. Let $W = \oplus W_i$, $i \in \mathbb{Z}$ be a graded vector space and $\P_1,\P_2$ be two Young diagrams, $|\P_1| = |\P_2| = l$. There is the following action of Σ_l on $\overset{l}{\otimes} W$:

$$\sigma(w_1 \otimes w_2 \otimes \ldots \otimes w_l) = (-1)^{\alpha}\, w_{\sigma^{-1}(1)} \otimes w_{\sigma^{-1}(2)} \otimes \ldots w_{\sigma^{-1}(l)}$$

where α is the number of pairs (i,j), $1 \leqslant i < j \leqslant l$, such that $\sigma(i) > \sigma(j)$ and $(-1)^{\deg w_i} = (-1)^{\deg w_j} = -1$. Consider the diagonal embedding $\Sigma_l \to \Sigma_l \times \Sigma_l$, $g \mapsto (g,g)$. Let $I_l(W)$ be the representation of $\Sigma_l \times \Sigma_l$ induced from that of Σ_l on $\overset{l}{\otimes} W$. Define the functor $F_{\P_1,\P_2}$ by the formula

$$F_{\P_1,\P_2}(W) = \mathrm{Hom}_{\Sigma_l \times \Sigma_l}(\Pi_{\P_1,\P_2};\ I_l(W)) \qquad (4.8.2)$$

where $\Pi_{\P_1,\P_2}$ means the exterior tensor product of irreducible representations corresponding to $\P_1,\P_2$.

We shall omit the symbol ε in the formulas.

Let $\bar{H}.(A,k)$ denote the reduced homology of the augmented algebra A : $\bar{H}.(A,k) = H_1(A,k) \oplus H_2(A,k) \oplus \ldots$; $H_i(A,k) = \mathrm{Tor}_i^A(k,k)$.

Theorem 4.8.3. The space $H.(\mathfrak{gl}(A), \rho(\P_1,\P_2))$ is a free $H.(\mathfrak{gl}(A),k)$-module generated by the graded space $F_{\P_1,\P_2}(\bar{H}.(A,k))$.

In particular, $H.(\mathfrak{gl}(A), \mathfrak{gl}(k)) \xrightarrow{\sim} H.(\mathfrak{gl}(A,k)) \otimes H.(A,k)$.

Proof. First we find $H.(\mathfrak{gl}(A), \mathfrak{gl}(k))$. We shall prove that the $\mathfrak{gl}$-invariant subcomplex in $C.(\mathfrak{gl}(A), \mathfrak{gl}(k))$ is isomorphic to $C.(A,k) \otimes S^{\cdot}(C^{cycl}_{\cdot}(A))$. Indeed, define a homomorphism

$$\varphi : C.(\mathfrak{gl}(A), \mathfrak{gl}(k)) \to C.(A,k) \qquad (4.8.3)$$

as follows:

$$\varphi((m_o \otimes a_o) \wedge \ldots \wedge (m_{k-1} \otimes a_{k-1}) \otimes n) = \sum tr(m_{\sigma o} \ldots m_{\sigma(k-1)}) sgn(\sigma) \otimes$$

$$\otimes (a_{\sigma o} \otimes \ldots \otimes a_{\sigma(k-1)}),$$

where $m_o, \ldots, m_{k-1}$ are elements of $\mathfrak{gl}(k)$. It is easily seen that is the morphism of the complexes. We have constructed a map (4.3)

$$C.(\mathfrak{gl}(A), k) \longrightarrow S^{\cdot}(C^{cycl}_{\cdot}(A)) \qquad (4.8.4)$$

Consider a commutative diagram

$$\begin{array}{ccc} C.(\mathfrak{gl}(A), \mathfrak{gl}(k)) \otimes C.(\mathfrak{gl}(A), k) & \longrightarrow & C.(A,k) \otimes S^{\cdot}(C^{cycl}_{\cdot}(A)) \\ \uparrow & \nearrow & \\ C.(\mathfrak{gl}(A), \mathfrak{gl}(k)) & & \end{array} \qquad (4.8.5)$$

The upper row is the tensor product of (4.8.3) and (4.8.4). The vertical arrow is the natural $C.(\mathfrak{gl}(A))$-comodule structure of $C.(\mathfrak{gl}(A), \mathfrak{gl}(k))$. The composition of these two arrows being restricted to the subcomplex $C.(\mathfrak{gl}(A), \mathfrak{gl}(k))^{inv}$ gives the desired isomorphism. This fact may be easily deduced from the invariant theory. Thus $H.(\mathfrak{gl}(A), \mathfrak{gl}(k)) \xrightarrow{\sim} H.(\mathfrak{gl}(A)) \otimes H.(A,k)$.

The homology comultiplication is obtained from one on the standard complex. We shall need the operation

$$\mu : C.(\mathfrak{gl}(A), \overset{s}{\otimes}\mathfrak{gl}(k)) \longrightarrow \overset{s}{\otimes} C.(\mathfrak{gl}(A), \mathfrak{gl}(k)).$$

Compose this with the map $\overset{s}{\otimes}\varphi$. We obtain a homomorphism

$$\theta : C.(\mathfrak{gl}(A), \overset{s}{\otimes}\mathfrak{gl}(k)) \longrightarrow \overset{s}{\otimes} C.(A,k).$$

Now recall that $\overset{s}{\otimes}\mathfrak{gl}(k) = (\overset{s}{\otimes} V) \otimes (\overset{s}{\otimes} V^*)$. The latter is a $(\Sigma_s \times \Sigma_s)$-module; the symmetric group Σ_s (embedded into $\Sigma_s \times \Sigma_s$ diagonally) acts by permutation of the factors in $\overset{s}{\otimes}\mathfrak{gl}(k)$.

The map θ clearly commutes with this diagonal action (if we put that Σ_s acts at $C.(A,k)$ by permutation of factors). If we compose θ with an automorphism given by a nondiagonal element of $\Sigma_s \times \Sigma_s$ we obtain just a new map. It is easy to see that θ may be extended to a map

$$\hat{\theta} : C.(\mathfrak{gl}(A), \overset{s}{\otimes}\mathfrak{gl}(k)) \longrightarrow I_s(\overset{s}{\otimes} C.(A,k));$$

this commutes with the natural action of $\Sigma_s \times \Sigma_s$.

Taking a composition

$$C.(\mathfrak{gl}(A), \overset{s}{\otimes}\mathfrak{gl}(k))^{inv} \longrightarrow C.(\mathfrak{gl}(A), \overset{s}{\otimes}\mathfrak{gl}(k)) \longrightarrow C.(\mathfrak{gl}(A), \overset{s}{\otimes}\mathfrak{gl}(k)) \otimes$$

$$\otimes\, C.(\mathfrak{gl}(A)) \twoheadrightarrow I_s(\overset{s}{\otimes} C.(A,k)) \otimes S^{\cdot}(C_{\cdot}^{cycl}(A))$$

we obtain (once more using the invariant theory) an isomorphism

$$C.(\mathfrak{gl}(A), \overset{s}{\otimes}\mathfrak{gl}(k))^{inv} \longrightarrow I_s(\overset{s}{\otimes} C.(A)) \otimes S^{\cdot}(C_{\cdot}^{cycl}(A)).$$

The Kunneth isomorphisms imply

$$H.(\mathfrak{gl}(A), \overset{s}{\otimes}\mathfrak{gl}(k)) \xrightarrow{\sim} I_s(H.(A,k)) \otimes H.(\mathfrak{gl}(A)).$$

Now look how the representation $\overset{s}{\otimes}\mathfrak{gl}(k)$ is decomposed into a direct sum of irreducible ones. This decomposition contains $\rho(\P_1,\P_2)$ where $|\P_1| = |\P_2| \leqslant s$. Put in correspondence to a pair (i, j) a homomorphism $u_{ij} : \overset{s}{\otimes}\mathfrak{gl}(k) — \overset{s-1}{\otimes}\mathfrak{gl}(k)$. Identify $\overset{s}{\otimes}\mathfrak{gl}(k)$ with $(\overset{s}{\otimes} V) \otimes \overset{s}{\otimes} V^*)$; then

$$u_{ij}(v_1 \otimes v_2 \otimes \ldots \otimes v_s \otimes v_1^* \otimes \ldots \otimes v_s^*) = (v_1 \otimes \ldots \otimes \hat{v}_i \otimes \ldots \otimes v_s \otimes v_1^* \otimes \ldots \otimes$$

$$\otimes\, \hat{v}_j^* \otimes \ldots \otimes v_s^*). \langle v_i, v_j^*\rangle.$$

Put $S = \bigcap_{i,j} \ker u_{ij}$. It is easily deduced from H.Weyl's results (cf. [64]) that S is a direct sum of the representations $\rho(\P_1,\P_2)$ where $|\P_1| = |\P_2| = s$ and the multiplicity of $\rho(\P_1,\P_2)$ is equal to the product of the dimensions of the irreducible representations of Σ_s corresponding to $\P_1,\P_2$. In other words, $\Sigma_s \times \Sigma_s$ and $\mathfrak{gl}(k)$ generate the centralizers for each other. It is easily seen from the what was said above that

$$H.(\mathfrak{gl}(A), S) \simeq I_s(\bar{H}.(A,k)) \otimes H.(\mathfrak{gl}(A)).$$

To show this it suffices to write down the operators u_{ij} on the space $I_s(H.(A,k))$; we omit it. The last equality implies Theorem 4.8.3.

Remark 4.8.4. This theorem might be deduced from Proposition 3.3.1. Namely, construct an algebra $A \oplus M$ where M is an 1-dimensional vector space and the product of (a_o, m_o) and (a_1, m_1) is equal to $(a_o a_1;\ \varepsilon(a_o)m_1 + m_o\varepsilon(a_1))$. Write down the Hochschild-Serre spectral sequence of $\mathfrak{gl}(A \oplus M)$ with respect to the ideal $\mathfrak{gl}(M)$. The second term is isomorphic to the limit; it is equal to $H.(\mathfrak{gl}(A), \Lambda^{\cdot}(\mathfrak{gl}(k) \otimes M))$. Let $m_1,\ldots,m_s$ be a basis in M. Take a polygrading of $\mathfrak{gl}(A \oplus M)$ by means of Z^s: m_i is of polydegree $(0,\ldots,1,\ldots,0)$ where 1 stands on the i-th place. Then it is easily verified that the space

$H.(gl(A), \overset{s}{\otimes} gl(k))$ is isomorphic to the term in $H.(gl(A \oplus M))$ of polydegree $(1, 1, \ldots, 1)$. But $K^+(A \oplus M) = K^+(A) \oplus H.(A,k) \otimes \tilde{M}$ where $M = \oplus(\overset{s}{\otimes} M/\mathrm{Im}(1-\tau))$.

Let D be an algebra without unit. Theorem 4.8.3. permits to calculate $H.(gl(D))$. The Lie algebra $gl(k)$ acts on the standard complex $C.(gl(D))$ and this one is the direct sum of the isotypical components $C.(gl(D))^{\rho}$, $\rho = \rho(\P_1, \P_2)$. $C.(gl(D))^{\rho} = \mathcal{K}.(D, \rho) \otimes \rho$ where $\mathcal{K}.(D, \rho) \simeq \mathrm{Hom}_{gl(k)}(\rho, C.(gl(D)))$. Denote the homology of $K(D, \rho)$ by $H.(gl(D), \rho)$.

Let $\tilde{D} = D \oplus k$ be the algebra obtained from D by adjoining a unit and $\varepsilon : \tilde{D} \to k$ the natural augmentation.

<u>Proposition 4.8.5</u>. $H.(gl(D), \rho) = H.(gl(D), gl(k); \varepsilon\rho^*)$.

This follows directly from the definition of the relative homology.

<u>Example 4.8.6</u>. Let $D = x\,k[x]$; we consider it as a graded algebra, $\deg x = 1$. Then $D = k[x]$; $H_0(k[x], k) \simeq k$; $H_1(k[x], k) \simeq k$; $H_i(k[x], k) = 0$, $i > 1$. Applying Proposition 4.8.5 and Theorem 4.8.3 we see that $H_k(gl(x\,k[x]), \rho(\P_1, \P_2)) = 0$ if $\P_1 \neq \P_2^*$ and $k \neq |\P_1|$; if $k = |\P_1|$, then the space $F_{\P,\P^*}(H.(k[x], k))$ is one-dimensional, where $\P^*$ is dual diagram. As usually, the information on such homology spaces may be used to obtain the Macdonald-Kac type identities. These express the equality of the Euler characteristic of the Lie algebra homology to this of its standard complex. More precisely, the complex $C.(gl(D))$ is the direct sum of the components $\mathcal{K}.(\rho, i) \otimes \rho$, $i \in \mathbb{N}$, where $\mathcal{K}.(\rho, i) \otimes \rho$ is the part of $\mathcal{K}.(D, \rho) \otimes \rho$ consisting of the chains of degree i. Write down the Poincaré series

$$\varphi(t) = \sum \rho \cdot t^i \chi(\mathcal{K}.(\rho, i))$$

where χ means the Euler characteristic. The function φ is an element of $R[[t]]$ where R is the representations ring of gl; φ may be factorized in a usual way:

$$\varphi(t) = \prod_{i=1}^{\infty} (1 - t^i gl(k) + t^{2i} \Lambda^2 gl(k) - \ldots) \qquad (4.8.6)$$

If we write down similarly the Poincaré series for the Lie algebra homology of $gl(x\,k[x])$ we obtain

$$\varphi(t) = \prod_{i=1}^{\infty} (1 + t^i)\Big(\sum_{\P : k = |\P|} (-1)^k t^k \rho(\P, \P^*)\Big) \qquad (4.8.7)$$

Note that the algebra $H^*(sl_\infty(x\,k[x]))$ is generated by $H^1(sl_\infty(x\,k[x])) \simeq (sl_\infty(k))^*$. The system of relations in $H^*(sl_\infty(xk[x]))$

s the space $P \subset \Lambda^2 H^1(\mathfrak{sl}_\infty(x\, k[x]))$, where $P \cong (\mathfrak{sl}_\infty(k))^*$ and the map $\mathfrak{sl}_\infty(k)^* \to (\Lambda^2 \mathfrak{sl}_\infty(k))^*$ is dual to natural map $\Lambda^2 \mathfrak{sl}_\infty(k) \to \mathfrak{sl}_\infty(k)$.

This may be deduced from the usual Macdonald-Kac identities for he current algebras $\mathfrak{sl}_n$, $n \to \infty$. Therefore we call the equality 4.8.6) = (4.8.7) the stable Macdonald-Kac identity. We shall not disuss such identities for another algebras D; note that they have inteesting combinatorial consequences.

Now we pass to the computation of the Lie algebra homology of $\mathfrak{gl}(A)$ with values in the adjoint representation. Let ¶ be a Young iagram and ¶(V) be the tensor functor: $¶(V) = P_¶(\otimes^{|¶|} V)$, where ¶ is the projection onto the isotypical component corresponding to ¶.

Let $\sigma \in \Sigma_s$; represent it as a product of the cycles: $\sigma = \sigma_1 \ldots \sigma_r$. ut $L(\sigma) = \overset{r}{\otimes} H.(A,A)$ and $L = \bigoplus_{\sigma \in \Sigma_s} L(\sigma)$. We write $h = h_{\sigma_1} \otimes \ldots \otimes h_{\sigma_r}$ or a monomial in $\overset{r}{\otimes} H.(A,A)$. Consider the action of Σ_s on the spae L:

$$\nu h = h_{\nu\sigma_1\nu^{-1}} \otimes \ldots \otimes h_{\nu\sigma_r\nu^{-1}} \in L(\nu\sigma\nu^{-1}).$$

<u>Theorem 4.8.7</u>. $H.(\mathfrak{gl}(A), ¶(\mathfrak{gl}(A))$ is a free $H.(\mathfrak{gl}(A), k)$-module enerated by the graded space $\mathrm{Hom}_{\Sigma_{|¶|}}(¶, L)$.

<u>Remark</u>: In the last formula the Young diagram and the corresponing representation of the symmetric group are denoted by the same etter ¶.

<u>Proof</u>. Consider the algebra $k \oplus M$, where M is a vector space ith a basis $m_1, \ldots, m_s$, $m_i m_j = 0$, $1 \leqslant i, j \leqslant s$. The K^+-functors f $A \otimes (k \oplus M)$ may be computed in two different ways: with use either roposition 3.3.1 or Proposition 5.5.1. Proposition 3.3.1 claims that $K^+_\cdot(A \oplus (A \otimes M)) \simeq K^+_\cdot(A) + \mathrm{Nil}^+_\cdot(A \otimes M[-1])$. We have

$$\mathrm{Nil}^{(i)}_\cdot(A \otimes M[-1]) \simeq H.(A,A) \otimes \left(\otimes^i M / \mathrm{Im}(1 - \tau)\right)$$

see Chapter 3).

Write down the Hochschild-Serre spectral sequence of the algebra $(A \oplus (A \otimes M))$ with respect to the ideal $\mathfrak{gl}(A \otimes M)$. It degenerates n the second term which is equal to $H.(\mathfrak{gl}(A), \Lambda^\cdot(M \otimes \mathfrak{gl}(k)))$. Introuce a polygrading of $A \oplus (A \otimes M)$ by the lattice $\mathbb{Z}^s$ as in Remark .8.4: $\deg A = (0,0,\ldots,0)$; $\deg m_i = (0,\ldots,1,0,\ldots,0)$ where the nit stands at the ith place. The Lie algebra $\mathfrak{gl}(A \oplus (A \otimes M))$ inherits the polygrading. The homology groups $H.(\mathfrak{gl}(A), \otimes^s \mathfrak{gl}(A))$ are somorphic to the $(1,1,\ldots,1)$ - polyhomogeneous components of $H.(\mathfrak{gl}(A \oplus (A \otimes M)))$. This implies that $H.(\mathfrak{gl}(A), \otimes^s \mathfrak{gl}(A)) \simeq$

$\simeq H.(\mathfrak{gl}(A)) \otimes L$. This clearly implies the statement of Theorem 4.8.7.

Corollary 4.8.8.

1) $H.(\mathfrak{gl}(A), S^{\cdot}(\mathfrak{gl}(A))) \xrightarrow{\sim} S^{\cdot}(\bigoplus_{i=0} [H_0(A,A] \oplus \dots \oplus H_i(A,A)]) \otimes H^{\cdot}(\mathfrak{gl}(A))$

2) $H.(\mathfrak{gl}(A), \Lambda^{\cdot}(\mathfrak{gl}(A))) \xrightarrow{\sim} S^{\cdot}(\bigoplus_{i=0} (H_i(A,A) \oplus H_{i-2}(A,A) + \dots]) \otimes H^{\cdot}(\mathfrak{gl}(A))$

(here the right hand sides mean the symmetric algebras of the graded spaces; the grading is usual: $\deg h = 1$ for $h \in H_1(A,A)$).

Note that $H.(\mathfrak{gl}(A), S^{\cdot}(\mathfrak{gl}(A))) \xrightarrow{\sim} H.(\mathfrak{gl}(A \otimes k[\xi]))$ where $k[\xi]$ is a commutative superalgebra with one generator, $\mathfrak{gl}(A \otimes k[\xi])$ is a Lie superalgebra (cf. [39] for the definition of the Lie superalgebra homology). On the other hand, $H.(\mathfrak{gl}(A), \Lambda^{\cdot}(\mathfrak{gl}(A))) \xrightarrow{\sim} H.(\mathfrak{gl}(A \otimes x(k[x]/(x^2)))$.

Remark 4.8.9. It is interesting to compare Corollary 4.8.8 with the well known case $A = k$. Say, $H.(\mathfrak{gl}(k), S^{\cdot}(\mathfrak{gl}(k)) \xrightarrow{\sim} H.(\mathfrak{gl}(k)) \otimes S^{\cdot}(\mathfrak{gl}(k))^{inv}$. Thus we have computed (in some sense) the analogues of the invariant polynomials on $\mathfrak{gl}$ for an arbitrary A (cf. Remark after Proposition 5.4.1).

Chapter 5. Operations in additive K-theory

5.1. In this Subsection we describe multiplications in additive K-theory. Begin with the construction of exterior multiplications, i.e. of the operations connecting $K^+_{\cdot}(A)$, $K^+_{\cdot}(B)$ and $K^+_{\cdot}(A \otimes B)$. Compute at first the K^+-functors of the tensor product of two free algebras. We suppose for simplicity that $\operatorname{char} k = 0$.

It will be convenient for us to deal sometimes with the algebras without unit. Put for such an algebra $K^+_{\cdot}(A) = \ker (K^+_{\cdot}(\tilde{A}) \longrightarrow K^+_{\cdot}(k))$ where $\tilde{A}$ is the algebra obtained by adjoining a unit. Note that $K^+_{\cdot}(A) = K^+_{\cdot -1}(k \longrightarrow A)$.

Proposition 5.1.1. Let R_1, R_2 be two free algebras without unit. Then

$$K^+_1(R_1 \otimes R_2) \simeq K^+_1(R_1) \otimes K^+_1(R_2) = R_1/[R_1,R_1] \otimes R_2/[R_2,R_2];$$

$$K^+_2(R_1 \otimes R_2) \simeq K^+_1(R_1 \otimes R_2);$$

$$K^+_i(R_1 \otimes R_2) = 0, \quad i > 2.$$

Proof. Put $R = R_1 \otimes R_2$. According to Theorem 1.2.5 there exists a spectral sequence converging to $K^+_{\cdot}(R)$ which has the second term equal to $H.(R,R) \otimes K^+_{\cdot}(k)$. In our case $H.(R,R) = H.(R_1,R_1) \otimes H.(R_2,R_2)$, i.e. $H_i(R,R) = 0$ when $i > 2$. This spectral sequence is periodical with period 2 because the differentials acting from $H_o(R,R) \otimes K^+_{2i+1}(k)$ to $H_1(R,R) \otimes K^+_{2i-1}(k)$ and from $H_o(R,R) \otimes K^+_{2i+1}(k)$ to $H_1(R,R) \otimes K^+_{2i-1}(k)$ do not depend on i. Form a complex with use these differentials:

$$0 \to H_o(R,R) \xrightarrow{B} H_1(R,R) \xrightarrow{B} H_2(R,R) \to 0 \qquad (5.1.1)$$

It is easy to see (cf. 1.3) that this is the tensor product of the complexes $0 \to H_o(R_i,R_i) \xrightarrow{B_i} H_1(R_i,R_i) \to 0$, $i = 1,2$. The operators B_i were computed in the proof of Lemma 2.3.1; in our case (i.e. when char $k = 0$) they are bijective. It follows that the complex (5.1.1) is acyclic. So our spectral sequence calculating $K^+_{\cdot}(R)$ degenerates in the third term; $E^3_{ij} = 0$, if $i \geqslant 1$ or $j > 2$; $E^3_{01} = H_o(R_1,R_1) \otimes H_o(R_2,R_2)$; $E^3_{02} = \operatorname{coker}(H_o(R_1,R_1) \otimes H_o(R_2,R_2) \to H_o(R_1,R_1) \otimes H_1(R_2,R_2) \oplus H_1(R_1,R_1) \otimes H_o(R_2,R_2))$; but B_1, B_2 are bijective, and it is easy to see that $E^3_{02} \simeq E^3_{01}$.

<u>Proposition 5.1.2</u>. Let A, B be algebras without unit. There is an exact sequence

$$\dots \to K^+_i(A \otimes B) \to \bigoplus_{\alpha+\beta=i+1} K^+_\alpha(A) \otimes K^+_\beta(B) \xrightarrow{\sigma} \bigoplus_{\alpha+\beta=i-1} K^+_\alpha(A) \otimes K^+_\beta(B) \to K^+_{i-1}(A \otimes B) \to \dots \qquad (5.1.2)$$

The map σ is given by the formula $\sigma = S \otimes 1 + 1 \otimes S$, S is the Bott periodicity map (cf. 1.2).

<u>Proof</u>. Let $R_1 \to A$ and $R_2 \to B$ be free resolutions of A and B over k. Then there is the spectral sequence converging to $K^+_{\cdot}(A \otimes B)$ with the first term $\tilde{K}^+_{\cdot}(R_1 \otimes R_2)$ (see Lemma 2.2.1). We may apply Proposition 5.1.1 (with the obvious changes in the graded case); $\tilde{K}^+_i(R_1 \otimes R_2) \simeq R_1/[R_1,R_1] \otimes R_2/[R_2,R_2]$ if $i = 1, 2$ and $\tilde{K}^+_i(R_1 \otimes R_2) = 0$ if $i > 2$. Compute the second term of the spectral sequence. We must find the homology of the differential acting in $K^+_i(R_1 \otimes R_2)$, $i = 1, 2$. Both complexes have homology equal to $K^+_{\cdot}(A) \otimes K^+_{\cdot}(B)$; there is a differential in the second term; all the differentials in the higher terms are obviously equal to zero. So our exact sequence (5.1.2) is constructed of the second term and the limit of the spectral sequence in a usual way; the map σ is equal to the differential acting in E^2.

To compute it we give another description of the periodicity homomorphism. Let C be an algebra without unit and $R \to C$ be its resolution. Write down the spectral sequence with the first term $\tilde{H}.(R,R)$ (see Lemma 2.2.1). Then $\tilde{H}_o(R,R) \simeq \tilde{H}_1(R,R)$; the homology groups of this complex are exactly the K^+-functors of C. The differential in the second term is the Bott homomorphism (the verification is easy). This gives the proof of Proposition 5.1.2.

The formula (5.1.2) shows that there are two operations in additive K-theory:

a) the exterior multiplication $K_i^+(A) \otimes K_j^+(B) \longrightarrow K_{i+j}^+(A \otimes B)$;

b) the exterior comultiplication $K_i^+(A \otimes B) \longrightarrow K_\alpha^+(A) \otimes K_\beta^+(B)$, $\alpha + \beta = i + 1$. The first was introduced by Loday and Quillen in [43] and the second by Connes in [13]. It is easily deduced from the definitions that the multiplication is associative and the comultiplication is coassociative. The analogous construction gives the operations in the relative additive K-functors:

$$K_i^+(A_1 \to A_2) \otimes K_j^+(B_1 \to B_2) \longrightarrow K_{i+j-1}^+(A_1 \otimes B_1 \longrightarrow A_2 \otimes B_2);$$

$$K_{i+j}^+(A_1 \otimes A_2 \longrightarrow B_1 \otimes B_2) \longrightarrow K_i^+(A_1 \to A_2) \otimes K_j^+(B_1 \to B_2).$$

If A is commutative then the multiplication $A \otimes A \longrightarrow A$ is a homomorphism; composition $K_\cdot^+(A) \otimes K_\cdot^+(A) \longrightarrow K^+(A \otimes A) \longrightarrow K_\cdot^+(A)$ turns $K_\cdot^+(A)$ into a skew-commutative graded algebra.

Proposition 5.1.3. 1) The multiplication $K_i^+(k) \otimes K_j^+(k) \longrightarrow K_{i+j}^+(k)$ is zero;

2) the comultiplication $K_{i+j+1}^+(k) \longrightarrow K_i^+(k) \otimes K_j^+(k)$ induces a coalgebra structure on $K_\cdot^+(k)$; the dual maps turn $(K_\cdot^+(k))^*$ into an algebra of polynomials generated by an element from $K_3^+(k)^*$;

3) let A be a k-algebra; then the map

$$K_i^+(A) \longrightarrow K_i^+(A \otimes k) \longrightarrow K_{i-2}^+(A) \otimes K_3^+(k) \xrightarrow{\sim} K_{i-2}^+(A)$$

is the Bott homomorphism.

These statements are deduced straightforward from the definitions; the proof of the statement 3) uses the description of the periodicity homomorphism used in the proof of Proposition 5.1.2.

Let A be a Hopf algebra. Then the comultiplication $A \longrightarrow A \otimes A$ is a ring homomorphism; hence it induces the map $K^+(A) \longrightarrow K^+(A \otimes A)$. Composing it with the exterior comultiplication we see that there is a structure of a skew commutative coalgebra on $K_{\cdot-1}^+(A)$ (note that it is necessary to shift the grading; for instance, the comultiplication sends $K_1^+(A)$ to $K_1^+(A) \otimes K_1^+(A)$; this map $A/[A,A] \to A/[A,A] \otimes A/[A,A]$

s induced by the comultiplication $A \xrightarrow{\Delta} A \otimes A$).

The preceding construction may be generalized in such a way. Let be a Hopf algebra and B a coalgebra with an action of A. This means hat there is an algebra homomorphism $\varphi : B \to A \otimes B$ such that the iagram

$$\begin{array}{ccc} A \otimes B & \xrightarrow{id_A \otimes \varphi} & A \otimes A \otimes B \\ \varphi \uparrow & & \uparrow \Delta \otimes id_B \\ B & \xrightarrow{\varphi} & A \otimes B \end{array}$$

s commutative. A standard example arises when B is a function ring t a G-space X and A is dual to the group algebra of G; φ is induced by $G \times X \to X$. In this situation we may define a homomorphism

$$K_i^+(B) \to K_i^+(A \otimes B) \to \bigoplus_{\alpha+\beta=i+1} K_\alpha^+(A) \otimes K_\beta^+(B)$$

This means that $K_\cdot^+(B)$ is a $K_\cdot^+(A)$-comodule.

Note that all previous results may be generalized immediately to the case of graded algebras.

Let $k[\xi]$ be a $\mathbb{Z}$-graded free skew commutative algebra with a generator ξ of degree one. Then $A \otimes k[\xi]$ is also a graded algebra. The space $\tilde{K}_i^+(A \otimes k[\xi]) = \tilde{K}_{i,\cdot}^+(A \otimes k[\xi])$ is a graded space; define a Hopf algebra structure on $k[\xi]$: $\Delta(\xi) = \xi \otimes 1 + 1 \otimes \xi$.

Proposition 5.1.4. 1) Let A be a k-algebra. Then

$$\tilde{K}_i^+(A \otimes k[\xi]) \simeq K_i^+(A) \oplus \left(\bigoplus_{\alpha=0}^{i-1} H_\alpha(A,A)\right)$$

2) The dual graded space $\tilde{K}_{\cdot+1,\cdot}^+(k[\xi])^*$ is an algebra with the base $1, y, y^2, \ldots; \eta_1, \eta_2, \ldots;$ $y^i \in \tilde{K}_{2i+1,0}^+(k[\xi])^*$; $\eta_i \in \tilde{K}_{i,i}^+(k[\xi])^*$.

The multiplication is given by the formulas

$$\eta_i\, y^k = \eta_i \eta_j = 0, \quad k > o; \quad y^i \cdot y^j = y^{i+j}.$$

The elements η_i form a base in the kernel of the map

$$\tilde{K}_i^+(k[\xi])^* \to K_i^+(k)^*.$$

2)$'$ The multiplication in $\tilde{K}^+(k[\xi])$ is described by the following formulas:

$$y_{2i+1}\, y_{2j+1} = y_{2i+1}\, \eta^j = 0; \quad \eta^i \eta^j = \eta^{i+j}.$$

Here the base $\langle y_{2i+1};\ \eta^j \rangle$ is dual to one $\langle y^i;\ \eta_j \rangle$ in $\tilde{K}^+(k[\xi])^*$.

To prove 1) compute the spectral sequence from Proposition 2.2.3.

The complexes in the second term are by Proposition 1.3.1. the truncated complexes $(H.(A,A); B) \otimes (\tilde{H}.(k[\xi], k[\xi]); B)$.

). But it is clear that in the second tensor factor $\tilde{H}_i(k[\xi], k[\xi])$ there is a base formed by the homology classes of the chains $\omega_0 = \xi\otimes\ldots\otimes\xi$, $\omega_1 = \xi\otimes\ldots\otimes\xi\otimes 1$; $B\omega_1 = 0$; $B\omega_0 = (i+1)\omega_1$; an easy computation shows that the spectral sequence for $\ker(\tilde{K}^+_\cdot(A \otimes k[\xi]) \to K^+_\cdot(A))$ degenerates in the third term equal to $E^3_{ij} = 0$, $i > 0$; $E^3_{oj} = \bigoplus_{\alpha=0}^{j} H_\alpha(A,A)$. The additive structure of $\tilde{K}^+_\cdot(k[\xi])$ is the particular case of 1). The chain corresponding to η^i is $\xi\otimes\xi\otimes\ldots\otimes\xi \in C_i^{cycl}(k[\xi])$. The formulas for the multiplication and the comultiplication follow from the direct description of these operations given below (Proposition 5.1.7).

Remark. Consider a Lie superalgebra $\mathfrak{gl}(k[\xi])$. Calculate the Hochschild-Serre spectral sequence of the pair $(\mathfrak{gl}(k[\xi]), \mathfrak{gl}(k))$. The second term is $H.(\mathfrak{gl}(k)) \otimes H.(\mathfrak{gl}(k[\xi]); \mathfrak{gl}(k))$. The relative homology $H.(\mathfrak{gl}(k[\xi]); \mathfrak{gl}(k))$ is isomorphic to the space of the invariant polynomials on $(\mathfrak{gl}(k))^*$. These form a free commutative algebra generated by the elements of degree 1, 2, 3,...; thus $H.(\mathfrak{gl}(k[\xi])) \simeq \Lambda^\cdot(\xi_1, \xi_3, \ldots) \otimes S^\cdot(\bar{\eta}_1, \bar{\eta}_2, \ldots)$; $\xi_{2i+1} \in H_{2i+1}(\mathfrak{gl}(k[\xi]))$ are odd and $\eta_i \in H_i(\mathfrak{gl}(k[\xi]))$ are even. The result on $\tilde{K}^+_\cdot(k[\xi])$ follows from this calculation.

Remark 5.1.5. Introduce a differential ∂ acting in $k[\xi]$: $\partial(\xi) = 1$. So $\tilde{K}^+_{i,j}(k[\xi])$ may be regarded as a complex; it is easily seen that the differential in it is equal to zero for $j > 1$; $\partial(\eta^1) = y_1$. So we may compute the Massey operation $\langle\partial, \partial, \eta^2\rangle$, i.e., regard $\partial\eta^2$ as a differential of a chain from the standard complex for $\tilde{K}^+_{\cdot,\cdot}$ and apply ∂ to this chain. It may be shown that $\langle\partial, \partial, \eta^2\rangle = y_3$. Analogously, $\langle\partial, \partial, \partial, \eta^3\rangle = y_5$; etc. Clearly these are the differentials in the spectral sequence $E^1_{\cdot\cdot} = \check{K}^+_{\cdot\cdot}(k[\xi]) \Rightarrow \tilde{K}^+_\cdot(k[\xi])$ applied to the elements η_i; $\tilde{K}^+_\cdot(k[\xi]) = 0$ because $H.(k[\xi]; \partial) = 0$ (Lemma 2.2.1). So the differentials in this spectral sequence are the "higher derivations" with respect to the variable ξ.

The Hopf algebra $k[\xi]$ acts naturally on $A \otimes k[\xi]$. Therefore $\tilde{K}^+_\cdot(A \otimes k[\xi])$ is a $\tilde{K}^+_\cdot(k[\xi])$-comodule. Furthermore, there is a natural map $(A \otimes k[\xi]) \otimes k[\xi] \to A \otimes k[\xi]$ whence $\tilde{K}^+_\cdot(A \otimes k[\xi])$ is a $\check{K}^+_\cdot(k[\xi])$-module and hence a $\tilde{K}^+(k[\xi])^*$-module. Now we describe these structures; represent $\tilde{K}^+_\cdot(A \otimes k[\xi])$ by the following table:

	j=0		j=1		j=2	
i=4	$K_4^+(A)$	$\longleftarrow$	$H_3(A,A)$	$\longleftarrow$	$H_2(A,A)$ ...	
i=3	$K_3^+(A)$	$\longleftarrow$	$H_2(A,A)$	$\longleftarrow$	$H_1(A,A)$	...
i=2	$K_2^+(A)$	$\longleftarrow$	$H_1(A,A)$	$\longleftarrow$	$H_o(A,A)$	0
i=1	$K_1^+(A)$	$\longleftarrow$	$H_o(A,A)$		0	

The entry (i, j) contains the space $\tilde{K}^+_{i,j}(A[\xi])$.

Proposition 5.1.6. 1) The operator y^1 acts by zero on all the components $H.(A,A)$; the map $y^1 : K^+_{\cdot}(A) \to K^+_{\cdot-2}(A)$ is the periodicity homomorphism; the operator y^k is the k-th power of the operator y^1.

2) The operator η_1 acts from the space in the entry (i, j) to one in the entry (i - l + 1; j - 1). If $j > 1$, then $\eta_1 : H_{i-j}(A,A) \to H_{i-j+1}(A,A)$ coincides with the differential B from 1.2. If j=1, then $\eta_1 : H_{i-1}(A,A) \to K_i^+(A)$ is induced by the projection $C.(A,A) \to C^{cycl}_{.+1}(A)$ (see 1.2).

3) The derivation ∂ of the algebra $k[\xi]$ induces the map $\tilde{K}_i^+(k[\xi] \otimes A) \to \tilde{K}_i^+(k[\xi] \otimes A)$ equal to η_1.

4) The operators $y_{2i+1} \in \tilde{K}^+_{2i+1}(k[\xi])$ act by zero.

5) The operator η^1 acts from the space in the entry (i,j) to one in the entry (i+1, j+1). If $j > 1$ then η^1 is identical; if $j = 1$ then $\eta^1 : K_1^+(A) \to H_1(A,A)$ coincides with the map in the exact sequence from 1.2.8; the operator η^l is the l-th power of the operator η^1.

The usual arrows in the table denote the operator η_1 and the dotted ones the operator η_2. The fact that the operator B generalizing the usual de Rham differential may be obtained as a multiplication operator by an element from $K_1^+(k[\xi])^*$ seems interesting to us.

We do not give the full proof of Proposition 5.1.6. Consider for example the homomorphism η_1. First of all the definitions imply directly that η_1 coincides with the action of the derivation ∂ (the statement 3) of Proposition); in particular, $(\eta_1)^2 = 0$. It is easy to see that η_1 and η^1 commute with each other. Suppose that we have already proved the statement 5). Then the equality $\eta_1\eta^1 = \eta^1\eta_1$

implies that η^1 maps the complex in the i-th row of the table to one in the (i+1)-th row. It is clear also that the map $H_i(A,A) \to$ $\to K^+_{i+1}(A)$ is induced by the homomorphism $C_{\cdot+1}(A,A) \longrightarrow C^{cycl}_{\cdot}(A)$. We have shown in Chapter 1 that the operator B may be decomposed as in the upper row of the following diagram:

$$H_i(A,A) \longrightarrow K^+_{i+1}(A) \xrightarrow{\eta^1} H_{i+1}(A,A)$$

$$H_i(A,A) \xrightarrow[\sim]{\eta^1} H_i(A,A) \xrightarrow{\eta_1} H_{i+1}(A,A)$$

The equality $B = \eta_1$ follows from the commutativity of this diagram. The statement 5) may be deduced from the direct formulas for the multiplication and comultiplication discussed in Proposition 5.1.7. below.

As in Remark 5.1.5, construct the differential algebra $A \otimes k[\xi]$, $\partial(a \otimes \xi) = a$, $\partial(a \otimes 1) = 0$. Then there is a spectral sequence $E^1_{\cdot\cdot} = \tilde{K}^+_{\cdot}(A \otimes k[\xi]) \Rightarrow 0$. The differential in the first term is the multiplication operator by η_1; the differentials in the higher terms are the "higher multiplications by η_1", i.e., the Massey operations $\langle \eta_1, \ldots, \eta_1, \alpha \rangle$, $\alpha \in \tilde{K}^+_{\cdot}(A \otimes k[\xi])$. Thus the first operation is the multiplication $\eta_1 : H_i(A,A) \longrightarrow H_{i+1}(A,A)$; the next one acts from the kernel of η_1 to the cokernel of $\eta_1 : H_{i+2}(A,A) \longrightarrow H_{i+3}(A,A)$, etc. All this situation may be drawn schematically:

$$0 - H_0(A,A) \to H_1(A,A) \to H_2(A,A) \to H_3(A,A) \to H_4(A,A) \to \ldots$$

The usual arrows denote the maps $\alpha \mapsto \eta_1\alpha$; the dotted ones denote the maps $\alpha \longmapsto \langle \eta_1, \eta_1, \alpha \rangle$ and $\alpha \mapsto \langle \eta_1, \eta_1, \eta_1, \alpha \rangle$. It is natural to call such a thing the "de Rham complex of A". The K^+-functors are obtained by the truncations of it; this relates our definition to the D-cohomology theory (cf. [2]).

Note also that as we had seen the relative additive K-functors $\tilde{K}^+_i(A \longrightarrow A \otimes k[\xi])$ are equal to $H_i(A,A) \oplus \ldots \oplus H_0(A,A)$. Therefore there is a spectral sequence $E^1_{ij} = \tilde{K}^+_{i,j}(A \longrightarrow A \otimes k[\xi]) \Rightarrow K^+_{i+j}(A \longrightarrow 0)$; which coincides with the spectral sequence from Theorem 1.2.4.

Now we shall give a direct description of the exterior multiplication. Fix some notations. Let C be a linear space; put

$$sh_l : (\overset{i_0}{\otimes} C) \otimes \ldots \otimes (\overset{i_{l-1}}{\otimes} C) \longrightarrow \overset{i_0+\ldots+i_{l-1}}{\otimes} C; \quad sh_l(c_0 \otimes \ldots \otimes c_{i_0-1},$$

$c_{i_0} \otimes \ldots \otimes c_{i_0+i_1-1}, \ldots, c_{i_0+\ldots+i_{l-2}} \otimes \ldots \otimes c_{i_0+\ldots+i_{\ell-1}-1}) =$

$$= \sum \operatorname{sgn}(\sigma)\, c_{\sigma^{-1}0} \otimes \ldots \otimes c_{\sigma^{-1}(i_0+\ldots+i_{\ell-1}-1)};$$

the sum is taken over all $\sigma \in \Sigma_{i_0+\ldots+i_{s-1}}$ such that $\sigma(k) < \sigma(k_1)$ if for some s $i_0+\ldots i_s < k < k_1 < i_0+\ldots+i_{s+1}-1$. Furthermore, define an operator

$$N : \overset{i}{\otimes} C \longrightarrow \overset{i}{\otimes} C,\ N(c_0 \otimes \ldots \otimes c_{i-1}) = \frac{1}{i}\sum_{k=0}^{i-1} (-1)^{(i-1)(k-1)} c_k \otimes \ldots$$

$$\ldots \otimes c_i \otimes \ldots \otimes c_0 \otimes \ldots \otimes c_{i-1}.$$

Let $\alpha = a_0 \otimes \ldots \otimes a_{i-1} \in C_i^{cycl}(A)$; $\beta = b_0 \otimes \ldots \otimes b_{j-1} \in C_j^{cycl}(B)$. Put $\alpha \cup \beta = sh_2(N((a_0 \otimes 1) \otimes \ldots \otimes (a_{i-1} \otimes 1));\ N((1 \otimes b_0) \otimes \ldots \otimes (1 \otimes b_{j-1}))$.

Proposition 5.1.5. The operation $\cup$ induces a well defined homomorphism $C^{cycl}_{\cdot}(A) \otimes C^{cycl}_{\cdot}(B) \longrightarrow C^{cycl}_{\cdot}(A \otimes B)$. The induced map in homology is the exterior multiplication.

Proof. The verification that $\cup$ is a morphism of complexes is straightforward. To show that the induced map coincides with the multiplication it suffices to show that it is so when A, B are free. We omit the details.

Remark 5.1.6. The definition of the exterior multiplication might be given in an another way with use Theorem 4.1.2. This way is analogous to the definition of the multiplication in K-theory due to Loday (cf. [41]).

The direct formula for the comultiplication may be given using Connes' approach with cyclic objects. We have

$$K^+_{i+1}(A) \xrightarrow{\sim} H_i(\Lambda^\circ, A^\natural)$$

(cf. Appendix): now, if char $k = 0$, $H.(\Lambda^\circ, A^\natural)$ may be calculated using an arbitrary Δ°-acyclic resolution of A . For example, $R^\natural$ is such a resolution.

Now, the map $R_A \otimes R_B \longrightarrow R_{A\otimes B}$ induces a map $R^\natural_A \otimes R^\natural_B \longrightarrow R^\natural_{A\otimes B}$. This is a chain map of acyclic resolutions. This shows that the comultiplication constructed above is the usual homological operation

$$H.(\Lambda^\circ, A^\natural) \otimes H.(\Lambda^\circ, B^\natural) \longrightarrow H.(\Lambda^\circ, A^\natural \otimes B^\natural) =$$

$$= H.(\Lambda^\circ, (A \otimes B)^\natural).$$

(dual to Yoneda multiplication). This enables one to give an explicit formula.

For instance

$$\Delta((a_0 \otimes b_0) \otimes (a_1 \otimes b_1)) = a_0 a_1 \otimes (b_0 \otimes b_1) + (a_0 \otimes a_1) \otimes (b_0 b_1).$$

Proposition 5.1.7. Let γ be a cycle in $\bar{C}_n^{cycl}(A \otimes B)$ (see Proposition 2.4.2). Then $\Delta\gamma$ is a cycle in $\bar{C}_{\cdot}^{cycl}(A) \otimes \bar{C}_{\cdot}^{cycl}(B)$ representing the coproduct of the homology class of γ in $K_{n-1}^{+}(k \to A)$.

The proof is simple; we omit the details. A similar formula is contained in [13].

We finish this Subsection by the following example. Let $A = k[x_1,\dots,x_n]$. It is a commutative cocommutative Hopf algebra: $\Delta(x_i) = x_i \otimes 1 + 1 \otimes x_i$. Describe the operations introduced above on $K_{\cdot}^{+}(k \to A)$. We have seen in Corollary 1.4.2 that $K_j^{+}(k \to A) = \Omega^i/d\Omega^{i-1}$, $i \geqslant 0$; $K_0^{+}(k \to A) = A/k$. Let $\omega \in \Omega^i$; we write $\tilde{\omega}$ for the corresponding class in $\Omega^i/d\Omega^{i-1}$. Then the formula for the $\cup$-multiplication together with the formula for the operator μ (see the proof of Theorem 1.4.1) show immediately that

$$\tilde{\omega}_1 \cup \tilde{\omega}_2 = \widetilde{\omega_1 \cdot d\omega_2}$$

The results of Chapter 6 show that the same formula gives the exterior product in the cohomology of the truncated de Rham complexes which is equal to $K^{+}(k \to k[X])$, X smooth, and in the cohomology of the Hodge complexes for an arbitrary X (cf. Chapter 6).

Now describe the comultiplication. There is one on $\Omega^{\cdot}$: $x_i \longmapsto x_i \otimes 1 + 1 \otimes x_i$, $dx_1 \mapsto dx_1 = dx_i \otimes 1 + 1 \otimes dx_i$; it is easy to see that the induced operation in $\Omega^{\cdot}/d\Omega^{\cdot}$ is the comultiplication defined above. A similar construction may be carried out for an arbitrary noetherian commutative algebra. We omit the details.

5.2. Now we want to discuss the operations in $H_{\cdot}(\mathfrak{gl}(A), k)$ when A is a commutative algebra.

Let τ be a representation of the Lie algebra $\mathfrak{gl}_n(k)$; $N = \dim\tau$. Then a Lie algebra homomorphism $\mathfrak{gl}_n(A) \to \mathfrak{gl}_N(A)$ is defined. Consider the induced morphism

$$C_{\cdot}(\mathfrak{gl}_n(A))^{inv} \to C_{\cdot}(\mathfrak{gl}_N(A))^{inv}.$$

The equivalent representations give the same morphisms.

We suppose $\tau = \P_1 \oplus \dots \oplus \P_k$, $\P_i$ are the irreducible representations corresponding to some Young diagrams. We shall characterize τ by a set of the Young diagrams or (that is the same) by a non-negative function f on the set $\mathfrak{h}$ of the Young diagrams having finite support; f associates to a diagram the multiplicity of the corresponding irreducible representation. We write $\tau = \tau(f)$.

It is easy to show that there exists a morphism of the complexes $\mathcal{K}_.(A,n) \to \mathcal{K}_.(A,N)$ (cf. 4.4) such that the following diagram is commutative:

$$\begin{array}{ccc} \mathcal{K}_i(A,n) & \longrightarrow & \mathcal{K}_i(A,N) \\ \downarrow & & \downarrow \\ C_i(\mathfrak{gl}_n(A))^{inv} & \longrightarrow & C_i(\mathfrak{gl}_N(A))^{inv} \end{array}$$

Indeed, the set of Young diagrams is fixed and the number n varies; then N depends on n polynomially (cf. 4.4). Fix an i; then for a sufficiently large n the vertical maps are bijective. We define the upper arrow to make the whole diagram commutative. Then the map $\mathcal{K}_i(A,n) \to \mathcal{K}_i(A,N)$ depends on n polynomially; thus, we may define it for an arbitrary n.

So, we associate to an arbitrary non-negative integer finitely supported function on $\mathfrak{h}$ a morphism of complexes

$$\theta(f) : \mathcal{K}.(A,n) \to \mathcal{K}.(A, N(f,n));$$

here $N(f,n)$ is the polynomial on n; for $n \in \mathbb{Z}_+$, $n >> 0$ $N(f,n)$ is the dimension of the representation $\tau(f)$ of $\mathfrak{gl}_n(k)$.

Fix a finite subset $X \subset \mathfrak{h}$. Consider the functions f with support in X. Then $\theta(f)$ may be regarded as a linear map depending on the parameters $f(x)$, $x \in X$. It is clear that this dependence is also polynomial (the number $N(f,n)$ depends on f linearly). This allows to define a map $\theta(f)$ for an arbitrary k-valued finitely supported function f on $\mathfrak{h}$. Now let f be such that $N(f,n) = n$, $\forall n$. Then $\theta(f)$ maps the complex $\mathcal{K}.(A,n)$ to itself; such $\theta(f)$ generate an algebra of transformations of the complex $\mathcal{K}.(A,n)$. It is clear that $\theta(f)$ acts on the homology groups of the complex $\mathcal{K}.(A,n)$.

Define a map $\rho(f,n)$ to make the following diagram commutative:

$$\begin{array}{ccc} \mathcal{K}.(A,n) & \longrightarrow & \mathcal{K}.(A, N(f,n)) \\ \downarrow & & \downarrow \\ S^{\cdot}(C_{\cdot}^{cycl}(A)) & \xrightarrow{\rho(f,n)} & S^{\cdot}(C_{\cdot}^{cycl}(A)) \end{array}$$

The vertical arrows are defined in 4.4. Let $\theta(f)$ correspond to a usual (not virtual) representation of $\mathfrak{gl}_n(k)$. Then for a sufficiently large n the vertical arrows are bijective. Thus, we really may define $\rho(f)$. In the general case the map $\rho(f)$ depends rationally on n. Composing it with the maps $C^{cycl}_{\cdot}(A) \to S^{\cdot}(C^{cycl}_{\cdot}(A))$, $S^{\cdot}(C^{cycl}_{\cdot}(A)) \to$ $\to C^{cycl}(A)$ we obtain the operations in additive K-theory. We shall give a detailed exposition of the ring generated by the operations $\rho(f,n)$ in $S^{\cdot}(C^{cycl}_{\cdot}(A))$ elsewhere.

Now describe briefly a method of studying the operations $\rho(f,n)$. Let $\tau: \mathfrak{gl}_n(k) \to \mathfrak{gl}_N(k)$ be a finite-dimensional representation. Consider the map

$$I_s(\tau) : [\overset{s}{\otimes}\mathfrak{gl}_n]^{inv} \longrightarrow [\overset{s}{\otimes}\mathfrak{gl}_N]^{inv}.$$

It commutes with the natural action of Σ_s at the invariant space. Furthermore the direct sum $\overset{\infty}{\underset{s=1}{\oplus}}(\overset{s}{\otimes}\mathfrak{gl}_n)$ is the free algebra and $\underset{s=1}{\oplus}(\overset{s}{\otimes}\mathfrak{gl}_n)^{inv}$ is subalgebra. The map

$$\underset{s}{\oplus} I_s(\tau) = I(\tau) : (\underset{s}{\oplus} \overset{s}{\otimes}\mathfrak{gl}_n)^{inv} \longrightarrow (\underset{s}{\oplus}\overset{s}{\otimes}\mathfrak{gl}_N)$$

is the homomorphism of algebras. Restrict it to the subalgebra $S^{\cdot}(\mathfrak{gl}_n)^{inv}$ of symmetrical invariants. By the Harish-Chandra theorem, the restriction of the polynomials to the Cartan subalgebra gives the isomorphism of $S^{\cdot}(\mathfrak{gl}_n)^{inv}$ onto the algebra $k[\delta_1,\dots,\delta_n]$ of symmetrical polynomials on n variables. Thus we obtain a map $J(\tau) : k[\delta_1, \dots, \delta_n] \to k[\delta_1,\dots,\delta_N]$.

<u>Lemma 5.2.1</u>. The map $I(\tau)$ is uniquely defined by $J(\tau)$. Indeed, consider the homomorphism $J(\tau)$. Suppose that we know $I_s(\tau)$ for all $s \leqslant i$; define a map $\alpha : \overset{s}{\otimes}\mathfrak{gl}_n \to \overset{s-1}{\otimes}\mathfrak{gl}_n$ as follows:

$$(m_o \otimes \dots \otimes m_{s-1}) = [m_o, m_1] \otimes m_2 \otimes \dots \otimes m_{s-1}.$$

Clearly $I_s(\tau)\alpha = \alpha I_{s-1}(\tau)$. It is easy to show that there exist no more then one map $I_i(\tau) : (\overset{s}{\otimes}\mathfrak{gl}_n)^{inv} \longrightarrow (\overset{s}{\otimes}\mathfrak{gl}_N)^{inv}$ such that:

1) $I_i(\tau)$ commutes with Σ_s;

2) $I_i(\tau)$ coincides with $I(\tau)$ on $S^i(\mathfrak{gl}_n)^{inv}$;

3) $I_i(\tau)$ is compatible with the multiplication in $\underset{s}{\oplus}(\overset{s}{\otimes}\mathfrak{gl}_n)^{inv}$.

This proves the Lemma.

Clearly we may suppose in the previous construction that τ is a virtual representation. Namely, consider the diagram

$$\begin{array}{ccc} S^i(\mathfrak{gl})^{inv} & \longrightarrow & S^i(\mathfrak{gl})^{inv} \\ \uparrow & & \uparrow \\ S^i(\mathfrak{gl}_n)^{inv} & \xrightarrow{J(\tau)} & S^i(\mathfrak{gl}_n)^{inv} \end{array}$$

For a sufficiently large n and a usual representation τ the vertical maps are bijective and we may define the upper horizontal map to make the diagram commutative. This map depends on the parameters rationally; therefore we may continue it "analytically" and obtain a definition for a virtual τ. So we may associate to an arbitrary virtual representation τ an endomorphism of the graded algebra $k[\sigma_1, \sigma_2, \ldots]$, $\deg \sigma_i = i$. In turn, such an endomorphism determines an operation

$$\mu : S^{\cdot}(C^{cycl}_{\cdot}(A)) \to S^{\cdot}(C^{cycl}_{\cdot}(A)).$$

It is easy to construct the operations which are the Hopf algebra endomorphisms, i.e., $\mu C^{cycl} \subset C^{cycl}$. For this it is necessary that the corresponding map $k[\sigma_1, \ldots] \to k[\sigma_1, \ldots]$ would be also a Hopf algebra endomorphism. Say it more precisely. The algebra $k[\sigma_1, \sigma_2, \ldots]$ consists of the invariant polynomials over $\mathfrak{gl}(k)$. The operation $\mathfrak{gl}(k) \oplus \mathfrak{gl}(k) \longrightarrow \mathfrak{gl}(k)$ turns it into a coalgebra. The elements σ_i are primitive with respect to the comultiplication. Then it is easy to see that the operations $C^{cycl}_{\cdot}(A) \to C^{cycl}_{\cdot}(A)$ may be determined by the set of numbers a_i, $\sigma_i \mapsto a_i \sigma_i$.

Now describe the operations of Adams and Grothendieck following the scheme of [40], [53].

5.3. Let $R(\mathfrak{gl}_n(k))$ be the virtual representation ring of $\mathfrak{gl}_n(k)$. There is a natural restriction map $R(\mathfrak{gl}_n(k)) \to R(\mathfrak{gl}_{n-1}(k))$. Denote by $R(\mathfrak{gl})$ the inverse limit of the system $R(\mathfrak{gl}_n(k))$.

Let τ and ρ be the usual representations of $\mathfrak{gl}_n(k)$ and τ_A, ρ_A - the compositions $C_{\cdot}(\mathfrak{gl}(A))^{inv} \xrightarrow{\sim} S^{\cdot}(C^{cycl}_{\cdot}(A)) \to C^{cycl}(A)$. Then $(\tau \oplus \rho)_A = \tau_A + \rho_A$. Therefore we may define a map

$$\alpha_n : R(\mathfrak{gl}_n(k)) \to \operatorname{Hom}(C_{\cdot}(\mathfrak{gl}_n(A)))^{inv}, C^{cycl}(A)) \to$$

$$\to \operatorname{Hom}(\operatorname{Prim} C_{\cdot}(\mathfrak{gl}_n(A))^{inv}, C^{cycl}(A)).$$

Passing to the inverse limits we obtain

$$\alpha : \varprojlim R(\mathfrak{gl}_n(k)) \simeq R(\mathfrak{gl}) \to \operatorname{End} C^{cycl}_{\cdot}(A).$$

Remark 5.3.1. There is a homological comultiplication in additive K-theory generalizing these maps:

$$H_{i+j-1}(\mathfrak{g} \otimes A, k) \to K_i^+(U(\mathfrak{g})) \otimes K_j^+(A)$$

for an arbitrary Lie algebra $\mathfrak{g}$. In our case $\mathfrak{g} = \mathfrak{gl}$; $i = 1$.

Lemma 5.3.2. Let $\rho_1, \ldots, \rho_l \in R(\mathfrak{gl}_n(k))$ and $f \in k[x_1, \ldots, x_l]$. Then

$$\alpha_n(f(\rho_1, \ldots, \rho_l)) = \sum_{i=1}^{l} \left(\frac{\partial f}{\partial x_i}\right)(\mathrm{rk}\,\rho_1, \ldots, \mathrm{rk}\,\rho_l)\,\alpha_n(\rho_i).$$

The proof may be easily deduced from the definitions, and we leave this to the reader.

Denote the standard n-dimensional representation of $\mathfrak{gl}_n(k)$ by 1_n; let Λ^i the exterior power functor; denote $\lambda^i = \Lambda^i(1_n - n)$. For an arbitrary polynomial we may define an element $f(\lambda^1, \ldots, \lambda^k) \in R(\mathfrak{gl})$. Note that $\mathrm{rk}\,\lambda^i = 0$; so the polynomial gives the same endomorphism of $C_{\cdot}^{cycl}(A)$ as its linear part.

Recall that the Grothendieck operations are by definition such γ^l, $l \in \mathbb{N}$, that

$$\sum_{l=0}^{\infty} \gamma^l t^l = \sum_{l=0}^{\infty} \lambda^l (t/(1-t))^l, \qquad \gamma^0 = 1.$$

In the following statement the complex $C_{\cdot}^{cycl}(A)$ is regarded as a sub-complex in $D_{\cdot}(A)$ (cf. 1.2.8).

Proposition 5.3.3. The action of γ^l on $C_{\cdot}^{cycl}(A)$ is given by the following formula:

$$\gamma^l(a_0 \otimes \ldots \otimes a_{i-1}) = \frac{(-1)^l}{l} \sum_{0<\rho_0<\ldots<\rho_{l-1}<l} sh_l((a_0 \otimes \ldots \otimes a_{\rho_0-1}), \ldots, (a_{\rho_{l-1}} \otimes \ldots \otimes a_{l-1}))$$

Proof. We must compute directly the maps

$$(\overset{i}{\otimes}\mathfrak{gl}_n)^{inv} \to (\overset{i}{\otimes}\mathfrak{gl}_N)^{inv}, \quad N = \binom{n}{k}, \quad \mathfrak{gl}_N = \mathrm{End}(\Lambda^k(k^n)),$$

induced by $m \mapsto m \wedge 1 \wedge \ldots \wedge 1 + \ldots + 1 \wedge 1 \wedge \ldots \wedge m$, $m \in \mathfrak{gl}_n$. The following formula is valid:

$$\mathrm{tr}((\overbrace{m_0 \wedge 1 \ \ldots \wedge 1}^{k\ \text{times}} + \ldots + 1 \wedge 1 \wedge \ldots \wedge m_0) \ldots (m_{i-1} \wedge 1 \wedge \ldots \wedge 1 + \ldots + 1 \wedge \ldots \wedge m_{i-1})) =$$

$$= \sum_{l=1}^{k} \frac{\binom{n}{k}}{\binom{n}{l}} \binom{k}{l} \sum_{\substack{I_0, \ldots, I_{l-1} \neq \emptyset \\ \{0,1,\ldots,i-1\} = I_0 \sqcup \ldots \sqcup I_{l-1}}} \mathrm{tr}(m_{I_0} \wedge m_{I_1} \wedge \ldots \wedge m_{I_{l-1}}) =$$

$$= \sum_{l=0}^{k} \binom{n-1}{k-1} \Big(\sum_{\substack{I_0,\dots,I_{l-1} \neq \emptyset \\ \{0,1,\dots,i-1\} = I_0 \sqcup \dots \sqcup I_{l-1}}} \frac{(-1)^i}{i} \operatorname{tr}(m_{I_0} \dots m_{I_{l-1}}) + \text{(decomposable}$$

form).

Here for an arbitrary subset $I \subseteq \{0,\dots,i-1\}$ we denote $m_I = m_{p_0} \dots m_{p_r}$, $I = \{p_0,\dots,p_r\}$, $p_0 < \dots < p_r$. Using this formula we may calculate the action of the exterior powers on $C^{cycl}_{\bullet}(A)$ and then the action of the Grothendieck operations. The details are left to the reader.

According to Lemma 5.3.1 all the operations are the linear combinations of γ^l. In particular, let Q_k be the Newton polynomials and $\psi^k = Q_k(\lambda^1,\dots,\lambda^k)$ the Adams operations. Note that the linear part of Q_k is equal to $(-1)^k k \lambda^k$; so $\psi^k = (-1)^k k \lambda^k$ in End $C^{cycl}_{\bullet}(A)$. But (if $\lambda_t = \sum \lambda^i t^i$ etc.)

$$\lambda_t = \gamma_{t/1-t} = \sum_{l,j} (-1)^j \binom{l+j-1}{l-1} \gamma^l t^{j+l} = \sum_{k,l} (-1)^{k-l} \gamma^l \binom{k-1}{l-1} t^k;$$

thus

$$(-1)^k k \lambda^k = \sum_l ((-1)^l l \gamma^l) \frac{k}{l} \binom{k-1}{l-1} = \sum_l ((-1)^l l \gamma^l) \binom{k}{l}$$

This provides a formula for the Adams operations. Denote by I the partition $\{0,\dots,i-1\} = I_0 \sqcup \dots \sqcup I_{k-1}$ (the order of the sets $I_0,\dots$ is essential). Let $\sigma_I \in \Sigma_i$ be a permulation such that

1) $i_1 \in I_{p_1}$, $i_2 \in I_{p_2}$, $p_1 < p_2 \Rightarrow \sigma_I i_1 < \sigma_I i_2$;

2) $\forall p$, $I_p = \{i_0,\dots,i_q\}$, $i_0 < i_1 < \dots < i_q$, $\sigma_I i_q = \sigma_I i_{q-1} + 1 = \dots = \sigma_I i_0 + q$.

Then we have

<u>Proposition 5.3.4</u>. $\psi^k(a_0 \otimes \dots \otimes a_{i-1}) = \sum_I \operatorname{sgn}(\sigma_I)\,(a_{\sigma_I(0)} \otimes \dots \otimes a_{\sigma_I(i-1)})$ in $C^{cycl}_{\bullet}(A) \subsetneqq D_{\bullet}(A)$. The sum is taken over all the partitions of $\{0,\dots,i-1\}$ into a disjoint union of k subsets. Note that the formula for $(-1)^k k \gamma^k$ is exactly the same, but the sets I_p are assumed to be non-empty.

<u>Proposition 5.3.5</u>. $\psi^k(\alpha \cup \beta) = \psi^k(\alpha) \cup \psi^k(\beta)$; $\psi^k \psi^l = \psi^{kl}$.

The proof is straightforward from 5.3.4.

<u>Proposition 5.3.6.</u> The Adams operation ψ^k being restricted to the subcomplex $F_nC^{cycl}(A)$ has the eigenvalues lying in the set $k, k^2, \dots, k^n$.

<u>Proof.</u> Note that $\gamma^l = 0$ on $F_nC^{cycl}(A)$, $l > n$.

$$\psi^k = \sum \binom{l}{i}(-1)^l(l\gamma^l);$$

thus $\psi^k|_{F^nC^{cycl}}$ is a polynomial of degree n on k: $\psi^k = \sum_{l=1}^{n} k^l \theta^l$. All the θ^l commute with each other; therefore we have

$$(\psi^k - k)\dots(\psi^k - k^n) = \sum_{i=0}^{n} \psi^{k^i} \sigma_{n-i}(k,\dots,k^n)(-1)^{i-n} = \sum_{l=1}^{n} \Big(\sum_{i=0}^{n} k^{li} \times$$

$$\times\, \sigma_{n-i}(k,\dots,k^n)(-1)^{n-i}\Big) = 0.$$

This ends the proof. We shall see in 6.5 that the eigenvalues of ψ^k in K_n^+ are equal to k^l, $[\frac{n}{2}] + 1 \leqslant l \leqslant n$, if A is a commutative noetherian algebra which is a quotient of a polynomial algebra by an ideal which is locally generated by a regular system of elements.

Chapter 6. Additive K-functors of the commutative noetherian algebras.

6.1. Let k be a field of characteristic zero, A be a commutative noetherian algebra over k, $X = \operatorname{Spec} A$; let also B be some polynomial algebra and $B \xrightarrow{\pi_0} A$ an epimorphism. Denote its kernel by I_0. We denote by $\Omega_B^\bullet$ the de Rham complex of B; let $F^\bullet$ be the Hodge filtration on $\Omega_B^\bullet$:

$$F^n\Omega_B^j = I_0^{n-j}\Omega_B^j, \quad n \geqslant j; \quad F^n\Omega_B^j = \Omega_B^j, \quad 0 \leqslant n \leqslant j.$$

The Leibnitz formula for the de Rham differential shows that $F^n\Omega_B^\bullet$ are subcomplexes of $\Omega_B^\bullet$. The cohomology groups of $\Omega_B^\bullet/F^{n+1}\Omega_B^\bullet$ are denoted by $H^\bullet_{cris}(X;n)$. It is well known (cf. [24]) that they do not depend on a choice of B and π_0. Put also $H^\bullet_{cris}(X) = H^\bullet(\varprojlim \Omega_B/F^{n+1}\Omega_B)$. These are the de Rham cohomology groups of a formal neighbourhood of X in $\operatorname{Spec} B$. If $k = \mathbb{C}$ then they are isomorphic to the usual cohomology groups of the space of closed points of X as a complex variety (maybe singular).

<u>Theorem 6.1.1.</u> 1) There are the functorial maps

$$\chi_{n,i} : K^{+}_{n+1}(A) \longrightarrow H^{n-2i}_{cris}(X;\ n-i)$$

for all n,i, such that $0 \leqslant 2i \leqslant n$; the diagrams

$$\begin{array}{ccc} K^{+}_{n+1}(A) & \xrightarrow{\ \chi_{n,i}\ } & H^{n-2i}_{cris}(X;\ n-i) \\ \Big\downarrow S & & \Big\downarrow \\ K_{n-1}(A) & \xrightarrow{\ \chi_{n-2,i-1}\ } & H^{n-2i}_{cris}(X;\ n-i-1) \end{array}$$

are commutative.

2) Assume that $A \cong B/I_o$, where B is a polynomial algebra and I_o is locally generated by a regular sequence. Then

$$\bigoplus_{0\leqslant 2i\leqslant n} \chi_{n,i} : K_{n+1}(A) \longrightarrow \bigoplus_{0\leqslant 2i\leqslant n} H^{n-2i}_{cris}(X;\ n-i)$$

is an isomorphism.

3) There is a functorial with respect to A isomorphism:

$$\chi : H^{\cdot}(A) \xrightarrow{\sim} H_{cris}(X)$$

(cf. 1.5), if A is finitely generated.

We shall prove this in 6.3-6.5. In the next Subsection we give some definitions which we use in the proof (and also in 6.6).

6.2. First of all we want to know the additive K-functors of a skew commutative free differential graded algebra P. Define the de Rham complex of P.

Let $P = \bigoplus_{n=-\infty}^{\infty} P_n$ be a skew commutative differential graded algebra; $\partial : R_n \longrightarrow R_{n-1}$, $n \in \mathbb{Z}$. Consider $F_P = P \otimes P$ as a right P-module. We denote an element $q \otimes p$ by the symbol $dq \cdot p$. Put $\Omega^1_P = F_P/\mathcal{K}$ where $\mathcal{K}$ is a submodule generated by the elements $d(pq) - dp\cdot q - (-1)^{\deg p \deg q}\, dq\cdot p$ for all homogeneous $p,q \in P$. Consider the tensor algebra of the graded module Ω^1_P over the graded ring P. Take its quotient by the ideal generated by the elements $\omega_1\omega_2 - (-1)^{\deg\omega_1 \deg\omega_2 + 1}\omega_2\omega_1$ for all homogeneous $\omega_1, \omega_2 \in \Omega^1_P$. We obtain a $\mathbb{Z}\oplus\mathbb{Z}$-graded skew commutative algebra $\Omega^{\cdot}_P$. More precisely, $\Omega^{\cdot}_P = \bigoplus_{\mathbb{Z}\ni i;\, j\geqslant 0} (\Omega^j_P)_i$ where the index i corresponds to the grading induced by one of P and the index j corresponds to the grading by number of factors from Ω^1_P. We call an element $\omega \in (\Omega^j_P)_i$ bihomogeneous and write $g_0(\omega) = i$, $g_1(\omega) = j$. For all bihomogeneous $\omega_0, \omega_1 \in \Omega_P$ we have

$$\omega_0 \omega_1 = (-1)^{g_0(\omega_0)g_1(\omega_1)+g_1(\omega_0)g_0(\omega_1)} \omega_1 \omega_0.$$

Now assume that the graded algebra P is the (skew) symmetric algebra of a graded space $V = \bigoplus_{i=-\infty}^{\infty} V_i$.

In this case, $\Omega_P^{\bullet}$ is the free $(\mathbb{Z} \oplus \mathbb{Z})$-graded skew-commutative algebra generated by the $(\mathbb{Z} \oplus \mathbb{Z})$-graded space $V \oplus dV$; $V_{i,o} = V_i$, $i \in \mathbb{Z}$; $V_{i,j} = 0$, $j \neq 0$; $(dV)_{i,1} \xrightarrow{\sim} V_i$; $(dV)_{i,j} = 0$, $j \neq 1$.

Furthermore, $F_P \cong P \otimes P$; so F_P is a complex. It is evident that $\mathcal{K}$ is a subcomplex. Therefore $\Omega_P^{\bullet}$ is a complex in a natural way. Moreover, the tensor algebra of Ω_P^1 and the ideal considered earlier are also the complexes. So there is a differential ∂' of bidegree $(-1, 0)$ acting on $\Omega_P^{\bullet}$. We have

$$\partial'(\omega_0 \omega_1) = \partial'\omega_0 \cdot \omega_1 + (-1)^{g_0(\omega_0)} \omega_0 \partial'\omega_1.$$

Let also d' be a map of bidegree $(0, 1)$ determined by conditions

$$d'(\omega_0\omega_1) = d'\omega_0 \cdot \omega_1 + (-1)^{g_1(\omega_0)}\omega_0 d'\omega_1, \quad \omega_0, \omega_1 \text{ are bihomogeneous;}$$

$$d'(dp) = 0, \quad dp \in \Omega_P^1; \quad d'p = dp \in \Omega_P^1, \quad p \in P.$$

So we obtain a $\mathbb{Z} \oplus \mathbb{Z}$ -graded skew-commutative algebra Ω_P together with the derivations ∂' of bidegree $(-1, 0)$ and d' of bidegree $(0, 1)$.

Define a quotient bicomplex $\overline{C}^{cycl}_{\cdot,\cdot}(P)$ of $C^{cycl}_{\cdot,\cdot}(P)$: $C^{cycl}_{n,m}(P) = (\overset{m}{\otimes}(P/k)/\mathrm{im}(1-\tau))_n$ (cf. 2.2). Let $\overline{C}^{cycl}_{\cdot}(P)$ be the simple complex associated to $\overline{C}^{cycl}_{\cdot,\cdot}(P)$; denote $K_n^+(k \to P) = H_{n+1}(\overline{C}^{cycl}_{\bullet}(P))$. Assume that either $P_n = 0$ for $n < 0$ or $P_n = 0$ for $n > 0$, $P_o = k$ and $P_{-1} = 0$. Then the graded version of 1.4.2 gives the quasi-isomorphism $\mu\colon \overline{C}^{cycl}_{\cdot}(P) \xrightarrow{\sim} \bigoplus_{n=0} (\Omega_P^n/d\Omega_P^{n-1})[-1-n]$; if $P_n = 0$ for $n < 0$, then the maps

$$C^{cycl}_{\cdot}(k) \longrightarrow C^{cycl}_{\cdot}(P) \longrightarrow \overline{C}^{cycl}_{\bullet}(P)$$

induce the long homology exact sequence.

6.3. It will be more convenient to change a little the definition of the de Rham complex $\Omega_P^{\cdot}$. Namely, let $\overline{\Omega}_P^{\cdot} = \Omega_P^{\cdot}$ as a vector space. We declare the product of bihomogeneous ω_0, ω_1 equal to

$(-1)^{g_0(\omega_0)\, g_1(\omega_1)}\omega_0\omega_1$; furthermore, let $g(\omega) = g_1(\omega) - g_0(\omega)$; $\bar{d}\omega = d'\omega$; $\bar{\partial}\omega = (-1)^{g_1(\omega)}\partial'\omega$. We have $\bar{d}\bar{\partial} + \bar{\partial}\bar{d} = 0$; the differential $\bar{\partial} + \bar{d}$ and the grading g turn $\bar{\Omega}^{\cdot}_P$ into a differential graded algebra. Denote by $G^k\bar{\Omega}_P$ the space of g-homogeneous elements. $\bar{\Omega}^j_P$ will denote the space of j-forms as above.

The differential $\bar{d} + \bar{\partial}$ is cohomological with respect to g. The values of g may be integer (not necessary $\geqslant 0$).

In our case (i.e., when $P = S(V)$) the graded algebra $\bar{\Omega}_P$ with the grading g is isomorphic to the free skew commutative graded algebra generated by the graded space $V' \oplus dV'$; $V'^i = V_{-i}$; $(dV)^i \xrightarrow{\sim} V_{1-i}$, $i \in \mathbb{Z}$.

Now return to the situation of 6.1. Consider a free resolution of A in the category of comm. differential graded algebras. Namely, let $P = (\bigoplus_{n=0}^{\infty} P_n;\ \partial : P_n \longrightarrow P_{n-1})$ be a differential graded algebra; assume that there is an epimorphism $\P : P \longrightarrow A$ such that:

a) $\P$ is a quasi-isomorphism;

b) the graded algebra P is isomorphic to the symmetric algebra of a graded space $V = \bigoplus_{n=0}^{\infty} V_n$.

The existence of such a resolution may be proved exactly in the same way as in Lemma 2.1.1. Furthermore, if A is fin. generated, we may choose P such that $\dim V_n < \infty$ for all n.

Consider the differential graded algebra $\bar{\Omega}^{\cdot}_P$ with the grading g and the cohomological differential $\bar{d} + \bar{\partial}$. If $\dim V_0 = m_0$, then $G^{m_0+1}\bar{\Omega}^{\cdot}_P = G^{m_0+2}\bar{\Omega}^{\cdot}_P = \ldots = 0$. We put $(P')^i = P_{-i}$.

Let $\bar{\Omega}^{\cdot}_P \longrightarrow P'$ be a differential graded algebra homomorphism: $p \longmapsto q$, $dp \longmapsto 0$. Composition with $\P$ gives the homomorphism $\bar{\Omega}^{\cdot}_P \longrightarrow A$. Put $I = \ker(\bar{\Omega}^{\cdot}_P \longrightarrow P')$; $J = \ker(\bar{\Omega}^{\cdot}_P \longrightarrow A)$; $B = P_0 = S(V_0)$; $I_0 = \ker(B \longrightarrow A)$.

<u>Definition 6.3.1.</u> $H^i(X,n) = H^i(\bar{\Omega}^{\cdot}_P / I^{n+1})$, $i \in \mathbb{Z}$.

The above considerations give the following

<u>Proposition 6.3.2.</u> There is a functorial with respect to A isomorphism:

$$K^+_{n+1}(A) \xrightarrow{\sim} H^n(X,n) \oplus H^{n-2}(X;\, n-1) \oplus \ldots \oplus H^{2-n}(X;\, 1) \oplus H^{-n}(X;\, 0).$$

The considered gradings, differentials, etc. are represented by the following diagram:

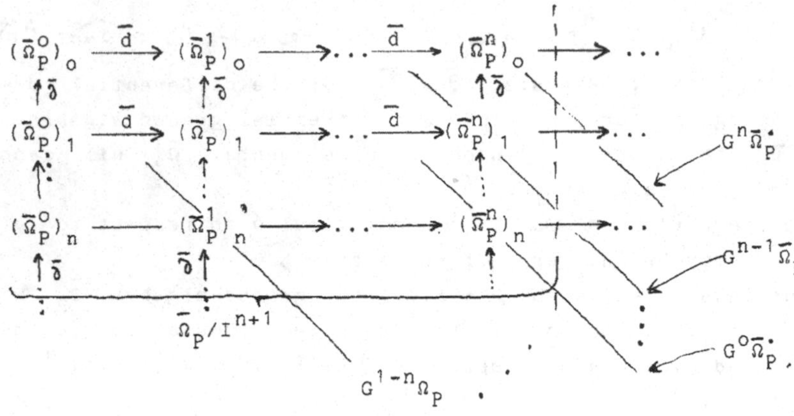

Consider now the embedding $B \to P$. Note that $F^{n+1}\Omega_B^{\cdot}$ is the $(n+1)$th power of the differential ideal $F^1\Omega_B^{\bullet} = \ker(\Omega_B^{\bullet} \to A)$.

Lemma 6.3.3. The induced map

$$\Omega_B^{\bullet}/F^{n+1}\Omega_B^{\cdot} \longrightarrow \bar{\Omega}_P^{\cdot}/J^{n+1}$$

is a quasi-isomorphism for all $n \geqslant 0$.

Proof. Let T_+ be the set of free generators of P whose degree is positive. The following isomorphism is valid for the differential ideal J^{n+1} regarded as a complex with differential $\bar{d}$:

$$J^{n+1} \xrightarrow{\sim} \bigoplus_{k+l+m=n+1} (I_o^k\, \Omega_B^l) \otimes \left(\bigoplus_{\{t_1,\ldots,t_m\}\subset T_+} (C(t_1) \underset{k}{\otimes} \cdots \underset{k}{\otimes} C(t_m)) \right)$$

where $C(t)$ is the de Rham complex of the graded algebra $k[t]$. The Künneth formula implies that the cohomology of J^{n+1} with respect to $\bar{d}$ is isomorphic to one of $F^{n+1}\Omega_B^{\cdot}$. Now, the spectral sequence computing the cohomology of J^{n+1} with respect to $\bar{d} + \bar{\partial}$ degenerates in the first term. Therefore $F^{n+1}\Omega_B^{\cdot} \xrightarrow{\sim} (J^{n+1}, \bar{d}+\bar{\partial})$ is a quasi-isomorphism. The map $\Omega_B^{\cdot} \to \bar{\Omega}_P^{\cdot}$ is also a quasi-isomorphism, and Lemma 6.3.3 follows from the five-lemma.

Consider now the obvious maps $\bar{\Omega}_P^{\cdot}/I^{n+1} \to \bar{\Omega}_P^{\cdot}/J^{n+1}$ $(I \subset J)$. Lemma 6.3.3 together with Proposition 6.3.2 give the homomorphisms

$$\chi_{n,i} : K_{n+1}^+(A) \to H^{n-2i}(X;\, n-i) \to H_{cris}^{n-2i}(X;\, n-i).$$

Their compatibility with the Bott morphisms is obvious. This gives the proof of the statement 1) in Theorem 6.1.1.

6.4. Now we prove the statement 2. We must show that under the assumption of Theorem 6.1.1, 2) $\bar{\Omega}_P^{\cdot}/I^{n+1} \to \bar{\Omega}_P^{\cdot}/J^{n+1}$ is a quasi-isomorphism.

Consider an ideal $L \subset \bar{\Omega}^{\cdot}_P$ generated by the elements db, $b \in B$. L is a differential ideal with respect to $\bar{d}$ and $\bar{\partial}$; $\bar{\Omega}^{\cdot}_{P/B} = \bar{\Omega}^{\cdot}_P/L$ by definition.

Consider a map

$$\bar{\Omega}^n_{P/B} \to (I^n_o/I^{n+1}_o)[-n]; \quad dp_o \dots dp_{n-1}\, p_n \mapsto \delta p_o \dots \delta p_{n-1} \cdot \P p_n$$

where $\delta p_i = \partial p_i \in I_o$ if $\deg p_i = 1$ and $\delta p_i = 0$ if $\deg p_i > 1$. If I_o is locally generated by a regular sequence, this map is a quasi-isomorphism (it has been shown in [48]). So, the spectral sequence of the double complex $\bar{\Omega}^o_{P/B} \to \dots \to \bar{\Omega}^n_{P/B} \to 0$ (i.e. $\bar{\Omega}^{\cdot}_P/L + I^{n+1}$) degenerates and we obtain $\bar{\Omega}^{\cdot}_P/L + I^{n+1} \xrightarrow{\sim} B/I^{n+1}_o$. Consider now the filtration by powers of $(L/L \cap I^{n+1})$. The i-th graded quotient is isomorphic to $\Omega^i_B \otimes_A (\bar{\Omega}^o_{P/B} \to \dots \to \bar{\Omega}^{n-i}_{P/B})$; we see that the spectral sequence corresponding to our filtration collapces; this gives the desired statement.

6.5. Now we pass to the proof of the statement 3). It suffices to show that $\varprojlim \bar{\Omega}^{\cdot}_P/I^n \xrightarrow{\sim} \varprojlim \bar{\Omega}^{\cdot}_P/J^m$ is a quasi-isomorphism. First note that $\varprojlim_m \bar{\Omega}^{\cdot}_P/(I^n + J^m) = (\bar{\Omega}^{\cdot}_P/I^n)^\wedge$ for any fixed n. The completion means here a usual one of a B-module in the I_o-adic topology. In fact, fix an $m \in \mathbb{Z}$; $G^m(\bar{\Omega}^{\cdot}_P/I^n) = \bigoplus_{i=0}^{n} (\bar{\Omega}^i_P/I^n)_{i+m}$; but J is generated by I_o and some other bihomogeneous elements having either $g_o > 0$ or $g_1 > 0$. So for $M \gg 0$

$$I^M_o\, G^m(\bar{\Omega}^{\cdot}/I^n) \subseteq G^m(J^M/(J^M \cap I^n)) \subseteq I^{M-p}_o\, G^m(\bar{\Omega}^{\cdot}/I^n)$$

where $p = \max(m+n, n)$.

For any $k \geqslant 0$ $(\bar{\Omega}^k_P, \bar{\partial})$ is a complex of free finite rank B-modules. Therefore

$$H.(\hat{\bar{\Omega}}^k_P) \simeq H.(\bar{\Omega}^k_P \otimes_B \hat{B}) \simeq H.(\bar{\Omega}^k_P) \otimes_B \hat{B} \simeq H.(\bar{\Omega}^k_P)^\wedge .$$

But it is clear that $I_o\, H.(\bar{\Omega}^k_P) = 0$. So $H.(\bar{\Omega}^k_P) \xrightarrow{\sim} H.(\hat{\bar{\Omega}}^k_P)$. The spectral sequence of the bicomplex $\bar{\Omega}^{\cdot}_P/I^n$ gives now a quasi-isomorphism

$$\bar{\Omega}^{\cdot}_P/I^n \xrightarrow{\sim} \varprojlim_m \bar{\Omega}^{\cdot}_P/(I^n + J^m).$$

Note also that the projections $\varprojlim_m \bar{\Omega}^{\cdot}_P/(I^{n+1} + J^m) \to \varprojlim_m \bar{\Omega}^{\cdot}_P/(I^n + J^m)$ are surjective again.

Consider now the complex $\varprojlim_n \varprojlim_m \bar{\Omega}^{\cdot}_P/(I^n + J^m)$. On one hand $H^{\cdot}(\varprojlim_n \varprojlim_m \bar{\Omega}^{\cdot}_P/(I^n + J^m)) = H^{\cdot}(\varprojlim_m \varprojlim_n \bar{\Omega}^{\cdot}_P/(I^n + J^m)) = H^{\cdot}(\varprojlim_m \Omega_P/J^m)$ since $I \subseteq J$; on the other hand we have seen that $\varprojlim_m \bar{\Omega}^{\cdot}_P/(I^n + J^m)$ and Ω_P/I^n are quasi-isomorphic inverse systems of complexes such that the maps in the inverse systems are surjective. The first circumstance shows that the total projective limits are quasi-isomorphic; the second one implies that these inverse systems have no higher projective limits. This ends the proof.

Corollary 6.5.1. Consider the decomposition from Proposition 6.3.1. The Adams operation ψ^k multiplies a summand $H^{n-2i}(X;\ n-i)$ by k^{n-i+1}.

Proof. Lemma 6.2.1 implies that there is a sequence

$$C^{cycl}_{\cdot}(k) \longrightarrow C^{cycl}_{\cdot}(P) \longrightarrow \bigoplus_{n=0} (\bar{\Omega}^n_P/d\bar{\Omega}^{n-1}_P)[-n-1]$$

giving a long homology exact sequence. So every homology class from $H^{n-2i}(X;\ n-i)$ may be represented by a cycle lying in $\overset{n-i+1}{\otimes} P/\mathrm{im}(1-\tau) \subset C^{cycl}(P)$ which is antisymmetric under all permutations of the tensor factors. Now our statement follows from the direct formula for the Adams operation given by Proposition 5.3.4 (with the evident changes in the graded case).

Corollary 6.5.2. Let X satisfy the assertion of Theorem 6.1.1, 2). Then the spectrum of the Adams operator $\psi^k : K^+_n(A) \longrightarrow K^+_n(A)$ contains in the set $k^n, \ldots, k^{[\frac{n}{2}]+1}$.

It was shown in 5.3.6 that this implies:

Corollary 6.5.3. Let A be a ring which satisfies the assertion of Theorem 6.1.1, 2) and $F.$ be a filtration defined in 4.4. Then

$$F_n K^+_{2n}(A) = F_n K^+_{2n+1}(A) = 0$$

Remark 6.5.4. Let $n \geqslant 1$. Then the Adams operation ψ^k on K^+_{n+1} has no eigenvectors with the eigenvalue k. This is an additive analogue of a theorem due to Kratzer [40] which states the same for the algebraic K-functors. The eigenspaces of ψ^k corresponding to the eigenvalue k^2 are the cohomology groups $H^{1-n}(X;\ 1) = D_n(A,A)$ $(n > 0)$ where $D.$ means the homology of A as a commutative ring in sence of Quillen, Harrison and André (cf. [48]). It seems to be interesting to study the lower graded factor of the γ-filtration on the Quillen K-functor (cf. [40], [53]), and its interplay with the homology of commutative rings. Cf. also Remark A 6.8.

Remark 6.5.5. The cohomology groups $H^{\cdot}(X, *)$ and their connection with crystalline cohomology were considered by L. Illusie in a more general situation (cf. [28]).

6.6. The K^+-functions of Sullivan models and Waldhausen K-theory. The results and methods of the present Chapter have an unexpected application to rational homotopy theory.

Let X be a simply connected topological space and $A^{\cdot}(X)$ be its Sullivan model (cf. [5]). We assume as usually that $\dim H^i(X;k) < \infty$ for all i.

If we want to apply the definitions of 2.2. and 6.2-6.3 to $A^{\cdot}(X)$ we must change the grading: $A_i(X)' := A^{-i}(X)$, $i \leqslant 0$; the multiplication and the differential ∂ are the same as in $A^{\cdot}(X)$.

It will be convenient to consider not just the complex $C^{cycl}_{\cdot}(A.(X)')$ but the reduced cyclic complex $\bar{C}^{cycl}_{\cdot}(A.(X)')$: $\bar{C}^{cycl}_{nm} = \left(\otimes^{n}(A.(X)'/k)\right)_m / \mathrm{im}(1-\tau)$. Put $K^+_{n-1}(k \to A.(X)') = H_n(\bar{C}^{cycl}(A.(X)'))$. If $A^{\cdot}(X)$ is the minimal model then the Hochschild homology and the relative K^+-functors of $A.(X)'$ may be non-zero only in the non-positive dimensions. So it is natural to put

$$K^n_+(k \to A^{\cdot}(A)) = K^+_{-n}(k \to A.(X)');$$

$$H^n(A^{\cdot}(X), A^{\cdot}(X)) = H_{-n}(A.(X)', A.(X)')$$

(do not mix with the Hochschild cohomology).

Theorem 6.6.1. 1) (Chen, [65])

$$H^{\cdot}(A^{\cdot}(X), A^{\cdot}(X)) \xrightarrow{\sim} H^{\cdot}(X^{S^1}; k);$$

$$2)\ K^{\cdot+1}_+(k \to A^{\cdot}(X)) \xrightarrow{\sim} \bar{H}^{\cdot}(ES^1 \times_{S^1} X^{S^1}; k).$$

Here X^{S^1} is the free loop space of X; ES^1 is the total space of the universal principal S^1-bundle; $\bar{H}^{\cdot}$ in the right hand side means the reduced singular cohomology.

Proof. Lemma 2.2.1 shows that we may restrict ourselves to the case when $A^{\cdot}(X)$ is the minimal model of X. There is a well known minimal model of X^{S^1} due to Sullivan and Vigue (cf. [55]). This is the algebra $\bar{\Omega}^{\cdot}_{A.(X)'}$ with the differential o and the grading

$$h(\omega) = g_0(\omega) + g_1(\omega)$$

(see 6.3).

Now to prove 1) we apply the standard method suggested in [48]. We sketch the proof: since char $k = 0$, the Hochschild complex of an arbitrary differential graded algebra $\mathfrak{A}$ is a Hopf algebra over $\mathfrak{A}$; so there is the canonical isomorphism between $C.(\mathfrak{A},\mathfrak{A})$ and the symmetric algebra over $\mathfrak{A}$ of the quotient complex of all decomposable elements; the latter is the Harrison complex of $\mathfrak{A}$. But the Harrison complex is quasi-isomorphic to $\mathfrak{A}\oplus \Omega^1_{\mathfrak{A}}[-1]$; this together with the Künneth formulas gives the desired statement (in our case $\mathfrak{A} = \bar{\Omega}^{\cdot}_{A.(X)'}$). The complete proof contains in [65].

Prove 2). There is the standard action of S^1 on X^{S^1}; at the level of models it is given by the homomorphism ($k[\xi]$ is the minimal model of S^1):

$$\bar{\Omega}^{\cdot}_{A.(X)'} \longrightarrow \bar{\Omega}^{\cdot}_{A.(X)'} \otimes k[\xi];$$

$\omega \mapsto \omega + \xi\cdot \bar{d}\omega$. Now, $H^{\cdot}(ES^1 \underset{S^1}{\times} X^{S^1})$ is equal to the cohomology of the following simplicial topological space:

$$X^{S^1} \underset{d_1}{\overset{d_0}{\leftleftarrows}} X^{S^1}\times S^1 \;\overleftarrow{\overleftarrow{\leftarrow}}\; X^{S^1}\times S^1\times S^1 \;\overleftarrow{\overleftarrow{\overleftarrow{\leftarrow}}}\; \cdots$$

Here $d_0(x, \theta_0,\dots,\theta_{n-1}) = (x\theta_0,\theta_1,\dots,\theta_{n-1})$; for $0 < i < n$ $d_i(x, \theta_0,\dots,\theta_{n-1}) = (x,\dots,\theta_{i-1}\cdot\theta_i,\dots)$; ...; $d_n(x, \theta_0,\dots,\theta_{n-1}) = (x, \theta_0,\dots,\theta_{n-1})$ where $x\in X^{S^1}$, $\theta_i \in S^1$.

Therefore, $H^{\cdot}(ES^1 \underset{S^1}{\times} X^{S^1})$ is isomorphic to the cohomology of the following cosimplicial complex:

$$\bar{\Omega}^{\cdot}_{A.(X)'} \underset{d_1}{\overset{d_0}{\rightrightarrows}} \bar{\Omega}^{\cdot}_{A.(X)'} \otimes k[\xi_0] \;\overrightarrow{\overrightarrow{\rightarrow}}\; \bar{\Omega}^{\cdot}_{A.(X)'} \otimes k[\xi_0, \xi_1] \quad \cdots$$

Here $d_i\omega = \omega$, $i > 0$; $d_0\omega = \omega + \xi_0\bar{d}\omega$;

$$d_i\xi_j = \begin{cases} \xi_{j+1}, & i \leqslant j; \\ \xi_j + \xi_{j+1}, & i = j+1; \\ \xi_j, & i > j+1 \end{cases}$$

The associated cohomological complex admits a decreasing filtrations; its n-th term is spanned by the monomials $\omega\otimes\xi_{i_0}\dots\xi_{i_k}$, $k \geqslant n$. This filtration is finite in each dimension, therefore the corresponding spectral sequence converges to $H^{\cdot}(ES^1 \underset{S^1}{\times} X^{S^1})$. Computing its first term we see that the projection of our complex onto the quotient complex

$$\bar{\Omega}^{\cdot}_{A.(X)'} — \bar{\Omega}^{\cdot}_{A.(X)'} \otimes \xi_o — \bar{\Omega}^{\cdot}_{A.(X')} \otimes \xi_o \xi_1 — \dots$$

(in which the differential sends $\omega \otimes \xi_o \dots \xi_{k-1}$ to $(-1)^{h(\omega)+1}\bar{d}\omega \otimes \xi_o \dots \xi_k$) is a quasi-isomorphism. Thus $H^{\cdot}(ES^1 \underset{S^1}{\times} X^{S^1}) \xrightarrow{\sim}$ $\xrightarrow{\sim} H^{\cdot}(\ker(d : \bar{\Omega}^{\cdot}_{A.(X)'} \to \bar{\Omega}^{\cdot}_{A.(X)'})$; therefore $\bar{H}^{\cdot}(ES^1 \underset{S^1}{\times} X^{S^1}) \xrightarrow{\sim}$ $\xrightarrow{\sim} H^{\cdot-1}(\bar{\Omega}^{>o}_{A.(X)'}/d\bar{\Omega}^{\cdot}_{A.(X)'})$. This (together with the statement proved in the end of 6.2) gives the proof of 2).

The statement obtained above is close to Theorem A 9.5. In view of Theorem A 9.4, it provides a tool for computing rational Waldhausen algebraic K-theory of the simply connected spaces.

Corollary 6.6.2. Let X be formal (cf. [5]). Then

$$K_n(X) \underset{\mathbb{Z}}{\times} \mathbb{Q} \to K_n(\mathbb{Z}) \underset{\mathbb{Z}}{\times} \mathbb{Q} + K^{n+1}_{+}(\mathbb{Q} \to H^{\cdot}(X;\ \mathbb{Q}))^*, \qquad n > 0.$$

For instance, X may be an H-space, or a compact Kahler manifold, or a compact oriented riemannian symmetric space.

Corollary 6.6.3. Let $H^{\cdot}(X;\ \mathbb{Q}) \simeq \mathbb{Q}[v_1,\dots,v_m]/I$ where I is an ideal generated by a regular sequence of homogeneous elements. Put $\mathbb{Q}[v_1,\dots,v_m] = B$; consider the complexes $\Omega_B/F^{n+1}\Omega_B$:

$$0 \to \bar{\Omega}^o_B/I^{n+1} \xrightarrow{\bar{d}} \bar{\Omega}^1_B/I^n\bar{\Omega}^1_B \to \dots \to \bar{\Omega}^n_B/I\ \bar{\Omega}^n_B \to 0$$

Then

$$K_n(X) \underset{\mathbb{Z}}{\otimes} \mathbb{Q} \xrightarrow{\sim} K_n(\mathbb{Z}) \underset{\mathbb{Z}}{\otimes} \mathbb{Q} + (\bigoplus_{2b-a+c=n} H^a(\bar{\Omega}^{\cdot}_B/F^{b+1}\bar{\Omega}^{\cdot}_B)^c)^*, \qquad n > 0.$$

In fact, Bousfield and Gugenheim have shown that the assertion of Corollary 6.6.3 implies formality of X (cf.[5]). So our statement follows from Theorem 6.1.1, 2) (or rather from its evident graded version).

For a simply connected space X, denote

$$\varphi(X;z) = \sum_{i=1}^{\infty} (\dim(K_i(X) \underset{\mathbb{Z}}{\otimes} \mathbb{Q}) - \dim(K_i(\mathbb{Z}) \underset{\mathbb{Z}}{\otimes} \mathbb{Q}))\ z^n.$$

6.6.4. Examples.
$$\varphi(S^{2n+1};\ z) = z^{2n+1}(1 - z^{2n})^{-1};$$
$$\varphi(S^{2n};\ z) = z^{2n}(1 - z^{4n-2})^{-1}$$

(Hsiang-Staffeldt);

$$\varphi(K(Z,\ sn);\ z) = z^{2n}(1 - z^{2n})^{-1}$$

(Burghelea).

Corollary 6.6.3 yields also a formula for $\varphi(\mathbb{C}P^m \times \ldots \times \mathbb{C}P^m ; z)$. We leave this to the reader as an exercise.

If X has rational homotopy type of $K(2u_1+1, \mathbb{Z}) \times \ldots \times K(2u_m+1, \mathbb{Z}) \times K(2v_1, \mathbb{Z}) \times \ldots \times K(2v_l, \mathbb{Z})$, $u_i, v_j > 0$, then

$$\varphi(X;z) = \frac{z}{z+1} \prod_{j=1}^{m} \frac{1+z^{2u_j+1}}{1-z^{2u}} \prod_{i=1}^{l} \frac{1+z^{2v_i-1}}{1-z^{2v}} - 1$$

(Burghelea).

The verifications may be easily done; we omit the details.

If $X = \Sigma Y$, Y connected, then the differentials in the reduced cyclic complex are all equal to zero; so, we obtain a theorem which is due to Hsiang and Staffeldt:

$$K.(X) \otimes_{\mathbb{Z}} \mathbb{Q} \xrightarrow{\sim} K.(\mathbb{Z}) \times_{\mathbb{Z}} \mathbb{Q} \oplus T(\tilde{H}.(Y; \mathbb{Q}))/[,]$$

Remark 6.6.5. One has to pay attention on the distinction in the indexes in Theorems 6.6.1 and A 9.5.

Chapter 7. Characteristic classes

7.1. Chern character for K_o. Let A be an algebra over k. Connes defined in [13] the Chern character $ch_{o,n}: K_o(A) \longrightarrow K^+_{2n+1}(A)$. We recall his construction just briefly.

Let E be a finitely generated projective A-module; it is an image of an idempotent $e \in End_A(A^N) = M_N(A)$. We obtain a map $k[x]/(x^2 - x) \longrightarrow M_N(A)$ which sends x to e. But $k[x]/(x^2 - x) \simeq k \oplus k$; therefore we may choose the elements $u_n \in K^+_{2n+1}(k[x]/(x^2 - x))$, $Su_n = u_{n-1}$, and obtain the elements in $K^+_{2n+1}(M_n(A)) \xrightarrow{\sim} K^+_{2n+1}(A)$. For example, we may regard these elements as the homology classes represented by cycles

$$\alpha(e) = e + \sum_{k=1}^{n} (2k - 1)!!(2^k e^{\otimes(2k+1)} - 2^{k-1} e^{\otimes 2k}) \cdot (-1)^k$$

in $L_{2n}(M_N(A))$. Connes shows that we obtain in such a way a well defined map

$$ch_{o,n} : K_o(A) \longrightarrow K^+_{2n+1}(A)$$

such that the following diagram is commutative:

$$\begin{array}{ccc} K_0(A) & \xrightarrow{ch_{0,n}} & K^+_{2n+1}(A) \\ & \underset{ch_{0,n-1}}{\searrow} & \downarrow S \\ & & K^+_{2n-1}(A) \end{array}$$

7.2. <u>Additive K-functors of the group algebras</u> ([32], [19]).

<u>Theorem 7.2.1.</u> Let G be a group. Then

$$H_n(G,k) \oplus H_{n-2}(G,k) \oplus H_{n-4}(G,K) \oplus \dots$$

is a direct summand in $K^+_{n+1}(k[G])$.

<u>Proof.</u> Consider the bicomplex $L_{..}(k[G])$. The map $g_0 \otimes \dots \otimes g_n \mapsto$ $\to g_0 \otimes \dots \otimes g_{n-1} \otimes g_0 \dots g_n$ gives an isomorphism

$$L_{..}(k[G]) \xrightarrow{\sim} \tilde{L}_{..}(k[G]); \quad \tilde{L}_{ij}(k[G]) = L_{ij}(k[G]), \quad i,j \geqslant 0;$$

the differentials in $\tilde{L}_{..}(k[G])$ are as follows: $\tilde{\partial}' : \tilde{L}_{ij} \to \tilde{L}_{i,j-1}$, $\tilde{\delta} : \tilde{L}_{ij} \to \tilde{L}_{i-1,j}$; $\tilde{\partial} = \tilde{\delta}$ for i odd and $\tilde{\partial} = \tilde{b}$ for i even; $\tilde{\partial} =$ $= 1 - \tilde{\tau}$ for i odd and $\tilde{\partial}' = 1 + \tilde{\tau} + \dots + \tilde{\tau}^j$ for i even; if $\alpha =$ $= g_0 \otimes \dots \otimes g_n$, then

$$\tilde{b}(\alpha) = \sum_{k=0}^{n-2} (-1)^k g_0 \otimes \dots \otimes g_k g_{k+1} \otimes \dots \otimes g_n + (-1)^{n-1} g_0 \otimes \dots \otimes g_{n-2} \otimes g_n;$$

$$\tilde{\delta} = -\tilde{b} + \tilde{d}_0; \quad \tilde{\tau}(\alpha) = g_1 \otimes \dots \otimes g_{n-1} \otimes (g_0 \dots g_{n-1})^{-1} g_n \otimes g_0^{-1} g_n g_0 (-1)^n;$$

$$\tilde{d}_0(\alpha) = g_1 \otimes \dots \otimes g_{n-1} \otimes g_0^{-1} g_n g_0.$$

We see that there is a direct summand in $\tilde{L}_{..}$ generated by the elements having $g_n = 1$. It is easy to see that this direct summand is isomorphic to $L_{..}(G) \otimes_{k[G]} k$ where $L_{..}(G)$ is the following bicomplex of free left $k[G]$-modules:

$$\begin{array}{ccccccc} & & \downarrow \delta' & & \downarrow b' & & \downarrow \delta' \\ \dots & \xrightarrow{1-\tau'} & \overset{n+1}{\otimes} k[G] & \xrightarrow{N'} & \overset{n+1}{\otimes} k[G] & \xrightarrow{1-\tau'} & \overset{n+1}{\otimes} k[G] \\ & & \downarrow \delta' & & \downarrow b' & & \downarrow \delta' \\ \dots & \xrightarrow{1-\tau'} & k[G] & \xrightarrow{N'} & k[G] & \xrightarrow{1-\tau'} & k[G] \end{array}$$

if $\alpha = g_0 \otimes \dots \otimes g_n$, then

$$b'(\alpha) = \sum_{k=1}^{n-1} (-1)^{k-1} g_0 \otimes \dots \otimes g_k g_{k+1} \otimes \dots \otimes g_n + (-1)^n g_0 \otimes \dots \otimes g_{n-1};$$

$$\delta' = -b' + d_0'; \; d_0' \alpha = g_0 g_1 \otimes \dots \otimes g_{n-1} \otimes g_n; \; \tau'(\alpha) = (-1)^n g_0 g_1 \otimes g_2 \otimes \dots \otimes (g_1 \dots g_n)^{-1}$$

Lemma 7.2.2. The bicomplex $L_{..}(G)$ is a free resolution of the complex

$$\ldots \longrightarrow 0 \longrightarrow k \longrightarrow 0 \longrightarrow k \qquad (7.2.1)$$

Proof. Consider the projection of $L_{..}$ onto the subcomplex (7.2.1): $p(L_{ij}) = 0$, $j \geqslant 1$; $p(g) = 1$, $g \in L_{2i,o}$; $p(g) = 0$, $g \in L_{2i+1,o}$. Clearly, p is an epimorphism of complexes. Furthermore, it is a quasi-isomorphism. In fact, the columns at even places are the free resolutions of k; the columns at odd places are contractible (they are obtained from the simplicial abelian groups with differential $d_1 - d_2 + \ldots$). Our statement follows from the comparison of spectral sequences.

This gives the proof of Theorem 7.2.1.

7.3. Chern character for the higher algebraic K-functors. Let $K_n(A)$ be Quillen K-functors (cf. [21]).

Theorem 7.3.1. There exist the homomorphisms $ch_{n,i} : K_n(A) \longrightarrow K^+_{n+2i+1}(A)$ which are functorial with respect to A; the following diagrams are commutative:

$$\begin{array}{ccc} K_n(A) & \xrightarrow{ch_{n,i}} & K^+_{n+2i+1}(A) \\ & \searrow^{ch_{n-1,i}} & \downarrow S \\ & & K^+_{n+2i-1}(A) \end{array}$$

Proof. The desired maps are given by the composition

$$K_n(A) = \pi_n(BGL(A)^+) \xrightarrow{h} H_n(BGL(A)^+) \longrightarrow H_n(BGL(A)) =$$

$$= H_n(GL(A), \mathbb{Z}) \longrightarrow H_n(GL(A), k) \longrightarrow K^+_{n+1+2i}(k[GL(A)]) \longrightarrow$$

$$\longrightarrow K^+_{n+1+2i}(M_\infty(A)) \overset{\sim}{\longleftarrow} K^+_{n+1+2i}(A).$$

Here h is the Hurewicz homomorphism, $BGL(A)$ is the classifying space of the group $GL(A)$ and $BGL(A)^+$ is its quillenization (see [21]); i is the embedding given by Theorem 7.2.1.

The above Theorem was obtained by Connes in [13] for $n = 1$ and in [32], [19] in the general case. We show in the next subsection that our construction for $i > 1$ it is "the higher derived functor" of the Connes' map.

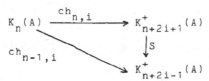

7.4. Chern character and Gersten-Swan construction. Here we suppose for simplicity that k is a field of characteristic zero. Let A be an algebra without unit. Denote by $\tilde{A}$ the algebra obtained from A by adjoining a unit. We shall write $K_i(A) = \ker(K_i(\tilde{A}) \longrightarrow K_i(k))$, $GL(A) = \ker(GL(\tilde{A}) \longrightarrow GL(k))$, etc.

Let R be a free simplicial resolution of A in the category of algebras without unit. The following statement was proved by (cf. [ ]) for the case $k = \mathbb{Z}$; our case follows easily from 's proof together with the Waldhausen theorem about K-theory of generalized free products (cf. [62]).

Theorem 7.4.1. The simplicial group $GL(R)$ coincides with its commutant. The sequence

$$BGL(R) \longrightarrow BGL(A) \longrightarrow BGL(A)^+$$

is a fibration up to homotopy.

Thus $\pi_i(GL(R)) = \pi_i(\Omega BGL(R)) \simeq K_{i+2}(A)$, $i > 0$; $\pi_0(GL(R)) \simeq St(A)$. We see that the functors $K_i(A)$ may be defined as the derived functors of the functor GL (the definition of Gersten-Swan). We want to compute in these terms the Chern character defined in 7.3.

In order to do this consider the complexes $F'(R) = \ker(F'(R) \longrightarrow F'(k))$, $F'(A) = \ker(F'(\mathbf{R}) \longrightarrow F'(A))$. The simplicial structure of R turns $F'(R)$ to be a bicomplex. The Lemma 2.3.1 implies that for any $m \geqslant 0$ $H_i(F'(R_m)) = 0$, $i > 0$. On the other hand, it is clear that $F'_n(R) \xrightarrow{\sim} F'_n(A)$ for all n. We denote the image of the differential $F'_{n+1}(R) \longrightarrow F'_n(R)$ by $B_n(R)$ and its kernel by $Z_{n+1}(R)$. The spectral sequence of the bicomplex $\bigoplus_{n'>n} F'_{n'}(R)$ shows that

$$B_n(R) \xleftarrow{\sim} F'_{n+1}(R)/Z_{n+1}(R) = F'_{n+1}(R)/B_{n+1}(R)$$

has the homology equal to $F'_{n+1}(A)/B_{n+1}(A)$ in dimension zero and to $K^+_{n+i+2}(A)$ in dimension $i > 0$. In other words, the sequence of complexes

$$\begin{array}{ccccc}
\cdots \longrightarrow & B_n(R_1) & \longrightarrow & B_n(R_0) \\
 & \downarrow & & \downarrow \\
\longrightarrow & 0 & \longrightarrow & B_n(A) \\
 & \downarrow & & \downarrow \\
\cdots \longrightarrow & F_{n+1}(A)' & \longrightarrow & Z_n(A)
\end{array}$$

induces a homology long exact sequence. We may write down exactly the same sequence for $M_\infty(R)$, $M_\infty(A)$.

Now let $g \in GL(R)$. Consider an element

$$\alpha(g) = \sum_{k=1}^{l+1} (k-1)!((g \otimes g^{-1} \otimes \ldots \otimes g \otimes g^{-1}) \quad (k \text{ times}) , \qquad (7.4.1)$$

$$\alpha(g) \in \bigoplus_{k=1}^{i+1} F'_{i+1-k,i+k}(M_\infty(R)).$$

An elementary verification shows that $\alpha(g) \in Z_{2i+1}(M_\infty(R)) = B_{2i+1}(M_\infty(R))$. Therefore we obtain a homomorphism of simplicial sets (but not of simplicial groups) $GL(R) \to B_{2i+1}(M_\infty(R))$ which induces the homomorphisms of homotopy groups $ch'_{n,i} : K_n(A) \to K^+_{n+1+2i}(A)$, $n > 2$.

Note also that the construction of 7.3 being applied to $\widetilde{A}$ gives in the obvious way the Chern character $K_n(A) \to K^+_{n+2i+1}(A)$ denoted again by $ch_{n,i}$.

Proposition 7.4.2. $ch_{n,i} = ch'_{n,i}$

Sketch of the proof. In the first place, it is easy to verify that for a $g \in GL(A)$ the formula (7.4.1) gives the cycle representing $ch_{1,i}[g]$ where $[g]$ is the class of g in $K_1(A) = GL(A)/E(A)$.

Recall that the functor from the category of simplicial abelian groups to the category of complexes which associates to a simplicial abelian group $C = (C_n,\ n \geq 0)$ the complex $(C.;\ d_0 - d_1 + \dots + (-1)^n d_n)$ admits a right adjoint denoted here by $\mathcal{L}$. (This is a little modification of the well known Dold-Puppe construction; cf. [25]). We regard the functors Ω, B to be realized in the category of simplicial sets as the well known Kan's functors G, W (cf. [70]); then Ω is left adjoint to B; the construction of 7.2-7.3 enables one to write down the canonical map of complexes

$$\varphi : \mathbb{Z}BGL(A) \longrightarrow C^{(i)}(A)$$

where $C_1^{(i)}(A) = Z_{2i+1}(M_\infty(A))$, $C_{1+k}^{(i)}(A) = F'_{2i+1+k}(M_\infty(A))$, $k > 0$, $C_k^{(i)}(A) = 0$, $k < 0$; the map φ induces the homomorphism $H.(GL(A);\mathbb{Z}) \to K^+_{.+2i+1}(A)$ such that the composition

$$K.(A) \xrightarrow{h} H.(GL(A);\ \mathbb{Z}) \to K^+_{.+2i+1}(A)$$

is equal to $ch_{n,i}$. The adjunction property of $\mathcal{L}$ shows that there is a map of simplicial sets

$$\mathbb{Z}BGL(A) \longrightarrow \mathcal{L}(C^{(i)}_{.}(A))$$

and whence a map

$$\Omega\mathbb{Z}BGL(A) \longrightarrow \Omega\mathcal{L}(C^{(i)}_{.}(A)).$$

Furthermore, it is easy to see that for any complex $C.$ such that

$C_k = 0$, $k < 0$ there exists the canonical weak homotopy equivalence $\Omega\mathcal{L}(C.) \to \mathcal{L}(C.[1])$. Thus, we obtain a map

$$\Omega BGL(A)^+ \longrightarrow |\Omega ZBGL(A)| \longrightarrow |\mathcal{L}(C.[1])|$$

where $|\ |$ is the geometric realisation functor . Now consider the diagram

$$\begin{array}{ccccc}
|\Omega BGL(R)| & \longrightarrow & |\Omega BGL(A)| & \longrightarrow & \Omega BGL(A)^+ \\
\downarrow \wr & & \downarrow \wr & & \\
|GL(R)| & \longrightarrow & |GL(A)| & & \downarrow \\
\downarrow |\varphi'| & & \downarrow |\varphi''| & & \\
|B_{2i+1}(R)| & \longrightarrow & |Z_{2i+1}(A)| & \longrightarrow & |\mathcal{L}(C^{(i)}[1])|
\end{array}$$

φ', φ'' associate to $g \in GL$ the cycle $\alpha(g)$ given by (7.4.1). It is easy to see that this diagram is homotopy commutative; the adjunction maps $\Omega BGL \longrightarrow GL$ are weak homotopy equivalences; the horizontal sequences are fibrations up to homotopy. This gives the proof of 7.4.2.

Remark 7.4.3. Karoubi and Connes use the non-commutative de Rham complex to define $K^+_\cdot(A)$ (cf. [12], [35]). In these terms one may say that $ch_{n,i}$ are the maps in the higher derived functors which are induced by the map $g \mapsto (g^{-1}\,dg)^{2i-1}$.

7.5. Regulator map for the relative algebraic K-theory of a two-sided nilpotent ideal. Let A and B be rings. Suppose that there are two homomorphisms $i : A \longrightarrow B$, $j : B \longrightarrow A$, $j \circ i = id_A$ and that $I = \ker(j)$ is a nilpotent two-sided ideal of B. We write $B = A + I$. Following standard notation we write also $K_i(A+I,I) = \ker(K_i(A+I) \longrightarrow K_i(A))$ and $K^+_i(A+I,I) = \ker(K^+_i(A+I) \longrightarrow K^+_i(A))$. Note that $K^+_i(A+I,I) = K^+_{i-1}(A \xrightarrow{i} A+I) =$ $= K^+_j(A+I \xrightarrow{j} A)$ (in notation of 2.4).

Theorem 7.5.1. 1) There exists a functorial map

$$\rho : K_i(A+I,I) \underset{\mathbf{Z}}{\otimes} \mathbf{Q} \longrightarrow K^+_i(A+I,I) \underset{\mathbf{Z}}{\otimes} \mathbf{Q} \qquad (7.5.1)$$

2) Let $i = 1,2$. Then ρ is an isomorphism.
3) Assume that either $i = 3$ or $I^2 = 0$. Then ρ is an epimorphism.
4) If A is the ring of integers in a number field, then ρ is an isomorphism.

If we consider $\mathbf{Q}$-algebras, then Theorem 7.5.1 gives precise information about the algebraic K-functors because of the following statement:

Theorem 7.5.2. The maps

$$K.(A+I,I) \otimes_{\mathbb{Z}} \mathbb{Q} \longrightarrow K.(A+I \otimes_{\mathbb{Z}} \mathbb{Q},\ I \otimes_{\mathbb{Z}} \mathbb{Q}) \otimes_{\mathbb{Z}} \mathbb{Q} \longleftarrow$$

$$\leftarrow K.(A+I \otimes_{\mathbb{Z}} \mathbb{Q},\ I \otimes_{\mathbb{Z}} \mathbb{Q})$$

are isomorphisms.

Corollary 7.5.3. (C.Soulé, [70]). Let $\mathcal{O}$ be the ring of integers in a number field F. Then

$$\mathrm{rk}\ K_i(\mathcal{O}[t]/(t^n)) = \mathrm{rk}\ K_i(\mathcal{O}) + \begin{cases} 0, & i \text{ even} \\ (n-1)\cdot[F:\mathbb{Q}], & i \text{ odd.} \end{cases}$$

Proof. We may apply the statement 2) of Theorem 6.1.1. Let $X = \mathrm{Spec}\ (\mathbb{Q}[t]/(t^n))$. The cohomology groups $H^{\cdot}(X;m)$ may be computed using the following complex:

$$0 \longrightarrow \mathbb{Q}[t]/(t^{nm}) \longrightarrow \mathbb{Q}[t]/(t^{n(m-1)})\cdot dt \longrightarrow 0,$$

$t^k \longmapsto k\cdot t^{k-1}\,dt$; we see that $H^0(X;\ m) \simeq \mathbb{Q}^n$ and $H^i(X;\ m) = 0,\ i > 0$. Thus, $K_i^+(\mathbb{Q}[t]/(t^n);\ (t^n)) = 0$ for i even and $\mathbb{Q}^{n-1}$ for i odd. But it is well known that $H_i(\mathcal{O},\mathcal{O}) \otimes \mathbb{Q} \simeq H_i(F,F) = 0,\ \ i > 0$. So we have $H_i(\mathcal{O}[t]/(t^n),\ \mathcal{O}[t]/(t^n)) \otimes \mathbb{Q} \xrightarrow{\sim} F \otimes H_i(\mathbb{Q}[t]/(t^n),\ \mathbb{Q}[t]/(t^n))$ and therefore

$$K^+(\mathcal{O}[t]/(t^n) \otimes_{\mathbb{Z}} \mathbb{Q} \xrightarrow{\sim} F \otimes K^+(\mathbb{Q}[t]/(t^n)).$$

Now applying the statement 4) of Theorem 7.5.1 we obtain the desired statement.

Corollary 7.5.4. (S.Bloch, [4]). Let A be a commutative $\mathbb{Q}$-algebra and $\mathfrak{a}$ an artinian commutative local algebra over $\mathbb{Q}$. Then

$$K_2(A \otimes_{\mathbb{Q}} \mathfrak{a}) \xrightarrow{\sim} K_2(A) \oplus \Omega^1_{A\otimes\mathfrak{a}}/(d(A\otimes\mathfrak{a}) + \Omega^1_A).$$

Indeed, the spectral sequence of Theorem 1.2. shows that for any commutative $\mathbb{Q}$-algebra B $K_2^+(B) \xrightarrow{\sim} \Omega^1_B/dB$.

Corollary 7.5.5. Let $\mathfrak{a}$ be an algebra without unit over $\mathbb{Q}$ such that $\mathfrak{a}^n = 0$ for some n. Then

$$(F^2_\gamma/F^3_\gamma)K_i(\mathfrak{a}) \xrightarrow{\sim} D_{i-2}(\mathfrak{a},\mathfrak{a}), \quad i > 2. \tag{7.5.2}$$

Here F_γ is the gamma-filtration (see [53]) and $D.$ is the Harrison-Quillen homology.

Proof. It follows from the construction of the map ρ given below that it commutes with the Adams operations (see Chapter 5). So our statement follows from Remark 6.5.4.

Remark 7.5.6. The Cartan-Milnor-Moore theorem (cf. [46]) implies that $K_i(A) \otimes_Z \mathbb{Q} = \mathrm{Prim}_i\, H.(GL(A), \mathbb{Q})$. Consider a filtration introduced by Soulé:

$$F_n(K_i(A) \otimes_{\mathbb{Z}} \mathbb{Q}) = \mathrm{im}\, H_i(GL_n(A), \mathbb{Q}) \cap \mathrm{Prim}\, H.(GL(A), \mathbb{Q}).$$

Soulé conjectured that for $i \geqslant 2n+1$ $F_n(K_i(A) \otimes_{\mathbb{Z}} \mathbb{Q}) = 0$. But the computation from Subsection 4.6 together with the statement 4) of Theorem 7.5.1 shows that in the general case this conjecture fails. If $\mathfrak{A} = \mathbb{Z} + I$ where I is an infinitely generated free abelian group and $I^2 = 0$, then for any $p > 0$ there exists such $n > 0$ that $F_n(K_{np}(\mathfrak{A}) \otimes_{\mathbb{Z}} \mathbb{Q}) \neq 0$.

There is another conjecture which is due to Beilinson [2]: if A is a regular commutative ring then $K_i(A) = F^n_\gamma K_i(A)$ (modulo torsion) for $i \geqslant \mathrm{Sup}(2n-1, 1)$. Corollary 7.5.5 shows that the assertion of regularity is actually essential.

Remark 7.5.7. The statement 3) of Theorem 7.5.1 is close to the results obtained by C.Kassel in [37], [39].

Proof of Theorem 7.5.1. First of all we need a generalization of the main statement on invariants (see 4.3).

Let A be a ring, P be a bimodule over A and N a natural number. We denote by $M_N(P) = M_N(\mathbb{Z}) \otimes_{\mathbb{Z}} P$ the abelian group of $N \times N$ - matrices with entries in P. We write $E_{ij}(p)$ (respectively $e_{ij}(a)$) for a matrix with the unique non-zero entry at the place (i,j) equal to p (respectively to a); $e_{ij} = e_{ij}(1)$. Let $m \in GL_N(A)$, $p^{(i)} \in M_N(P)$. Put

$$m*(p^{(o)} \otimes \ldots \otimes p^{(n-1)}) = mp^{(o)} m^{-1} \otimes \ldots \otimes m\, p^{(n-1)} m^{-1} \qquad (7.5.3)$$

This formula gives a $GL_N(A)$-module structure on $\overset{n}{\otimes} M_N(P)$.

Now let $\sigma \in \Sigma_n$; denote

$$P^\sigma = (\overset{n}{\otimes} P)/\langle p_o \otimes \ldots \otimes p_i a \otimes \ldots \otimes p_{\sigma(i)} \otimes \ldots \otimes p_{n-1} - p_o \otimes \ldots \otimes p_i \otimes \ldots$$
$$\ldots \otimes a p_{\sigma(i)} \otimes \ldots \otimes p_{n-1}\rangle,$$

$p_i \in P$, $a \in A$. For example, let e be the identity element of Σ_n. Then $P^e = \overset{n}{\otimes}(P \otimes_{A \otimes A^o} A)$. If t is the standard cycle $0 \to 1 \to \ldots \to n-1$ (with $n-1 \to 0$), then

$$P^t = ((P \otimes_A \ldots \otimes_A P) \otimes_{A\otimes A^\circ} A).$$

We have the maps

$$\varkappa_\sigma : (\overset{n}{\otimes} M_N(P))_{GL_N(A)} \longrightarrow P^\sigma;$$

$$\varkappa_\sigma(p^{(o)} \otimes \ldots \otimes p^{(n-1)}) = \sum_{0 \leqslant i_o, \ldots, i_{n-1} \leqslant N} p^{(o)}_{i_o, i_{\sigma(o)}} \otimes \ldots \otimes p^{(n-1)}_{i_{n-1}, i_{\sigma(n-1)}}$$

(it is easy to see that they are well defined).

These maps give a homomorphism

$$\varkappa : (\overset{n}{\otimes} M_N(P))_{GL_N(A)} \longrightarrow \bigoplus_{\sigma \in \Sigma_n} P^\sigma.$$

The following statement has been obtained by Kassel for $n = 1,2$ (cf. [37], [39]). We use his method in the proof.

Proposition 7.5.8. Assume that $N > 2n$. Then $\varkappa$ is an isomorphism.

Proof. Let $p_o \otimes \ldots \otimes p_{n-1} \in P^\sigma$. Put

$$\zeta_\sigma(p_o \otimes \ldots \otimes p_{n-1}) = E_{o,\sigma(o)}(p_o) \otimes \ldots \otimes E_{n-1,\sigma(n-1)}(p_{n-1}).$$

We obtain the map $\zeta : \bigoplus_\sigma P^\sigma \longrightarrow (\overset{n}{\otimes} M_N(P))_{GL_N(A)}$. It suffices to show that:

a) ζ is well defined;

b) ζ is onto;

c) $\varkappa \circ \zeta = \mathrm{id}$.

Let $\alpha = E_{i_o j_o}(p_o) \otimes \ldots \otimes E_{i_{n-1}, j_{n-1}}(p_{n-1})$. We write $r_i(\alpha) = \mathrm{card}\{m \in [0,n-1] : i_m = i\}$; $s_j(\alpha) = \mathrm{card}\{m \in [0,n-1] : j_m = j\}$. For arbitrary elements $\beta, \beta' \in \overset{n}{\otimes} M_N(P)$ we shall write $\beta \equiv \beta'$ if $\beta - \beta' = 0$ in $(\overset{n}{\otimes} M_N(P))_{GL_N(A)}$.

Lemma 7.5.9. 1) Suppose that there exists $i \in [1,N]$ such that $r_i(\alpha) \neq s_i(\alpha)$. Then $\alpha \equiv 0$.

2) The abelian group $(\overset{n}{\otimes} M_N(P))_{GL_N(A)}$ is generated by the elements α such that for all i either $s_i(\alpha) = r_i(\alpha) = 1$ or $s_i(\alpha) = r_i(\alpha) = 0$.

Proof. Assume that there exists an i such that $s_i(\alpha) = 0$, $r_i(\alpha) = r \neq 0$. We claim that $\alpha \equiv 0$. We use induction on r. Choose $i' \notin \{i_o, \ldots, i_{n-1}; j_o, \ldots, j_{n-1}\}$. Let $i = i_{m_1} = \ldots = i_{m_r}$; replace i

by i' in α:

$$\alpha' = \dots \otimes E_{i',j_{m_1}}(p_{m_1}) \otimes \dots \otimes E_{i',j_{m_r}}(p_{m_r}) \otimes \dots$$

Let $r = 1$. Then

$$(1 + e_{ii'}) * \alpha' = (1 + e_{ii'}) * (E_{i_o,j_o}(p_o) \otimes \dots \otimes E_{i',j_{m_1}}(p_{m_1}) \otimes \dots$$

$$\dots \otimes E_{i_{n-1},j_{n-1}}(p_{n-1})) = \alpha' + \alpha ;$$

this implies $\alpha \equiv 0$. Now for an arbitrary r.

$$(1 + e_{ii'}) * \alpha' = \alpha + \alpha' + \sum_{l=1}^{2^r-2} \alpha_l$$

where $0 < r_i(\alpha_l) < r$ for all l; the inductive assumption shows that $\alpha \equiv 0$.

Assume now that $r_i(\alpha) \geqslant s_i(\alpha)$ and $r_i(\alpha) > 1$. Fix an m such that $i_m = i$. Consider an element

$$\alpha'' = E_{i_o j_o}(p_o) \otimes \dots \otimes E_{i' j_m}(p_m) \otimes \dots \otimes E_{i_{n-1},j_{n-1}}(p_{n-1}).$$

Note that we replace i by i' only in one tensor factor; $\alpha'' \neq \alpha'$. It is easily seen that

$$(1 + e_{ii'}) * \alpha'' = \alpha + \alpha'' + \sum_{l=1}^{s_i(\alpha)} \alpha_l^{(o)} + \sum \alpha_l^{(1)}$$

where

$$r_i(\alpha_\ell^{(o)}) = r_i(\alpha) - 1,\ s_i(\alpha_\ell^{(o)}) = s_i(\alpha) - 1;\ r_j(\alpha_\ell^{(o)}) = r_j(\alpha),\ j \neq i,i';$$

$$r_{i'}(\alpha_\ell^{(o)}) = s_{i'}(\alpha_\ell^{(o)}) = 1;\quad r_i(\alpha_\ell^{(1)}) - s_i(\alpha_\ell^{(1)}) > r_i(\alpha) - s_i(\alpha).$$

Again proceeding by induction we see that

$$\alpha \equiv \sum \beta_l^{(o)} + \sum \beta_l^{(1)}$$

where $s_j(\beta_l^{(o)}) = r_j(\beta_l^{(o)}) = 0$ or 1 for all j; $r_i(\beta_l^{(1)}) > s_i(\beta_l^{(1)}) = 0$. This together with the analogous computation for the case $r_i(\alpha) \leqslant s_i(\alpha)$ proves Lemma 7.5.9.

Now prove the statement a). If $\sigma = \sigma_1 \dots \sigma_k$ where σ_l are cycles, then it is clear that $P^\sigma \xrightarrow{\sim} \bigotimes_{l=1}^{k} P^{\sigma_\ell}$; so it suffices to show that the map

$$\zeta_t : P^t = (P \otimes_A \dots \otimes_A P) \otimes_{A \otimes A^\circ} A \longrightarrow (\tilde{\otimes} M_N(P))_{GL_N(A)}$$

is well defined where t is the standard cycle. We want to show that for $a \in A$

$$E_{01}(p_o) \otimes \dots \otimes E_{n-2,n-1}(p_{n-2}) \otimes E_{n-1}(ap_{n-1}) \equiv$$

$$\equiv E_{01}(p_o) \otimes \dots \otimes E_{n-2,n-1}(p_{n-2}a) \otimes E_{n-1,n}(p_{n-1})$$

(other cases are obtained by the cyclic permutation of tensor factors). We use induction on n. Let $n = 1$. Then $0 \equiv (1 + e_{01}(a)) * E_{10}(p) =$

$$= (1 + e_{01}(a))\, E_{10}(p)(1 - e_{01}(a)) = E_{10}(p) + E_{00}(ap) - E_{11}(pa) - E_{01}(apa) \equiv$$

$$\equiv E_{00}(ap) - E_{11}(pa) = E_{00}(ap) - \begin{pmatrix} 0 & 1 \\ 1 & 0 \end{pmatrix} * E_{00}(pa) \equiv E_{00}(ap - pa);$$

now let $n > 1$. We have

$$(1 + e_{n-1,o}(a)) * (E_{01}(p_o) \otimes \dots \otimes E_{n-2,n-1}(p_{n-2}) \otimes E_{00}(p_{n-1})) =$$

$$= (E_{01}(p_o) + E_{n-1,1}(ap_o)) \otimes E_{12}(p_1) \otimes \dots \otimes (E_{n-2,n-1}(p_{n-2}) -$$

$$- E_{n-2,0}(p_{n-2}a)) \otimes (E_{00}(p_{n-1}) + E_{n-1,0}(ap_{n-1})) = E_{01}(p_o) \otimes \dots \otimes E_{n-2,n-1}(p_{n-2}) \otimes$$

$$\otimes E_{00}(p_{n-1}) - E_{01}(p_o) \otimes \dots \otimes E_{n-2,n-1}(p_{n-2}) \otimes E_{n-1,0}(p_{n-2}a) \otimes$$

$$\otimes E_{00}(p_{n-1}) + E_{01}(p_o) \otimes \dots \otimes E_{n-2,n-1}(p_{n-2}) \otimes E_{n-1,0}(ap_{n-1}) +$$

$$+ E_{n-1,1}(ap_o) \otimes E_{12}(p_1) \otimes \dots \otimes E_{n-2,n-1}(p_{n-2}) \otimes E_{00}(p_{n-1}) -$$

$$- E_{01}(p_o) \otimes \dots \otimes E_{n-3,n-2}(p_{n-3}) \otimes E_{n-2,0}(p_{n-2}a) \otimes E_{n-1,0}(ap_{n-1}) +$$

$$+ E_{n-1,1}(ap_o) \otimes E_{12}(p_1) \otimes \dots \otimes E_{n-2,n-1}(p_{n-2}) \otimes E_{n-1,0}(ap_{n-1}) -$$

$$- E_{n-1,1}(ap_o) \otimes \dots \otimes E_{n-2,0}(p_{n-2}a) \otimes E_{00}(p_{n-1}) - E_{n-1,1}(ap_o) \otimes$$

$$\otimes E_{12}(p_1) \otimes \dots \otimes E_{n-2,0}(p_{n-2}a) \otimes E_{n-1,0}(ap_{n-1}).$$

But by Lemma 7.5.9 the last four summands are equal to zero in $(\tilde{\otimes} M_N(P))_{GL_N(A)}$. For example, for the last one $s_o = 2$, $r_o = 0$. Thus,

$$E_{01}(p_o) \otimes \dots \otimes E_{n-2,n-1}(p_{n-2}) \otimes E_{n-1,0}(ap_{n-1}) \equiv -E_{n-1,1}(ap_o) \otimes \dots$$

$$\dots \otimes E_{n-2,n-1}(p_{n-2}) \otimes E_{00}(p_{n-1}) + E_{01}(p_o) \otimes \dots \otimes E_{n-3,n-2}(p_{n-3}) \otimes E_{n-2,0}(p_{n-2}a) \otimes$$

$$\otimes E_{00}(p_{n-1}).$$

But this is valid for all $a \in A$ and $p_i \in P$. Replace p_{n-2} by $p_{n-2}a$ and a by the unit. We obtain

$$E_{01}(p_o) \otimes \ldots \otimes E_{n-2,n-1}(p_{n-2}a) \otimes E_{n-1,0}(p_{n-1}) \equiv -E_{n-1,1}(p_o) \otimes \ldots$$

$$\ldots \otimes E_{n-2,n-1}(p_{n-2}a) \otimes E_{00}(p_{n-1}) + E_{01}(p_o) \otimes \ldots \otimes E_{n-3,n-2}(p_{n-3}) \otimes$$

$$\otimes E_{n-2,0}(p_{n-2}a) \otimes E_{00}(p_{n-1});$$

thus,

$$E_{01}(p_o) \otimes \ldots \otimes E_{n-2,n-1}(p_{n-2}) \otimes E_{n-1,0}(ap_{n-1}) - E_{01}(p_o) \otimes \ldots$$

$$\ldots \otimes E_{n-2,n-1}(p_{n-2}a) \otimes E_{n-1,0}(p_{n-1}) \equiv (-E_{n-1,1}(ap_1) \otimes \ldots \otimes E_{n-2,n-1}(p_{n-2}) +$$

$$+ E_{n-1,1}(p_1) \otimes \ldots \otimes E_{n-2,n-1}(p_{n-2}a)) \otimes E_{00}(p_{n-1}).$$

Applying the inductive assumption we obtain the desired statement.

To prove b) note that an element $\alpha = E_{i_o j_o}(p_o) \otimes \ldots \otimes E_{i_{n-1}, j_{n-1}}(p_{n-1})$ such that $s_i(\alpha) = r_i(\alpha) = 0$ or 1 for all i is conjugated by a permutation matrix from $GL_N(\mathbb{Z})$ to an element $E_{0,\sigma(o)}(p_o) \otimes \ldots \otimes E_{n-1,\sigma(n-1)}(p_{n-1}) = \zeta_\sigma(p_o \otimes \ldots \otimes p_{n-1})$ for some σ. Now we may apply the statement b) from Lemma 7.5.9. The statement c) is obvious.

Now we shall construct the map ρ. Note that our method is similar to one suggested by Burghelea in [8] and by Hsiang and Staffeldt in [27].

Clearly it suffices to prove all the statements of Theorem 7.5.1 for the case when I is a $\mathbb{Q}$-algebra (without unit). Indeed, then in the general case we shall put ρ to be a composition $K.(A+I,I) \otimes_{\mathbb{Z}} \mathbb{Q} \xrightarrow{\sim}$ $\xrightarrow{\sim} K.(A+I \otimes_{\mathbb{Z}} \mathbb{Q},\ I \otimes_{\mathbb{Z}} \mathbb{Q}) \longrightarrow K^+_.(A+I \otimes_{\mathbb{Z}} \mathbb{Q},\ I \otimes_{\mathbb{Z}} \mathbb{Q}) = K^+_.(A+I,I) \otimes_{\mathbb{Z}} \mathbb{Q}$; the first isomorphism will be given by Theorem 7.5.2 (see below).

So let I be a $\mathbb{Q}$-algebra without unit; $I^n = 0$. Consider a simplicial algebra $A + P$ such that:

a) $P^n = 0$;

b) for any k, P_k is a free bimodule over $A \otimes_{\mathbb{Z}} \mathbb{Q}$;

c) there is an epimorphic quasi-isomorphism $P \longrightarrow I$.

It is easy to see that such P does exist. For example, we may apply the standard method of constructing free resolutions and obtain such a P that for all i

$$P_i \simeq (A \otimes_{\mathbb{Z}} \mathbb{Q}) *_{\mathbb{Z}} T(V_i)/J^n$$

where V_i is a free abelian group, $T(V_i)$ is its tensor algebra and J is the two sided ideal generated by V_i.

Now recall the definition of Waldhausen K-theory (cf. [63]). For a simplicial ring R let $\widetilde{GL}(R)$ be the monoid of matrices which are invertible up to homotopy, i.e., whose image in $M_\infty(\pi_0 R)$ lies in $GL(\pi_0 R)$. The fundamental group of $B\widetilde{GL}(R)$ is isomorphic to $GL(\pi_0 R)$ and we may apply Quillen + construction with respect to the perfect normal subgroup $E(\pi_0 R)$. Put

$$K_i(R) = \pi_i(B\widetilde{GL}(R)^+), \quad i > 0;$$

Waldhausen has shown that if $R \longrightarrow R'$ is a quasi-isomorphism, then the induced maps $K_i(R) \to K_i(R')$ are isomorphisms for all i.

On the other hand, we have seen in 2.4 that

$$K_1^+(A+I, I) = H_1(C_\bullet^{cycl}(P,A)).$$

So we have to compare Waldhausen K-functors of the simplicial ring $A+P$ (they are isomorphic to $K.(A+I)$) with the homology groups of the bicomplex $C_\bullet^{cycl}(P,A)$.

A matrix from $M_\infty(A+P)$ is invertible up to homotopy iff it is invertible iff its image under the projection $M_\infty(A+P) \to M_\infty(A)$ lies in $GL(A)$. Thus,

$$\widetilde{GL}(A+P) = GL(A+P) = GL(A) \ltimes M_\infty(P); \tag{7.5.4}$$

here $M_\infty(P)$ is the simplicial group of infinite matrices over P with the group law given by the formula

$$(m_1,m_2) \longmapsto m_1 + m_2 + m_1 m_2; \tag{7.5.5}$$

$GL(A)$ acts on $M_\infty(P)$ by conjugations; $GL(A+P)$ is equal to the semidirect product corresponding to this action; thus,

$$BGL(A+P) \simeq EGL(A) \times_{GL(A)} BM_\infty(P). \tag{7.5.6}$$

We want to find a suitable realization of $BM_\infty(P)$. Roughly speaking, this is the spatial realization of the standard Lie algebra homology complex $C.(\mathfrak{gl}(P); \mathbb{Q})$ with its differential graded coalgebra structure. More precisely, fix an $i \geqslant 0$. Since $M_\infty(P_i)^n = 0$, it is a direct limit of the filtered system of finite dimensional subalgebras P_i^α (a subalgebra is finite dimensional iff it is finitely generated). Consider the Lie algebra cohomology complexes $C^\bullet(\mathrm{Lie}\, P_i^\alpha)$; they are finite dimensional differential graded algebras. Apply to them Sullivan spatial realization functor <> (cf. [5]). Put

$$\langle C.(M_{\infty}(P_i))\rangle = \varinjlim_{\alpha} \langle C^{\cdot}(\mathrm{Lie}\, P_i^{\alpha})\rangle.$$

This is the direct limit of the filtered system of pointed simplicial sets enjoying Kan extension property; therefore

$$\pi_1(\langle C.(M_{\infty}(P_i))\rangle) = \varinjlim_{\alpha} \pi_1(\langle C^{\cdot}(\mathrm{Lie}\, P_i^{\alpha})\rangle).$$

But $\langle C^{\cdot}(\mathrm{Lie}\, P_i^{\alpha})\rangle$ is nothing other than the classifying space of P_i^{α} (regarded as a discrete group with operation (7.5.5)). In fact, it follows from the theorem proved by A.Haefliger in [67] and E.Kuzmin in [68], that the group cohomology $H^1(P_i^{\alpha}; \mathbb{Z})$ is equal to the Lie algebra cohomology $H^1(\mathrm{Lie}\, P; \mathbb{Q})$ for $1 > 0$. Furthermore, Kuzmin's proof clearly shows that the transgressions in the Hochschild-Serre spectral sequences coincide: $H^{\cdot}((P_i^{\alpha})^r/(P_i^{\alpha})^{r+1};\ H^{\cdot}(P_i^{\alpha})^{r+1}; \mathbb{Z}) \Rightarrow H^{\cdot}((P_i^{\alpha})^r; \mathbb{Z})$; $H^{\cdot}(\mathrm{Lie}(P_i^{\alpha})^r/\mathrm{Lie}(P_i^{\alpha})^{r+1};\ H^{\cdot}(\mathrm{Lie}(P_i^{\alpha})^{r+1}; \mathbb{Z}) \Rightarrow H^{\cdot}(\mathrm{Lie}(P_i^{\alpha})^r; \mathbb{Z})$, $r \in \mathbb{N}$;

$H^1(\mathrm{Lie}(P_i^{\alpha})^r; \mathbb{Z}) := H^1(\mathrm{Lie}(P_i^{\alpha})^r; \mathbb{Q})$ for $1 > 0$ and $H^0(\mathrm{Lie}(P_i^{\alpha})^r; \mathbb{Z}) := \mathbb{Z}$.

Proceeding by induction on r we obtain that Postnikov invariants of the differential graded algebra $C^{\cdot}(\mathrm{Lie}(P_i^{\alpha}); \mathbb{Q})$ are the same as Postnikov invariants of the classifying space BP_i^{α}. Both spaces are rational and nilpotent. This shows that

$$BP_i^{\alpha} \simeq \langle C^{\cdot}(\mathrm{Lie}(P_i^{\alpha}))\rangle;$$

so

$$BM_{\infty}(P_i) \simeq \langle C.(M_{\infty}(P_i))\rangle;$$

these spaces are connected; therefore we obtain from the realization lemma that

$$BM_{\infty}(P) \simeq \langle C.(M_{\infty}(P))\rangle. \qquad (7.5.7)$$

Note that this argument enables one to prove Theorem 7.5.2. In fact, we see that in the case when I is a $\mathbb{Q}$-algebra the space $\mathrm{fibre}(BGL(A)^+ \longrightarrow BGL(A+I)^+)$ is rational; if I is arbitrary, then the theorem of Haefliger and Kuzmin easily shows that $BM_{\infty}(P \otimes_{\mathbb{Z}} \mathbb{Q})$ is the O-localization of $BM_{\infty}(P)$; now the fibration homology spectral sequence shows that $\mathrm{fibre}(BGL(A)^+ \longrightarrow BGL(A + I \otimes_{\mathbb{Z}} \mathbb{Q})^+)$ is the O-localization of $\mathrm{fibre}(BGL(A)^+ \longrightarrow BGL(A+I)^+)$.

Return to the case when I is a $\mathbb{Q}$-algebra. Put

$$<C.(M_\infty P)_{GL(A)}> = \varinjlim <\bar{C}.(\mathrm{Lie}\ P_i^\alpha)^*>;$$

$$\bar{C}.(\mathrm{Lie}\ P_i) := C.(\mathrm{Lie}\ P_i^\alpha)/(C.(\mathrm{Lie}\ P_i^\alpha) \cap (\sum_g \mathrm{im}(g-1));$$

the sum is taken over all $g \in GL(A)$. There is a map

$$<C.(M_\infty(P))> \longrightarrow <C.(M_\infty(P))_{GL(A)}>.$$

Note that the operation of Whithney sum is strictly commutative and associative in $C.(M_\infty(P))_{GL(A)}$; in fact, $\alpha \oplus \beta$ is conjugated to $\beta \oplus \alpha$ and $\alpha \oplus (\beta \oplus \gamma)$ is conjugated to $(\alpha \oplus \beta) \oplus \gamma$ by means of the permutation matrices from $GL(\mathbb{Z})$. Therefore, $BGL(A)^+ \times <C.(M_\infty(P))_{GL(A)}>$ is an H-space; thus, the universal property of the Quillen's + construction implies that there is a map making the following diagram homotopy commutative:

$$\begin{array}{ccc}
BGL(A+I) & = & BGL(A+P) \\
\Big\downarrow & & \| \\
 & & EGL(A) \times_{GL(A)} BGL(M_\infty(P)) \\
 & & \downarrow \\
 & & BGL(A) \times <C.(M_\infty(P))_{GL(A)}> \\
 & & \downarrow \\
BGL(A+I)^+ & \dashrightarrow & BGL(A)^+ \times <C.(M_\infty(P))_{GL(A)}>
\end{array} \qquad (7.5.8)$$

On the other hand, Proposition 7.5.8 implies that

$$C.(M_\infty(P))_{GL(A)} \simeq S^\cdot(C_\cdot^{cycl}(P,A)) = \varinjlim S^\cdot(C_\cdot^\beta);$$

$C_\cdot^\beta$ is the filtered system of subcomplexes of $C_\cdot^{cycl}(P,A)$ having finite dimensional homology. It is easily seen that

$$<C.(M_\infty(P))_{GL(A)}> = \varinjlim <S^\cdot(C_\cdot^\beta)^*>;$$

the right hand side is again a direct limit of filtered system of Kan pointed simplicial sets; therefore $\pi_1(<C.(M_\infty(P))_{GL(A)}>) \simeq K_1^+(A+I,I)$. So we obtained the maps

$$K_1(A+I) \longrightarrow K_1(A) \oplus K_1^+(A+I,I);$$

the restriction to $K.(A+I,I)$ is the desired map ρ.

It is clear that ρ is an isomorphism when $l=1$. In order to investigate the cases $l = 2,3$ one has to consider the spectral sequence converging to $H.(BGL(A+P);\mathbb{Q})$. We have seen that $GL(A+P) = GL(A) \ltimes M_\infty(P)$ and that

$$H.(BGL(A+P);\mathbb{Q}) \simeq \mathbb{H}.(GL(A); \Lambda^\cdot M_\infty(P.))$$

is equal to the hyperhomology of $GL(A)$ with values in the standard

Lie algebra complex $\Lambda^{\cdot}M_{\infty}(P.)$ in which the differential $d_0 - d_1 + \dots \pm d_n : P_n \to P_{n-1}$ induces a structure of a bicomplex. We may consider the projection

$$H.(GL(A); \Lambda^{\cdot}M_{\infty}(P.)) \xrightarrow{\P} H.(GL(A); \Lambda^{\cdot}M_{\infty}(P.)_{GL(A)});$$

the operation of Whithney sum induces a coalgebra structure on the both sides and ¶ is a coalgebra homomorphism. The construction of the map ρ (together with the Cartan-Milnor-Moore theorem) shows that ρ is equal to the restriction of ¶ to the primitive part.

Consider the simple complex $(\Lambda M_{\infty} P).$ associated to the bicomplex $\Lambda^{\cdot}M_{\infty} P.$ We have the spectral sequence

$$E^1_{ij} = H_i(GL(A), (\Lambda M_{\infty} P)_j) \Rightarrow H.(BGL(A+P); \mathbb{Q})$$

There is the analogous spectral sequence

$$\tilde{E}^1_{ij} = H_i(GL(A), ((\Lambda M_{\infty} P)_j)_{GL(A)}) \Rightarrow H.(GL(A); \mathbb{Q}) \otimes$$

$$\otimes H.((\Lambda^{\cdot}M_{\infty} P.)_{GL(A)});$$

$\check{E}$.. degenerates in the second term. Now apply the following statement which is due to C.Kassel ([39]):

<u>Lemma 7.5.10</u>. Let A be a ring and P a free bimodule over A. Then the map

$$H_1(GL(A); M_{\infty} P) \to H_1(GL(A); (M_{\infty} P)_{GL(A)})$$

is an isomorphism.

So, we have $E^1_{11} \xrightarrow{\sim} \tilde{E}^1_{11}$. The desired statement follows now from the comparison of spectral sequences $E, \tilde{E}$. We leave the details to the reader.

Now assume that $I^2 = 0$. Then we may consider the Hochschild-Serre spectral sequence:

$$E^2_{ij} = H_i(GL(A); H_j(M_{\infty} P)) = H_i(GL(A); \Lambda^{\cdot}M_{\infty}(P)) \Rightarrow H_{i+j}(GL(A+P)).$$

There is a map $H_0(GL(A); \Lambda^{\cdot}M_{\infty}(P)) \to H.(GL(A+P))$; its restriction to the primitive elements splits the homomorphism ρ.

To prove the statement 4) of Theorem 7.5.1 recall the following statement (it was proved for the case $\mathcal{O} = \mathbb{Z}$; the proof in the general case is similar).

<u>Lemma 7.5.11</u>. (Farrell and Hsiang, [69]). Let $\mathcal{O}$ be the ring of integers in a number field F. Consider the algebraic group $G_n = \mathrm{Res}_{F/Q} SL_n$. Let ρ be an irreducible non-trivial representation of $G(\mathbb{R})$. Then for $n \geq 5$, $1 \leq (n-1)/4$

$$H^1(SL_n(\mathcal{O});\ \rho) = 0.$$

The Hochschild-Serre spectral sequence shows that $H^1(GL_n(\mathcal{O});\rho) = 0$, $1 < \frac{n-1}{4}$, $n \geq 5$; now it is enough to check that $\Lambda^{\cdot} M_n(P.) \otimes \mathbb{R}$ admits a decomposition into a direct sum of irreducible representations of $G(\mathbb{R})$; so, the map

$$H.(GL(\mathcal{O});\ \Lambda^{\cdot} M_{\infty}(P.)) \otimes \mathbb{R} \to H.(GL(\mathcal{O});\ \Lambda^{\cdot} M_{\infty}(P.)_{GL(\mathcal{O})}) \otimes \mathbb{R}$$

is a coalgebra isomorphism; so it is an isomorphism being restricted to the primitive part. This ends the proof of Theorem 7.5.1.

Remark 7.5.12. There is a generalization of the map ρ. Assume that $k = \mathbb{Z}$. Let the $\hat{\mathcal{L}}.$ be the complex introduced in 1.5; put $\check{L}. = \mathrm{filt}^0 \hat{\mathcal{L}}.$; $\check{K}^+_{\cdot}(B) = H_{\cdot,-2}(\check{L}.(B))$ for any ring B. Then one may define a functorial map (it is obtained from the Chern character 7.3).

$$K_l(A+I,I) \xrightarrow{\rho_l} \check{K}^+_l(A+I,I), \quad l > 0.$$

The construction involves the technique of May coalgebras over operads (cf. [25]) instead of the technique of Sullivan theory. Let $l = 1$. Then $\check{L}_{l-2}(B) = \prod_{i=1}^{\infty} (\overset{i}{\otimes} B)$; there is a map $B/[B,B] \to \check{K}^+_1(B)$; the image of b^k B under this map becomes to be divisible by k. For instance, $b^2 + 2b \otimes b$ is the boundary of $b^2 \in \check{L}_0$; so $-b \otimes b = b^2/2$ in $\check{K}_1$. It may be shown that $\check{K}^+_1(M_n(B)) \cong \check{K}^+_1(B), \forall B$; let $1+g \in GL_n(B)$, $g \in M_n(I)$. There is an element $\log(1+g) \in \check{K}_1(M_n(A+I),\ M_n(I))$; $\log(1+g) = \sum_{k=1}^{\infty} \frac{(-g)^k}{k}$; it may be shown that this gives a well defined map

$$\rho_1 = \log : \check{K}_1(A+I,I) \to \check{K}^+_1(A+I,I).$$

There is an explicit formula for log:

$$\log(1+g) = \sum_{k=0}^{\infty} k!((g \otimes \bar{g})^{\otimes(k+1)} - (g \otimes \bar{g})^{\otimes k} \otimes g),$$

where $1+\bar{g} = (1+g)^{-1}$; $\log(1+g) \in \prod_{k=1}^{\infty} (\overset{k}{\otimes} M_{\infty}(A+I)) = \check{L}_{-1}(A+I)$; a straightforward verification shows that this is a cycle.

It is likely that the maps $\rho.$ are closely related to the various maps which are constructed for studying deformations of algebraic K-theory (S.Bloch, H.Maazen, J.Stienstra, C.Soulé). We shall discuss these questions elsewhere.

Appendix. Cyclic objects

The notion of a cyclic object in a category was introduced by Connes in [14]. It enables one to give another (comparing with Chapter 2) translation of the basic definitions of Chapter 1 to the invariant language of homological algebra and to define the bivariant K^+-functors. The cyclic objects are also a combinatorial analogue of the topological spaces on which the topological group S^1 acts. Subsection A1 contains a brief recapitulation of the homology of a small category. In A2, A3 we introduce the notion of a cyclic object in a category and give some other definitions. After this we study the connection of these notions with additive K-theory (following Connes). In A4, A5 we prove some simple facts on the homology of small categories and give one of the possible interpretations of the analogy between the cyclic sets and the topological spaces with an action of S^1. Then we compute the additive K-functors of the twisted group algebras (A6). This is a generalization of the theorem which is due to Burghelea and Fiedorowicz [9]. In A7 we prove the statement of Connes about the isomorphism of the category Λ with its dual. Point A8 is devoted to the generalisation of the results of Chapter 2. We define the relative additive K-functors over an arbitrary scalar ring in terms of the derived functors on the category of algebras. As an application we give the proofs of some statements of Chapter 3. In Subsection A9 we give a brief sketch of homotopy theory of cyclic sets and of some related questions (according to the papers by Burghelea and Fiedorowicz [7], [9]).

Almost all small categories considered below have the set of objects numerated by nonnegative integers. So we fix the notations [i], [j], etc. for an object of an arbitrary small category and M_i, M_j etc. for the value of a functor M on an object [i].

For the functors M, N from a small category to a category with tensor products we denote by $M \boxtimes N$ the tensor product of functors. This is the functor whose values on the objects and morphisms of C are the tensor products of those for M and N.

A1. Homology of small categories. Let $\mathcal{C}$ be a small category and $\mathfrak{A}$ an abelian category*) Let $\mathcal{C}^{\circ}$ be the category opposite to $\mathcal{C}$. The category of contravariant functors from $\mathcal{C}$ to $\mathfrak{A}$ is abelian. We denote it by $\mathfrak{A}^{\mathcal{C}^{\circ}}$. Recall that for $M : \mathcal{C}^{\circ} \to \mathfrak{A}$ the homology groups $H.(\mathcal{C}^{\circ}, M)$ of the category $\mathcal{C}$ with values in M are defined as the left derived functors of the right exact functor $\varinjlim : \mathfrak{A}^{\mathcal{C}^{\circ}} \longrightarrow \mathfrak{A}$ (cf. [23]).

*) with enough projectives

Let Z be a projective object in $\mathfrak{A}$ and $[j]$ an object in $\mathcal{C}$. The functor $Z^{(j)} : [i] \mapsto \hom_{\mathcal{C}}([i],[j])Z$ is projective as an object of $\mathfrak{A}^{\mathcal{C}^\circ}$. Indeed, Yoneda's lemma implies $\mathrm{Hom}_{\mathfrak{A}^{\mathcal{C}^\circ}}(Z^{(j)},M) = \mathrm{Hom}_{\mathfrak{A}}(Z,M_j)$.

Let $\mathcal{C}_o$ be the subcategory in $\mathcal{C}$ such that $\mathrm{Ob}\,\mathcal{C}_o = \mathrm{Ob}\,\mathcal{C}$ and $\mathrm{Mor}\,\mathcal{C}_o = \{1_{[i]} : [i] \in \mathrm{Ob}\,\mathcal{C}\}$. Let $I : \mathfrak{A}^{\mathcal{C}^\circ} \to \mathfrak{A}^{\mathcal{C}_o^\circ}$ be the forgetting functor. For an M in $\mathrm{Ob}\,\mathfrak{A}^{\mathcal{C}_o^\circ}$ let FM be an object in $\mathfrak{A}^{\mathcal{C}^\circ}$:

$$(FM)_i = \coprod_{[i]\in \mathrm{Ob}\,\mathcal{C}} \hom_{\mathcal{C}}([i],[j]) \times M_j .$$

The functor F is left adjoint to I. Particularly, let M be an object of $\mathfrak{A}^{\mathcal{C}^\circ}$ such that M_i is projective for any $[i]$. Then we receive a projective cover $FIM \to M$. Having a pair of adjoint functors F,I we may now construct a standard Maclane bar resolution of M. Namely, $E_{\mathcal{C}^\circ}(M)$ is a simplicial object of $\mathfrak{A}^{\mathcal{C}^\circ}$; for $n \in \mathbb{Z}_+$, $[i]_{-1} \in \mathrm{Ob}\,\mathcal{C}$ $E_{\mathcal{C}^\circ}(M)_n$ is an object of $\mathfrak{A}^{\mathcal{C}^\circ}$ such that

$$(E_{\mathcal{C}^\circ}(M)_n)_{i_{-1}} = \coprod_{[i]_{-1}\to\dots\to[i]_n} M_{i_n};$$

$$d_k([i]_{-1} \xrightarrow{f_{-1}} [i]_o \xrightarrow{f_o} \dots \xrightarrow{f_{n-1}} [i]_n;\ m) = ([i]_{-1} \xrightarrow{f_{-1}} [i]_o \to \dots$$
$$\to i_{k-1} \xrightarrow{f_k f_{k-1}} [i]_{k+1} \to \dots \xrightarrow{f_{n-1}} [i]_n;\ m),\quad 0 \leqslant k < n;$$

$$d_n([i]_{-1} \to \dots \to [i]_{n-1}:m) = ([i]_{-1} \xrightarrow{f_{-1}} \dots \xrightarrow{f_{n-2}} [i]_{n-1};\ mf_{n-1});$$

$$s_k([i]_{-1} \to \dots \to [i]_n;m) = ([i]_{-1} \to \dots \xrightarrow{f_{k-1}} [i]_k \xrightarrow{1_{[i_k]}} [i]_k \to [i]_{k+1} \to \dots$$
$$\dots \to [i]_n;\ m),\quad 0 \leqslant k \leqslant n.$$

Now put $\underrightarrow{\mathrm{holim}}_{\mathcal{C}^\circ} M := \underrightarrow{\lim}_{\mathcal{C}^\circ} E_{\mathcal{C}^\circ}(M)$. It is a simplicial object of $\mathfrak{A}$:

$$(\underrightarrow{\mathrm{holim}}_{\mathcal{C}^\circ} M)_n = \coprod_{[i]_o\to\dots\to[i]_n} M_{i_n} ;$$

if $x = ([i]_o \xrightarrow{f_o} \dots \xrightarrow{f_{n-1}} [i]_n;\ m) \in \underrightarrow{\mathrm{holim}}_n$, then

$d_o x = ([i]_1 \to \dots \to [i]_n;\ m)$; $d_k x = ([i]_0 \to [i]_1 \to \dots \to [i]_{k-1} \xrightarrow{f_k f_{k-1}} [i]_{k+1} \to \dots$
$\dots \to [i]_n;\ m)$, $0 < k < n$; $d_n x = ([i]_o \to \dots \to [i]_{n-1};\ mf_{n-1})$; $s_k x = ([i]_0 \to \dots$
$\dots \to [i]_k \xrightarrow{1_{[i_k]}} i_k \to \dots \to [i]_n;\ m)$, $0 \leqslant k \leqslant n$.

Now note that the definitions of $E_{\mathcal{C}^\circ}(M)$ and $\underrightarrow{\mathrm{holim}}_{\mathcal{C}^\circ} M$ are reasonable for the functors $\mathcal{C}^\circ \to \mathfrak{A}$, where $\mathfrak{A}$ is an arbitrary category with direct sums (particularly the category (Sets)).

Let $\mathfrak{A}$ be the category of modules over a commutative ring k, M an object of $\mathfrak{A}^{\mathcal{C}^\circ}$, N an object of $\mathfrak{A}^{\mathcal{C}}$. Define $M \otimes_{k[\mathcal{C}]} N$ as a quotient module of $\bigoplus_{[i]\in Ob\mathcal{C}} M_i \otimes_k N_i$ by the submodule generated by

$$mf \otimes n - m \otimes fn, \quad m \in M_i, \quad n \in N_j, \quad f \in hom_{\mathcal{C}}([i],[j]).$$

Let $k^\natural$ be the constant functor $\mathcal{C}^\circ \to \mathfrak{A}$: $k_i = k, \forall [i]$. Then $\varinjlim_{\mathcal{C}^\circ} M = M \otimes_{k[\mathcal{C}]} k^\natural$. The functor $\otimes_{k[\mathcal{C}]}$ is right exact with respect to both arguments. The derived functors are denoted by $Tor_\cdot^{k[\mathcal{C}]}(M,N)$. Note that since $k_i = k$ are projective in $\mathfrak{A}$ for all $[i]$ it is easy to see that $H_\cdot(\mathcal{C}^\circ, M) = H_\cdot(\operatorname{holim}_{\overrightarrow{\mathcal{C}^\circ}} M)$ for an arbitrary object M of $\mathfrak{A}^{\mathcal{C}^\circ}$.

A2. <u>The categories Λ_p, $1 \leqslant p \leqslant \infty$</u>. For any $j \in \mathbb{Z}_+$ let $[j]$ be the set of all integers with the standard ordering $<$ and t_j an automorphism of $[j]$: $t_j(i) = i+1$, $i \in \mathbb{Z}$. Denote by Λ_∞ a category in which objects are $[j]$ for all $j \geqslant 0$ and $hom_{\Lambda_\infty}([i],[j])$ is the set of all nondecreasing maps $f : [i] \longrightarrow [j]$ such that $t_j^{j+1} f = f\, t_i^{i+1}$.

Let now r be a fixed natural number. We say that the two elements $f, f' \in hom_{\Lambda_\infty}([i],[j])$ are equivalent if $f = t_j^{k(j+1)} f'$ $(= f'\, t_i^{k(i+1)})$. It is evident that if f is equivalent to f' then f g is equivalent to f'g and hf is equivalent to hf' for all g,h. Therefore we may define a category Λ_r whose objects are the same as ones in Λ_∞ and morphisms are the equivalence classes of ones in Λ_∞.

Put $\Lambda = \Lambda_1$. This category was introduced by Connes in [14].

For any $r \in [1, \infty]$ consider a subcategory of Λ_r consisting of equivalence classes of such $f : [i] \longrightarrow [j]$ that $f^{-1}(\{0,1,\dots,j\}) = \{0,1,\dots,i\}$. It is clear that this category is isomorphic to the category Δ of isomorphism classes of totally ordered non-empty finite sets and nondecreasing maps. We denote it by the same letter.

An element $g \in hom_{\Lambda_r}([i],[j])$ admits unique decomposition $g = f \cdot \sigma$ where $\sigma \in Aut_{\Lambda_r}([i])$, $f \in hom_\Delta([i],[j])$. Here

$$Aut_{\Lambda_r}([i]) \simeq \mathbb{Z}/(i+1)r\mathbb{Z}, \quad r < \infty \ ; \quad Aut_{\Lambda_\infty}([i]) \simeq \mathbb{Z}.$$

These groups are generated by t_i. Now for the standard generators $d_l^{(i)} : [i-1] \longrightarrow [i]$, $s_l^{(i)} : [i+1] \longrightarrow [i]$ we have in Λ_∞

$$t_i d_o^{(i)} = d_1^{(i)} t_{i-1}; \dots; \ t_i d_{i-1}^{(i)} = d_i^{(i)} t_{i-1}; \ t_i d_i^{(i)} = d_o^{(i)} ; \qquad \text{(A2.1)}$$

$$t_i s_0^{(i)} = s_1^{(i)} t_{i+1}; \ldots ; \ t_i s_{i-1}^{(i)} = s_i^{(i)} t_{i+1}; \ \ t_i s_i^{(i)} = s_0^{(i)} t_{i+1}^2 \qquad \text{(A2.2)}$$

When $r < \infty$, we have a supplementary relation in Λ_r :

$$t_i^{r(i+1)} = 1 \qquad \text{(A2.3)}$$

Definition A2.1. A cyclic object of a category $\mathfrak{A}$ is a functor $M : \Lambda^0 \to \mathfrak{A}$; an r-cyclic object of $\mathfrak{A}$ is a functor $M : \Lambda_r^0 \to \mathfrak{A}$, $1 \leqslant r \leqslant \infty$.

So, an r-cyclic object of $\mathfrak{A}$ is the same as a simplicial object M together with the automorphisms $t_i \in \mathrm{Aut}_{\mathfrak{A}}(M_i)$, $i \geqslant 0$, such that (A2.1), (A2.2) and (for $r < \infty$) (A2.3) hold.

Examples. A2.2. Let A be an associative k-algebra. Construct a cyclic object $A^\natural$ in the category of k-modules:

$$A_i = \overset{i+1}{\otimes} A; \ (a_0 \otimes \ldots \otimes a_i)d_l = a_0 \otimes \ldots \otimes a_{l-1}a_l \otimes \ldots \otimes a_i, \quad 0 < l \leqslant i;$$

$$(a_0 \otimes \ldots \otimes a_i)d_0 = a_1 \otimes \ldots \otimes a_i a_0; \ (a_0 \otimes \ldots \otimes a_i)t = a_1 \otimes \ldots \otimes a_i \otimes a_0;$$

$$(a_0 \otimes \ldots \otimes a_i)s_l = a_0 \otimes \ldots \otimes 1 \otimes a_l \otimes \ldots \otimes a_i .$$

More generally, let $\alpha \in \mathrm{Aut}\, A$. Define an r-cyclic k-module $A_\alpha^{\natural, r}$ where $\alpha^r = 1$: $(A_\alpha^{\natural, r})_i = \overset{i+1}{\otimes} A$; s_l and d_l (for $l > 0$) as above;

$$(a_0 \otimes \ldots \otimes a_i)d_0 = a_1 \otimes \ldots \otimes a_i \alpha(a_0);$$

$$(a_0 \otimes \ldots \otimes a_i)t = a_1 \otimes \ldots \otimes a_i \otimes \alpha(a_0).$$

The corresponding simplicial module has the homology groups equal to $H.(A, A_\alpha)$.

A2.3. Let N be a group and $g \in N$. Construct an ∞-cyclic set $C(N,g) : C(N,g)_i = N^{i+1}$; for $x = (g_{-1}, \ldots, g_{i-1})$ put

$$xd_l = (g_{-1}, \ldots, g_{l-1}\, g_l, \ldots, g_{i-1}) \quad \text{for } 0 \leqslant l < i;$$

$$xd_i = (g_{-1}, \ldots, g_{i-2}); \ \ xt = (g_{-1}\, g_0, g_1, \ldots, g_{i-1}, (g_0 \ldots g_{i-1})^{-1} g);$$

$$xs_l = (g_{-1}, \ldots, g_{l-2}, g_{l-1}, 1, \ldots, g_{i-1}) \quad \text{for all } l.$$

Here $xt^{i+1} = (g_{-1} g, g^{-1} g_0 g, \ldots, g^{-1} g_{i-1} g)$. So, if $g^r = 1$, then we obtain an r-cyclic object. The corresponding simplicial set is the total space of the universal fibration over the classifying space. Put $\mathfrak{P}(N,g) = k \cdot C(N,g)$. This is an ∞-cyclic free $k[N]$-module; it was used in 7.3 for $g = 1$ (in this case we denote it by $\mathfrak{P}(N)$).

A3. Homology of the categories Λ_r. Put for $r < \infty$

$$N = \sum_{i=0}^{r(n+1)-1} (t_n \cdot (-1)^n)^i .$$

Consider the diagrams

$$\begin{array}{ccccccc}
 & & \downarrow\delta & & \downarrow b & & \downarrow\delta \\
\dots \xrightarrow{1-(-1)^n t} & & k\cdot\hom_{\Lambda_r}([n],\cdot) & \xrightarrow{N} & k\cdot\hom_{\Lambda_r}([n],\cdot) & \xrightarrow{1-(-1)^n t} & k\ \hom([n],\cdot) \\
 & & \downarrow\delta & & \downarrow b & & \downarrow\delta \\
\dots \longrightarrow & & k\cdot\hom_{\Lambda_r}([n-1],\cdot) & \xrightarrow{N} & k\cdot\hom_{\Lambda_r}([n-1],\cdot) & \longrightarrow & k\cdot\hom_{\Lambda_r}([n-1],\cdot) \\
 & & \downarrow\delta & & \downarrow b & & \downarrow\delta \\
 & & \vdots & & \vdots & & \vdots \\
 & & \downarrow & & \downarrow & & \downarrow \\
\xrightarrow{1+t} & & k\cdot\hom_{\Lambda_r}([1],\cdot) & \xrightarrow{N} & k\cdot\hom_{\Lambda_r}([1],\cdot) & \xrightarrow{r+t} & k\cdot\hom_{\Lambda_r}([1],\cdot) \\
 & & \downarrow\delta & & \downarrow b & & \delta\downarrow \\
\xrightarrow{1-t} & & k\cdot\hom_{\Lambda_r}([0],\cdot) & \xrightarrow{N} & k\cdot\hom_{\Lambda_r}([0],\cdot) & \xrightarrow{1-t} & k\cdot\hom_{\Lambda_r}([0],\cdot)
\end{array} \qquad (A3.1)$$

$(r < \infty)$;

$$\begin{array}{ccccccc}
\xrightarrow{b} & k\cdot\hom_{\Lambda_\infty}([n],\cdot) & \xrightarrow{b} & k\cdot\hom_{\Lambda_\infty}([n-1],\cdot) & \xrightarrow{b} \dots \xrightarrow{b} & k\cdot\hom_{\Lambda_\infty}([0],\cdot) \\
 & \Big\downarrow 1-(-1)^n t & & \Big\downarrow 1-(-1)^{n-1} t & & \Big\downarrow 1-t \\
\xrightarrow{\delta} & k\ \hom_{\Lambda_\infty}([n],\cdot) & \xrightarrow{\delta} & k\cdot\hom_{\Lambda_\infty}([n-1],\cdot) & \xrightarrow{\delta} \dots \xrightarrow{\delta} & k\cdot\hom_{\Lambda_\infty}([0],\cdot)
\end{array} \qquad (A3.2)$$

Here $\delta = d_0 - d_1 + \dots + (-1)^n d_n$, $b = d_1 - d_2 + \dots + (-1)^{n-1} d_n$ (acting by right multiplication).

Lemma A3.1. The diagrams (A3.1), (A3.2) are bicomplexes. The corresponding simple complexes are the projective resolutions of the constant r-cyclic objects $k_1^{\natural, r}$, $r \leq \infty$.

Proof. The differentials in the rows anticommute with those in the columns. It may be shown exactly in the same way as in proof of Lemma 1.2.1. Furthermore, we have seen that $k\cdot\hom_{\Lambda_r}([n],\cdot)$ are free $k[\mathrm{Aut}_{\Lambda_r}([n])]$-modules; so the differentials including t are those of the standard resolutions of the cyclic groups of order r ($\leq \infty$). So, the first terms of the corresponding spectral sequences are equal to $k\cdot\hom_{\Lambda_r}([n],\cdot)/\mathrm{im}(1\pm t) = k\cdot\hom_\Delta([n],\cdot)$; but it is easy (and well known) that the complex

$$\to k\cdot\hom_\Delta([n],\cdot) \xrightarrow{\delta} k\cdot\hom_\Delta([n-1],\cdot) \xrightarrow{\delta} \dots \xrightarrow{\delta} k\ \hom_\Delta([0],\cdot) \to 0 \qquad (A3.3)$$

is a projective resolution of the cosimplicial module $k^\natural$.

For the case $r = 1$ cf. [14].

Corollary A3.2. The homology groups $H.(\Lambda_r^o ; M)$ may be computed by the following double complexes:

$$
\begin{array}{ccccccc}
 & & \vdots & & \vdots & & \vdots \\
\cdots & \xrightarrow{1-(-1)^n t} & M_n & \xrightarrow{N} & M_n & \xrightarrow{1-(-1)^n t} & M_n \\
 & & \downarrow \delta & & \downarrow b & & \downarrow \delta \\
\cdots & \xrightarrow{1-(-1)^{n-1} t} & M_{n-1} & \xrightarrow{N} & M_{n-1} & \longrightarrow & M_{n-1} \\
 & & \downarrow \delta & & \vdots & & \downarrow \delta \\
 & & \vdots & & \downarrow b & & \vdots \\
\cdots & \xrightarrow{1+t} & M_1 & \longrightarrow & M_1 & \xrightarrow{1+t} & M_1 \\
 & & \downarrow \delta & & \downarrow b & & \downarrow \delta \\
\cdots & \xrightarrow{1-t} & M_o & \xrightarrow{N} & M_o & \xrightarrow{1-t} & M_o
\end{array}
\qquad\qquad
\begin{array}{ccc}
\vdots & & \vdots \\
\downarrow b & & \downarrow \delta \\
M_n & \xrightarrow{1-(-1)^n t} & M_n \\
\downarrow b & & \downarrow \delta \\
M_{n-1} & \longrightarrow & M_{n-1} \\
\downarrow b & & \downarrow \delta \\
\vdots & & \vdots \\
M_1 & \xrightarrow{1+t} & M_1 \\
\downarrow b & & \downarrow \delta \\
M_o & \xrightarrow{1-t} & M_o
\end{array}
\qquad (A3.4)
$$

$$(r < \infty) \qquad\qquad\qquad (r = \infty)$$

Theorem A3.3. (Connes). $K^+_{i+1}(A) \simeq H_i(\Lambda^o, A^\natural)$.

Passing to the spectral sequences of the bicomplexes (A3.4) we obtain

Corollary A3.4. 1) There exist the spectral sequences for $r < \infty$:

$$E^2_{ij} = H_i(\Lambda^o_r, k^\natural) \otimes H_j(\Delta^o, M) \Rightarrow H_{i+j}(\Lambda^o_r, M)$$

2) $H_i(\Lambda^o_\infty, M) \xrightarrow{\sim} H_i(\Delta^o, M)$.

Indeed, the differential b is acyclic.

Corollary A3.5. Let M be a ∞-cyclic k-module. Then the automorphisms t_n^{n+1} act trivially on $H.(\Delta^o; M)$. In particular, α acts trivially on the Hochschild homology $H.(A, A_\alpha)$.

Definition A3.6. Let char $k = 0$. Put for a k-algebra A and an automorphism α such that $\alpha^r = 1$ $(r < \infty)$:

$$C^{cycl}_{n+1}(\mathbb{Z}/r\mathbb{Z}, \alpha; A) = C_n(A, A_\alpha)/\mathrm{im}(1 - \tau),$$

$\tau(a_o \otimes \ldots \otimes a_n) = (-1)^n a_1 \otimes \ldots \otimes a_n \otimes \alpha(a_o)$. Then δ induces a differential in $C^{cycl}(\mathbb{Z}/r\mathbb{Z}, \alpha; A)$; we denote the homology of the corresponding complex by $K^+_\cdot(\mathbb{Z}/r\mathbb{Z}, \alpha; A)$.

Definition A3.7. For an arbitrary scalar ring k and $\alpha \in \mathrm{Aut}\, A$, $\alpha^r = 1$, put $K^+_{n+1}(\mathbb{Z}/r\mathbb{Z}, \alpha; A) = H_n(\Lambda^o_r, A^{\natural, r}_\alpha)$. If $r = \infty$, then $K^+_{n+1}(Z, \alpha; A) = H_n(A, A_\alpha)$.

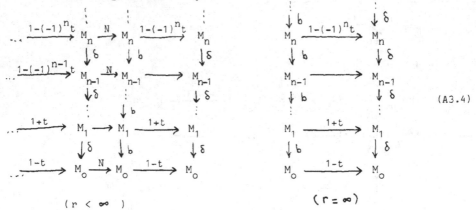

A4. Let $\mathcal{C}$ be a small category, $\mathcal{D}$ be its subcategory such that for any $[i] \in \mathrm{Ob}\,\mathcal{C}$ a subgroup $\Gamma_i \subseteq \mathrm{Aut}_{\mathcal{C}}[i]$ is fixed and for every $g \in \hom_{\mathcal{C}}([i],[j])$ there is uniquely defined decomposition

$$g = f \cdot \sigma, \quad \sigma \in \Gamma_i, \quad f \in \hom_{\mathcal{D}}([i],[j]) \qquad (A4.1)$$

In particular, write down (A4.1) for $g = \sigma \cdot f$, $f \in \hom_{\mathcal{D}}([i],[j])$, $\sigma \in \Gamma_j$:

$$\sigma \cdot f = \sigma(f) \cdot f^*(\sigma) \qquad (A4.2)$$

Uniqueness of (A4.1) implies

$$g^*(f^*(\sigma)) = (fg)^*(\sigma); \quad (\sigma\tau)(f) = \sigma(\tau(f)); \qquad (A4.3)$$

$$\sigma(fg) = \sigma(f) \cdot (f^*\sigma)(g); \quad f^*(\sigma\tau) = \tau(f)(\sigma)\, f^*(\tau) \qquad (A4.4)$$

Thus, $\hom_{\mathcal{D}}([i],[j])$ is a Γ_j-set and $\Gamma.$ is a functor from $\mathcal{D}^o$ to (Sets).

Example A4.1. We have for the category Λ_r, $r \leqslant \infty$, $\Gamma_i = \mathbb{Z}/r(i+1)\mathbb{Z}$; for Λ_∞ $\Gamma_i = \mathbb{Z}$. For any $r \leqslant \infty$ we obtain a simplicial set $\Gamma.$ Let I be the standard simplicial interval: $I_n = \hom_\Delta([n],[1])$. There are two maps $* \overset{\partial_0}{\underset{\partial_1}{\rightrightarrows}} I$. The formulas (A2.1)-(A2.3) imply that for Λ_∞

$$\Gamma \simeq \left(\coprod_{i=-\infty}^{\infty} I_i\right) / \sim \qquad (A4.5)$$

where $I_i = I$ for all i and $I_i \ni \partial_1(*) \sim \partial_0(*) \in I_{i+1}$. If $r < \infty$, then also I_{r+i} is identified with I_i.

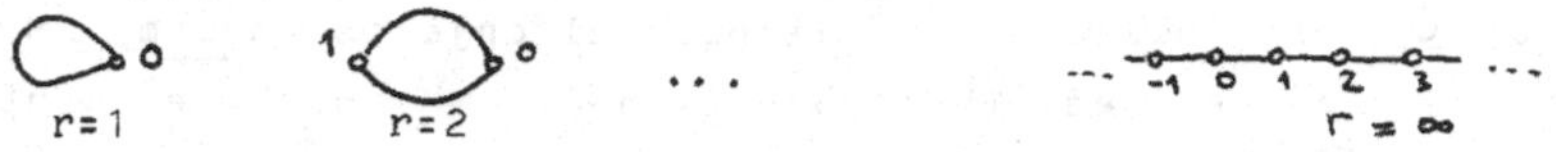

The isomorphism (A4.5) associates to an element $t^{a+(n+1)b} \in \Gamma_n$, $0 \leqslant a \leqslant n$, an element $\varphi_a \in (I_b)_n = \hom_\Delta([n],[1])$ such that $\varphi_a^{-1}(\{0\}) = \{0,1,\dots,n-a\}$.

Return to the general situation. Certainly, the multiplication in $\Gamma.$ does not commute with the maps f^*. Nevertheless, these structures are connected to each other in a certain way.

Let $[i]_0 \xrightarrow{f_0} [i]_1 \to \dots \xrightarrow{f_{n-1}} [i]_n$ be an element of $\underrightarrow{\mathrm{holim}}_{\mathcal{D}^o}(*)_n$ and $\sigma \in \Gamma_{i_n}$. We denote the element

$$[i]_0 \xrightarrow{(f_1 \dots f_{n-1}\sigma)^*(f_0)} [i]_1 \to \dots \xrightarrow{(f_{n-1}^*\sigma)(f_{n-2})} [i]_{n-1} \xrightarrow{\sigma(f_n)} [i]_n$$

by $\sigma([i]_0 \to \dots \to [i]_n)$. It is easily verified that this gives a group action of Γ_{i_n} on the set $\{[i]_0 \to \dots \to [i]_n\}$.

Now define $\partial_0, \partial_1 : \underrightarrow{\mathrm{holim}}_{\mathcal{D}^o} \Gamma \to \underrightarrow{\mathrm{holim}}_{\mathcal{D}^o}(*)$

$$\partial_1([i]_0 \to [i]_1 \to \ldots \to [i]_n, \delta) = [i]_0 \to \ldots \to [i]_n;$$

$$\partial_0([i]_0 \to [i]_1 \to \ldots \to [i]_n, \delta) = \delta([i]_0 \to \ldots \to [i]_n).$$

Consider the morphism

$$\mu : (\underrightarrow{\mathrm{holim}}_{\mathcal{D}^\circ} \Gamma)_{\partial_1} \times_{\underrightarrow{\mathrm{holim}}_{\mathcal{D}^\circ}(*)} {}_{\partial_0}(\underrightarrow{\mathrm{holim}}_{\mathcal{D}^\circ} \Gamma) \to \underrightarrow{\mathrm{holim}}_{\mathcal{D}^\circ} \Gamma; \qquad \text{(A4.6)}$$

$$(\tau([i]_0 \to \ldots \to [i]_n), \delta) \times ([i]_0 \to \ldots \to [i]_n, \tau) \mapsto ([i]_0 \to \ldots \to [i]_n, \delta\tau).$$

There are the maps ∂_0, ∂_1 acting from the left hand side of (A4.6) to $\underrightarrow{\mathrm{holim}}_{\mathcal{D}^\circ}(*)$. It is evident that ∂_0, ∂_1 commute with μ; μ is associative. Furthermore, let ε be the map $\underrightarrow{\mathrm{holim}}_{\mathcal{D}^\circ} \Gamma \to \underrightarrow{\mathrm{holim}}_{\mathcal{D}^\circ} \Gamma$ such that

$$([i]_0 \to \ldots \to [i]_n, \delta) = (\delta([i]_0 \to \ldots \to [i]_n), \delta^{-1}).$$

It is clear that $\varepsilon\partial_0 = \partial_1\varepsilon$, $\varepsilon\partial_1 = \partial_0\varepsilon$. So we obtain

Lemma A4.2. There exists a simplicial category $\mathcal{G}.$ such that

1) All morphisms in $\mathcal{G}.$ are invertible;

2) The simplicial set of morphisms in $\mathcal{G}.$ is $\underrightarrow{\mathrm{holim}}_{\mathcal{D}^\circ} \Gamma$;

3) The simplicial set of objects in $\mathcal{G}.$ is $\underrightarrow{\mathrm{holim}}_{\mathcal{D}^\circ}(*)$.

Here ∂_0 and ∂_1 associate to a morphism its source and target and ε an inverse morphism. Now let $\mathcal{A}$ be a category with direct sums (possibly infinite) and the final object $*$ and let M be a functor from $\mathcal{D}^\circ$ to $\mathcal{A}$. In this case the simplicial objects $\underrightarrow{\mathrm{holim}}_{\mathcal{D}^\circ} M$ and $\underrightarrow{\mathrm{holim}}_{\mathcal{D}^\circ}(*)$ may be defined. The projection of M onto the constant functor induces a projection $\partial : \underrightarrow{\mathrm{holim}}_{\mathcal{D}^\circ} M \to \underrightarrow{\mathrm{holim}}_{\mathcal{D}^\circ}(*)$.

Proposition A4.3. The following conditions are equivalent.

1) There exists a functor $\tilde{M} : \mathcal{C}^\circ \to \mathcal{A}$ such that the diagram

$$\begin{array}{ccc} \mathcal{D}^\circ & \hookrightarrow & \mathcal{C}^\circ \\ & {\scriptstyle M} \searrow & \downarrow {\scriptstyle \tilde{M}} \\ & & \mathcal{A} \end{array}$$

is commutative;

2) There exists a map

$$\gamma : (\underrightarrow{\mathrm{holim}}_{\mathcal{D}^\circ} M)_{\partial} \times_{\underrightarrow{\mathrm{holim}}_{\mathcal{D}^\circ}(*)} {}_{\partial_0}(\underrightarrow{\mathrm{holim}}_{\mathcal{D}^\circ} \Gamma) \to \underrightarrow{\mathrm{holim}}_{\mathcal{D}^\circ} M$$

turning $\underrightarrow{\mathrm{holim}}_{\mathcal{D}^\circ} M$ to be a simplicial ~~co~~variant (handwritten above: ntra) functor from the simplicial category $\mathcal{G}.$ to the constant simplicial category $\mathcal{A}$.

Proof. If 1) holds, put

$$\gamma : (\sigma([i]_0 \to \ldots \to [i]_n); m) \times ([i]_0 \to \ldots \to [i]_n; d) \mapsto ([i]_0 \to \ldots \to [i]_n; m\sigma)$$

In turn, if we assume 2), then the action $M_i \times \Gamma_i \to M_i$ is defined at the level of $(\underrightarrow{\mathrm{holim}}_{\mathcal{D}^\circ})_0$. It is easy to see that it satisfies (A4.2).

The verification of the fact that γ actually turns $\underrightarrow{\mathrm{holim}}_{\mathcal{D}^\circ} M$ to be a simplicial functor is direct, and we omit it.

Proposition A4.3 provides one of the possible interpretations of the notion of a cyclic object. A group is a set of morphisms in a small category with one object and with invertible arrows. Passing to the standard simplicial realisation of S^1 we loose the multiplicative structure. Nevertheless, it may be partially reconstructed at the level of homotopy limits. In fact, we have seen that $\underrightarrow{\mathrm{holim}}_{\Delta^\circ} S^1$ is the simplicial set of morphisms in a simplicial category in which all arrows are invertible and the simplicial set of objects is equal to the canonical contractible simplicial set $\underrightarrow{\mathrm{holim}}_{\Delta^\circ}(*)$ (i.e., the classifying space of Δ). A simplicial set has a cyclic structure iff it's homotopy limit has a structure of a simplicial functor from $\underrightarrow{\mathrm{holim}}_{\Delta^\circ} S^1$ to (Sets).

One may see the relation of the cyclic objects to the spaces with action S^1 in Theorems 6.6.1 and A9.5.

<u>A5. Computation of functors Tor</u>. Let $\mathcal{C}, \mathcal{D}$ be the small categories satisfying the conditions of the beginning of A4. Consider the category $\mathcal{A}$ of modules over an arbitrary commutative ring k. Let M be an object of $\mathcal{A}^{\mathcal{C}^\circ}$ and N be an object of $\mathcal{A}^{\mathcal{D}^\circ}$. In this Subsection, we give a method for computing the functors $\mathrm{Tor}^{k[\mathcal{C}]}_{\cdot}(M,N)$ and deduce several consequences.

At first, we shall construct a resolution (not projective) of M in $\mathcal{A}^{\mathcal{C}^\circ}$. Consider the functor $\Phi^\circ : \mathcal{A}^{\mathcal{D}^\circ} \to \mathcal{A}^{\mathcal{C}^\circ}$:

$$(\Phi^\circ M)_i = M_i \otimes k[\Gamma_i]; \tag{A5.1}$$

$$(m \otimes \sigma)f = (m \cdot \sigma(f)) \otimes f^*(\sigma); \tag{A5.2}$$

$$(m \otimes \sigma)\tau = m \otimes \sigma\tau. \tag{A5.3}$$

Here $\sigma, \tau \in \Gamma_i$ and f is a morphism in $\mathcal{D}$. If $M = k\cdot \mathrm{hom}_{\mathcal{D}}(\cdot,[j])$, then $\Phi^\circ M = k\cdot \mathrm{hom}_{\mathcal{C}}(\cdot,[j])$.

Let $J : \mathcal{A}^{\mathcal{C}} \to \mathcal{A}^{\mathcal{D}}$, $J^\circ : \mathcal{A}^{\mathcal{C}^\circ} \to \mathcal{A}^{\mathcal{D}^\circ}$ be the forgetting functors. It is easy to see that:

1) Φ° is left adjoint to J°;

2) $\Phi^\circ M \otimes_{k[\mathcal{C}]} N \simeq M \otimes_{k[\mathcal{D}]} N.$ (A5.4)

Clearly, J^o is exact. So, Φ^o sends projective objects to projectives. Firthermore, the adjunction morphism is surjective and acts as follows: $\Phi^o J^o M \longrightarrow M$; $m \otimes \delta \mapsto m\delta$, $m \in M_i$.

Having these data we construct in the standard way Maclane bar-resolution. It is a simplicial object in $\mathcal{A}^{\mathcal{C}^o}$ denoted by $E_{\mathcal{C}^o,\mathcal{D}^o}(M)$: for $n \in \mathbb{Z}_+$, $[i] \in \mathrm{Ob}\,\mathcal{C}$

$$[E_{\mathcal{C}^o,\mathcal{D}^o}(M)_n]_i = M_i \otimes \overbrace{k[\Gamma_i] \otimes \ldots \otimes k[\Gamma_i]}^{(n+1)\text{ times}} ;$$

$$(m \otimes \delta_0 \otimes \ldots \otimes \delta_n)f = m\,\delta_0 \ldots \delta_n(f) \otimes \ldots \otimes \delta_n(f)^*(\delta_{n-1}) \otimes f^*(\delta_n).$$

The face and degeneracy maps are the same as in the standard resolution of M as a $k[\Gamma_i]$-module.

It is easily verified using (A4.2)-(A4.3) that the map $m \otimes \delta_0 \otimes \ldots \otimes \delta_n \mapsto m\delta_0 \ldots \delta_n \otimes \delta_1 \ldots \delta_n \otimes \ldots \otimes \delta_{n-1}\delta_n \otimes \delta_n$ gives an isomorphism $E_{\mathcal{C}^o,\mathcal{D}^o}(M) \xrightarrow{\sim} E'_{\mathcal{C}^o,\mathcal{D}^o}(M)$ where:

$$\begin{aligned} &1)\ E'_{\mathcal{C}^o,\mathcal{D}^o}(M)_n \text{ is equal to } M \otimes k[\Gamma]^{\otimes(n+1)} \text{ in } \mathcal{A}^{\mathcal{C}^o}; \\ &2)\ (m \otimes \delta_0 \otimes \ldots \otimes \delta_n)d_l = m \otimes \delta_0 \otimes \ldots \otimes \widehat{\delta_l} \otimes \ldots \otimes \delta_n,\quad 0 \leqslant l \leqslant n. \end{aligned} \qquad (A5.4)$$

<u>Proposition A5.1</u>. Let $\mathcal{P}.$ be a resolution of N in $\mathcal{A}^{\mathcal{C}^o}$, such that all $\mathcal{P}_n$ $(n \geqslant 0)$ are projective as the objects in $\mathcal{A}^{\mathcal{D}}$. Then

$$H.(E_{\mathcal{C}^o,\mathcal{D}^o}(M) \otimes_{k[\mathcal{C}]} \mathcal{P}.) \simeq \mathrm{Tor}_{\bullet}^{k[\mathcal{C}]}(M,N).$$

<u>Proof</u>. It suffices to show that for all objects L in $\mathcal{A}^{\mathcal{D}^o}$ and L' in $\mathcal{A}^{\mathcal{C}}$

$$JL' \text{ is projective in } \mathcal{A}^{\mathcal{D}} \Rightarrow \mathrm{Tor}_i^{k[\mathcal{C}]}(\Phi^o L, L') = 0,\quad i > 0.$$

But Φ^o is exact. So, if $Q.$ is a projective resolution of L in $\mathcal{A}^{\mathcal{D}^o}$, then $\Phi^o Q.$ is a projective resolution of $\Phi^o L$ in $\mathcal{A}^{\mathcal{C}^o}$. Therefore we have (using (A5.4))

$$H_i(\Phi^o Q. \otimes_{k[\mathcal{C}]} L') \simeq H_i(Q. \otimes_{k[\mathcal{D}]} L') \simeq \mathrm{Tor}_i^{k[\mathcal{D}]}(L,L') = 0,\quad i > 0.$$

<u>Corollary A5.2</u>. Let $\mathcal{S}.$ be the simplicial category constructed in Lemma A4.2 and M be an object in $\mathcal{A}^{\mathcal{C}^o}$. Then

$$H.(\mathcal{C}^o,M) \simeq H.(\mathcal{S}., \underrightarrow{\mathrm{holim}}_{\mathcal{D}^o} M).$$

<u>Proof</u>. Consider the standard resolution $E_{\mathcal{D}}(k^{\natural})$ of the constant functor $\mathcal{D} \to \mathcal{A}$ (see A1). It turns to be a complex of objects in $\mathcal{A}^{\mathcal{C}}$ and therefore we may apply to it Proposition A5.1.

In fact, put $\Psi(K) = \bigoplus_{[i] \in \mathrm{Ob}\mathcal{D}} K_i \times \hom_{\mathcal{D}}([i],\cdot)$ for an object K in $\mathcal{A}^{\mathcal{D}}$. We claim that for an object N in $\mathcal{A}^{\mathcal{C}}$ the object $\Psi(JN)$

in $\mathfrak{A}^{\mathfrak{D}}$ admits a structure of an object in $\mathfrak{A}^{\mathfrak{C}}$.

Let

$$\sigma(n,[i] \xrightarrow{f} [j]) = ((f^*\sigma)n;\ \sigma(f)),\quad n \in N_i,\quad \sigma \in \Gamma_j \qquad \text{(A5.5)}$$

Using (A4.2)-(A4.4) we obtain:

$$\sigma(g(n,f)) = \sigma(n,gf) = (((gf)^*\sigma)(n),\ \sigma(gf)) = (((gf)^*(\sigma))n,\ \sigma(gf)) =$$

$$= (((gf)^*(\sigma))n,\ \sigma g\cdot(g^*\sigma)f) = (\sigma g)(f^*g^*\sigma(n),\ (g^*\sigma)f) = (\sigma g)((g^*\sigma)(n,f));$$

$$\sigma(\tau(n,f)) = \sigma((f^*\tau)(n),\ \tau f) = ((\tau f)^*\sigma\cdot f^*\tau n,\ \sigma\tau f) =$$

$$= (f^*(\sigma\tau)\cdot n,\ \sigma\tau f) = (\sigma\tau)(n,f).$$

So, the formula (A5.5) gives the group action of Γ_j on $\Psi(JN)_j$ which satisfies (A4.2). Hence $\Psi(JN)$ is an object of $\mathfrak{A}^{\mathfrak{C}^\circ}$. It is easy to verify that the map $\Psi(JN) \longrightarrow N$, $(n,f) \mapsto f\cdot n$, is a morphism in $\mathfrak{A}^{\mathfrak{C}}$. But the resolution $E_{\mathfrak{C}}(JN)$ is obtained by iterating the functor ΨJ. So we conslude that $E_{\mathfrak{C}}(JN)$ is a simplicial object of $\mathfrak{A}^{\mathfrak{C}}$. We have (according to Lemma A5.1)

$$\mathrm{Tor}_{\cdot}^{k[\mathfrak{C}]}(M,\ k^{\natural}) = H.(E_{\mathfrak{C}^\circ,\mathfrak{D}^\circ}(M) \underset{k[\mathfrak{C}]}{\otimes} E_{\mathfrak{C}}(k^{\natural})) \simeq H.(\mathcal{L}.,\ \underrightarrow{\mathrm{holim}}_{\mathfrak{D}^\circ}(M)).$$

The verification of the second isomorphism is direct. We omit the details.

Another particular case of Proposition A5.1 is often useful. Let Δ_{inj} be the subcategory of injective maps in Δ.

We make the following assumption:

$$\left.\begin{array}{l} \mathfrak{D} = \Delta \quad \text{or} \quad \Delta_{inj}\ ; \\ \text{for } n \geqslant 0 \text{ there exists a group action} \\ \text{of } \Gamma_n \text{ on } [o,\ n] \text{ such that } \sigma(d_i) = d_{\sigma i}; \\ \text{There exist the homomorphisms } \varepsilon : \Gamma_n \longrightarrow \{\pm 1\}, \\ \quad \varepsilon(d_i^*\sigma)/\varepsilon(\sigma) = (-1)^{\sigma i - i} \end{array}\right\} \qquad \text{(A5.6)}$$

For example, if $\mathfrak{C} = \Lambda_r$, then the generator t_n of $\Gamma_n = \mathbb{Z}/r(n+1)$ acts on $[o,\ n]$ by cyclic permutations and $\varepsilon(t_n) = (-1)^n$.

Let (A5.6) be valid. Consider the complex $\mathcal{P}.$:

$$\ldots \longrightarrow k\cdot\hom_{\mathfrak{D}}([n],\cdot) \longrightarrow k\cdot\hom_{\mathfrak{D}}([n-1],\cdot) \longrightarrow\ldots\longrightarrow k\cdot\hom_{\mathfrak{D}}([1],\ ) \longrightarrow$$

$$\longrightarrow k\ \hom_{\mathfrak{D}}([0],\cdot) \longrightarrow 0$$

where the differential is $d_0 - d_1 + d_2 - \ldots \pm d_n$. Take an action of Γ_i on $k\cdot\hom([n],[i])$ such that

$$\sigma : f \mapsto \varepsilon(f^*\sigma)\cdot\sigma(f).$$

A straightforward computation shows that this is actually a group action which turns $\mathcal{P}.$ to be a complex of objects in $\mathfrak{A}^{\mathcal{C}}$. On the other hand, $\mathcal{P}.$ is a projective resolution of $k^{\natural}$ in $\mathfrak{A}^{\mathcal{D}}$. Applying the Proposition A5.1 we obtain

Corollary A5.4. Let (A5.6) be valid. For an object M in $\mathfrak{A}^{\mathcal{C}^{\circ}}$ consider a bicomplex $\mathcal{E}..(M)$:

$$\mathcal{E}(M)_{nk} = M_k \otimes (k[\Gamma_k])^{\otimes n}; \quad \partial'(m \otimes \sigma_0 \otimes \ldots \otimes \sigma_{n-1}) = \sum_{i=0}^{n-1} (-1)^i \times (m \otimes \sigma_0 \otimes \ldots$$

$$\ldots \otimes \sigma_i \otimes \ldots \otimes \sigma_{n-1} + (-1)^n m\, \sigma_{n-1}^{-1} \otimes \sigma_0 \sigma_{n-1}^{-1} \otimes \ldots \otimes \sigma_{n-2}\, \sigma_{n-1}^{-1} \varepsilon(\sigma_{n-1})$$

$$\partial''(m \otimes \ldots \otimes \sigma_{n-1}) = \sum_{i=0}^{k} (-1)^{i+n} m d_i \otimes d_i^* \sigma_0 \otimes \ldots \otimes d_i^* \sigma_{n-1}$$

The homology groups of the corresponding simple complex are isomorphic to $H.(\mathcal{C}^{\circ}, M)$.

Proof. Indeed, $\mathcal{E}..(M) \simeq E_{\mathcal{C}^{\circ}, \mathcal{D}^{\circ}}(M) \underset{k[\mathcal{C}]}{\otimes} \mathcal{P}.$

Proposition A5.5. Let $\mathcal{D} = \Delta$. Suppose that a map $M_1 \xrightarrow{f} M_2$ in $\mathfrak{A}^{\mathcal{C}^{\circ}}$ induces an isomorphism $H.(\Delta^{\circ}, M_1) \xrightarrow{\sim} H.(\Delta^{\circ}, M_2)$. Then $H.(\mathcal{C}^{\circ}, M_1) \xrightarrow{\sim}$ $\xrightarrow{\sim} H.(\mathcal{C}^{\circ}, M_2)$.

Proof. This statement follows from Corollary A5.2 (or A5.5 when (A5.6) holds). In fact, the spectral sequence of the bicomplex $E'_{\mathcal{C}^{\circ}, \mathcal{D}^{\circ}}(M_i) \underset{k[\mathcal{C}]}{\otimes} E_{\mathcal{D}}(k^{\natural})$ has the first term equal to $H.(\varinjlim_{\mathcal{D}^{\circ}} (M \otimes k[\Gamma]^{\otimes n})$.

Remark A5.6. Let B be an associative algebra (without unit) over a field k of characteristic zero. Consider the corresponding Lie algebra Lie(B) (see 4.1). There exists a pair of categories $\mathcal{C} \supset \Delta_{inj}$ such that $\Gamma_j = \Sigma_{j+1}$ for all j and (A5.6) holds. We may construct an object B^{Σ} in $\mathfrak{A}^{\mathcal{C}^{\circ}}$ such that $B_n^{\Sigma} = \overset{1+n}{\otimes} B$ and $H.(\mathcal{C}^{\circ}, B^{\Sigma}) = H.(\mathrm{Lie}(B), k)$. The double complex $\mathcal{E}(B^{\Sigma})$ may be extended to the right (as in 1.5); the horizontal complexes in the double complex obtained in such a way are the Tate complexes of the groups Σ_{j+1} with values in $\overset{j+1}{\otimes} A$. As a consequence we obtain a filtration on H.(Lie B, k). Its initial term is equal to the kernel of the map $\mathrm{Tor}_{.}^{U(\mathrm{Lie}\, B)}(k, k) \longrightarrow$ $\longrightarrow \mathrm{Tor}_{.}^{B \oplus k}(k, k)$. When k is arbitrary, this construction gives the homology groups which do not coincide with the standard ones.

A6. Additive K-functors of the twisted group algebras. Let k be a commutative ring, A be a k-algebra, G be a group and α a homomorphism $G \to \mathrm{Aut}\, A$, $g \mapsto \alpha_g$. We shall write $\alpha_{g(a)} = a^g$. Consider the algebra $A_{\alpha}[G]$ which is equal to $A \times G$ as a k-module and in hich the product is given by

$$(a\cdot g)\cdot(b\cdot h) = (ab^{g})\cdot(gh)$$

we write $a\cdot g = (a,g) \in A\times G$). The present subsection is devoted to omputation of $K_{\cdot}^{+}(A_{\alpha}[G])$.

Let $\widehat{G}$ denote the set of classes of conjugated elements of G; e write $[g]$ for the class containing $g \in G$. Denote by N_g the nor- alizing subgroup of g and by (g) its central subgroup generated y g.

heorem A6.1. There is a decomposition

$$K_{\cdot}^{+}(A_{\alpha}[G]) \xrightarrow{\sim} \bigoplus_{[g]\in\widehat{G}} K_{\cdot}^{+}(A;G;[g]) \tag{A6.1}$$

here $K^{+}(A;G;g)$ are some homology groups such that there exist the pectral sequences

$$E_{ij}^{2} = H_i(N_g/(g);\ K_j^{+}(\mathbf{Z}/\mathrm{ord}\ g\cdot\mathbf{Z},\ \alpha_g;\ A) \Rightarrow K_{i+j}^{+}(A;G;[g]).$$

roposition A6.2. Let $\mathrm{char}\ k = 0$ and $\mathrm{ord}\ g < \infty$. Then

$$K_{\cdot}^{+}(A;G;[g]) = \mathbb{H}_{\cdot}(N_g/(g);\ C_{\cdot}^{cycl}(A)/\mathrm{im}(1-\alpha_g)).$$

s a consequence we have

heorem A6.3. (Burghelea, Fiedorowicz, see [9]). Suppose that har $k = 0$. Then

$$K_{n+1}^{+}(k[G]) = \bigoplus_{\substack{[g]\in\widehat{G}\\ \mathrm{ord}\ g<\infty}} (H_n(N_g/(g);\ k) \oplus H_{n-2}(N_g/(g);k) \oplus\ldots) \oplus$$

$$\oplus \Big(\bigoplus_{\substack{[g]\in\widehat{G}\\ \mathrm{ord}\ g=\infty}} H_n(N_g/(g);\ k)\Big).$$

roof of Theorem A6.1. Consider the cyclic k-module $A_{\alpha}[G]^{\natural}$. Let $\widetilde{A_{\alpha}[G]}^{\natural}$ be another cyclic module:

$$\widetilde{A_{\alpha}[G]}^{\natural} \xrightarrow{\sim} (\overset{n+1}{\otimes} A) \otimes (\overset{n+1}{\otimes} k[G]); \tag{A6.3}$$

f $x = (a_o \otimes\ldots\otimes a_n) \otimes (g_o \otimes\ldots\otimes g_n)$, $a_i \in A$, $g_i \in G$, then

$$xd_i = (a_o \otimes\ldots\otimes a_{i-1}a_i \otimes\ldots\otimes a_n) \otimes (g_o \otimes\ldots\otimes g_{i-1}g_i \otimes\ldots\otimes g_n),\quad 1 \leqslant i \leqslant n-1;$$

$$xd_0 = (a_1^{g_o^{-1}} \otimes\ldots\otimes (a_n\ a_0^{g_n})^{g_o^{-1}}) \otimes (g_1 \otimes\ldots\otimes g_{n-1} \otimes g_o^{-1}g_n g_o);$$

$$xd_n = (a_o \otimes\ldots\otimes a_{n-1}a_n) \otimes (g_o \otimes\ldots\otimes g_{n-2} \otimes g_n);$$

$$xt = (a_1^{g_o^{-1}} \otimes\ldots\otimes a_n^{g_o^{-1}} \otimes a^{g_o^{-1}g_n}) \otimes (g_1 \otimes\ldots\otimes (g_o\ldots g_{n-1})^{-1}\ g_n \otimes g_o^{-1}g_n g_o);$$

$$xs_i = (a_o \otimes \ldots \otimes a_{i-1} \otimes 1 \otimes a_i \otimes \ldots \otimes a_n) \otimes (g_o \otimes \ldots \otimes g_{i-1} \otimes 1 \otimes g_i \otimes \ldots \otimes g_n),$$
$$0 \leqslant i \leqslant n.$$

The isomorphism (A6.3) acts as follows:

$$a_o \cdot g_o \otimes a_1 \cdot g_1 \otimes \ldots \otimes a_n \cdot g_n \mapsto (a_o \otimes a_1^{g_o} \otimes \ldots \otimes a_n^{g_o \ldots g_{n-1}}) \otimes (g_o \otimes \ldots \otimes g_{n-1} \otimes g_o \ldots g_n)$$

There is the decomposition

$$\widetilde{A_\alpha[G]}^\natural = \bigoplus_{[g] \in \hat{G}} M_{[g]};$$

a subobject $M_{[g]}$ is generated by elements for which $g_n \in [g]$. We have for $M_{[g]}$ (regarded as a simplicial module):

$$M_{[g]} \xrightarrow{\sim} \mathcal{P}(G) \boxtimes_{k[G]} (\bigoplus_{x \in [g]} A_{\alpha_x}^{\natural,\ \mathrm{ord}\ g}) \xrightarrow{\sim}$$

$$\xrightarrow{\sim} \mathcal{P}(G) \boxtimes_{k[G]} (k[G] \otimes_{k[N_g]} A_{\alpha_g}^{\natural,\ \mathrm{ord}\ g}) \underset{(\mathrm{quis})}{\xleftarrow{\sim}} \mathcal{P}(N_g) \boxtimes_{k[N_g]} A_{\alpha_g}^{\natural,\ \mathrm{ord}\ g}$$

$\mathcal{P}(G)$ was defined in the end of A2. Here we are interested in simplicial structure of $\mathcal{P}(G)$: this is one of the standard bar-resolution of the G-module k. The second isomorphism is equal to $1_{\mathcal{P}(G)} \boxtimes \theta$ and θ is given by the formula

$$\theta(x;\ a_o \otimes \ldots \otimes a_n) = y \otimes a_o^{y^{-1}} \otimes \ldots \otimes a_n^{y^{-1}},$$

where $ygy^{-1} = x$; the third arrow is the natural embedding. But now Proposition A5.5 implies

<u>Lemma A6.4</u>. $K^+_{\cdot+1}(A_\alpha[G]) \xrightarrow{\sim} \bigoplus_{[g] \in \hat{G}} H_\cdot(\Lambda^o;\ \mathcal{P}(N_g, g) \boxtimes_{k[N_g]} A_{\alpha_g}^{\natural,\mathrm{ord}\ g})$.

In fact, we have seen that the embedding of the subobject generated by $(a_o \otimes \ldots) \otimes (g_o \otimes \ldots)$, $g_o, g_1, \ldots \in N_g$, to $M_{[g]}$ induces an isomorphism in $H_\cdot(\Delta^\circ, \cdot)$, whence in $H_\cdot(\Lambda^o, \cdot)$. But the considered subobject is clearly isomorphism to $\mathcal{P}(N_g, g) \boxtimes_{k[N_g]} A_{\alpha_g}^{\natural,\mathrm{ord}\,g}$.

Note that neither $A_{\alpha_g}^{\natural,\mathrm{ord}\,g}$ nor $\mathcal{P}(N_g, g)$ are cyclic k-modules. But the operators t_n^{n+1} coincide for both of them with the action of g; so $\mathcal{P}(N_g, g) \boxtimes_{k[N_g]} A_{\alpha_g}^{\natural,\mathrm{ord}\,g}$ is really an object in $\mathfrak{A}^{\Lambda^o}$.

Our next aim is to express the right hand side in Lemma A6.4 in terms of homology of categories.

Let N be a group and $g \in \mathrm{Cent}\ N$. We denote by $\mathcal{C}_N$ the category with one object corresponding to N. Consider the cartesian product

$\Lambda_\infty \times \mathcal{C}^0_N$. Introduce the supplementary relations $(t_n^{n+1}, 1) = (1, g)$. We obtain a new category denoted by $\Lambda^o_{N,g}$. The functors from $\Lambda^o_{N,g}$ to $\mathfrak{A}$ are in 1-1 correspondence with the ∞-cyclic objects in the category of left N-modules such that $mt_n^{n+1} = gm$, $m \in M_n$, $n \geqslant 0$. For example, $A_{\alpha_g}^{\natural, \mathrm{ord}\, g}$ is such an object.

The category $\Lambda_{N,g}$ satisfyes the conditions of Subsection A4: $\mathfrak{D} = \Delta$ and $\Gamma_n = \mathrm{Coker}\,(\mathbb{Z} \xrightarrow{\varphi} \mathbb{Z} \times N)$, $1 \overset{\varphi}{\mapsto} (n+1;\, g^{-1})$. There is the exact sequence

$$1 \longrightarrow \mathbb{Z}/(n+1)\mathrm{ord}\, g \cdot \mathbb{Z} \longrightarrow \Gamma_n \longrightarrow N/(g) \longrightarrow 1.$$

Moreover, $\Lambda_{N,g}$ satisfyes the conditions (A5.6). Indeed, put $\mathcal{E}(\sigma) = (-1)^{nl}$ for $\sigma = (l, x) \in \mathbb{Z} \times N \pmod{\mathrm{im}\, \varphi}$.

We have the following diagrams of categories and functors:

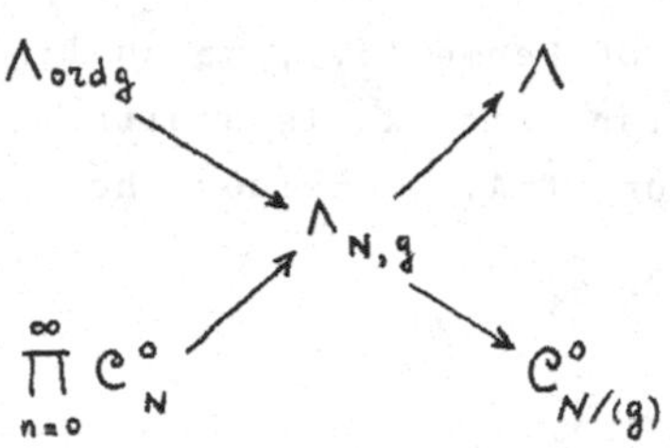

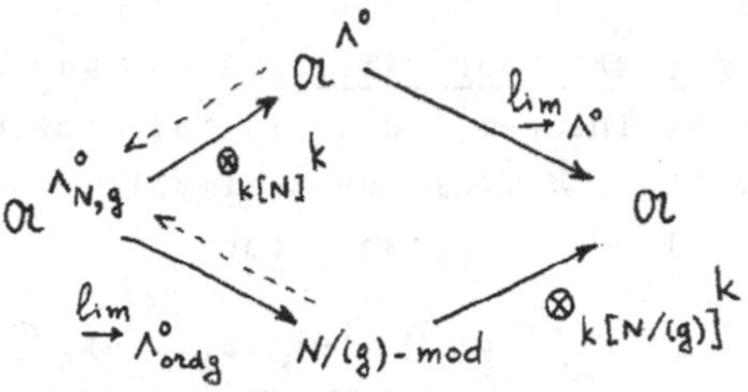

Here the upper functor in the first diagram sends $[i]$ to $[i]$ and $x \cdot f \cdot t_n^i$ to $f \cdot t_n^i$ where $x \in N$, $0 \leqslant i \leqslant n$ and $f \in \mathrm{hom}_\Delta([n],[m])$. The lower right functor sends $[n]$ to the single object of $\mathcal{C}^o_{N/g}$ and $x \cdot f \cdot t_n^i$ to $x \bmod (g)$.

The dotted arrows in the second diagram are the restriction functors; they are exact and right adjoint to the functors acting in the opposite directions. So we may apply the spectral sequences of the composition to the both ways $\mathfrak{A}^{\Lambda^o_{N,g}} \longrightarrow \mathfrak{A}$ in the second diagram. First, apply it to the upper way.

<u>Lemma A6.5</u>. $K^+_{\cdot+1}(A_\alpha[G]) \xrightarrow{\sim} \bigoplus_{[g] \in \hat{G}} H_\cdot(\Lambda^o_{N_g, g};\, A_{\alpha g}^{\natural, \mathrm{ord}\, g})$.

<u>Proof</u>. It suffices to show that

$$H_\cdot(\Lambda^o; \mathfrak{P}(N,g) \underset{k[N]}{\otimes} M) \xrightarrow{\sim} H_\cdot(\Lambda^o_{N,g};\, M)$$

for any M from $\mathrm{Ob}\, \mathfrak{A}^{\Lambda^o_{N,g}}$. But $\mathfrak{P}(N,g) \underset{k}{\otimes} M$ is an object in $\mathfrak{A}^{\Lambda^o_{N,g}}$ whose simplicial homology is isomorphic to $H_\cdot(\Delta^o, M)$. So Lemma A5.5 shows that the homology groups of $\Lambda_{N,g}$ with values in $\mathfrak{P}(N,g) \otimes_k M$ are isomorphic to those with values in M. Therefore the spectral sequence of the composition $\lim_{\to \Lambda^o_{N,g}} = \lim_{\to \Lambda^o} \circ (\otimes_{k[N]} k)$ applied to $\mathfrak{P}(N,g) \underset{k}{\otimes} M$ gives the desired statement. Indeed, since $\mathcal{P}(N,g) \boxtimes_k M$ consists of free N-modules, its higher derived functors of $(\otimes_{k[N]} k)$ are clearly equal to zero.

<u>Proposition A6.6</u>. Let M be an object of $\mathfrak{A}^{\Lambda^\circ_{N,g}}$. There exists a spectral sequence

$$E^2_{ij} = H_i(N/(g),\ H_j(\Lambda^\circ_{\mathrm{ord}\ g};\ M)) \Longrightarrow H_{i+j}(\Lambda^\circ_{N,g};\ M).$$

<u>Proof</u>. This follows from Grothendieck theorem applied to the composition $\varinjlim_{\Lambda^\circ_{N,g}} = (\underset{k[N/(g)]}{\otimes} k) \circ \varinjlim_{\Lambda^\circ_{\mathrm{ord}\ g}}$.

This ends the proof of Theorem A6.1.

Consider the case $G = \mathbb{Z}$. Let t be a generator of G; we write α for α_t and $A_\alpha[t,t^{-1}]$ for $A_\alpha[\mathbb{Z}]$ (see Chapter 3).

<u>Computation of $\mathrm{Nil}^+_\cdot(A,A_\alpha)$</u>. As a consequence of Lemma A6.5, we shall prove Theorem 3.2.2 in the case when the scalar ring k is arbitrary. Let $\mathcal{B}.(A)$ be a simplicial bar-resolution of A. Consider the simplicial abelian groups

$$D^{(k)}_\cdot = \underbrace{(\mathcal{B}.(A)_\alpha \boxtimes_A \cdots \boxtimes_A \mathcal{B}.(A)_\alpha)}_{k\ \text{times}} \underset{A\otimes A^\circ}{\otimes} A.$$

We write $x_1 \otimes\ldots\otimes x_k = (x_1 \otimes_A \ldots \otimes_A x_k) \underset{A\times A^\circ}{\otimes} 1$, $x_i \in \mathcal{B}.(A)$. There is an automorphism $\bar{t}_k$ of $D^{(k)}_\cdot$ which sends $x_1 \otimes\ldots\otimes x_k$ to $x_2 \otimes\ldots\otimes x_k \otimes x_1$, $x_i \in \mathcal{B}.(A)$. We have $\bar{t}^k_k = 1$;

$$\mathrm{Nil}^+(A,A_\alpha) = \bigoplus_{k=1}^{\infty} \mathbb{H}.(\mathbb{Z}/k\mathbb{Z},\ D^{(k)}_\cdot).$$

Put $\tilde{\alpha}(a_{-1} \otimes a_0 \otimes\ldots\otimes a_n) = \alpha(a_{-1})\otimes\ldots\otimes\alpha(a_n)$, $a_i \in A$. Then the map, which sends $x_1 \otimes\ldots\otimes x_k$ to $x_1 \otimes x_2 \otimes\ldots\otimes \tilde{\alpha}^{k-1}x_k$, establishes an isomorphism

$$\varphi : D^{(k)} \xrightarrow{\sim} (\mathcal{B}.(A) \boxtimes_A \cdots \boxtimes_A \mathcal{B}.(A)_{\alpha^k}) \underset{A\otimes A^\circ}{\otimes} A;$$

it is easy to see that

$$(\varphi \bar{t}_k \varphi^{-1})(x_1 \otimes\ldots\otimes x_k) = \tilde{\alpha}^{-1}x_1 \otimes\ldots\otimes\tilde{\alpha}^{-1}x_{k-1} \otimes\tilde{\alpha}^{k-1}x_k.$$

We see that for all $k \in \mathbb{N}$ there is a $k[\mathbb{Z}/k\mathbb{Z}]$-module isomorphism of $D^{(k)}$ with the following complex $\tilde{D}^{(k)}$:

$$\tilde{D}_n^{(k)} = (A^{\natural,\infty}_{\alpha^k})_{k(n+1)-1}, \quad n \geq 0;$$

$$\delta : \tilde{D}_n^{(k)} \longrightarrow \tilde{D}_n^{(k-1)}; \quad \delta = \partial_0 - \partial_1 + \ldots + (-1)^n \partial_n;$$

$$x\partial_i = x d_i d_{i+n} \ldots d_{i+(k-1)n} \quad \text{for} \quad x \in (A^{\natural,\infty}_{\alpha^k})_{k(n+1)-1}.$$

The action of $\mathbb{Z}/k\mathbb{Z}$ on $\tilde{D}^{(k)}_{.}$ is the following:

$$x\,\tilde{t}_k = x\,\alpha^{-1}\,t^{n+1}$$

where t is the corresponding map in Λ_∞.

We shall reformulate this in terms of a suitable subcategory of Δ. Let ${}_k\Delta \subset \Delta$; $\mathrm{Ob}({}_k\Delta) = \{[k(n+1)-1] : n \geq 0\}$. Put ${}_k[n] = [k(n+1) - 1]$; $\mathrm{hom}_{{}_k\Delta}({}_k[n], {}_k[n']) = \{ f \in \mathrm{hom}_\Delta([k(n+1) - 1], [k(n'+1) - 1]) : f(i + n + 1) = f(i + n' + 1), \ 0 \leq i < (k-1)(n+1)\}$.

Denote by $A^{\natural,\infty}_{\alpha^k}\big|_{{}_k\Delta^\circ}$ the restriction of the functor $A^{\natural,\infty}_{\alpha^k} : \Lambda^\circ_\infty \to$ k-mod to the subcategory ${}_k\Delta^\circ \subset \Delta^\circ \subset \Lambda^\circ_\infty$. Then

$$H_.(\tilde{D}^{(k)}_.) \xrightarrow{\sim} H_.({}_k\Delta^\circ, A^{\natural,\infty}_{\alpha^k}\big|_{{}_k\Delta^\circ}).$$

Indeed, there is the isomorphism

$$G : {}_k\Delta \xrightarrow{\sim} \Delta;$$

$${}_k[n] \longmapsto [n];$$

if $f \in \mathrm{hom}_{{}_k\Delta}({}_k[n], {}_k[n'])$, then

$$f(\{0,\ldots, n\}) \subseteq \{0,\ldots, n'\}, \quad \text{and}$$

$$Gf = f\,|\,\{0,\ldots, n\} : \{0,\ldots, n\} \to \{0,\ldots, n'\}.$$

This isomorphism sends ∂_i to d_i.

Now let $\mathcal{C}$ be a small category satisfying the property of the beginning of A4. Assume that $D = \Delta$ and $\Gamma_0 = 1$. Moreover, suppose that for all $f \in \mathrm{Mor}_k\Delta$ and $\sigma \in \Gamma_.$: $\sigma^* f \in \mathrm{Mor}_k\Delta$. Denote by ${}_k\mathcal{C}$ the subcategory of $\mathcal{C}$ generated by ${}_k\Delta$ and $\Gamma_{k(n+1)-1}$, $n \geq 0$. Then ${}_k\mathcal{C}$ satisfies the property of the beginning of A4.

Let $\mathcal{A}$ be an abelian category, $M \in \mathrm{Ob}\,\mathcal{A}^{\mathcal{C}^\circ}$, and let $M|_k\mathcal{C}^\circ$ be the restriction of the functor M to the subcategory ${}_k\mathcal{C}^\circ$.

<u>Lemma A6.7</u>. $H_.({}_k\Delta^\circ, M|_k\Delta^\circ) \xrightarrow{\sim} H_.(\mathcal{C}^\circ, M)$.

<u>Proof</u>: It is well known that

$$H.({}_k\Delta^\circ,\ M|_{k}\Delta^\circ) \xrightarrow{\sim} H.(\Delta^\circ,\ M)$$

for any M. Note that for all $n \geqslant 0$

$$\hom_{\mathcal{C}}(\cdot,\ [n]) \cong \Phi^\circ \hom_\Delta(\cdot,\ [n]).$$

Here the both sides are regarded as the objects in $\mathfrak{A}^{{}_k\mathcal{C}^\circ}$ and the functor Φ° (cf. (A5.1)-(A5.3)) is the functor $\mathfrak{A}^{{}_k\Delta^\circ} \to \mathfrak{A}^{{}_k\mathcal{C}^\circ}$ which is left adjoint to the restriction functor. Now take a resolution P. of M in $\mathfrak{A}^{\mathcal{C}^\circ}$ each term of which is isomorphic to $\oplus A_n \times \hom_{\mathcal{C}}(\cdot,[n])$. Then for all $i \geqslant 0$ $P_n|_{k}\mathcal{C} \cong \Phi^\circ(Q_n|_{k}\Delta^\circ)$ where $\mathbb{L}_i \varinjlim_{{}_k\Delta^\circ}(Q_n|_{k}\Delta^\circ) =$ $= 0,\ i > 0$ (since $\mathbb{L}_i \varinjlim_{\Delta^\circ} Q_n = 0,\ i > 0$). This implies that $P.|_{k}\mathcal{C}^\circ$ is an acyclic resolution of $M|_{k}\mathcal{C}^\circ$ in $\mathfrak{A}^{{}_k\mathcal{C}^\circ}$. On the other hand, $\varinjlim_{{}_k\mathcal{C}^\circ}(P.|_{k}\mathcal{C}^\circ) \xrightarrow{\sim} \varinjlim_{{}_k\Delta^\circ}(P.|_{k}\mathcal{C}^\circ) \xrightarrow{\sim} \varinjlim_{\Delta^\circ}P. \xrightarrow{\sim} \lim_{\mathcal{C}^\circ} P.$, whence

$$H.({}_k\mathcal{C}^\circ,\ M|_{k}\mathcal{C}^\circ) \xrightarrow{\sim} H.(\mathcal{C}^\circ,\ M),\quad \text{Q.E.D.}$$

We want to apply the previous Lemma to the case which arises in Lemma A6.5 for $G = \mathbb{Z}$, $g = k \in Z$, $k \neq 0$, $N_g = Z$, $\mathcal{C} = \Lambda_{N_g,g} = \Lambda_{\mathbb{Z},k}$. Note that ${}_k\Lambda_{\mathbb{Z},k}$ is isomorphic to the cartesian product $\mathcal{C}_{\mathbb{Z}/k\mathbb{Z}} \times$ $\times\ {}_k\Lambda_\infty$ where $\mathcal{C}_{\mathbb{Z}/k\mathbb{Z}}$ is the category with one object corresponding to the group $\mathbb{Z}/k\mathbb{Z}$. If $*$ is the single object of $\mathcal{C}_{Z/kZ}$, then the isomorphism acts as follows:

$$\mathcal{C}_{Z/kZ} \times {}_k\Lambda_\infty \xrightarrow{\sim} {}_k\Lambda_{\mathbb{Z},k};$$

$$(*,\ {}_k[n]) \mapsto {}_k[n];$$

$$(g^i,\ f) \mapsto (\alpha^{-1}\ t^{n+1})^i.f,$$

where g is the generator of $\mathbb{Z}/k\mathbb{Z}$ and $f \in \mathrm{Mor}\ {}_k\Lambda$.

Now Lemma A6.7 implies that for M in $\mathrm{Ob}\,\mathfrak{A}^{{}_k\Lambda_{Z,k}}$

$$H.(\Lambda^\circ_{\mathbb{Z},k};\ M) \xleftarrow{\sim} H.({}_k\Lambda^\circ_{\mathbb{Z},k};\ M|_{k}\Lambda_{\mathbb{Z},k}) \xrightarrow{\sim} H.(Z/kZ,\ D^{(k)}_{.}).$$

Thus,

$$\mathrm{Nil}^+(A,A_\alpha) \xrightarrow{\sim} \bigoplus_{k=1}^{\infty} H.(Z/kZ,\ D^{(k)}_{.}) \xrightarrow{\sim} \bigoplus_{k=1}^{\infty} H.(\Lambda^\circ_{\mathbb{Z},k};\ A^{\natural,\infty}_{\alpha^k}).$$

Comparing with Lemma A6.5 we have

$K^+_{\cdot+1}(A_\alpha[t,t^{-1}]) \simeq H.(\mathbb{Z}, L.(A)) \oplus Nil^+(A,A_\alpha) \oplus Nil^+_{\cdot}(A,A_{\alpha^{-1}})$.

This proves Theorem 3.2.2.

Corollary A6.8. Let char k = 0. Then

$$K^+_{\cdot+1}(A_\alpha[t,t^{-1}]) \xrightarrow{\sim} H.(\mathbb{Z}, C^{cycl}_{\cdot+1}(A)) \oplus \bigoplus_{k\neq 0} H.(A,A_{\alpha^k})/\mathrm{im}(1-\alpha_*);$$

$$K^+_{\cdot+1}(A_\alpha[t]) \xrightarrow{\sim} K^+_{\cdot+1}(A) \oplus \bigoplus_{k>0} H.(A,A_{\alpha^k})/\mathrm{im}(1-\alpha_*).$$

Corollary A6.9. There is an exact sequence

$$0 \to \bigoplus_{k\geq 0}[\bigoplus_{i\geq 1}({}_kH_{n-2i}(A,A)) \oplus H_n(A,A)] \to Nil^+_n(A,A) \to$$

$$\to \bigoplus_{k\geq 0}\bigoplus_{i\geq 1} H_{n-2i+1}(A,A)/kH_{n-2i+1}(A,A) \to 0$$

for any $n \geq 0$.

Remark A6.8. All results of the present work may be reformulated for the functors $K^+(\mathbb{Z}/r\mathbb{Z}, \alpha; A)$. Here we give a brief discussion of the analogues of the main statements of Chapter 6.

Let k be a commutative ring and X an affine noetherian scheme over Spec k; let α be an automorphism of X such that $\alpha^r = 1_X$, $r < \infty$. For an arbitrary cosimplicial affine scheme V and $\alpha \in \mathrm{Aut}\, V$ we put

$$K^+(Z/rZ, \alpha; V) = K^+(\mathbb{Z}/r\mathbb{Z}; \alpha'; k[V]),$$

where k[V] is the coordinate simplicial ring of V and α' is the automorphism of k[V] induced by α. The definition of the K^+-functors of a simplicial ring is obvious; compare with Subsection 2.2.

We may use a construction dual to that of A1 and form homotopy inverse limit on the category of schemes (this is a category with direct products). We obtain a cosimplicial affine noetherian scheme

$$\mathrm{holim}_{Z/r\mathbb{Z}} X: \quad \ldots \;\substack{\leftarrow\\\leftarrow\\\leftarrow\\\leftarrow}\; \mathrm{Hom}((\mathbb{Z}/r\mathbb{Z})^2, X) \;\substack{\leftarrow\\\leftarrow\\\leftarrow}\; \mathrm{Hom}(\mathbb{Z}/r\mathbb{Z}, X) \;\substack{\leftarrow\\\leftarrow}\; X$$

(Certainly, we might act directly in the category of commutative k-algebras which admits direct sums $\otimes_k$).

Furthermore, let X^α be the fixed point subscheme (i.e. $X^\alpha = \operatorname{Spec}(k[X]/I)$, I is the ideal generated by all elements $\alpha'(f) - f$, $f \in k[X]$). Finally, note that we may construct the cohomology groups $H^\cdot(\mathbb{Z}/r\mathbb{Z}, \alpha; A)$ attached to $K^+_\cdot(\mathbb{Z}/r\mathbb{Z}, \alpha; A)$, exactly as in 1.5.

Theorem A6.9. 1) There is a natural isomorphism

$$K^+_\cdot(\mathbb{Z}/r\mathbb{Z}, \alpha; X) \xrightarrow{\sim} K^+_\cdot(\mathbb{Z}/r\mathbb{Z}, 1; \underleftarrow{\operatorname{holim}}_{\mathbb{Z}/r\mathbb{Z}} X).$$

If r is invertible in k, then

$$K^+_\cdot(\mathbb{Z}/r\mathbb{Z}, \alpha; X) \xleftarrow{\sim} K^+_\cdot(\underleftarrow{\operatorname{holim}}_{\mathbb{Z}/r\mathbb{Z}} X).$$

2) Let k be a field of characteristic zero, X of finite type

$$H^\cdot(\mathbb{Z}/r\mathbb{Z}, \alpha; X) \xrightarrow{\sim} H^\cdot_{cris}(X^\alpha).$$

The proof will be given elsewhere.

A7. Duality in the categories Λ_r. The following statement is due to Connes.

Lemma A7.1. There are the isomorphisms $\Lambda_r \xrightarrow{\sim} \Lambda_r^o$, $1 \leqslant r \leqslant \infty$.

Proof. Let $f : [m] \longrightarrow [n]$ be a morphism in Λ_∞. Put

$$\tilde{f}(i) = \min\{j : f(j) = i\}, \quad \text{if} \quad i \in \operatorname{Im} f;$$

$$\tilde{f}(i) = f(i'),\ i' = \max\{i'' : i'' \leqslant i,\ i'' \in \operatorname{Im} f\}, \text{ if } \quad i \notin \operatorname{Im} f.$$

It is easy to see that $\widetilde{fg} = \tilde{g}\tilde{f}$ and $\tilde{\tilde{f}} = 1$. Furthermore, for $r < \infty$, $\tilde{f}$ is equivalent to $\tilde{f}'$, if f is equivalent to f'. This proves the Lemma.

It is easy to see that $\tilde{d}_i = s_{i-1}$, $i > 0$; $\tilde{t} = t^{-1}$; $\tilde{d}_o = s_{n-1}t^{-1}$. The above Lemma enables one to regard a cyclic k-module $A^\natural$ as an object in $\mathcal{O}\!\mathcal{L}^\Lambda$; let f be in $\hom_\Delta([m],[n])$, $a_o \otimes \ldots \otimes a_m \in A_m$; then

$$f(a_o \otimes \ldots \otimes a_m) = (a_{i_{-1}} \ldots a_{i_o - 1}) \otimes (a_{i_o} \ldots a_{i_1 - 1}) \otimes \ldots \otimes (a_{i_{m-1}} \ldots a_{i_m - 1})$$

where $i_{-1} = 0$, $i_m = n+1$ and $i_l - i_{l-1} = \operatorname{card} f^{-1}(\{l\})$, $0 \leqslant l \leqslant m$; the product taken over empty set of indices is equal to unit by definition.

A8. The results of Chapter 2.3 may now be generalized to the case of an arbitrary scalar ring k. We suppose that all the considered algebras are projective k-modules. Having a morphism $A \longrightarrow B$ define $K^+_\cdot(A \to B)$ as the homology of bicomplex $\Omega.(A \to R)$ (see Remark 2.4.2) where R is a simplicial free resolution of B over A. We have seen proving 2.3.1 that the projection $F'(R) \to \Omega.(A \to R)$ includes itself to a distinguished triangle in the derived category:

$$F_{\bullet}^{!}(A) \to F_{\bullet}^{!}(R) \to \Omega.(A \to R)$$

Furthermore, Lemma 2.2.1 may be reformulated in obvious way in the case when R is a simplicial k-algebra. So $F'(A) \xrightarrow{\sim} L.(A)$, $F'(R) \xrightarrow{\sim} F'(B) \xrightarrow{\sim} L.(B)$ and we have

Proposition A8.1. All the statements of Theorem 2.4.1 hold for the functors $K_{\bullet}^{+}(A)$ and $K_{\bullet}^{+}(A \to B)$.

Consider now a cyclic k-module $B_A^{\natural}$ for any $A \to B$: $(B_A^{\natural})_n = (B \otimes_A \ldots \otimes_A B) \otimes_{A \otimes A^{\circ}} A$, where the number of factors B is equal to $n+1$; the operators d_i, s_i and t act by the formulae (A2.1), (A2.2).

Proposition A8.2. $K_{\bullet}^{+}(A \to B) \simeq H.(\Lambda, R_A^{\natural}/A_A^{\natural})$.

(The right hand side denotes the hyperhomology of Λ with values in a complex of objects in $\mathfrak{A}^{\Lambda}$).

Proof. Note that $R_{n,A}^{\natural}/A_A^{\natural}$ are projective cosimplicial modules. In fact, let T_n be the set of the free generators of R_n. We have an isomorphism of the cosimplicial k-modules:

$$\bigoplus_{m=0}^{\infty} \bigoplus_{(t_o,\ldots,t_m) \in T_n} (\bigotimes^{m+1} A) \times \hom_{\Delta}([m],\cdot) \xrightarrow{\sim} (R_n^{\natural})_A/A_A^{\natural}.$$

At the direct summand corresponding to $(t_o,\ldots,t_m)$ this isomorphism is given by the following formula:

$$(a_o \otimes \ldots \otimes a_m;\ f : [m] \to [m']) \mapsto (t_o a_o \ldots t_{i_o} a_{i_o} \otimes \ldots \otimes t_{i_{m'-1}+1}\, a_{i_{m'-1}+1} \ldots t_m a_m)$$

where $i_k - i_{k-1} = \operatorname{card} f^{-1}(\{k\})$; $i_{-1} = -1$; $i_{m'} = m$; the product over the empty set is equal by definition to the unit.

Therefore the homology groups $H.(\Lambda, R_A^{\natural}/A_A^{\natural})$ may be computed using the resolution $E'_{\Lambda^{\circ},\Delta^{\circ}}(k^{\natural})$ (see A5). Note also that $H.(\Lambda, R_A^{\natural}/A_A^{\natural}) = H.(E. \otimes_{k[\Lambda]} (R_A^{\natural}/A_A^{\natural}))$ where $E. \to k^{\natural}$ is an arbitrary resolution such that $(E.)_n$ is a free $\mathbb{Z}/(n+1)\mathbb{Z}$-module complex for all $[n] \in \operatorname{Ob}\Lambda$. In fact, if M is an object of $\mathfrak{A}^{\Lambda^{\circ}}$ and N an object of $\mathfrak{A}^{\Lambda}$, then $H.(E'_{\Lambda^{\circ},\Delta^{\circ}}(M) \otimes_{k[\Lambda]} N) = \oplus \operatorname{Tor}_{\bullet}^{\mathbb{Z}/(n+1)\mathbb{Z}}(M_n, N_n)$; so for a resolution of such type $\operatorname{Tor}_i^{k[\Lambda]}(E_j, (R_A^{\natural}/A_A^{\natural})_k) = 0$, $i > 0$, $j,k \geqslant 0$.

Now we shall write down such a resolution E.; we shall see that for all $A \to B$ the complex $E. \otimes_{k[\Lambda]} (B_A^{\natural}/\operatorname{Im}(A_A^{\natural}))$ turns out to be equal to the complex $\Omega.(A \to B)$.

Let $E_i = k\cdot\hom_\Lambda(\cdot, [0])$, $i \equiv 0 \pmod 2$; $E_i = k\cdot\hom_\Lambda(\cdot, 1)/L.$, $i \equiv 1 \pmod 2$; L_n is the subspace in $k\cdot\hom_\Lambda([n],[1])$ generated by the elements $d_o s_1^{n+1}$ and $(-s_o^{i_o+i_1-1} s_1^{i_2-1} + s_o^{i_o-1} s_1^{i_1+i_2-1} + s_o^{i_1-1} s_1^{i_o+i_2-1} t^{-i_o}) t^l$ for all $i_o, i_1 > 0$; $l \geqslant 0$. It is easy to see that $g \in L_n \Rightarrow gf \in L_n$ for $f : [n] \to [m]$. So E_i are well defined. Let $\partial(f) = d_1 \cdot f$, $f \in E_{2l}$; $\partial(f) = s_1(1-t)f$, $f \in E_{2l+1}$. It is verified directly that the map from $k[\mathbb{Z}/(n+1)\mathbb{Z}]$ sending t^i to $s_1^{n-1} t^i \in k\cdot\hom_\Lambda([n],[1])$ induces a bijection $k[\mathbb{Z}/(n+1)\mathbb{Z}] \xrightarrow{\sim} k\cdot\hom_\Lambda([n],[1])/L_n$. So $(E_o)_n \simeq (E_1)_n \simeq (E_2)_n \simeq \dots$ are free $k[\mathbb{Z}/(n+1)\mathbb{Z}]$-modules of rank one; it is easily seen that under this isomorphism the differential ∂ is exactly the same as in the standard resolution for the cyclic group; furthermore, $E. \otimes_{k[\Lambda]} B_A^\natural / A_A^\natural \simeq \Omega.(A \to B)$. This proves A8.2.

An example of using these ideas for the direct calculations is the proof of the rest of the statements of Chapter 3 (which were not proved there). So, to prove Theorem 3.2.3 it suffices to take a free simplicial bimodule resolution P of M; then $T_A(P)$ is the free simplicial resolution of $T_A(M)$ over A and the proof follows straightforward from the Theorem A8.1. Indeed, $\Omega.(A \to T_A(P)) \simeq \mathrm{Nil}^+_.(A;P)$.

<u>Proof of Theorem 3.2.1</u>. As in 3.2, we consider the free simplicial augmented resolutions R^B, R^A of the augmented algebras B, A over C. Then $R^A *_C R^B$ is such a resolution for $A \underset{C}{*} B$. The Proposition A8.2 shows that we may use the complex $E'_{\Lambda^o,\Delta^o}(k^\natural) \otimes_{k[\Lambda]} (R^A *_C R^B)_C^\natural / C_C^\natural$ (cf. A5). Write $R = R^A \underset{C}{*} R^B$. Consider the direct complement to $E'_{\Lambda^o,\Delta^o}(k^\natural) \otimes_{k[\Lambda]} (R_A^{A\natural} + R_A^{B\natural})$; denote it by $\mathcal{E}$. Then

$$\mathcal{E}_{kn} = \bigoplus_{m \geqslant 2} (\mathbb{Z}/m\mathbb{Z})^n \times (R^A *_C R^B)^{(m)}$$

where the superscript (m) means the grading by the length of a monomial in the free algebra. $\mathcal{E}$ is a bisimplicial k-module. The face and degeneracy operators with respect to n act as in the standard complex computing the group homology $H.(\mathbb{Z}/m\mathbb{Z}, (R^A \underset{C}{*} R^B)^{(m)})$ (compare with (A5.4)); let us describe these operators with respect to k.

Let $t_o c_o \dots t_m c_m$ $R_k^{(m+1)}$; $\sigma_i \in \mathbb{Z}/(m+1)\mathbb{Z}$; $f : [k]' \to [k]$ in Δ. Suppose that

$$(t_i a_i) f = \sum_{\alpha_i \in A_i} r_{\alpha_i}, \quad r_{\alpha_i} \text{ are the monomials in } R^{(m_{\alpha_i})};$$

A_i are some index sets. Then

$$((\sigma_o \times \ldots \times \sigma_{n-1}) \times (t_o c_o \ldots t_m c_m))f = \sum_{(\alpha_o,\ldots,\alpha_m)\in A_o \times \ldots \times A_m} (\bar{f}^*_{\alpha_o,\ldots,\alpha_m} \sigma_o \times$$

$$\times \ldots \times \bar{f}^*_{\alpha_o,\ldots,\alpha_m} \sigma_{n-1}) \times (r_{\alpha_o} \ldots r_{\alpha_m})$$

where

$$\bar{f}_{\alpha_o,\ldots,\alpha_m} \in \hom_\Delta([\textstyle\sum m_{\alpha_i} - 1], [m]); \quad \text{card } \bar{f}^{-1}_{\alpha_o,\ldots,\alpha_m}(i) = m_{\alpha_i}$$

Now consider the submodules $\mathcal{E}'' \subset \mathcal{E}' \subset \mathcal{E}$; $\mathcal{E}'$ is generated by the elements $(\sigma_o \times \ldots \times \sigma_{n-1}) \times (p_o q_o \ldots p_r q_r)$ where p_i are the monomials in $\bar{R}^A$ and q_i are the monomials in $\bar{R}^B$ and $\sigma_i(p_o q_o \ldots p_r q_r)$ is again equal to $p'_o q'_o \ldots p'_r q'_r$; $\mathcal{E}''$ is generated by the elements $(\sigma_o \times \ldots \times \sigma_{n-1}) \times (p_o q_o \ldots p_r q_r)$ such that $\sigma_i(p_o q_o \ldots p_r q_r) = p_o q_o \ldots p_r q_r$. It is easy to see that $\mathcal{E}'$ is a bisimplicial submodule of $\mathcal{E}$ isomorphic to $\bigoplus_r C.(\mathbf{Z}/(r+1)\mathbf{Z}; \ (\otimes_C^{r+1}(\bar{R}^A \otimes_C \bar{R}^B)) \underset{C\otimes C^o}{\otimes} C))$. We want to prove that $\mathcal{E}' \hookrightarrow \mathcal{E}$ is a quasi-isomorphism. In order to do this note that $\mathcal{E}''$ is a simplicial submodule in $\mathcal{E}$ with respect to the first simplicial structure; the inclusion $\mathcal{E}'' \hookrightarrow \mathcal{E}'$ is certainly a quasi-isomorphism with respect to this structure (by Shapiro lemma). So, $\mathcal{E}' \hookrightarrow \mathcal{E}$ is a quasi-isomorphism. This proves the Theorem 3.2.1.

Exactly the same method applyed to the bicomplex $\mathcal{L}((A + M)^\natural)$ (cf. A5) gives the proof of the Theorem 3.3.1. We omit the details.

Remark A8.3. The method of proof of Theorem A8.2 may be convenient also in the case of characteristic zero. It helps to study another derived functors on the category of k-algebras. For example we may take a functor $A \to A^{ab}$ where A^{ab} means the maximal commutative quotient. It may be shown that the derived functors of this functor (in sense of [47]; cf. 2.1) admit a categorical interpretation in spirit of the Theorem 3.3. They may be computed as the homology groups of a certain small category satisfying the conditions of A4 if $\mathfrak{Z} = \Delta$ and $\Gamma_j = \Sigma_{j+1}$, $j \geqslant 0$. This is the same category which arises in the computation of the Lie algebra homology of Lie(B), see Remark A5.6. It might be said that the constructions discussed here and in the Remark A5.6 are dual to each other in sense of A7.

A9. We have restricted ourselves by studying homology properties of the cyclic objects in an abelian category. Now we finish this paper reviewing some results on homotopy theory of cyclic sets and topological spaces and some related questions.

Theorem A9.1. (Connes, [14]). The classifying space of small category Λ is homotopy equivalent to $BS^1 = \mathbb{C}P^\infty$.

Theorem A9.2. (Burghelea, Fiedorowicz, [9]). Let X be a cyclic topological space. There are the homotopy fibrations

$$S^1 \longrightarrow \underrightarrow{\mathrm{holim}}_{\Delta^\circ} X \longrightarrow \underrightarrow{\mathrm{holim}}_{\Lambda^\circ} X;$$

$$\underrightarrow{\mathrm{holim}}_{\Delta^\circ} X \longrightarrow \underrightarrow{\mathrm{holim}}_{\Lambda^\circ} X \longrightarrow BS^1.$$

The following statement may be proved exactly in the same way as Theorem A9.2; it may also be obtained using the methods of the end of the Subsection A4. Let $\mathcal{C}$, $\mathcal{D}$, Γ be the same as in A4.

Theorem A9.3. Let X be a functor from to the category of topological spaces. Then there is a homotopy fibration

$$\underrightarrow{\mathrm{holim}}_{\mathcal{D}^\circ}\Gamma \longrightarrow \underrightarrow{\mathrm{holim}}_{\mathcal{D}^\circ} X \longrightarrow \underrightarrow{\mathrm{holim}}_{\mathcal{C}^\circ} X.$$

In particular, let $B\mathcal{C}$, $B\mathcal{D}$ denote the classifying spaces of the corresponding categories. Then there is a homotopy fibration

$$\underrightarrow{\mathrm{holim}}_{\mathcal{D}^\circ}\Gamma \longrightarrow B\mathcal{D} \longrightarrow B\mathcal{C}$$

Now let X be a simply connected topological space, $\tilde{X}$ a simplicial set corresponding to X and $\Omega\tilde{X}$ the Kan's free simplicial loops group for X. Let K. denote Waldhausen's algebraic K-functor.

Theorem A9.4. (D. Burghelea, [7]).

$$K.(X) \underset{\mathbb{Z}}{\times} \mathbb{Q} \longrightarrow K.(\mathbb{Z}) \underset{\mathbb{Z}}{\times} \mathbb{Q} + K^+_{.-1}(\mathbb{Q} \longrightarrow \mathbb{Q}[\Omega\tilde{X}]) \qquad (A9.1)$$

We finish by the statement about the right hand side of (A9.1).

Theorem A9.5. (Goodwillie-Dwyer-Jones, [7]). Let X^{S^1} be the free loop space with the natural action of S^1 and ES^1 be the total space of the universal S^1-bundle. Then

$$H.(\mathbb{Q}[\Omega\tilde{X}, \Omega\tilde{X}]) \simeq H.(X^{S^1}; \mathbb{Q});$$

$$K^+(\mathbb{Q} \to \mathbb{Q}[\Omega\tilde{X}]) \to H.(ES^1 \underset{S^1}{\times} X^{S^1}; \mathbb{Q})$$

10. Symmetric objects. In this Subsection we define the symmetric ategory and the symmetric objects. We show that the (co-)homology of hese objects generalize the Lie algebra (co-)homology. We discuss this eneralization in the framework of the theory of May coalgebras over he operads. One of the consequences is that the objects generalizing he Lie algebra cohomology in characteristic p possess a natural tructure of a graded module over the Steenrod algebra.

Define the category $\mathcal{S}$. Put $\mathrm{Ob}\,\mathcal{S} = \{[n] : n \geqslant 0,\ n \in Z\}$; let $n] = \{0, \dots, n\}$; we define $\hom_{\mathcal{S}}([n],[m])$ as the set of all famiies $\{f; <_0, \dots, <_m\}$ where $<_k$ is a total ordering of $f^{-1}(k)$. ut

$$\{g; <_0, \dots, <_l\} \circ \{f; <_0, \dots, <_m\} = \{gf; \prec_0, \dots, \prec_l\}, \qquad \text{(A10.1)}$$

here for all $i \in \{0, \dots, l\}$, $p, q \in (gf)^{-1}(\{i\})$:

$$p \prec_i q, \text{ if either } f(p) <_i f(q) \text{ or } \\ f(p) = f(q) \text{ and } p <'_{f(p)} q. \qquad \text{(A10.2)}$$

The condition (A10.2) determines $<_i$ uniquely. It is easy to see hat the composition (A10.1) satisfies the category axioms.

Let Δ be the subcategory of $\mathcal{S}$ such that:

$\mathrm{Ob}\,\Delta = \mathrm{Ob}\,\mathcal{S}$;

$\hom_\Delta([n], [m]) = \{(f; <_0^{(o)}, \dots, <_m^{(o)})\}$,

here f is nondecreasing with respect to the standard orderings on n], [m] and the $<_k^{(o)}$ are restrictions of the standard ordering to $^{-1}(\{k\})$.

Furthermore, for $n \geqslant 0$ we have

$$\Gamma_n = \mathrm{Aut}_{\mathcal{S}}([n]) \xrightarrow{\sim} \Sigma_{n+1},$$

nd for $f \in \hom_{\mathcal{S}}([m], [n])$ there are uniquely determined elements $\in \hom_\Delta([m], [n])$, $\sigma \in \Gamma_n$: $f = g \cdot \sigma$. So, $\mathcal{S}$ satisfies the propery of the beginning of A4. In particular, $\Gamma_\cdot = \Sigma_{\cdot+1}$ is a simpliial set.

Furthermore, let $\mathcal{S}_{inj}$ denote the subcategory of such $\{f, <_0, \dots,$ $_m\}$ that f is injective and $\mathcal{S}_{sur}$ denote the subcategory of such $\{f, <_0, \dots, <_m\}$ that f is surjective. Clearly, $\mathcal{S}_{sur}$ and $\mathcal{S}_{inj}$ satisfy the roperty (A4.1). Note also that $\mathcal{S}$ and $\mathcal{S}_{inj}$ satisfy (A5.6) if we ut $\varepsilon(\sigma) = \mathrm{sgn}\,\sigma$. Category $_{sur}$ is generated by the morphisms

$s_0, \ldots, s_{n-1} : [n] \longrightarrow [n-1]$, $\sigma : [n] \longrightarrow [n]$, $n \geqslant 0$; category $\mathcal{S}_{inj}$ is generated by the morphisms $d_0, \ldots, d_n : [n] \longrightarrow [n+1]$, $\sigma : [n] \longrightarrow [n]$, $n \geqslant 0$.

Let $\mathfrak{A}$ be a category.

Definition A10.1. A functor $\mathcal{S}_{sur} \longrightarrow \mathfrak{A}$ is called a Σ-object of $\mathfrak{A}$; a functor $\mathcal{S}^{\circ}_{ing} \longrightarrow \mathfrak{A}$ is called a Σ°-object of $\mathfrak{A}$; a functor $\mathcal{S}_{sur} \rightarrow \mathfrak{A}$ is called a symmetric object of $\mathfrak{A}$.

Example A10.2. Let A be an associative algebra (possibly without unit) over a commutative ring k. We define a Σ-object $A^{\natural}$ and a Σ°-object $A^{\flat}$ in the category of k-modules.

Define $A^{\natural}$. Put $A^{\natural}_n = A^{\otimes(n+1)}$; let $\varphi = \{f; <_0, \ldots, <_m\} \in \hom_{\mathcal{S}}([n], [m])$. Put

$$\varphi(a_0 \otimes \ldots \otimes a_n) = a_{i_{00}} \ldots a_{i_{0l_0}} \otimes \ldots \otimes a_{i_{m0}} \ldots a_{i_{ml_m}}$$

where for $s = 0, \ldots, m$ $f^{-1}(\{s\}) = \{i_{s0}, \ldots, i_{sl_s}\}$ and $i_{s0} <_s i_{s1} <_s \ldots <_s i_{sl_s}$. If A is an algebra with unit, then A may be extended to a functor from $\mathcal{S}$ to k-mod.

Define $A^{\flat}$. Put

$$(a_0 \otimes \ldots \otimes a_n)d_i = \sum_{j \neq i} a_0 \otimes \ldots \otimes a_j a_i \otimes \ldots \otimes a_i \otimes \ldots \otimes a_n .$$

Then $xd_jd_i = xd_id_{j-1}$, $i < j$; put also

$$(a_0 \otimes \ldots \otimes a_n)\sigma = a_{\sigma 0} \otimes \ldots \otimes a_{\sigma n}, \quad \sigma \in \Sigma_{n+1} .$$

It is easy to see that the relations between d_i and σ are the same as in $\mathcal{S}$.

Example 10.3. Let $\mathfrak{A}$ be a category with products and with final object $*$ (i.e., the category of sets, of simplicial sets, of topological spaces, of augmented complexes, ...). Let $\mathcal{O}$ be an operad in $\mathfrak{A}$ (cf. [ ]). Assume that a morphism $\Delta : * \longrightarrow \mathcal{O}(2)$ is given. Let γ denote the multiplication in the operad $\mathcal{O}$. Construct the morphisms $s_i : \mathcal{O}(n) \longrightarrow \mathcal{O}(n+1)$, $0 \leqslant i > n-1$. Let $e : * \longrightarrow \mathcal{O}(1)$ be the unit in $\mathcal{O}$. We define s_i to be equal to the following composition:

$$\mathcal{O}(n) = \mathcal{O}(n) \times * \times \ldots \times * \xrightarrow{\;1_{\mathcal{O}(n)} \times \underbrace{e \times \ldots \times e}_{i-1} \times \Delta \times e \times \ldots\;}$$

$$\longrightarrow \mathcal{O}(n) \times \mathcal{O}(1) \times \ldots \times \mathcal{O}(2) \times \mathcal{O}(1) \times \ldots \longrightarrow \mathcal{O}(n+1).$$

We say that $\mathcal{O}$ is the operad with associative coproduct if in the following diagram $s_o s_o = s_1 s_o$:

$$\mathcal{O}(1) \xrightarrow{s_o} \mathcal{O}(2) \overset{s_o}{\underset{s_1}{\rightrightarrows}} \mathcal{O}(3).$$

If $\mathcal{O}$ is an operad with associative coproduct, then the morphisms s_i together with the right action of Σ_n on $\mathcal{O}(n)$ determine a symmetric object of $\mathcal{Q}$ which is denoted by $\mathcal{O}^{\natural}$:

$$\mathcal{O}^{\natural}(\mathcal{O}^{\natural})_n = \mathcal{O}(n+1).$$

An important example arises if $\mathcal{O}$ is the operad of natural transformations of the chain complex functor:

$$\mathcal{O}(n) = \mathrm{Hom}_{fun}(C, C^{xn}),$$

where $C : \Delta^o\text{-Sets} \longrightarrow (\mathrm{Com})$ is the functor associating to a simplicial set its chain complex. Then Δ is represented by a cycle in $\mathcal{O}(2)_o$ which is the Alexander-Whithney coproduct. We shall return to this example below.

Since $\mathcal{G}_{inj}$ satisfies (A5.6), one may compute the homology of this category using A5.

Proposition A.10.4. Let char $k = 0$. Then

$$H.(\mathcal{G}^o_{inj}, A^b) \xrightarrow{\sim} H._{x1}(\mathrm{Lie}\ A, k).$$

Proof. Consider the bicomplex from We see that the horizontal homology of this bicomplex vanishes in all dimensions except of zero. The corresponding spectral sequence collapses, and we obtain

$$H.(\mathcal{G}^o_{inj}, A^b) \xrightarrow{\sim} H_{\cdot+1}(\Lambda^{\cdot} A, d_o - d_1 + \dots + (-1)^n d_n)$$

where

$$d_i(a_o \wedge \dots \wedge a_n) = \sum_{j \neq i} a_o \wedge \dots \quad a_j a_i \wedge \dots \wedge \hat{a}_i \wedge \dots \wedge a_n ;$$

$$(a_o \wedge \dots \wedge a_n) \cdot \sum_{i=0}^{n} (-1)^i d_i = \sum_{i=0}^{n} \sum_{j>i} (-1)^{i+j-1} \times$$

$$\times\ a_j a_i \wedge \dots \wedge \hat{a}_i \wedge \dots \wedge \hat{a}_j \quad \dots \wedge a_n +$$

$$+ \sum_{j=0}^{n} \sum_{i<j} (1)^{i+j} a_i a_j \wedge \dots \wedge \hat{a}_i \wedge \dots \wedge \hat{a}_j \wedge \dots \wedge a_n =$$

$$= \sum_{i<j} (-1)^{i+j}[a_i, a_j] \wedge \ldots \wedge \hat{a}_i \wedge \ldots \wedge \hat{a}_j \wedge \ldots \wedge a_n.$$

This gives the proof of A10.4.

Let A be an algebra and R be its free differential graded (or simplicial) resolution in the category of algebras without unit. Consider the functor $A \mapsto A^{ab}$, which associates to an algebra its maximal commutative quotient algebra. Then R^{ab} is a complex (moreover, it is a commutative differential graded (or simplicial) algebra). Put

$$\mathbb{L}_i A^{ab} = H_i(R^{ab}), \quad i \geqslant 0.$$

Proposition A10.5. Let char $k = 0$. Then

$$H.(\mathcal{G}_{sur}, A^{\natural}) \xrightarrow{\sim} \mathbb{L}.\, A^{ab}.$$

Proof. The complex $R.^{\natural}$ of Σ-objects is Δ_{sur}-acyclic in each dimension (compare with A.8). So, using
we have

$$H.(\mathcal{G}_{sur}, A^{\natural}) \xrightarrow{\sim} H.(E_{\mathcal{G}^{\circ}_{sur}, \Delta^{\circ}_{sur}}(k^{\natural}) \otimes_{\mathcal{G}_{sur}} R.^{\natural}) \xrightarrow{\sim}$$

$$\xrightarrow{\sim} H.(k^{\natural} \otimes_{\mathcal{G}_{sur}} R.^{\natural}) \xrightarrow{\sim} H.(R^{ab}), \quad \text{Q.E.D.}$$

We see from A10.4 that in the case when k is arbitrary the homology groups $H.(\mathcal{G}^{\circ}_{inj}, A^{\flat})$ provide a reasonable version of $H_{\bullet+1}(\text{Lie } A, k)$. However, the (co-)homology invariants of the object $A^{\natural}$ seem to be of greater importance. We shall explain this below.

Now we shall construct a double complex connecting the structures of $A^{\flat}$ and $A^{\natural}$.

Put

$$\mathcal{T}^n_m(A) = \begin{cases} \text{Hom}(\Sigma^n_{m+1}, A^{\otimes(m+1)}), & n \geqslant 0; \\ A^{\otimes(m+1)} \times \Sigma^{-(n+1)}_{m+1}, & n < 0; \end{cases}$$

define

$$d : \mathcal{T}^n_m(A) \longrightarrow \mathcal{T}^{n+1}_m(A);$$

$$\partial : \mathcal{T}^n_m(A) \longrightarrow \mathcal{T}^n_{m-1}(A);$$

if $n \geqslant 0$, $\varphi \in \text{Hom}(\Sigma^n_{m+1}, A^{\otimes(m+1)})$, then

$$(d\varphi)(\sigma_0,\ldots,\sigma_n) = \sum_{i=0}^{n-1} (1)^i \varphi(\sigma_0,\ldots,\hat{\sigma}_i,\ldots,\sigma_n) + \qquad \text{(A10.3)}$$
$$+ (-1)^n \operatorname{sgn}(\sigma_n)\, \sigma_n^{-1}(\varphi(\sigma_0\sigma_n^{-1},\ldots,\sigma_{n-1}\,\sigma^{-1});$$

f $n < -1$, $x \in A^{\otimes(m+1)}$, then

$$d(x \otimes \sigma_1 \otimes \ldots \otimes \sigma_{-1-n}) = \sum_{i=1}^{-2-n} (-1)^i\, x \otimes \sigma_1 \times\ldots\times \hat{\sigma}_i \times\ldots\times \qquad \text{(A10.4)}$$
$$\times\, \sigma_{-1-n} + (-1)^{-1-n}\, x\, \sigma_{-1-n}^{-1} \times \sigma_1\sigma_{-1-n}^{-1} \times\ldots\times \sigma_{-2-n}\, \sigma_{-1-n}^{-1};$$

f $n = 0$, then

$$dx = \sum_{\sigma\in\Sigma_{m+1}} \operatorname{sgn}(\sigma)\, x\, \sigma := N\, x; \qquad \text{(A10.5)}$$

f $n \geqslant 0$, then

$$(\partial\varphi)(\sigma_0,\ldots,\sigma_{n-1}) = (-1)^n \sum_{i=0}^{m-1} (-1)^i\, s_i\, \varphi(s_i^*\, \sigma_0,\ldots,s_i^*\, \sigma_{n-1}); \qquad \text{(A10.6)}$$

f $n < 0$, then

$$\partial(x \times \sigma_1 \times\ldots\times \sigma_{-1-n}) = (-1)^n \sum_{i=0}^{m} (-1)^i\, x d_i \times d_i^*\sigma_1 \times\ldots \qquad \text{(A10.7).}$$
$$\ldots\, d_i^*\, \sigma_{-1-n}.$$

The action of the maps s_i, d_i on $A^{\otimes(\cdot+1)}$ was defined in Examp-
e A10.2; the maps d_i^*, s_i^* correspond to the simplicial set $\Gamma. =$
$= \Sigma_{\cdot+1}$ which was defined in A4.

The double complex $\mathcal{T}_\cdot(A)$ is represented by Fig. A10.1.

$$\begin{array}{ccccccc}
\downarrow & & \downarrow & & \downarrow & & \downarrow \\
A^{\otimes(m+1)} \times \Sigma_{m+1} & \xrightarrow{d} & A^{\otimes(m+1)} & \xrightarrow{d=N} & A^{\otimes(m+1)} & \xrightarrow{d} & \mathrm{Hom}(\Sigma_{m+1}, A^{\otimes(m+1)}) \longrightarrow \ldots \\
\vdots & & \vdots & & \vdots & & \vdots \\
A^{\otimes 3} \times \Sigma_3 & \xrightarrow{d} & A^{\otimes 3} & \xrightarrow{d=N} & A^{\otimes 3} & \xrightarrow{d} & \mathrm{Hom}(\Sigma_3, A^{\otimes 3}) \longrightarrow \ldots \\
\downarrow\partial & & \downarrow\partial & & \downarrow\partial & & \downarrow\partial \\
A^{\otimes 2} \times \Sigma_2 & \xrightarrow{d} & A^{\otimes 2} & \xrightarrow{d=N} & A^{\otimes 2} & \xrightarrow{d} & \mathrm{Hom}(\Sigma_2, A^{\otimes 2}) \longrightarrow \ldots \\
\downarrow\partial & & \downarrow\partial & & \downarrow\partial & & \downarrow\partial \\
A \times \Sigma_1 & \xrightarrow{d} & A & \xrightarrow{d=N} & A & \xrightarrow{d} & \mathrm{Hom}(\Sigma_1, A) \longrightarrow \ldots
\end{array}$$

Fig. A10.1.

Lemma A10.6. For all $n \in \mathbb{Z}$, $m \geqslant 0$: $\partial d + d\partial = 0$.

Proof. Note that $\{\mathcal{T}_m^{-(n+1)}\}_{n\geqslant 0}$ is isomorphic to $\mathcal{E}(A^b)_{nm}$ from Corolary A5.4 (the differentials differ by a sign). So $\partial d + d\partial = 0$ in the left half of Fig. A10.1.

Let M be an object in $\mathcal{O}\!\iota^{\mathcal{G}^{sur}}$ where $\mathcal{O}\!\iota$ is an arbitrary abelian category. Define $\partial_n : M_n \to \mathrm{Hom}(\Sigma_{n+1}, M_n)$:

$$(\partial_n m)(\sigma) = \mathrm{sgn}\sigma\ \sigma^{-1}m.$$

Put $dm = \sum_{i=0}^{n-1} (-1)^i s_i m$; $(d'\varphi)(\sigma) = \sum_{i=0}^{n-1} (-1)^i s_i \varphi(s_i^* \sigma)$,

$m \in M_n$, $\varphi \in \mathrm{Hom}(\Sigma_{n+1}, M_n)$.

Sublemma A10.7. $\partial_n \cdot d = d' \cdot \partial_n$.

Proof. Note that $\sigma(s_i) = s_{\sigma i}$, $\mathrm{sgn}(s_i^*\sigma)/\mathrm{sgn}(\sigma) = (-1)^{\sigma i - i}$.
Thus,

$$(d'\partial_n m)(\sigma) = \sum_{i=0}^{n-1} (-1)^i \,\mathrm{sgn}(s_i^*\sigma)\, s_i (s_i^*\sigma)^{-1} m =$$

$$= \sum_{i=0}^{n-1} (-1)^{\sigma i}\, \mathrm{sgn}(\sigma)\, s_i \cdot s_{\sigma i}^*(\sigma^{-1})m = \sum_{j=0}^{n-1} (-1)^j\, \mathrm{sgn}\sigma \cdot \sigma^{-1}(s_j)\, s_j^*(\sigma^{-1})m =$$

$$= \sum_{j=0}^{n-1} (-1)^j\, \mathrm{sgn}\sigma\, s_j \sigma^{-1} m = (\partial_n d)(m)(\sigma), \quad \text{Q.E.D.}$$

The Sublemma implies that for any Σ-object M, the complex $\mathcal{F}_n(M) = \{m \in M_n : \sigma m = \mathrm{sgn}(\sigma)\cdot m,\ \forall\ \sigma \in \Sigma_{n+1}\}$ with the differential $\sum_{i=0}^{n-1} (-1)^i s_i$ is well defined. The bicomplex $\{\mathcal{T}_m^n\}_{n>0}$ is isomorphic to $\mathcal{T}_n(\mathrm{Hom}(E'_{\mathcal{G}^o_{sur},\ \Delta^o_{sur}}(k^\natural), A^\natural))$ (cf. A5). Thus, $d\partial + \partial d = 0$ also in the right half of Fig. A10.1.

It remains to show that

$$N \sum_{i=0}^{n} x d_i (-1)^i = \sum_{i=0}^{n-1} (-1)^i s_i N x, \quad x \in A^{\otimes(n+1)}.$$

Define the following morphisms in $\mathcal{G}$. For $i < j$ put

$$s_{ij}^+(l) = \begin{cases} l, & l \leqslant i; \\ l-1, & l > i,\ l \neq j; \\ i, & l = j; \end{cases} \qquad s_{ij}^-(l) = \begin{cases} l, & l < i; \\ l-1, & l > i; \\ j-1, & l = i. \end{cases}$$

Put also $i <_i j$ for s_{ij}^+ and $j <_{j-1} i$ for s_{ij}^-. We obtain two elements $s^{\pm}_{ij} \in \hom_{\mathcal{G}}([n],[n-1])$. Every element $f \in \hom_{\mathcal{G}}([n],[n-1])$ admits either a decomposition $f = \tau s^{+}_{ij}$, $\tau \in \Sigma_m$, or $f = \tau s^{-}_{ij}$, $\tau \in \Sigma_m$. This decomposition is unique. On the other hand, f has the standard decomposition $f = g\sigma$, $g \in \mathrm{Mor}\,\Delta$; it is easy to see that

$g = s_{\tau i}$ and $\mathrm{sgn}(\sigma)/\mathrm{sgn}(\tau) = (-1)^{j-1-\tau i}$ if $f = \tau s^{+}_{ij}$;

$g = s_{\tau(j-1)}$ and $\mathrm{sgn}(\sigma)/\mathrm{sgn}(\tau) = (-1)^{i-1-\tau(j-1)}$ if $f = \tau s^{-}_{ij}$.

Thus,

$$\partial Nx = (-1)^m \sum_{i=0}^{m-1} (-1)^i \,\mathrm{sgn}\sigma \cdot s_i \sigma x = (-1)^m \sum_{i<j} (-1)^{-\tau i + j - 1 + \tau i} \,\mathrm{sgn}\,\tau\, x$$

$$\tau s^{+}_{ij} x + (-1)^m \sum_{i<j} (-1)^{-\tau(j-1)+i-1+\tau(j-1)} \,\mathrm{sgn}\,\tau \cdot \tau \cdot s^{-}_{ij}\, x =$$

$= -N\partial x$, Q.E.D.

This ends the proof of Lemma A10.6.

Remark A10.8. The construction of the Σ°-object analogous to b and of the bicomplex analogous to $\mathcal{T}(A)$ may be carried out for any Σ-object of an abelian category $\mathcal{A}$ (not only for $A^{\natural}$).

Remark A10.9. Let $\mathrm{char}\, k = 0$. We have seen in Chapter 5 that when $\mathbb{Q} \subseteq k$) for any Young diagram ¶ (possibly infinite) one may define the Σ-vector spaces $F_{¶}S$, $F^{¶}S$ and the homology groups

$$H_{\cdot+1}(gl_{¶}(A)) = H_{\cdot}(\mathcal{F}(F_{¶}S \otimes_k A^{\natural}));$$

$$H_{\cdot+1}(gl^{¶}(A)) = H_{\cdot}(\mathcal{F}(F^{¶}S \otimes_k A^{\natural}));$$

If ¶ is an infinite diagram with n cells in each column, then

$$H_{\cdot}(gl^{¶}(A)) = H_{\cdot}(gl_n(A));$$

If ¶ is a diagram with the single non-empty column of height $n+1$, then

$$H_{\cdot}(gl_{¶}(A)) = H_{\cdot}(gl_n(A)).$$

But

$$\mathcal{T}(A)^l = \bigoplus_{n-m=l} \mathcal{T}^n_m(A);$$

$$\mathcal{T}^l_+(A) = \bigoplus_{n-m=l} \mathcal{T}^n_m(A); \qquad \text{(A10.9)}$$

$$\mathcal{T}^l_-(A) = \bigoplus_{\substack{n-m=l \\ n\geq 0}} \mathcal{T}^n_m(A); \quad \hat{\mathcal{T}}^l_-(A) = \prod_{\substack{n-m=l \\ n\geq 0}} \mathcal{T}^n_m(A).$$

Now we begin to discuss the relation of the complex $\mathcal{T}.(A)$ to May theory.

Fix an acyclic operad $\mathcal{O}$ in the category of augmented complexes of k-modules (cf. [25]). Recall that the complex X is called a May algebra over $\mathcal{O}$ if there is a morphism of operads $\mathcal{O} \longrightarrow \underline{\mathrm{Hom}}(X^{\otimes n}, X)$ (cf. [25]). This means that there are the operations $\mathcal{O}(n) \otimes X^{\otimes n} \longrightarrow$ $\longrightarrow X$ satisfying certain natural conditions. Dually, a complex X. is called a May coalgebra over $\mathcal{O}$ if there is a morphism of operads $\mathcal{O} \longrightarrow \mathrm{Hom}(X, X^{\otimes n})$, i.e., if there is a collection of operations $\mathcal{O}(n) \otimes X \longrightarrow X^{\otimes n}$ with some natural properties. For our purposes, it is convenient to consider the topological May coalgebras. A topological May coalgebra is a complex X. which is full in a topology connected with the given non-increasing filtration together with the collection of continuous operations $\mathcal{O}(n) \hat{\otimes} X — X^{\hat{\otimes} n}$, $n \in \mathbb{N}$, satisfying the natural properties. We denote the category of May algebra over $\mathcal{O}$ by May-alg, the category of May coalgebras over $\mathcal{O}$ by May-coalg, the category of topological May coalgebras over $\mathcal{O}$ by May-coalgt and the category of complexes of k-modules by Com(k). There is an inclusion of categories May-coalg $\xrightarrow{I}$ May-coalgt: a May coalgebra X. is regarded as a topological May coalgebra with respect to the trivial filtration $X \supset 0 \supset 0 \ldots$ Note that here, in the framework of May coalgebras theory, we regard all our complexes as homological ones (possibly non-bounded in both sides).

There is a forgetting functor F : May-coalgt $\longrightarrow$ Com(k). It has a right adjoint (constructed by Smirnov in the dual case); if X. is a complex, put

$$GX = \prod \underline{\mathrm{Hom}}_{\Sigma_n}(\mathcal{O}(n), X^{\otimes n});$$

the operation μ_n for GX, being restricted to a direct factor, is equal to

$$\mathcal{O}(k) \hat{\otimes} \mathrm{Hom}_{\Sigma_n}(\mathcal{O}(n), X^{\hat{\otimes} n}) \longrightarrow \prod_{n_1+\ldots+n_k=n} \underline{\mathrm{Hom}}_{\Sigma_{n_1} \times \ldots \times \Sigma_{n_k}}(\mathcal{O}(n_1) \otimes \ldots \otimes \mathcal{O}(n_k),$$

$$X^{\hat{\otimes} n} \hat{\otimes} \ldots \hat{\otimes} X^{\hat{\otimes} n}) \longrightarrow (\prod \underline{\mathrm{Hom}}_{\Sigma_1}(\mathcal{O}(1), X^{\otimes 1}))^{\hat{\otimes} k}.$$

It is easy to see that

$$\mathrm{Hom}_{\text{May-coalgt}}(Y, GX) \xrightarrow{\sim} \mathrm{Hom}_{\mathrm{Com}(k)}(FY, X)$$

for any topological May coalgebra Y.

Smirnov's construction is a May-theoretic generalization of the ymmetric algebra functor (and hence also of the exterior algebra functor). We want to modify the differential in GX in order to obtain he generalization of the Lie algebra complex.

Let A be an associative k-algebra without unit. Assume that $\mathcal{O}$ s an operad with associative coproduct (Example A10.2). Then we form complex $\mathcal{F}(\mathrm{Hom}_k(\mathcal{O}^{\natural}, A^{\natural}))$. Here $\mathcal{O}^{\natural}$ is regarded as a homological omplex of Σ-objects in the category of k-modules; the bivariant unctor $\mathcal{F}$ from the category of Σ-objects to the category of complexes was defined in the proof of Sublemma A10.7. Put

$$ {}_{\mathcal{O}}\hat{\mathfrak{J}}^{-}(A) = \mathcal{F}(\mathrm{Hom}_k(\mathcal{O}^{\natural}, A^{\natural}))^{\wedge}, \qquad (A10.10)$$

here the completion is taken with respect to the decreasing filtraion

$$\mathrm{filt}^n \mathcal{F} = \bigoplus_{n'>n} \mathrm{Hom}_{\Sigma_{n'}}(\mathcal{O}(n'), A^{\otimes n'}).$$

he complex ${}_{\mathcal{O}}\hat{\mathfrak{J}}^{-}(A)$ coincide with the "symmetric coalgebra" G(A[-1]) s a graded space (or even as a bigraded space). The differential in ${}_{\mathcal{O}}\hat{\mathfrak{J}}^{-}(A)$ is equal to the sum of the differential d in G(A[-1]) with n additional differential ∂ which acts from the component $\mathrm{Hom}(\mathcal{O}(n), A^{\otimes n})$ to the component $\mathrm{Hom}(\mathcal{O}(n-1), A^{\otimes(n-1)})$. Note that just the ame formula (A10.10) determines the May coalgebra ${}_{\mathcal{O}}\hat{\mathfrak{J}}^{-}(A)$ in the ase when A is a differential graded algebra. In this case, $A^{\natural}$ is egarded as a complex of Σ-objects of k-mod.

We shall show that the construction of ${}_{\mathcal{O}}\hat{\mathfrak{J}}^{-}$ is the direct generalization of the construction of $\hat{\mathfrak{J}}^{-}$ given by (A10.8). In order to o this, we shall now construct an operad $\mathcal{O}^{s}$. First, we shall recall the construction of the operad in the category of sets and of the perad in the category of simplicial sets which was suggested by J. May (cf. [ ]).

Let M be a free monoid with infinitely many generators. Consider an operad $\mathcal{O}^{M}(n) = \mathrm{Hom}_{sets}(M^n, M)$. This is an operad in the category of sets. Consider a suboperad $\mathcal{O}^{o}(n) \to \mathcal{O}^{M}(n)$ consisting of ransformations $M^n \text{ — } M$ of the type

$$(m_1, \ldots, m_n) \mapsto m_{\sigma^{-1}1} \ldots m_{\sigma^{-1}n}, \quad \sigma \in \Sigma_n, \quad n \in \mathbb{N}.$$

$\mathcal{O}^{o}(n)$ is an operad in the category of sets; $\mathcal{O}^{o}(n) \xrightarrow{\sim} \Sigma_n$, $n \in \mathbb{N}$.

Now let $\mathcal{O}^S(n)$ be the simplex with the set of vertices numerated by Σ_n: $\mathcal{O}^S(n) \in \Delta^\circ$ Sets; $\mathcal{O}^S_k(n) = \Sigma_n^{k+1}$; the face maps are given by the deleting of components, and the degeneracy maps are given by the repeating of components. We may define the operation $\mathcal{O}^S_k(n) \times \times \mathcal{O}^S_k(l_1) \times \ldots \times \mathcal{O}^S_k(l_n) \to \mathcal{O}^S_k(l_1+\ldots+l_n)$ componentwise: if $\sigma_o,\ldots,\sigma_k \in \mathcal{O}^S_k(n)$, $\tau_{io},\ldots,\tau_{ik} \in \mathcal{O}^S_k(l_i)$, then

$$\gamma((\sigma_o,\ldots,\sigma_k);\ (\tau_{10},\ldots,\tau_{1k}),\ldots,(\tau_{no},\ldots,\tau_{nk})) =$$

$$= (\gamma^o(\sigma_o;\tau_{10},\ldots,\tau_{no}),\ldots,\gamma^o(\sigma_k;\tau_{1k},\ldots,\tau_{nk})) \in \mathcal{O}^S_k(l_1+\ldots+l_n).$$

Here γ^o is the operation in $\mathcal{O}^o$. Now construct the maps of complexes

$$\mathcal{O}^S(n) \otimes \mathcal{O}^S(l_1) \otimes\ldots\otimes \mathcal{O}^S(l_n) \xrightarrow{EZ} \mathcal{O}^S(n) \boxtimes \mathcal{O}^S(l_1) \boxtimes \ldots \boxtimes \mathcal{O}^S(l_n) \longrightarrow$$

$$\longrightarrow \mathcal{O}^S(l_1+\ldots+l_n); \qquad (A10.9).$$

the first map is the Eilenberg-Zilber homomorphism. Since EZ is associative and commutes with permutations, it is not hard to verify that the operation (A10.9) (together with the natural Σ_n-module structures on $\mathcal{O}^S(n)$) turn $\mathcal{O}^S$ to be an acyclic operad in the category of augmented complexes. Certainly, $\mathcal{O}^S$ is an operad with associative comultiplication $\Delta \in \mathcal{O}(2)_o$; Δ is represented by the chain $e \in \mathcal{O}(2)_o = \Sigma_2$.

<u>Lemma A10.10</u>. ${}_{\mathcal{O}^S}\hat{\mathcal{T}}^-(A) \xrightarrow{\sim} \hat{\mathcal{T}}^-(A)$ for any associative algebra A.

The proof is straightforward.

<u>Conjecture A10.11</u>. For any simplicial set K, the chain complex C.(K) has the functorial structure of a May coalgebra over $\mathcal{O}^S$, in which Δ acts via Alexander-Whithney comultiplication.

If A10.11 held, the calculations below might be simplified. However, we may apply a result by Smirnov, Hinich and Schechtman, which may be stated as follows.

<u>Proposition A10.12</u>. There exists an acyclic operad $\mathcal{O}$ in the category of augmented complexes, such that:

a) for any $n \in \mathbb{N}$, $\mathcal{O}(n)$ is the complex of free Σ_n-modules; $\mathcal{O}_{<0}(n) = 0$;

b) $\mathcal{O}$ has an associative comultiplication Δ;

c) for any simplicial set K, the chain complex C.(K) has the functorial structure of a May coalgebra over $\mathcal{O}$, in which Δ acts via the Alexander-Whithney coproduct.

The proof contain in Smirnov's paper [ ].

Fix an operad $\mathcal{O}$ which has the properties a)-c) of A10.12. Then we may construct the spatial realization functor as it has been done in Sullivan theory, cf. [ ].

Definition A10.13. Let X be a topological May coalgebra over $\mathcal{O}$. Pute

$$\langle X\rangle_n = \mathrm{Hom}_{\text{May-coalgt}}(C.(\Delta^n),\ X.),\quad n \geqslant 0.$$

Then the face and degeneracy maps between the standard simplices induce on $\langle X\rangle$ the structure of a simplicial set. We call this set the spatial realization of X.

Example A10.14. Let X be a complex and GX a symmetric May coalgebra generated by X. Since G is right adjoint to the forgetting functor, we have

$$\langle GX\rangle_n = \mathrm{Hom}_{\mathrm{Com}(k)}(C.(\Delta^n),\ X);$$

we see that GX is the Dold-Puppe transform of X.

Now let A be an associative algebra over k; let $\tilde{A}$ be an algebra obtained from A by adjoining a unit. Consider the monoid $(1 + A)^x \in \tilde{A}$ where the operation is multiplication in A. We construct, as usually, the simplicial set $B(1 + A)^x$:

$$(B(1 + A)^x)_{n+1} = \underbrace{(1 + A) \times \dots \times (1 + A)}_{n+1 \text{ times}};$$

$$(y_0,\dots,y_n)d_i = (y_0,\dots,y_{i-1}y_i,\dots,y_n),\quad 0 < i < n;$$

$$(y_0,\dots,y_n)d_0 = (y_1,\dots,y_n);\ (y_0,\dots,y_n)d_{n+1} = y_0,\dots,y_{n-1};$$

$$(y_0,\dots,y_n)s_i = (y_0,\dots,1,y_i,\dots,y_n),\quad 0 \leqslant i \leqslant n+1.$$

Proposition A10.15. There is the canonical isomorphism of simplicial sets

$$\langle {}_{\mathcal{O}}\hat{\mathcal{J}}^-(A)\rangle \xrightarrow{\sim} B(1 + A)^x.$$

Proof. Let A be a differential graded algebra and K a simplicial set. It is easy to see that a homomorphism in May-coalgt from $C.(K)$ to ${}_{\mathcal{O}}\hat{\mathcal{J}}^-(A)$ is uniquely determined by a homomorphism of graded modules $\varphi: X \longrightarrow A[-1]$ such that

$$d\varphi - \varphi d = \mu(\varphi \otimes \varphi)\Delta \qquad \text{(A10.11)}$$

where Δ is the Alexander-Whithney comultiplication in $C.(K)$ and μ is the multiplication $A \otimes A \longrightarrow A$. Note that both sides of (A10.11) have degree -1. In our case, i.e., when $A = A_0$ and $K = \Delta_n$, φ associates to every of Δ^n the element of A; let a_{ik} be the image of the segment between the i-th and k-th vertices of Δ^n under the map φ. Clearly, we must verify (A10.11) only at the level of two-chains of $C.(\Delta^n)$; let $(i\ j\ k)$ be the 2-simplex of Δ^n with vertices i, j, k, $i < j < k$. Then (A10.15) implies

$$a_{ij} - a_{ik} + a_{jk} = -a_{ij}a_{jk}, \qquad \text{(A10.12)}$$

i.e., $(1 + a_{ij})(1 + a_{jk}) = 1 + a_{ik}$.

Thus, $\langle {}_{\mathcal{O}}\hat{\mathcal{T}}^{-}(A)\rangle_n = (y_{ij})_{0\leq i<j\leq n} : y_{ij} \in 1+A, \quad y_{ij}y_{jk} = y_{ik}$ for all $i < j < k$. The i-th face map is obtained by deleting of the elements a_{ij}, a_{ji} for all $j \neq i$; the i-th degeneracy map is obtained by repeating of the row (a_{ij}) and of the column (a_{ki}). We see that the map $(y_{ij}) \mapsto (y_{01}, \dots, y_{n-1,n})$ establishes the isomorphism A10.15.

Remark A10.16. Certainly, one may easily carry out the construction of the whole Tate complex ${}_{\mathcal{O}}\mathcal{T}^{\cdot}_{\cdot}(A)$ when $\mathcal{O}$ is an arbitrary operad with associative comultiplication and A a differential graded algebra.

Using the dual version of the bicomplex $\hat{\mathcal{T}}^{-}$, one may define the cohomology of an algebra A. Consider the following double complex:

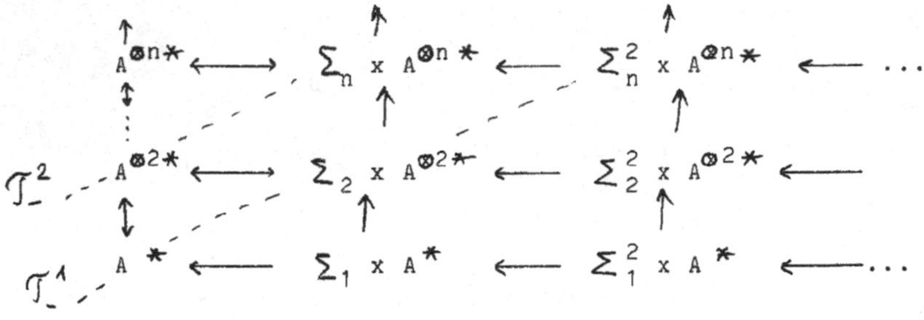

$$\mathcal{T}^{1}_{-}(A) = \bigoplus_{n-m=1} \Sigma_n^m \times A^{\otimes n *};$$

if $(\varphi s_i)(x) = \varphi(s_i x)$, $(\varphi\sigma)(x) = \varphi(\sigma x)$, $\varphi \in A^{\otimes n*}$, $x \in A^{\otimes n*}$, $\sigma \in \Sigma_n$, then

$$d(\sigma_0 \times \dots \times \sigma_{m-1} \times \varphi) = \sum_{i=0}^{n-1} (-1)^i \times \varphi s_i \times s_i\sigma_0 \times \dots \times s_i\sigma_{m-1};$$

$$\partial(\sigma_o \times \ldots \times \sigma_{m-1} \times \varphi) = (-1)^n \sum_{i=0}^{m-2} (-1)^i \sigma_o \times \ldots \times \hat{\sigma}_i \times \ldots \times \sigma_{m-1} \times \varphi +$$

$$(-1)^{m-1} \operatorname{sgn}(\sigma_{m-1}) \sigma_o \sigma_{m-1}^{-1} \times \ldots \times \sigma_{m-2} \sigma_{m-1}^{-1} \times \varphi \sigma_{m-1}^{-1}).$$

hen $\partial d + d\partial = 0$. If multiplication in A is zero, then $\mathcal{T}^{\bullet}_{-}(A)$ is he free May algebra of the complex A[-1]. In the general case, the ame formulas endow $\mathcal{T}^{\bullet}_{-}(A)$ with the structure of a May algebra over he operad $\mathcal{O}^S$. As a consequence, we have the following. Put $L^{\bullet}_{-}(A) = H(\mathcal{T}^{\bullet}_{-}(A))$.

Proposition A10.16. The cohomology module $HL^{\bullet}_{-}(A)$ has the ca- onical structure of a graded skew-commutative algebra. If char k = p, hen $HL_{-}(A)$ has the canonical structure of a graded module over the teenrod algebra A_p. This structure is related to the multiplication ia the Cartan formulas.

Indeed, the result by Hinich and Schechtman ([25]) shows that hese statements are valid for the cohomology of an arbitrary May al- ebra over an acyclic operad.

Example A10.17. $HL^i_{-}(\mathbb{Z}/p) \xrightarrow{\sim} \bigoplus_{n-m=i} H_m(\Sigma_n, (\mathbb{Z}/p)^-)$, where $(\mathbb{Z}/p)^-$ s the Σ_n-module corresponding to the character $\operatorname{sgn}: \Sigma_n \to (\mathbb{Z}/p)^*$. he multiplication in $HL^i(\mathbb{Z}/p)$ is the composition

$$H_m(\Sigma_n, (\mathbb{Z}/p)^-) \otimes H_{m'}(\Sigma_{n'}, (\mathbb{Z}/p)^-) \longrightarrow$$

$$\to H_{m+m'}(\Sigma_n \times \Sigma_{n'}, (\mathbb{Z}/p)^-) \to H_{m+m'}(\Sigma_{n+n'}, (\mathbb{Z}/p)^-).$$

ere the first map is the exterior multiplication; the second map is nduced by the homomorphism $\Sigma_n \times \Sigma_{n'} \to \Sigma_{n+n'}$. The Steenrod powers cts from $H.(\Sigma_n, (\mathbb{Z}/p)^-)$ to $H.(\Sigma_{pn}, (\mathbb{Z}/p)^-)$.

We may make the following conclusion. For an associative algebra , we have constructed the complexes $\mathcal{T}^+_{\bullet}(A)$, $\mathcal{T}^-_{\bullet}(A)$, $\mathcal{T}^{\bullet}_{+}(A)$ and $\mathcal{T}^{\bullet}_{-}(A)$ whose (co)homology $HL^+_{\bullet}(A)$, $HL^-_{\bullet}(A)$, $HL^{\bullet}_{+}(A)$, $HL^{\bullet}_{-}(A)$ satis- y the following properties:

a) There are the maps $HL^+_{\bullet}(A) \longrightarrow H.(\text{Lie } A, k)$, $H^{\bullet}(\text{Lie } A, k) \longrightarrow HL^{\bullet}_{+}(A)$; if char k = 0, they are isomorphisms; if char k = 0, here are the maps $HL^-_{\bullet}(A) \longrightarrow H.(\text{Lie } A, k)$ and $H^{\bullet}(\text{Lie } A, k) \longrightarrow HL^{\bullet}_{-}(A)$;

b) $\mathcal{T}^{\bullet}_{-}(A)$ is a May algebra over an acyclic operad $\mathcal{O}^S$, and $\hat{\mathcal{T}}^-_{\bullet}$ s a topological May coalgebra;

c) The similar constructions may be worked out starting from an rbitrary acyclic augmented operad $\mathcal{O}$ with associative comultiplication; we have for the May coalgebra ${}_{\mathcal{O}}\hat{\mathcal{T}}^{-}$:

$$<{}_{\mathcal{O}}\hat{\mathcal{T}}_{.}^{-}> \xrightarrow{\sim} B(1 + A)^{x}$$

(the "Lie theorem");

d) The cohomology module $HL^{\cdot}_{-}(A)$ is a skew-commutative graded algebra; if char $k = p$, it is a graded module over the Steenrod algebra A_p;

e) The complexes $\mathcal{T}_{.}^{+}$, $\mathcal{T}_{.}^{-}$ may be involved to the single Tate complex; similarly for $\mathcal{T}^{\cdot}_{+}$, $\mathcal{T}^{\cdot}_{-}$.

We finish this work by a brief discussion of the opportunity which is provided by these constructions to the theory of characteristic classes.

Let A be an algebra without unit. Then

$$GL(A) \xrightarrow{\sim} GL(k) \times (1 + M_{\infty}(A))^{*};$$

$$BGL(A) \xrightarrow{\sim} EGL(k) \times_{GL(k)} B(1 + M_{\infty}(A))^{*};$$

we have the inclusion of monoids

$$(1 + M_{\infty}A)^{*} \longrightarrow (1 + M_{\infty}A)^{x};$$

thus, there is a map

$$BGL(A) \longrightarrow EGL(k) \times_{GL(k)} B(1 + M_{\infty}(A))^{x} =$$

$$= EGL(k) \times_{GL(k)} <{}_{\mathcal{O}}\hat{\mathcal{T}}_{.}^{-}(M_{\infty}(A))> \longrightarrow \qquad (A10.\quad)$$

$$\longrightarrow BGL(k) \times <{}_{\mathcal{O}}\hat{\mathcal{T}}_{.}^{-}(M_{\infty}(A))_{GL(k)}> \longrightarrow BGL(k)^{+} \times <{}_{\mathcal{O}}\hat{\mathcal{T}}_{.}^{-}(M_{\infty}(A))_{GL(k)}>;$$

since the operation of Whithney sum is strictly associative and commutative at the level of $GL(k)$-coinvariants, the target of the composition (A10.) is an H-space, and we obtain the maps

$$BGL(A)^{+} \longrightarrow BGL(k)^{+} \times <{}_{\mathcal{O}}\hat{\mathcal{T}}^{-}(M_{\infty}(A))_{GL(k)}>;$$

$$BGL(A)^{+} \longrightarrow <{}_{\mathcal{O}}\hat{\mathcal{T}}^{-}(M_{\infty}(A))_{GL(k)}>$$

The May coalgebra ${}_{\mathcal{O}}\hat{\mathcal{T}}^{-}(M_{\infty}(A))_{GL(k)} =: Z(A)$ may be computed using Theorem 7.5. We have

$$Z_{\cdot}(A) \subset \prod_{m_0+\ldots+m_k=n} \underline{\mathrm{Hom}}_{\mathbb{Z}/m_0\times\ldots\times\mathbb{Z}/m_k}(\mathcal{O}(m_0+\ldots+m_k),$$

$$A[-1]^{\otimes(m_0+\ldots+m_k)});$$

$Z.(A)$ comprises all such homomorphisms which commute also with he permutations $f^{*}\sigma$, where σ is an arbitrary permutation in Σ_{k+1} nd f_o is the morphism from $[m_o + \ldots + m_k - 1]$ to $[k]$ in Δ such that ard $f_o^{-1}(\{i\}) = m_i$, $0 \leqslant i \leqslant k$.

The differential in the complex $Z.(A)$ is just the same as one n $_{\mathcal{O}}\hat{\mathfrak{T}}_{\cdot}^{-}(A)$.

In particular, $\langle Z.(A)\rangle$ maps to the Dold-Puppe transform of the omplex $\prod \mathrm{Hom}_{\mathbb{Z}/n}(\mathcal{O}(n), A[-1]^{\otimes n})$ which is quasi-isomorphic to the omplex $\mathrm{filt}^{o}\,\hat{\mathfrak{L}}.(A)$ (cf. 1.5). On the other hand, since the Chern haracters $K_n(A) \longrightarrow H_n(A,A)$, $K_n(A) \longrightarrow K^{+}_{n+2i+1}(A)$ (cf.7.2, 7.3) are efined at the level of chains and commute with the Bott homomorphisms, hey determine the maps $K_n(A) \xrightarrow{\widehat{chn}} H_{n-1}(\mathrm{filt}^{o}\hat{\mathfrak{L}}(A))$. It may be hown that the compositions $K_n(A) \longrightarrow \P_n(\langle Z.(A)\rangle) \longrightarrow H_{n-1}(\mathrm{filt}^{o}\hat{\mathfrak{L}}.(A))$ oincide with ch_n.

This explains the structure of "the primitive part" of the map $GL(A)^{+} \longrightarrow \langle Z.(A)\rangle$. The computation of the homotopy groups of $\langle Z.(A)\rangle$ eems to be rather complicated. For instance, when k is a field of haracteristic $p > 0$, these homotopy groups depend essentially on he multiplication and Steenrod powers in the cohomology of the dual omplex $Z^{\cdot}(A)$ (which is a May algebra).

References

1. Anderson D.W.: Relationship among K-theories. Springer Lecture Notes N 341, 1973, pp. 57-72.

2. Beilinson A.A.: Higher regulators and the values of L-functions. Modern Problems of Mathematics, VINITI, vol. 24, 1984, pp. 181-239.

3. Bloch S.: The dilogarithm and extensions of Lie algebras. Springer Lecture Notes N 854, 1981, pp. 1-24.

4. Bloch S.: K_2 of Artinian $\mathbb{Q}$-algebras, with application to algebraic cycles. Comm. Alg., 1975, v.3, N 5, pp. 405-428.

5. Bousfield A.K., Gugenheim V.K.A.M.: On PL de Rham theory and rational homotopy type. Memoirs of the American Mathematical Society, v.8, N 179, 1976.

6. Bousfield A.K., Kan D.M.: Homotopy limits, completions and localizations. Springer Lecture Notes N 304, 1975.

7. Burghelea D.: Cyclic homology and the K-theory of spaces, I and II, preprint, 1984.

8. Burghelea D.: Some rational computations in Waldhausen K-theory. Comment. Math. Helv., v.54, N 2, 1979, pp.185-198.

9. Burghelea D., Fiedorowicz Z.: Cyclic homology of group rings, preprint, 1984.

10. Cartan H., Eilenberg S.: Homological algebra, Princeton Univ. Press, 1956.

11. Cartier P.: Homologie cyclique: Rapport sur des travaux recentes de Connes, Karoubi, Loday, Quillen,... Sem. Bourbaki, 36e annee, 1983-1984, N 621, pp. 1-24.

12. Connes A.: Non commutative differential geometry. Part I: The Chern character in K-homology. Preprint IHES, oct. 1982.

13. Connes A.: Non commutative differential geometry. Part II: De Rham homology and non-commutative algebra. Preprint IHES, may 1983.

14. Connes A.: Cohomologie cyclique et foncteurs Ext^n. C.R.Acad. Sci. Paris, t. 296, Ser. I, pp. 953-958.

15. Connes A., Karoubi M.: Caractère multiplicatif d'un module de Fredholm. C.R. Acad. Sci. Paris. t. 299 (1984) 963-968.

16. Daletskii Yu.L., Tsygan B.L.: Formal differential geometric structures connected with the Lie superalgebras. Funct. Anal. and Appl. to appear.

17. Dennis R.K.: Algebraic K-theory and Hochschild homology, Conf. Evanston, 1976 (non published).

18. Feigin B.L., Tsygan B.L.: Cohomology of Lie algebras of generalized Jacobi matrices. Funct. Anal. and Appl., 17, N 2, 1983, pp. 86-87.

19. Feigin B.L., Tsygan B.L.: Additive K-theory and crystalline cohomology. Funct. Anal. and Appl., 19, N 2, 1985, pp. 52-62.

20. Gelfand I.M., Fuks D.B.: The cohomology of the Lie algebra of vector fields on the circle. Funct. Anal. and Appl., 2, N 4, 1968, pp. 92-93.

21. Gersten S.M.: Higher K-theory of rings. Springer Lect. Notes N 341, 1973, pp. 3-42.

22. Grayson D.R.: Higher algebraic K-theory, II (after Daniel Quillen). Springer Lect. Notes N 551, 1976, pp. 217-240.

23. Grothendieck A.: Théorie de la descente. Sem. Bourbaki N 105, 1959/1960.

24. Grothendieck A.: Crystals and the de Rham cohomology of schemes. In b.: Dix exposés sur la cohomologie des schemas, North Holland, 1968.

25. Hinich V.A., Schechtman V.V.: On the homotopy limit of homotopy algebras, these proceedings.

26. Hochschild G., Kostant B., Rosenberg : Differential forms on regular affine algebras. Trans. AMS, 102 (1962), pp. 383-408.

27. Hsiang W.C., Staffeldt R.E.: A model for computing rational algebraic K-theory of simply connected spaces. Inv. Math., 68 (1982), pp. 227-239.

28. Illusie L.: Complex cotangent et déformations. Springer Lect. Notes NN 239 (1971) and 287 (1972).

29. James G.: The representations theory of the symmetric groups. Springer Verlag, 1978.

30. Kac V.G.: Dedekind's η-function, classical Möbius function and the very strange formula. Adv. Math., 30, 1978, pp. 85-136.

31. Karoubi M.: Homologie cyclique des groupes et algebres. C.R.Acad. Sci. Paris, t. 297, Ser. I, 1983, pp. 381-384.

32. Karoubi M.: Homologie cyclique et K-théorie algébrique, I et II, C.R. Acad. Sci. Paris, Ser. I, 1983, pp. 447-450 et 513-516.

33. Karoubi M.: Homologie cyclique et regulateurs en K-théorie algébrique. C.R. Acad. Sci. Paris, v. 297, Ser. I, 1983, pp. 557-560.

34. Karoubi M.: Connexions, curboures et classes caracteristiques en K-théorie algébrique. Can. Math. Soc. Proc., vol. 2, part 1, 1982, pp. 19-27.

35. Karoubi M.: Homologie cyclique and classes caracteristiques, I, preprint, 1984.

36. Leites D.A., Fuks D.B.: On the Lie superalgebra cohomology. C.R. Bulg. AN, 1984, 37, N 10, pp. 1294-1296.

37. Kassel C.: K-théorie relative d'un idéal bilatére de carre nul. Etude homologique en basse dimension. Springer Lect. Notes N 854, 1981, pp. 249-262.

38. Kassel C., Loday J.L.: Extensions centrales d'algébres de Lie. Ann. Inst. Fourier, 32 (1982), 119-142.

39. Kassel C.: Homologie du groupe lineare generale et K-théorie stable. These, Strassbourg, 1981.

40. Kratzer C.: λ-structure en K-theorie algebrique. Comment. Math. Helv., 1980, 55, N 2, pp. 233-254.

41. Loday J.L.: K-theorie algebrique et representations des groupes. Ann. Sci. Ec. Norm. Sup., 9, 1976, pp. 309-377.

42. Loday J.L., Quillen D.G.: Homologie cyclique et homologie d'algebres de Lie des matrices. C.R. Acad. Sci. Paris, t.296, Ser.1, 1983, pp. 295-297.

43. Loday J.L., Quillen D.G.: preprint, 1983.

44. Loday J.L., Quillen D.G.: Cyclic homology and Lie algebras of matrices, Comment. Math. Helv., N 4 (1984), pp.559-587.

45. Macdonald I.G.: Symmetric functions and Hall polynomials. Oxford, Clarendon, 1979.

46. Milnor J., Moore J.: On the structure of Hopf algebras. Ann. Math. 1965, pp. 211-265.

47. Quillen D.G.: Homotopical algebra. Springer Lect. Notes N 43, (1967).

48. Quillen D.G.: On the (co)-homology of commutative rings. Proc. Symp. Pure Math., v. XVII (1970), pp. 65-87.

49. Quillen D.G.: Higher algebraic K-theory, I. Springer Lect. Notes N 341, 1973, pp. 85-149.

50. Quillen D.G.: On the cyclic homology of algebras. Preprint, Oxford, 1984.

51. Rowen H.R.: Polynomial identities in ring theory. N.Y., Academic Press, 1980.

52. Seminaire "Sophus Lie", 1954/1955, Paris, 1955.

53. Soule C.: Operations en K-théorie algebrique. Preprint C.N.R.S., 1983.

54. Soule C.: Regulateurs. Sem. Boubaki, Fevr. 1985.

55. Sullivan D., Vique-Poirrier M.: The homology theory of the closed geodesic problem. Differ. Geom., 11, N 4 (1976), 633-644.

56. Swan R.G.: Non abelian homological algebra and K-theory. Proc. Symp. Pure Math., v. XVII, pp. 88-123.

57. Tate J.: Residues of differentials on curves. Ann. Sci. Ec. Norm. Sup., Ser. 4, 1968, v. 1, N 1, pp. 149-159.

58. Tsygan B.L.: Homology of the matrix Lie algebras over rings and Hochschild homology. Uspekhi Math. Nauk, vol. 38 (1983), N 2, pp. 217-218.

59. Tsygan B.L.: On the homology of certain matrix Lie superalgebras. Funct. Anal. and Appl., to appear.

60. Tsygan B.L.: On the relative algebraic K-theory of a two-sided nilpotent ideal. Proc. Conf. MGU, to appear.

61. Van Est W.T.: Group cohomology and Lie algebra cohomology in Lie groups. Indag. Math., 1953, 15, pp. 484-504.

62. Waldhausen F.: Algebraic K-theory of generalized free products. Ann. Math., Ser. 2, 108, 1978, N 1, pp. 135-204; N 2, pp. 205-256.

63. Waldhausen F.: Algebraic K-theory of topological spaces I. Proc. Symp. Pure Math., v. XXXII, 1978, pp. 35-60.

64. Weyl H.: The classical groups. Their invariants and representations. Princeton Univ. Press, 1939.

65. Chen K.: Circular bar-construction. J. Alg., 1979, v.57, N 2, pp. 466-483.

66. Atiyah M.F., Patodi V.K., Singer I.M.: Spectral asymmetry and Riemannian geometry. Math. Proc. Camb. Phil. Soc., 1975, v. 77, N 1, pp. 43-65; v.78, N 3, pp. 405-432; 1976, v. 79, N 1, pp. 71-99.

67. Haefliger A.: The homology of nilpotent Lie groups. Preprint.

68. Kuzmin E.N.: On the Lie groups and algebras cohomology. Uspekhi Math. Nauk, 198 , v. , N 2, pp.

69. Farrell F.T., Hsiang W.C.: On the rational homotopy groups of the diffeomorphism groups of spheres, discs and aspherical manifolds. Proc. Symp. Pure Math., v. XXXII, part I, pp. 325-338.

70. Soulé C.: Rational K-theory of the dual numbers of a ring of algebraic integers. Springer Lect. Notes N 854, 1981, pp. 403-409.

CYCLIC HOMOLOGY OF ALGEBRAS WITH QUADRATIC RELATIONS, UNIVERSAL ENVELOPING ALGEBRAS AND GROUP ALGEBRAS

B. L. Feigin

B. L. Tsygan

0. Introduction.

Cyclic homology of rings, introduced in [5], [17] (see definition 1.1) was an object of study in a series of recent papers: as additive analogues of Quillen's K-functors ([12], [17]) as a non-commutative generalization of de Rham cohomology ([5], [10], [11]), in connection with the index of elliptic operators ([5]), characteristic classes in algebraic K-theory ([5], [10], [11]), with problems of smooth topology. It turned out that in applications to K-theory and topology knowing the cyclic homology of twisted group algebras is important. The problem of calculating cyclic homology of universal enveloping algebras of Lie algebras arises also in applications to the rational topology and constructing operations in cyclic homology of commutative rings.

Calculating cyclic homology of universal enveloping algebras is performed by a method suitable for a wider class of algebras. Consider let an algebra A be determined by its generators $\{x_i\}$ with relations

$$x_i x_j = \sum c_{ij}^{kl} x_k x_l + \sum b_{ij}^{k} x_k \quad , \quad (i, j) \notin S \qquad (0.1)$$

(here $\{x_i x_j\}_{(i,j)\in S}$ is a maximal system of monomials linearly independent modulo $\sum_j k\cdot x_j$ where k is the ground field. Sources of important examples of such algebras besides Lie algebra theory are topology and mathematical physics ([6, 14]). Studying cyclic homology of algebras with quadratic relations we encounter an interesting duality that we will briefly discuss. First we shall give several facts and constructions from Priddy's paper [14].

Consider together with A the algebra $\tilde{A}$ with generators $\{x_i\}$ and relations

$$x_i x_j = \sum c_{ij}^{kl} x_k x_l \quad , \quad (i, j) \notin S .$$

If A is finitely generated and the graded ring $\mathrm{Ext}_A^*(k, k)$ is generated by Ext^1, then A is called a Koszul algebra. If A and $\tilde{A}^+$ are Koszul algebras then $\tilde{A}^{++} \cong \tilde{A}$.

If A is a Koszul algebra then the graded algebra $\tilde{A}^+$ is determined by generators $\{\xi_i\}$ of degree -1 with relations

$$\xi_i\xi_j + \sum_{(k,l)\notin S} c_{kl}^{ij}\xi_k\xi_l = 0 \quad , \quad (i, j) \in S .$$

Consider the differential $\partial : \tilde{A}^+_\cdot \to \tilde{A}^+_{\cdot -1}$ making $\tilde{A}^+$ into a graded differential algebra such that

$$\partial\xi_i = \sum_{(k,l)\notin S} b_{kl}^i\xi_k\xi_l \quad .$$

$A^+ = (\tilde{A}^+, \partial)$ is called the cohomological Koszul complex and $H.(A^+) \cong \mathrm{Ext}_A^{-\cdot}(k, k)$. If char k = 0 and A = k[V] then $A^+ \simeq \Lambda^\cdot(V^*)$; if g is a Lie algebra and $A = U(g)$ then $A^+ = C^\cdot(g)$ is the standard Koszul complex; if char k > 0 and A is the Steenrod algebra then A^+ is the Van algebra etc. In all these cases A and $\tilde{A}^+$ are Koszul algebras. It is not difficult to generalize this construction to the case when A is a graded algebra satisfying certain reasonable finiteness conditions and $\{x_i\}$ are homogeneous generators.

Thus from a graded Koszul algebra determined by relations (0.1) we recover a graded differential algebra A^+. If $b_{kl}^i \equiv 0$ i.e. A is bigraded then the differential in A^+ vanishes i.e. A^+ is a bigraded algebra of the same form as A. If A^+ is a Koszul algebra then $A^{++} \cong A$.

Generalizing then cyclic cohomology of [just] algebras (see 1.1) we can determine cyclic homology and (dually) cyclic cohomology of graded differential algebras (1.3), (1.5).

Besides, by analogy with [5], [10] we can determine periodic cyclic (co)homology $HC^{per}_\cdot$, $HC^\cdot_{per}$ which satisfy $HC_i^{per} \cong HC_{i+2}^{per}$ for all i. Notice that it is this homology that serves as non-commutative generalization of de Rham cohomology; e.g. if A is a finitely generated algebra over C and X = Spec A then

$$HC_0^{per} \cong \bigoplus_{i=0}^{\infty} H^{2i}(X,\mathbb{C}); \; HC_1^{per}(A) \cong \bigoplus_{i=0}^{\infty} H^{2i+1}(X,\mathbb{C}) ,$$

where X is considered as a topological subspace of a complex affine space with complex topology. Set

$\overline{HC}.(A) = \ker(HC.(A) \to HC.(k))$; $\overline{HC}^\cdot(A) = \mathrm{coker}(HC^\cdot(k) \to HC^\cdot(A))$; $\overline{HC}^\cdot_{per}(A) = \mathrm{coker}(HC^\cdot_{per}(k) \to HC^\cdot_{per}(A))$.

The main result of Ch. 2 is the following Theorem (2.4.1). Let A and $\tilde{A}^+$ be Koszul algebras. Suppose that $\#\{k : \deg x_k \leq i\} < \infty$ for all $i \in \mathbb{Z}$ or $\#\{k : \deg x_k \geq i\} < \infty$ for all $i \in \mathbb{Z}$, and that either $b^i_{kl} \equiv 0$ or the grading of A is non-negative. Then the following sequence is exact:

$$\ldots \to \overline{HC}^{-2-2n}_{per}(A^+) \to \overline{HC}_n(A) \to \overline{HC}^{-1-n}(A^+) \to \overline{HC}^{-1-n}_{per}(A^+) \to \ldots$$

If char k = 0 then $\overline{HC}_n(A) \xrightarrow{\sim} \overline{HC}^{-1-n}(A^+)$.

Comments on the proof. Cyclic homology has two definitions: one in terms of the standard complex (1.1 and 1.3) and another one in terms of the derived functors (1.4). If R is a free graded differential resolution A and char k = 0 then $R/([R, R] + k)$ is a complex and

$$\overline{HC}.(A) \xrightarrow{\sim} H.(R/([R, R] + k) \qquad (0.2)$$

For $\overline{HC}.(A)$ the standard complex for char k = 0 is of the form $\overline{A}^{\otimes(\cdot+1)}/\mathrm{im}(1-\tau)$ where τ is the cyclic permutation of multiples multiplied by its sign and $\overline{A} = \ker(A \to k)$. In other words, the standard complex is isomorphic as a graded space to $T(\overline{A}[-1])/(k + (T(A[-1]), T(A[-1]))$ where T is the tensor algebra. More detailed analysis shows that there is a duality between the two definitions of $\overline{HC}.(A)$. To determine it we replace R, with a standard trick, by a particular free resolution, the cobar-construction of the Koszul co plex. Here the righthand side of (0.2) is isomorphic exactly to the standard complex calculating $\overline{HC}^{-1-\cdot}(A^+)$.

In Ch. 3 we calculate cyclic homology of universal enveloping algebras. Let g be a Lie algebra over k. Set

$$Y_{ij}(g) = \Lambda^{j-i}g \otimes S(g),\ j \geq i \geq 0;\ Y_{ij}(g) = 0,\ i > j \geq 0;$$

$$\delta : Y_{ij}(g) \to Y_{i,j-1}(g);\ d_* : Y_{ij}(g) \to Y_{i-1,j}(g);$$

δ is the differential in the standard complex for calculating the homolo $H.(g, S(g))$ and d_* the de Rham differential in $\Lambda g \otimes S(g) \xrightarrow{\sim} \Omega^{\cdot}_{S(g)}$. Then $Y_{\cdot\cdot}(g)$ is a bicomplex. Let $Y_{\cdot}(g)$ be the associated complex.

Theorem 3.1.1. Let char k = 0. Then there is a canonical isomorphism

$$HC_{\cdot}(U(g)) \xrightarrow{\sim} H_{\cdot}(Y_{\cdot}(g)) .$$

Another formulation: Let $W_{ij}(g) = \Lambda^{j-i}(g) \otimes S^i(g)$, i.e. $W_{\cdot\cdot}(g)$ is a subcomplex of $Y_{\cdot\cdot}(g)$. The Weyl algebra of g is the dual bicomplex.

Let $W_{\cdot}(g)$ be the associated complex. On it, there is a descending filtration:

$$F^m W_\cdot(g) = \bigoplus_{i<m} W_{i\cdot}(g) .$$

Corollary 3.1.2. Let char k = 0. Then

$$HC_\cdot(U(g)) \cong \bigoplus_{m\geq 0} H_{\cdot-2m}(W_\cdot(g)/F^m W_\cdot(g)) .$$

Notice that thanks to [5] or [15] if V is a vector space and char k = 0 then

$$HC_n(k[V]) \cong \Omega^n_{k[V]}/d\Omega^{n-1}_{k[V]} \oplus HC_n(k) \qquad (0.3)$$

Clearly (0.3) is exactly Theorem 3.1.1 for an Abelian g.

Notice that there exists a functorial homomorphism $H_\cdot(g) \overset{\varphi}{\to} HC_{\cdot-1}(U(g))$ where $H_\cdot(g)$ is homology of g with trivial coefficients. On the level of standard complexes it is determined by the formula

$$x_1 \wedge \dots \wedge x_n \mapsto \sum_{\sigma\in\Sigma_{n-1}} \operatorname{sgn}\sigma\cdot(x_{\sigma 1} \otimes \dots \; x_{\sigma(n-1)} \otimes x_n) .$$

Now let A be a commutative k-algebra. Consider the chain of maps

$$H_n(g \otimes A) \overset{\varphi}{\to} HC_{n-1}(U(g \otimes A)) \to HC_{n-1}(U(g) \otimes A) \to$$

$$\overset{\pitchfork}{\to} \bigoplus_{i+j=n-1} HC_i(U(g)) \otimes HC_j(A) ,$$

where $\pitchfork$ is a comultiplication in cyclic homology (see [5]). We get a homological operation

$$H_n(g \otimes A) \to \bigoplus_{i+j=n-1} HC_i(U(g)) \otimes HC_j(A) .$$

On the other hand, by [17], [12], $H_\cdot(gl(A))$ is a Hopf algebra and $HC_{\cdot-1}(A)$ the space of its primitive elements. Therefore there are multiplications

$$HC_{i_1-1}(A) \otimes \dots \otimes HC_{i_k-1}(A) \to HC_i(U(gl(k)) \otimes HC_{i_1+\dots+i_k-1}(A) .$$

For k = 1, i = 0 this is the way to get γ-operations in cyclic homology (notice that $HC_o(U(gl(k)) \cong R(gl(k)) \otimes k$ where R is the representation ring).

In Ch. 4 we calculate cyclic homology of the twisted group algebra recovered from a group G acting by automorphisms on a ring A. For the usual group algebras we get Burghelea's and Fiedorowicz's theorem [2]. Another interesting particular case arises for G = $\mathbb{Z}$. Recall that by [17] $HC_{\cdot-1}(A)$ are additive analogues of Quillen's K-functors. It is possible to determine additive analogues of Waldhausen's $Nil^+_\cdot$-functors ([18, 10])

so that the analogues of theorems of [18] hold. In particular, for $G = \mathbb{Z}$ Theorem 4.1.1 yields an analogue of Waldhausen's theorem on K-functors of Laurent extensions.

If A is a graded vector space and $a \in A$ a homogeneous element, set $\bar{a} = \deg a$. If $a_1, \ldots, a_n \in A$ then let $(a_1, \ldots, a_n) = a_1 \otimes \ldots \otimes a_n \in A^{\otimes n}$. Further if A is a graded algebra then set $[a, b] = ab - (-1)^{\bar{a}\bar{b}}ba$, $a, b \in A$.

1. Cyclic homology.

1.1. Cyclic homology of algebras. Let k be a commutative ring, A an associative algebra with unit over k. Consider the mappings

$$b : A^{\otimes(n+1)} \to A^{\otimes n} \quad ; \; \delta : A^{\otimes(n+1)} \to A^{\otimes n} \quad ;$$

$$\tau : A^{\otimes(n+1)} \to A^{\otimes(n+1)} ; \; N : A^{\otimes(n+1)} \to A^{\otimes(n+1)} \; ;$$

where

$$b(a_0,\ldots,a_n) = \sum_{0 \le i \le n-1} (-1)^i (a_0,\ldots,a_i a_{i+1},\ldots,a_n);$$

$$\delta(a_0,\ldots,a_n) = \sum_{0 \le i \le n-1} (-1)^{i+1} (a_0,\ldots,a_i a_{i+1},\ldots,a_n) + (a_1,\ldots,a_{n-1}, a_n a_0)$$

$$\tau(a_0,\ldots,a_n) = (-1)^n (a_1,\ldots,a_n a_0);$$

$$N = 1 + \tau + \tau^2 + \ldots + \tau^n .$$

Then the diagram

$$\begin{array}{ccccccccc}
 & \vdots & & \vdots & & \vdots & & \vdots \\
 & \downarrow b & & \downarrow \delta & & \downarrow b & & \downarrow \delta \\
\cdots \xrightarrow{N} & A^{\otimes(n+1)} & \xrightarrow{1-\tau} & A^{\otimes(n+1)} & \xrightarrow{N} & A^{\otimes(n+1)} & \xrightarrow{1-\tau} & A^{\otimes(n+1)} \\
 & \vdots & & \vdots & & \downarrow b & & \downarrow \delta \\
 & \vdots & & \vdots & & \vdots & & \vdots \\
 & \downarrow b & & \downarrow \delta & & \downarrow b & & \downarrow \\
\cdots \xrightarrow{N} & A^{\otimes 2} & \xrightarrow{1-\tau} & A^{\otimes 2} & \xrightarrow{N} & A^{\otimes 2} & \xrightarrow{1-\tau} & A^{\otimes 2} \\
 & \downarrow b & & \downarrow \delta & & \downarrow b & & \downarrow \delta \\
\cdots \xrightarrow{N} & A & \xrightarrow{1-\tau} & A & \xrightarrow{N} & A & \xrightarrow{1-\tau} & A
\end{array}$$

is a bicomplex, i.e. $(1-\tau)N = 0 = N(1-\tau)$, $\delta^2 = b^2 = 0$, $\delta(1-\tau) + (1-\tau)b = 0$, $N\delta + bN = 0$ (see [5, 17]). Denote this bicomplex by $C_{..}(A) : C_{ij}(A) = A^{\otimes(j+1)}$, $i, j \geq 0$. Let $C_{.}(A)$ be the associated complex. Set $HC_n(A) = H_n(C_{.}(A))$, $n \geq 0$. If $\mathbb{Q} \subseteq k$ then

$$HC_n(A) \xrightarrow{\sim} H_n(A^{\otimes(\cdot+1)}/\mathrm{im}(1-\tau)) .$$

Further set $C_n(A, A) = A^{\otimes(n+1)}$. The differential δ turns $C.(A, A)$ into a complex; set

$$H_n(A, A) = H_n(C.(A, A)) , \; n \geq 0 .$$

Here $H_n(A, A) \simeq \mathrm{Tor}_n^{A\otimes A^o}(A, A)$ (see [3]) where A^o is the algebra opposite to A. The groups $H_\bullet(A, A)$ are called Hochschild homology of A with coefficients in the module A and $HC_\bullet(A)$ cyclic homology of A.

1.2. <u>Cyclic objects</u>. Denote by $[n]$, $n \geq 0$ the set $\mathbb{Z}$ with the standard linear order and the automorphism $T_n : k \mapsto k + n + 1$. Let Λ_∞ be the category whose objects are $[n]$, $n = 0, 1, 2, \ldots$ and morphisms of $[n]$ to $[n']$ are non-decreasing maps $f : [n] \to [n']$, $T_{n'}f = fT_n$. Now fix an integer $r > 0$ and call two morphisms $f, f' : [n] \to [n']$ equivalent if $f = f' \cdot T_n^r$ $(= T_{n'}^r f')$. Denote by Λ_r the category of equivalence classes of morphisms of Λ_∞. The category $\Lambda = \Lambda_1$ is that introduced by Connes.

Let us describe the structure of the categories Λ_r more explicitly. Each category Λ_r contains a subcategory consisting of equivalence classes containing the morphism

$$f : [n] \to [n'] , \quad f(\{0,\ldots,n'\}) = \{0,\ldots,n\} .$$

Clearly this category is isomorphic to the category Δ of isomorphism classes of finite non-empty fully ordered sets and non-decreasing maps. We will assume Δ embedded in Λ_r. For each morphism $g : [n] \to [n']$ in Λ_r there is a unique factorization $g = f\sigma$, where $f \in \hom_\Delta([n], [n'])$ and $\sigma \in \mathrm{Aut}_{\Lambda_r}([n])$. Here $\mathrm{Aut}_{\Lambda_r}([n]) \simeq \mathbb{Z}/(n+1)r$ is a cyclic group of order $r(n+1)$ generated by the automorphism $t_n : [n] \to [n]$, $t_n(k) = k + 1$, $k \in \mathbb{Z}$. If s_i, d_i are the standard generators in Mor Δ (see [13]) then

$$\begin{gathered} t_n d_o = d_1 t_n \;;\; t_n d_1 = d_2 t_n \;;\; \ldots \;;\; t_n d_n = d_o \;; \\ t_n s_o = s_1 t_n \;;\; t_n s_1 = s_2 t_n \;;\; \ldots \;;\; t_n s_n = s_o t_n^2 . \end{gathered} \tag{1.1}$$

Let a be a category. We call a contravariant functor $\Lambda_r \to a$ an r-cyclic object. As usual , if M is such a functor we write M_n instead of $M([n])$ and for $f \in \hom_{\Lambda_r}([n], [n'])$ we write $f : M_{n'} \to M_n$ instead of $M(f)$. It is more convenient to us to write the symbol of this map to the right of the argument: $M_{n'} \ni m \mapsto mf \in M_{n'}$.

To determine an r-cyclic object of a is the same as to determine a simplicial object M of a together with the automorphisms $t_n : M_n \rightarrow M_n$ for which 1.1 holds and (if $r < \infty$) $t_n^{r(n+1)} = 1$ for $n \geq 0$.

1-cyclic objects are exactly Connes' cyclic objects.

Example 1.2.1. Let A be an associative algebra with unit over the commutative ring k; let $\alpha \in \mathrm{Aut}\, A$. Suppose that $\alpha^r = 1$. Consider an r-cyclic object in the category of k-modules: $(A_\alpha^{\natural,r})_n = A^{\otimes(n+1)}$ and if $x = (a_0, \ldots, a_n) \in A^{\otimes(n+1)}$ then

$$xd_0 = (a_1, \ldots, a_{n-1}, a_n \cdot \alpha a_0) ;$$

$$xd_i = (a_0, \ldots, a_{i-1}a_i, \ldots, a_n) , \ i = 1,\ldots,n ;$$

$$xt = (a_1, \ldots, a_n, \alpha a_0) ;$$

$$xs_i = (a_0, \ldots, a_{i-1}, 1, a_i, \ldots, a_n) , \ 0 \leq i \leq n .$$

Example 1.2.2. Let G be a group; $g \in G$, $g^r = 1$. Determine an r-cyclic G-module P(G; g, r) setting $P(G; g, r)_n = k[G]^{\otimes(n+1)}$ and if $x = (g_0, \ldots, g_n)$ then

$$xd_i = (g_0, \ldots, g_i g_{i+1}, \ldots, g_n) , \ 0 \leq i \leq n-1 ;$$

$$xd_n = (g_0, \ldots, g_{n-1}) ;$$

$$xt = (g_0 g_1, g_2, \ldots, g_{n-1}, (g_0 \ldots g_n)^{-1} g) ;$$

$$xs_i = (g_0, \ldots, g_i, 1, g_{i+1}, \ldots, g_n) , \ 0 \leq i \leq n .$$

The corresponding simplicial G-module is the standard free resolution of the G-module k. Set P(G) = P(G; 1, 1).

Let M be an r-cyclic object of an Abelian category a. Consider the following maps:

$$\delta : M_n \rightarrow M_{n-1} ; \ b : M_n \rightarrow M_{n-1} ; \ \tau : M_n \rightarrow M_n ;$$

$$N : M_n \rightarrow M_n ; \ \delta = \sum_{i=0}^{n} (-1)^i d_i ; \ b = \sum_{i=1}^{n} (-1)^{i+1} d_i ;$$

$$\tau = (-1)^n t \ ; \ N = 1 + \tau + \tau^2 + \ldots + \tau^{r(n+1)-1} \quad (r < \infty) .$$

It is easy to verify that the following diagrams are bicomplexes:

$$
\begin{array}{ccccccccc}
 & & r<\infty & & & & & & \\
 & \vdots & \vdots & \vdots & \vdots & & & & \\
 & \downarrow b & \downarrow \delta & \downarrow b & \downarrow \delta & & & & \\
\cdots \xrightarrow{N} & M_n & \xrightarrow{1-\tau} M_n & \xrightarrow{N} M_n & \xrightarrow{1-\tau} M_n & & & & \\
 & \downarrow b & \downarrow \delta & \downarrow b & \downarrow \delta & & & & \\
 & \vdots & \vdots & \vdots & \vdots & & & & \\
 & \downarrow b & \downarrow \delta & \downarrow b & \downarrow \delta & & & & \\
\cdots \xrightarrow{N} & M_0 & \xrightarrow{1-\tau} M_0 & \xrightarrow{N} M_0 & \xrightarrow{1-\tau} M_0 & & & &
\end{array}
\qquad
\begin{array}{ccccc}
 & & r=\infty & & \\
 & \vdots & \vdots & \vdots & \vdots \\
 & \downarrow & \downarrow & \downarrow b & \downarrow \delta \\
\cdots \to & 0 & \to 0 & \to M_n & \to M_n \\
 & \downarrow & \downarrow & \downarrow b & \downarrow \delta \\
 & \vdots & \vdots & \vdots & \vdots \\
 & \downarrow & \downarrow & \downarrow b & \downarrow \delta \\
\cdots \to & 0 & \to 0 & \to M_0 & \to M_0
\end{array}
$$

Denote these bicomplexes by $C_{..}(r, M)$ $(r \leqslant \infty)$. More exactly, $C_{ij}(r, M) = M_j$, $i, j \geqslant 0$ for $r < \infty$, $C_{ij}(r, M) = M_j$, $i = 0, 1$, $j \geqslant 0$, $C_{ij}(r, M) = 0$, $i > 1$, $j \geqslant 0$ for $r = \infty$.

Let $C_{\cdot}(r, M)$ be the complex associated with $C_{..}(r, M)$. The following statement has been proved in [6] for $r = 1$. The general case is treated similarly, see [10].

Proposition 1.2.3. Let $\mathcal{a}$ be an Abelian category with enough projective objects. Then

$$H_{\cdot}(C_{\cdot}(r, M)) \overset{\sim}{\to} \mathbb{L}_{\cdot}\varinjlim_{\Lambda_r^o} M .$$

Here $\mathbb{L}_{\cdot}\varinjlim$ denotes the left derived functors of the direct limit functor from the abelian category of r-cyclic objects of $\mathcal{a}$ to $\mathcal{a}$.

In the setting of Example 1.2.2 set

$$HC_n(A; \alpha, r) = H_n(C_{\cdot}(r, A_\alpha^{\natural, r})) .$$

In particular, $HC_{\cdot}(A) = HC_{\cdot}(A; 1, 1)$.

1.3. Cyclic homology of graded differential algebras. Let A be a graded k-algebra with unit. Consider the cyclic object $A^\natural$ in the category of graded k-modules: $A_n^\natural = A^{\otimes(n+1)}$ and if $x = (a_0, \dots, a_n) \in A_n^\natural$ and a_i are homogeneous elements of A then

$$xd_0 = (-1)^{\bar{a}_0 \cdot \sum_{1\leqslant i\leqslant n}(\bar{a}_i+1)} (a_1,\dots,a_n a_0) ;$$

$$xd_i = (-1)^{\sum_{0\leqslant l\leqslant i-1}\bar{a}_l} (a_0,\dots,a_{i-1}a_i,\dots,a_n) , \quad 1 \leqslant i \leqslant n ;$$

$$xs_i = (-1)^{\sum_{0\leqslant l\leqslant i-1}\bar{a}_l} (a_0,\dots,a_{i-1},1,a_i,\dots,a_n) ;$$

$$xt = (-1)^{n+(\bar{a}_0+1)\sum_{1\leqslant l\leqslant n}(\bar{a}_l+1)} (a_1,\dots,a_n,a_0) .$$

The bicomplex $C_{..}(A) := C_{..}(1, A^{\natural})$ is a bicomplex of graded k-modules.

Now suppose that there is given a differential $\partial : A_i \to A_{i-1}$ making A into a graded differential algebra. Consider the differential $C_{nm}(A)_k \xrightarrow{\partial} C_{nm}(A)_{k-1}$:

$$\partial(a_o,\dots,a_n) = \sum_{i=0}^{n} (-1)^{m-1+\sum_{l=0}^{i-1}(\bar{a}_l+1)} (a_o,\dots,\partial a_i,\dots,a_n) \qquad (1.2)$$

If β is a differential in $C_{..}(A)$ then clearly $\beta\partial + \partial\beta = 0$. Set $C_s(A) = \bigoplus_{m+n+k=s} C_{mn}(A)_k$; the differential $\beta + \partial$ makes $C_.(A)$ into a complex; homology is denoted by $HC.(A)$.

Now suppose that A is augmented by $\varepsilon : A \to k$. Let $\bar{A} = \ker \varepsilon$. Set $\bar{C}_s(A) = \bigoplus_{m+n+k=s} \bar{C}_{mn}(A)_k$; $\bar{C}_{mn}(A)_k = (\bar{A}^{\otimes(n+1)})_k$. Then $\bar{C}_.(A)$ is a subcomplex in $C_.(A)$. Set $\overline{HC}_.(A) = H_.(\bar{C}_.(A))$ and call it the reduced cyclic homology [12], [15].

Now we present another realization of cyclic homology. Let $D_{mn}(A)_k = (A^{\otimes(n+1)})_k$ for $n \geq m$ and $D_{mn}(A) = 0$ for $m > n$. Consider the operators

$$\delta : D_{mn}(A)_k \to D_{m,n-1}(A)_k \ ;$$

$$B : D_{mn}(A)_k \to D_{m-1,n}(A)_k \ ;$$

$$\partial : D_{mn}(A)_k \to D_{mn}(A)_{k-1} \ ;$$

$$\delta = d_o - d_1 + \dots + (-1)^n d_n ; \ B = (1-\tau)s_o N \ ,$$

where ∂ is determined by the formula 1.2. We have

$$B^2 = (1-\tau)s_o\left[N(1-\tau)\right]s_o N = 0 \ ;$$

$$\delta B = \delta(1-\tau)s_o N = -(1-\tau)bs_o N = -(1-\tau)N + (1-\tau)s_o bN$$

$$= -(1-\tau)s_o N\delta = -B\delta \quad ; \quad B\delta + \delta B = 0 \ .$$

Further since $\tau\partial = \partial\tau$, $N\partial = \partial N$, $s_o\partial + \partial s_o = 0$, the differentials δ, b, ∂ make $D_{..}(A)$. into a tricomplex. The associated complex is denoted by $D_.(A)$. There is a chain map $D_.(A) \to C_.(A)$ inducing the isomorphism $H_.(D_.(A)) \xrightarrow{\sim} H_.(C_.(A))$ if $\partial = 0$ or $A = \bigoplus_{i\geq 0} A_i$ (concerning the non-graded case see [15]).

The quotient $D'_.(A)$ of the complex $D_.(A)$ where $D'_{mn}(A)_k = \bar{A}^{\otimes n} \otimes A$,

$m \leq n$, where δ, B, ∂ are the same as in $D_{..}(A)_{.}$ is often convenient. The projection $D_{.}(A) \to D'_{.}(A)$ induces an isomorphism on homology with respect to $\partial + \delta + B$. The comparison of the spectral sequences shows that $H_{.}(D_{.}(A)) \xrightarrow{\sim} H_{.}(D'_{.}(A))$.

It is convenient to consider the quotient $E(A)$ of the complex $D'_{.}(A)$ modulo

$$\bigoplus_{j-1\geq 2} D'_{ij}(A) \oplus \bigoplus_{i} (D'_{i,i+1}(A) \cap \operatorname{im} \delta) \ ;$$

with the help of the isomorphism

$$\bar{A} \otimes A \xrightarrow{\sim} A \otimes \bar{A},\ (a,b) \mapsto (-1)^{(\bar{a}+1)\bar{b}}(\bar{b}, \bar{a}) \ ,$$

we see that $D'_{.}(A)/E_{.}(A)$ is isomorphic to the complex associated with the bicomplex $\Omega_{..}(A)$:

$$\Omega_{2i,j}(A) = \bar{A}_j \ ; \ \Omega_{2i+1,j}(A) = (\bar{\Omega}^1_A)_j, \text{ where } i \geq 0 \ ,$$

$$\bar{\Omega}^1_A := A \otimes \bar{A}/\operatorname{im} \tilde{\delta} \ ,$$

$$\tilde{\delta}(a,b,c) = (ab,c) - (a,bc) + (-1)^{\bar{c}(\bar{a}+\bar{b})}(ca,b)$$

for homogeneous a, b, $c \in \bar{A}$;

$$d : \Omega_{ij}(A) \to \Omega_{i-1,j}(A) \ ; \ \partial : \Omega_{ij}(A) \to \Omega_{i,j-1}(A) \ ;$$

$$d(a,b) = [a,b],\ i \equiv 1 \pmod 2 \ ; \ da = (1,a),\ i \equiv 0 \pmod 2 \ ;$$

$$\partial(a,b) = (\partial a,b) + (-1)^{\bar{a}}(a,\partial b),\ i \equiv 1 \pmod 2; \ \partial(a) = -\partial a,\ i \equiv 0 \pmod 2 \ .$$

1.4. <u>Cyclic homology as derived functors</u>. In [10], [15], [17], the cyclic homology of the tensor algebra was computed.

Theorem 1.4.1. Let V be a graded k-module and $A = T(V)$. Then

$$\overline{HC}_n(A) \xrightarrow{\sim} \Big(\bigoplus_{m=1}^{\infty} \bigoplus_{k+s=n} H_k(\mathbb{Z}/m\mathbb{Z}, V_s^{\otimes m}) \Big) \ ; \qquad (1.3)$$

$$HC_n(A) \xrightarrow{\sim} \overline{HC}_n(A) \oplus HC_n(k) \ . \qquad (1.4)$$

Here $V_s^{\otimes m}$ is the $(\mathbb{Z}/m)$-module of elements of degree s in $V^{\otimes m}$ on which the generator of $\mathbb{Z}/m$ acts via

$$\theta : (v_1,\dots,v_m) \mapsto (-1)^{\bar{v}_1(\bar{v}_2+\dots+\bar{v}_m)}(v_2,\dots,v_m,v_1) \ .$$

Sketch of the proof. Consider the spectral sequence of the bicomplex of the graded k-modules $\bar{C}_{..}(A)$. The homology with respect to the differential is isomorphic to $\operatorname{Tor}^{A\otimes A^{\circ}}_{.,.}(A, \bar{A})$ and the complex of the graded k-modules

$(\overline{A}^{\otimes(\cdot+1)}, b)$ is quasi-isomorphic to the complex

$$\ldots \to 0 \to V \otimes \overline{T(V)} \xrightarrow{\delta_0} \overline{T(V)} \to 0\ , \quad v \otimes x \overset{\delta_0}{\longmapsto} [v, x]\ .$$

The homology of $(\overline{A}^{\otimes(\cdot+1)}, b)$ is isomorphic to $\mathrm{Tor}^A_{\cdot+1,\cdot}(k, k)$; $\mathrm{Tor}^A_{1,\cdot}(k, k) \xrightarrow{\sim} V$, $\mathrm{Tor}^A_{i,\cdot}(k, k) = 0$, $i > 1$. Therefore the second term of the spectral sequence is isomorphic to the following one:

$$\begin{array}{cccccccc}
 & \vdots & & \vdots & & \vdots & & \vdots \\
 & 0 & & 0 & & 0 & & 0 \\
\ldots\ 0 & V^{\otimes>1} \cap \ker(\theta-1) & 0 & V^{\otimes>1} \cap \ker(\theta-1) & & 0 & & \ker(\theta-1) \cap V^{\otimes>1} \\
 & & \nearrow & & & & \nearrow & \\
\ldots\ 0 & V^{\otimes>1}/\mathrm{im}(\theta-1) & & 0 & V^{\otimes>1}/\mathrm{im}(\theta-1) & 0 & & V^{\otimes>0}/\mathrm{im}(\theta-1)
\end{array}$$

It can be directly verified that the differential $E^2_{2i,0} \to E^2_{2i-2,1}$ is mapped by this isomorphism onto the operator $N_\theta = 1 + \theta + \ldots + \theta^{\mathrm{ord}\ \theta-1}$. Further, the spectral sequence is degenerate at E^2 by dimensional considerations. This implies (1.3). In exactly the same way the term E^2 of the spectral sequence of the bicomplex $C_{\cdot\cdot}(A)$ of the graded k-modules is isomorphic to

$$\begin{array}{ccccc}
 & \vdots & \vdots & \vdots & \vdots \\
 & 0 & 0 & 0 & 0 \\
\ldots\ 0 & & \ker(\theta-1) \cap V^{\otimes\geq1} & 0 & \ker(\theta-1) \cap V^{\otimes\geq1} \\
 & & & \nearrow & \\
\ldots\ 0 & & V^{\otimes\geq0}/\mathrm{im}(\theta-1) & 0 & V^{\otimes\geq0}/\mathrm{im}(\theta-1)
\end{array}$$

As above the differential in E^2 is mapped by this isomorphism onto N_θ implying (1.4). Details see in [10] or [15].

It is clear from the proof that if A is a free augmented graded algebra then the projection $D'_\cdot(A) \to D'_\cdot(A)/E_\cdot(A)$ induces a map such that the sequence

$$0 \to HC_\cdot(k) \to HC_\cdot(A) \to \overline{HC}_\cdot(A) \to 0$$

is exact. The same applies to a graded differential algebra $A = \bigoplus_{i\geq0} A_i$ which is free as a graded algebra.

Now let $A = \bigoplus_{i\geq0} A_i$ be an augmented graded differential algebra. A free resolution of A is an augmented graded differential algebra R together with an epimorphism $\pi : R \to A$ such that

a) the graded algebra R is free;

b) π is a quasi-isomorphism.

Such a resolution always exists (if $\partial = 0$ in A then it exists without assuming the positiveness of the grading). If k is a field then $\overline{C}_{..}(R) \to \overline{C}_{..}(A)$, $C_{..}(R) \to C_{..}(A)$ are quasi-isomorphisms. Theorem 1.4.1 applied to R implies that if k is a field then

$$\overline{HC}_{\cdot}(A) \oplus HC_{\cdot}(k) \tilde{\to} HC_{\cdot}(A) \tag{1.5}$$

and

$$\overline{HC}_{\cdot}(A) \tilde{\to} H_{\cdot}(\Omega_{\cdot}(R)) \tag{1.6}$$

Notice that the differential d in $\Omega_{\cdot}(R)$ is acyclic for $i > 0$ if $k \supseteq \mathbb{Q}$. Therefore in this case

$$\overline{HC}_{\cdot}(A) \tilde{\to} H_{\cdot}(\overline{R}/[\overline{R}, \overline{R}]) \ . \tag{1.7}$$

This means that the functors $\overline{HC}_{\cdot}$ are derived functors in Quillen's sense [16] of the functor from the category of algebras to the category of complexes assigning to A the 2-periodic complex $\Omega_{\cdot}(A)$ (or the k-module $A/([A, A] + k)$ if $\mathbb{Q} \subseteq k$).

1.5. <u>Cyclic cohomology</u>. Let $A = \bigoplus_{n \in \mathbb{Z}} A_n$ be a graded differential algebra. Suppose that on A the additional grading $A_n = \bigoplus_{s \in \mathbb{Z}} A_{ns}$ is determined. Set

$$(A^{\otimes k})^* = \oplus_n (A^{\otimes k})^*_{ns} := \oplus_n (\oplus_s (A^{\otimes k})_{ns})^*$$

and

$$\overline{C}^{ij}(A) = (\overline{A}^{\otimes(j+1)})^* \ , \ i \geq 0, \ j \geq 0 \ ;$$

$$C^{ij}(A) = (A^{\otimes(j+1)})^* \ , \ i \geq 0, \ j \geq 0 \ .$$

If $x \in (A^{\otimes(j+1)})_n$ then set

$$(b\varphi)(x) = (-1)^{\overline{\varphi}+n}\varphi(bx) \ , \ (\delta\varphi)(x) = (-1)^{\overline{\varphi}+n}\varphi(\delta x) \ ,$$

$$(\tau\varphi)(x) = \varphi(\tau x) \ ; \ (\partial\varphi)(x) = (-1)^{\overline{\varphi}+n}\varphi(\partial x) \ ;$$

$$C^s(A) = \bigoplus_{i+j+k=s} C^{ij}(A)_k \ .$$

Set

$$\beta\varphi = \delta\varphi + (1-\tau)\varphi \quad \text{for} \quad \varphi \in C^{2i,\cdot}(A) \ ; \tag{1.8}$$

$$\beta\varphi = b\varphi + N\varphi \quad \text{for} \quad \varphi \in C^{2i+1,\cdot}(A) \ . \tag{1.9}$$

The differential $\beta + \partial : C^s(A) \to C^{s+1}(A)$ makes $C^{\cdot}(A)$, $\overline{C}^{\cdot}(A)$ into complexes. Set

$$HC^{\cdot}(A) = H^{\cdot}(C^{\cdot}(A)) \ ; \ \overline{HC}^{\cdot}(A) = H^{\cdot}(\overline{C}^{\cdot}(A)) \ .$$

Let $\overline{C}^{ij}_{per}(A) = (\overline{A}^{\otimes(j+1)})^*$, $j \geq 0$, $i \in \mathbb{Z}$ and $\overline{C}^{s}_{per}(A) = \bigoplus_{i+j+k=s} \overline{C}^{ij}_{per}(A)_k$. Denote by $\overline{HC}^{\cdot}_{per}(A)$ the cohomology of $\overline{C}^{\cdot}_{per}(A)$ with respect to the differential $\beta + \partial$ (formulae (1.8), (1.9), (1.2)). Notice that $\overline{HC}^{\cdot}_{per}(A) \cong \overline{HC}^{\cdot+2}_{per}(A)$.

Lemma 1.5.1. Let either $\partial = 0$ in A or $A = \bigoplus_{i\geq 0} A_i$. Suppose that A admits an additional grading $A_n = \bigoplus_s A_{ns}$, where $A_{\cdot,<0} = 0$ and $A_{\cdot 0} = k\cdot 1$. Then $\overline{HC}^{\cdot}_{per}(A) = 0$ if $\mathbb{Q} \subseteq k$.

Sketch of the proof. First let $\partial = 0$ in A. Then ψ is a derivation of degree zero with respect to the grading $A = \bigoplus_n A_n$. Set

$$(i_\psi\varphi)(a_0,\ldots,a_n) = (-1)^{\overline{\varphi}}\varphi(a_0,\ldots,a_{n-2},\psi(a_{n-1})a_n) \ ;$$

$$(L_\psi\varphi)(a_0,\ldots,a_n) = \sum_{i=0}^{n} \varphi(a_0,\ldots,\psi(a_i),\ldots,a_n) \ .$$

Let the graded k-modules $H^{\cdot}(A, A^*)$ be the cohomology of $(A^{\otimes(\cdot+1)^*}, \delta)$ and $B : H^{\cdot}(A, A^*) \to H^{\cdot-1}(A, A^*)$ the differential dual to the one introduced in 1.3. It can be shown that

$$L_\psi = i_\psi B + B i_\psi \ . \tag{1.10}$$

Now let $A_n = \bigoplus_s A_{ns}$; set $\psi x = sx$ for $x \in A_{\cdot s}$. It follows from (1.10) that the non-trivial cohomology of $(H^{\cdot}(A, A^*); B)$ can only be in degree 0 with respect to the grading by s. This easily implies that $\overline{HC}^{\cdot}_{per}(A) = 0$ if $\partial = 0$. For $\partial \neq 0$ our statement is easily derived from this since, thanks to our hypothesis on the grading, the corresponding spectral sequence converges to zero.

2. Cyclic homology of Koszul algebras.

2.1. <u>Cohomology of graded algebras</u>. Let $A = \bigoplus_{i\in\mathbb{Z}} A_i$ be a graded associative algebra with unit over k, $\varepsilon : A \to k$ an augmentation, $\overline{A} = \ker \varepsilon$. Set

$$B_{nm}(A) = (\overline{A}^{\otimes n})_m \ , \ n \geq 0 \ , \ m \in \mathbb{Z} \ ;$$

$$B_s(A) = \bigoplus_{m+n=s} B_{nm}(A) \ ;$$

$b : B_{nm}(A) \to B_{n-1,m}(A)$ is determined in 1.3:

$$-b(a_1,\dots,a_n) = \sum_{i=1}^{n-1} (-1)^{\sum_{l=1}^{i}(\bar{a}_l+1)} (a_1,\dots,a_ia_{i+1},\dots,a_n)$$

for homogeneous $a_1,\dots,a_n \in \bar{A}$. Let $\mathrm{Tor}^A_{nm}(k, k)$ be the homology of $B_{\cdot}(A)$ in dimension (n, m):

$$\mathrm{Tor}^A_s(k, k) = \bigoplus_{n+m=s} \mathrm{Tor}^A_{nm}(k, k) .$$

Further, set

$$B^{nm}(A) = (\bar{A}^{\otimes n})^*_m \; ; \; B^s(A) = \bigoplus_{m+n=s} B^{nm}(A) \; ;$$

$$\mathrm{Ext}^{nm}_A(k,k) = H^{\cdot}(B^{\cdot\cdot}(A))^{nm} \; ; \; \mathrm{Ext}^s_A = \bigoplus_{n+m=s} \mathrm{Ext}^{nm}_A(k,k)$$

For homogeneous $\varphi \in B^{nm}(A)$ and $a_1, \dots, a_{n+1} \in \bar{A}$

$$-(b\varphi)(a_1,\dots,a_{n+1}) = (-1)^{\bar{\varphi}+n} \sum_{i=1}^{n} (-1)^{\sum_{l=1}^{i}(\bar{a}_l+1)} \varphi(a_1,\dots,a_ia_{i+1},\dots,a_{n+1});$$

$$\Delta(a_1,\dots,a_n) = \sum_{i=1}^{n-1} (a_1,\dots,a_i) \otimes (a_{i+1},\dots,a_n) \; ;$$

$$(\varphi\cup\psi)(a_1,\dots,a_{m+n}) = (-1)^{(\bar{\psi}+m)\cdot\sum_{l=1}^{n}(\bar{a}_l+1)} \varphi(a_1,\dots,a_n)\psi(a_{n+1},\dots,a_{n+m}) .$$

The $\cup$-product turns $B^{\cdot}(A)$ into a graded differential algebra and Δ makes $B_{\cdot}(A)$ a graded differential coalgebra. Therefore $\mathrm{Ext}^{\cdot}_A(k, k)$ is a graded algebra and $\mathrm{Tor}^A_{\cdot}(k, k)$ a graded coalgebra.

2.2. <u>Homogeneous Koszul algebras</u>. Let $\{x_i\}_{i\in I}$ be a system of homogeneous generators of A. Let S be a subset of I × I. In this subsection we suppose that A is determined by relations

$$x_kx_l = \sum_{(i,j)\in S} c^{ij}_{kl}x_ix_j \; , \; (k, l) \notin S \qquad (2.1)$$

where c^{ij}_{kl} are elements of k. Suppose that either

$$\#\{k : \bar{x}_k \leq i\} < \infty \; , \; i \in \mathbb{Z} \qquad (2.2)$$

or

$$\#\{k : \bar{x}_k \geq i\} < \infty \; , \; i \in \mathbb{Z} . \qquad (2.3)$$

Then A is bigraded:

$$A_{rs} = \Big(\sum_{i_1,\dots,i_s} k\cdot x_{i_1} \dots x_{i_s}\Big) \cap A_r .$$

Let $\varepsilon : A \to k$ be an augmentation sending x_i to zero for all i and $\overline{A} = \ker \varepsilon$; $J = \overline{A}/\overline{A}^2 = \sum k \cdot x_i$.

Following [14] we call A a homogeneous Koszul algebra, HKA, if the ri Ext$_A^{\cdot}(k, k)$ is generated by $\mathrm{Ext}_A^{1,\cdot}(k, k)$. Notice that $\mathrm{Ext}_A^{1,\cdot}(k, k) \cong J^* \cong \mathrm{Ext}_A^{1,\cdot}(k, k)_1$ (the lower index corresponds to the grading induced by the grading $A_{\cdot s}$ of A). Therefore

$$\mathrm{Ext}_A^{\cdot}(k, k) = \bigoplus_{r \geq 0} \mathrm{Ext}_A^{r,\cdot}(k, k)_r$$

for an HKA A. Set $A_t^+ := \mathrm{Ext}_A^{-t}(k, k)$. We call A^+ the algebra dual to A.

It is not difficult to determine A^+ in terms of defining relations. In fact, let

$$x_i^* \in \overline{A}^*_{\cdot,1} \text{ , } x_i^*(x_j) = \delta_{ij} \text{ ; } (x_i x_j)^* \in \overline{A}^*_{\cdot,2} \text{ , } (i, j) \in S \text{ ;}$$

$$(x_i x_j)^*(x_k x_l) = \delta_{ik}\delta_{jl} \quad \text{for} \quad (k, l) \in S \text{ ;}$$

$$(x_i | x_j)^*(x_k x_l) = \delta_{ik}\delta_{jl} \text{ , } k, l \in I; \ (x_k | x_l)^* \in (\overline{A}^{\otimes 2}_{\cdot,2})^* \text{ .}$$

We have $bx_i^* = 0$. Denote the corresponding cohomology class by

$$\xi_i \in \mathrm{Ext}_A^{1,\overline{x}_i}(k, k)_1 \text{ .}$$

The family $\{\xi_i\}$ is a system of generators of A^+. All the relations among them follow from the conditions

$$[b(x_i x_j)^*] = 0 \quad , \quad (i, j) \in S$$

where $[\psi]$ denotes the cohomology class of the cocycle ψ. We have

$$x_i^* \cup x_j^* = (-1)^{(\overline{x}_i+1)(\overline{x}_j+1)} (x_i | x_j)^* \text{ , } i, j \in I \text{ ;}$$

$$-(b(x_i x_j)^*)(x_k x_l) = (-1)^{\overline{x}_i+\overline{x}_j+\overline{x}_k} (x_i x_j)^*(x_k x_l)$$

$$= (-1)^{\overline{x}_i+\overline{x}_j+\overline{x}_k} \begin{cases} \delta_{ik}\delta_{jl} \text{ , } k, l \in S \text{ ;} \\ c_{kl}^{ij} \text{ , } (k, l) \notin S \text{ .} \end{cases}$$

Therefore

$$-b(x_i x_j)^* = (-1)^{\overline{x}_j+(\overline{x}_i+1)(\overline{x}_j+1)} x_i^* \cup x_j^* +$$

$$+ \sum_{(k,l) \notin S} (-1)^{\overline{x}_i+\overline{x}_j+\overline{x}_k+(\overline{x}_k+1)(\overline{x}_l+1)} c_{kl}^{ij} x_k^* \cup x_l^* \quad \text{for} \quad (i,j) \in S.$$

We have got the following statement [14]).

Lemma 2.2.1. Let A be an HKA with generators $\{x_i\}$ and relations (2.1). Then A^+ is determined by generators $\{\xi_i\}$, $\bar{\xi}_i = -1 - \bar{x}_i$ and defining relations

$$(-1)^{\bar{\xi}_i(1+\bar{\xi}_j)}\xi_i\xi_j + \sum_{(k,l)\notin S}(-1)^{\bar{\xi}_k(1+\bar{\xi}_l)}c^{ij}_{kl}\xi_k\xi_l = 0,\ (i,j) \in S.$$

Lemma implies in particular, that if A and A^+ are HKAs then $A^{++} \simeq A$. In general, we have only a homomorphism $A^{++} \to A$.

2.3. Koszul algebras; Koszul complexes. Now let a graded algebra A over a field k be determined by homogeneous generators $\{x_i\}_{i\in I}$ and homogeneous relations

$$x_kx_l = \sum_{(i,j)\notin S} c^{ij}_{kl}x_ix_j + \sum_{i\in I} b^i_{kl}x_i\ ,\ (k,\ l) \notin S\ ,$$

where c^{ij}_{kl}, $b^i_{kl} \in k$ and either (2.2) of (2.3) holds. Let $\tilde{A}$ be a bigraded algebra with generators $\{x_i\}$ and relations

$$x_kx_l = \sum_{(i,j)\in S} c^{ij}_{kl}x_ix_j\ ,\ (k,\ l) \notin S\ .$$

If $\tilde{A}$ is an HKA then call A a Koszul algebra. Set $J = \tilde{A}_{\cdot,1}$.

In [14] a Koszul complex for A is constructed. Let

$$\partial_p : J^{\otimes n} \to \tilde{A}^{\otimes(n-1)}\ ,\quad p = 1,\ \dots,\ n-1\ ;$$

$$\partial_p(x_{i_1},\ \dots,\ x_{i_n}) = (x_{i_1},\ \dots,\ x_{i_p}x_{i_{p+1}},\ \dots,\ x_{i_n})\ .$$

Set

$$K_{nm}(A) = (\bigcap_{1\leq p\leq n-1} \ker\partial_p)_m\ ;\ K_s(A) = \bigoplus_{m+n=s} K_{nm}(A)\ .$$

Clearly, if A is a Koszul algebra then $K_s(A) \simeq \mathrm{Tor}^{\tilde{A}}_s(k,\ k)$. Now determine the differential $b:K_s(A)\to K_{s-1}(A)$ by setting $b^i_{kl} = 0$ for $(k,\ l) \in S$. If

$$x = \sum c_{i_1\dots i_n}(x_{i_1},\ \dots,\ x_{i_n}) \in K_{nm}(A)\ ,$$

then set

$$-bx = \sum c_{i_1..i_n}\sum_{1\leq l\leq n-1}(-1)^{\sum_{1\leq k\leq l}(\bar{x}_{i_k}+1)}(x_{i_1},\dots,\sum b^j_{i_li_{l+1}}x_j,\dots,x_{i_n})$$

$$\in K_{n-1,m}(A)\ .$$

We have $b^2 = 0$. In [14] it is shown that the homology of $K_\cdot(A)$ is isomorphic to $\mathrm{Tor}^A_\cdot(k,\ k)$ and the comultiplication in $K_\cdot(A) \simeq \mathrm{Tor}^{\tilde{A}}_\cdot(k,\ k)$

makes $K_{\cdot}(A)$ into a graded differential coalgebra.

Passing to the dual complex we easily get an explicit form of the cohomological Koszul complex calculating $Ext_A^{\cdot}(k, k)$.

Consider a graded differential algebra A^+ with generators $\{\xi_i\}_{i\in I}$, $\bar{\xi}_i = -1 - \bar{x}_i$ and defining relations

$$(-1)^{\bar{\xi}_i(1+\bar{\xi}_j)}\xi_i\xi_j + \sum_{(k,l)\notin S}(-1)^{\bar{\xi}_k(1+\bar{\xi}_l)}c_{kl}^{ij}\xi_k\xi_l = 0, (i,j) \in S$$

and the differential

$$(-1)^{\bar{\xi}_i+1}b\xi_i = \sum_{(k,l)\notin S}(-1)^{\bar{\xi}_k(1+\bar{\xi}_l)}b_{kl}^{i}\xi_k\xi_l .$$

Lemma 2.3.1. There are natural isomorphisms $H_n(A^+) \tilde{\to} Ext_A^{-n}(k, k)$ for all integers n.

Examples. Let V be a graded vector space and $A = T(V)$. Then A and A^+ are Koszul algebras. Set $V_n^+ = V_{-1-n}^*$, $n \in \mathbb{Z}$. We have $A^+ \tilde{\to} k \oplus V^+$, $(V^+)^2 = 0$. Conversely, if $A = k \oplus V$ where $V^2 = 0$ then $A^+ \tilde{\to} T(V)$.

b) If char $k = 0$ and $A = S(V)$ is a symmetric algebra of a graded space V then A and A^+ are Koszul algebras and $A^+ \tilde{\to} S(V^+)$. If g is a graded Lie algebra and $A = U(g)$ then A, A^+ are Koszul algebras and A is a standard cohomological complex $C^{\cdot}(g)$.

c) If A is the standard algebra then A and $\tilde{A}^+$ are Koszul algebras and A^+ is the Van algebra [14].

2.4. Formulation of the theorem. Let A be the same as in 2.3. Assume that either $b_{kl}^i = 0$ for all i, k, l or $A = \bigoplus_{i\geq 0} A_i$. Let A, $\tilde{A}^+$ be Koszul algebras such that either (2.2) or (2.3) hold.

Theorem 2.4.1. a) The sequence

$$\ldots \to \overline{HC}^{-2-n}(A^+) \to \overline{HC}_n(A) \to \overline{HC}^{-1-n}(A^+) \to \overline{HC}_{per}^{-1-n}(A^+) \to \ldots$$

is exact.

b) If char $k = 0$ then

$$\overline{HC}_n(A) \tilde{\to} \overline{HC}^{-1-n}(A^+) .$$

2.5. <u>Cobar-construction of a graded differential coalgebra</u>. Let $K = \bigoplus_{i\in\mathbb{Z}} K_i$ be a graded differential coalgebra. This means that there are given a differential $\partial : K_i \to K_{i-1}$, a counit $\varepsilon : K \to k$ and a comultiplication

$$\Delta : K_i \to \bigoplus_{j+k=i} K_j \otimes K_k \;, \quad \Delta c = \sum c^{l_1} \otimes c^{l_2} \;,$$

$$\sum (\partial c)^{l_1} \otimes (\partial c)^{l_2} = \sum \partial c^{l_1} \otimes c^{l_2} + ((-1)^{\bar{c}^{l_1}}) c^{l_1} \circ \partial c^{l_2} \;.$$

Suppose that ε is a split homomorphism of graded differential coalgebras. Let $\bar{K} = \ker \varepsilon$. Set

$$B_{st}(K) = (\bar{K}^{\otimes s}) \;;$$

$$b_* : B_{st}(K) \to B_{s+1,t}(K) \;; \quad \partial_* : B_{st}(K) \to B_{s,t-1}(K) \;;$$

defining for homogeneous elements c_i of $\bar{K}$

$$b_*(c_1,\dots,c_n) = \sum_{i=1}^{n} (-1)^{\sum_{1\leq k\leq i-1} (\bar{c}_k+1)} \sum (-1)^{\bar{c}_i^{l_1}} (c_1,\dots,c_{i-1},c_i^{l_1},c_i^{l_2},\dots,c_n) \;; \tag{2.4}$$

$$\partial_*(c_1,\dots,c_n) = \sum_{i=1}^{n} (-1)^{\sum_{1\leq k\leq i-1} (\bar{c}_k+1)} (c_1,\dots,\partial c_i,\dots,c_n) \;. \tag{2.5}$$

Further,

$$B_i(K) = \bigoplus_{t-s=i} B_{st}(K) \;;$$

$$(c_1,\dots,c_k)\cdot(c_{k+1},\dots,c_n) = (c_1,\dots,c_n) \;. \tag{2.6}$$

The differential $\partial_* + b_*$ together with the multiplication (2.6) make $B_\cdot(K)$ into a graded differential algebra. Notice that $B_\cdot(K)$ is free.

Let $A = \bigoplus_{i\geq 0} A_i$ be a Koszul algebra, $K_\cdot(A)$ its Koszul complex (see .3). Consider the algebra homomorphism $\pi : B_\cdot(K_\cdot(A)) \to A$ determined by

$$\pi(x) = 0 \quad \text{for} \quad x \in K_{i,\cdot}(A) \;, \; i > 1 \;;$$

$$\pi(x_i) = x_i \in A \quad \text{for} \quad x_i \in K_{1,\bar{x}_i}(A) \;.$$

If $\tilde{A}^+$ is a Koszul algebra then, since the homomorphism $A^{++} \to A$ is an isomorphism, π clearly is a quasiisomorphism. Therefore $B_\cdot(K_\cdot(A))$ is a free resolution of A.

Our aim is to calculate $\overline{HC}_\cdot(A)$ with the help of this resolution.

2.6. Cyclic homology of cobar-constructions. Let K be the same as in 2.5, $\bar{K} = \bigoplus_{i>0} K_i$, $R = B_\cdot(K)$. We have $\bar{R} \stackrel{\sim}{\to} \bar{K}^{\otimes>0}$, $\bar{\Omega}^{-1}(R) \stackrel{\sim}{\to} \bar{K}^{\otimes>0}$. In fact, any element of $R \otimes \bar{R}$ is uniquely represented modulo $\operatorname{im} \tilde{\delta}$ by an element of the form

$(c_1 \ldots c_{n-1}, c_n)$. Since the differential in the cobar-construction splits in two components, the differential ∂ in $\Omega(R)$ also splits into two components induced by ∂_* and b_* (see 2.5) respectively: $\partial = \partial' + \partial''$. From (2.4), (2.5) we get

$$-\partial''(c_o \ldots c_{n-1}, c_n) = \sum_{i=0}^{n-1} (-1)^{\sum_{k<i} (\bar{c}_k+1)} \sum (-1)^{\bar{c}_i^{l_1}} (c_o \ldots c_i^{l_1} c_i^{l_2} \ldots c_{n-1}, c_n)$$

$$+ \sum (-1)^{\sum_{k<n} (\bar{c}_k+1) + \bar{c}_n^{l_1}} (c_o \ldots c_{n-1}, c_n^{l_1} c_n^{l_2})$$

$$= \sum_{i=0}^{n-1} (-1)^{\sum_{k<i} (\bar{c}_k+1)} \sum (-1)^{\bar{c}_i^{l_1}} (c_o \ldots c_i^{l_1} c_i^{l_2} \ldots c_{n-1}, c_n)$$

$$+ \sum (-1)^{\sum_{k<n} (\bar{c}_k+1) + \bar{c}_n^{l_1}} (c_o \ldots c_{n-1} c_n^{l_1}, c_n^{l_2})$$

$$- \sum (-1)^{\bar{c}_n^{l_2}(\bar{c}_n^{l_1}+1+\sum_{k<n} (\bar{c}_k+1))} (c_n^{l_2} c_o \ldots c_{n-1}, c_n^{l_1})$$

(here we have made use of the explicit form for

$$\sum \tilde{\delta}(c_1 \ldots c_{n-1}, c_n^{l_1}, c_n^{l_2})) .$$

We see that under the identification

$$\bar{\Omega}_R^1(R) \cong \overline{K}^{\otimes >0} , \quad (c_o \ldots c_{n-1}, c_n) \mapsto (c_o, \ldots, c_n)$$

the definition of $-\partial''$ in $\bar{\Omega}_R^1$ is dual to the definition of δ given in 1.3. Further

$$\partial''(c_o \ldots c_n) = - \sum_{i=0}^{n} (-1)^{\sum_{k<i} (\bar{c}_k+1)} \sum (-1)^{\bar{c}_i^{l_1}} (c_o \ldots c_i^{l_1} c_i^{l_2} \ldots c_n) .$$

It means that under the identification $\bar{R} \cong \overline{K}^{\otimes >0}$,

$$c_o \ldots c_n \mapsto (c_o, \ldots, c_n)$$

the definition of $-\partial''$ in $\bar{R}$ is dual to the definition of b given in 1.3.

Now consider the differentials $d : \bar{R} \to \bar{\Omega}_R^1$, $d : \bar{\Omega}_R^1 \to \bar{R}$. We have

$$d(c_o\ldots c_{n-1},c_n) = [c_o\ldots c_{n-1},c_n] =$$

$$= c_o\ldots c_{n-1}c_n - (-1)^{(\bar{c}_n+1)\cdot\sum_{k<n}(\bar{c}_k+1)} c_nc_o\ldots c_{n-1} .$$

This is exactly dual to the definition of $1-\tau$ (see 1.3). Further

$$d(c_o\ldots c_{n-1}c_n) = (1,c_o\ldots c_n) = (c_o,c_1\ldots c_n)$$

$$+ (-1)^{\sum_{i>0}(\bar{c}_i+1)\cdot(\bar{c}_o+1)} (c_1\ldots c_n,c_o) + \tilde{\delta}(1,c_o,c_1\ldots c_n) = (c_oc_1,c_2\ldots c_n)$$

$$+ (-1)^{\sum_{i>1}(\bar{c}_i+1)(\bar{c}_o+1+\bar{c}_1+1)} (c_2\ldots c_nc_o,c_1)$$

$$+ \tilde{\delta}(1,c_o,c_1\ldots c_n) + \tilde{\delta}(c_o,c_1,c_2\ldots c_n) = \ldots \equiv$$

$$\equiv \sum_{k=0}^{n} (-1)^{\sum_{i>k}(\bar{c}_i+1)\cdot\sum_{i\leq k}(\bar{c}_i+1)} (c_{k+1}\ldots c_nc_o\ldots c_{k-1},c_k) \pmod{\operatorname{im}\tilde{\delta}} .$$

Therefore under our identifications this operator is dual to the operator N of 1.3.

Finally, it is obvious that the definition of $-\partial'$ is dual to the definition of ∂ in 1.3.

Thus we see that the construction of $\Omega_{.}(R)$ where $R = B_{.}(K)$ is completely dual to the construction of $C^{-2-\cdot}_{per}/C^{-2-\cdot}$ in 1.5. This implies (a) of Theorem 2.4.1. Part b) follows form a) and Lemma 1.5.1.

3. <u>Cyclic homology of universal enveloping algebras</u>. In this chapter we suppose that k is a field of characteristic zero.

3.1. Formulation of the theorem. Let g be a Lie algebra over k. Set

$$Y_{ij}(g) = \Lambda^{j-i}g \otimes S(g),\ j \geq i \geq 0;\ Y_{ij}(g) = 0,\ i > j \geq 0;$$

$$d_* : Y_{ij}(g) \to Y_{i-1,j}(g) ;$$

$$\delta : Y_{ij}(g) \to Y_{i,j-1}(g) ;$$

and if $x_1,\ldots,x_m,y_1,\ldots,y_m \in g$, $f = y_1 \ldots y_m \in S(g)$, $j - i = n$, then

$$d_*(x_1\wedge\ldots\wedge x_n\otimes f) = \sum_{k=1}^{m} (x_1\wedge\ldots\wedge x_n\wedge y_k)\otimes(y_1\ldots\hat{y}_k\ldots y_m) \tag{3.1}$$

$$\begin{aligned}\delta(x_1\wedge\ldots\wedge x_n\otimes f) = (-1)^i\Big(&\sum_{k,l=1}^{n} (-1)^{k+1}[x_k,x_l]\wedge\ldots\wedge\hat{x}_k\\ &\wedge\ldots\wedge\hat{x}_l\wedge\ldots\wedge x_n\otimes f + \sum_{k=1}^{n} (-1)^k x_1\wedge\ldots\wedge\hat{x}_k\wedge\ldots\wedge x_n\otimes[x_i,f]\ .\end{aligned} \tag{3.2}$$

Therefore d_* is the de Rham differential in $\Lambda g \otimes S(g) \simeq \Omega_{S(g)}$, δ the differential in the standard complex for calculating the homology of g with coefficients in $S(g)$. It is directly verified that $d_*\delta + \delta d_* = 0$. Therefore $Y_{\cdot\cdot}(g)$ is a bicomplex. Denote the associated complex by $Y_\cdot(g)$.

Theorem 3.1.1. There is a canonical isomorphism

$$\overline{HC}_\cdot(U(g)) \xrightarrow{\sim} H_\cdot(Y_\cdot(g))\ .$$

The proof will be given in 3.2 - 3.4.

Now we shall give another formulation. Let

$$W_{ij}(g) = \Lambda^{j-i}g \otimes S^i(g)\ ,\ j \geq i \geq 0\ ;\ W_{ij}(g) = 0\ ,\ i > j \geq 0\ .$$

Clearly, $W_{\cdot\cdot}(g)$ is a subcomplex in $Y_{\cdot\cdot}(g)$. Let

$$F^m W_\cdot(g) = \bigoplus_{i<m} W_{i,\cdot}(g)\ .$$

Corollary 3.1.2. There are canonical isomorphisms

$$HC_\cdot(U(g)) \xrightarrow{\sim} \bigoplus_{m\geq 0} H_{\cdot-2m}(W_\cdot(g)/F^m W_\cdot(g))\ .$$

The complex dual to $W_\cdot(g)$ is the Weyl algebra of a Lie algebra g. The relation of the cohomology of the truncated Weyl algebras with an additive analogue of the algebraic K-functor was discovered by Beilinson in [1].

Theorem 3.1.1 yields also

Proposition 3.1.3. There exists a spectral sequence

$$E^1_{ij} = H_{j-i}(g,\ S(g)) \Rightarrow HC_{i+j}(U(g))\ .$$

3.2. <u>Free graded commutative algebras</u>. Let A be a free graded (skew) commutative algebra with homogeneous generators $\{x_i\}$. Let Ω_A be a free graded commutative algebra with generators $\{x_i\}$ and $\{\xi_i\}$, $\bar{\xi}_i = \bar{x}_i - 1$. Consider the unique $d : \Omega_A \to \Omega_A$ of degree -1 such that $x_i \mapsto \xi_i$, $\xi_i \mapsto 0$ for all i. Any element of Ω_A can be expressed as a line combination of monomials of the form $da_o da_1 \dots da_{n-1} \cdot a_n$, where $a_i \in A$. Let Ω_A^n be the subspace in Ω_A generated by elements of this form, n fixed. Consider the maps

$$\gamma_n : \overline{A}^{\otimes(n+1)} \to \Omega_A^n / d\Omega_A^{n-1};\ (a_o, \dots, a_n) \mapsto da_o da_1 \dots da_{n-1} \cdot a_n .$$

It is easy to verify that $\gamma_n x = \gamma_n \tau x$ and that the mappings $\gamma_\cdot$ determine a homomorphism γ from the complex of graded spaces $(\overline{A}^{\otimes(\cdot+1)}/\mathrm{im}(1-\tau), \delta)$ to the complex of graded spaces

$$\dots \xrightarrow{o} \Omega_A^n / d\Omega_A^{n-1} \xrightarrow{o} \Omega_A^{n-1} / d\Omega_A^{n-2} \to \dots \to \Omega_A^o / k \to 0 .$$

Lemma 3.2.1. γ is a quasi-isomorphism.

For the proof of the non-graded case, see [15]. The general case is treated similarly.

Notice that the cohomology of d is trivial. Further

$\overline{A}^{\otimes(\cdot+1)}/\mathrm{im}(1-\tau) \overset{\sim}{\to} \overline{A}^{\otimes(\cdot+1)}/\ker N \overset{\sim}{\to} \mathrm{im}\, N = \ker(1-\tau)$

which yields

Lemma 3.2.2. There exists a functorial quasi-isomorphism of complexes of the graded spaces $(\overline{A}^{\otimes(\cdot+1)} \cap \ker(1-\tau);\ b)$ and

$$\dots \xrightarrow{o} d\Omega_A^n \xrightarrow{o} d\Omega_A^{n-1} \xrightarrow{o} \dots \xrightarrow{o} d\Omega_A^1 \xrightarrow{o} d\Omega_A^o \to 0 .$$

3.3. <u>Free cocommutative graded differential coalgebras</u>. Let A be the same as in 3.2. On A, there exists a unique comultiplication Δ sending x_i to $x_i \otimes 1 + 1 \otimes x_i$ which is consistent with the multiplication in A. Suppose, there is given a differential $\partial : \overline{A}_\cdot \to \overline{A}_{\cdot-1}$, which together with Δ makes A into a graded differential coalgebra. Then the comultiplication which sends x_i to $x_i \otimes 1 + 1 \otimes x_i$ and ξ_i to $\xi_i \otimes 1 + 1 \otimes \xi_i$ and which is compatible with the multiplication makes Ω_A a graded coalgebra.

Attributing to x_i and ξ_i degree 1 for all i, we get an additional grading on Ω_A. There exists a unique coderivation of Ω_A of degree 0 with respect to this grading, of degree 1 with respect to the initial grading of Ω_A that sends ξ_i to x_i and x_i to 0 for all i. Denote it by d_*. Further there exists a unique codifferentiation ∂' of the coalgebra Ω_A of degree -1 sending x to ∂x for $x \in A$ and such that $\partial' d_* + d_* \partial' = 0$. The differential ∂' makes $\Omega_A^{n+1}/d_*\Omega_A^{n+2}$ into a complex.

Further let $L_m(A) = \bigoplus_{i+j=m} (\overline{A}^{\otimes i})_j$ be a complex with the differential $b_* + \partial_*$ where b_* and ∂_* are determined in 2.5.

Lemma 3.3.1. There is a quasi-isomorphism

$$\mu : L_{\cdot}(A)/\mathrm{im}(1-\tau) \xrightarrow{\sim} \bigoplus_{n\geq 0} (\Omega_A^{n+1}/d_*\Omega_A^{n+2})_{\cdot -n} \tag{3.3}$$

Proof. Let $\partial = 0$ in A. If A' is a subalgebra of A generated by a finite number of $\{x_i\}$ then the quasi-isomorphism μ of the form (3.1) is obtained by passage to the dual algebras and applying Lemma 3.2.2. This isomorphism is functorial; passing to the inductive limit we get a quasi-isomorphism μ for A.

Now if $\partial \neq 0$ then clearly the quasi-isomorphism μ constructed above satisfies $\mu\partial' = \partial'\mu$. Now our statement follows from the comparison of the spectral sequences (clearly the corresponding sequences converge).

3.4. The end of the proof. Thanks to (1.7) there is a canonical isomorphism

$$\overline{HC}_{\cdot}(U(g)) \xrightarrow{\sim} H_{\cdot}(\bar{R}/[\bar{R}, \bar{R}]) ,$$

where R is a free resolution of $U(g)$. Taking for R the cobar-construction of the chain complex $C_{\cdot}(g)$ for calculating the homology of g with trivial coefficients we see that

$$\bar{R}/[\bar{R}, \bar{R}] \xrightarrow{\sim} L_{\cdot}(C_{\cdot}(g))/\mathrm{im}(1-\tau) \tag{3.4}$$

where $L_{\cdot}(\cdot)$ is introduced in 3.3. Now clearly $\Omega_{C_{\cdot}(g)} \simeq \Lambda^{\cdot}g \otimes S(g)$, the differential ∂' coincides with δ of (3.1) and the differential d_* of 3.3 with the differential d_* of 3.2. Taking into account that the cohomology with respect to d_* is trivial, we immediately get Theorem 3.1.1

Clearly the cobar-construction of $C_{\cdot}(g)$ is a resolution of $U(g)$ without any assumptions on finiteness of the dimension.

<u>Semisimple Lie algebras</u>. Let g be a semisimple Lie algebra over $\mathbb{C}$. Consider the spectral sequence of Proposition 3.1.3. We have

$$H_{\cdot}(g, S(g)) \simeq H_{\cdot}(g) \otimes S(g)_g ;$$

$$S(g)_g \xrightarrow{\sim} \mathbb{C}[\sigma_1,\dots,\sigma_n],\ \sigma_i \in S^{m_i}(g) ;$$

$$H_{\cdot}(g) \xrightarrow{\sim} \Lambda^{\cdot}[\eta_1,\dots,\eta_n],\ \eta_i \in H^{2m_i-1}(g) ;$$

where the differentials of higher terms of the spectral sequence send σ_i to η_i and η_i to 0 and are multiplicative with respect to the multiplication in

$$\Lambda^{\cdot}[\eta_1,\dots,\eta_n] \otimes \mathbb{C}[\sigma_1,\dots,\sigma_n] .$$

From that we easily derive an explicit form for the E^∞ term of the spectral sequence. Let

$$x = \eta_{l_1} \dots \eta_{l_p} \sigma_{r_1}^{s_1} \dots \sigma_{r_q}^{s_q},\ l_1 < \dots < l_p,\ r_1 < \dots < r_q,\ s_j > 0\ ,$$

be a monomial of $H_\cdot(g) \otimes S(g)$. Call $\sum_{i=1}^{p} (2m_{l_i} - 1)$ its order. We say that x is i-admissible if $i < m_{r_1} < m_{l_1}$.

Theorem 3.5.1. The space E^∞_{ij} has a basis consisting of homology classes of admissible monomials of order j. In particular $E^\infty_{ij} = 0$, $i > m_n$.

Therefore there is a filtration of $HC_n(U(g))$ such that for its associated graded quotients we have isomorphisms

$$(F^{i-1}/F^i)HC_n(U(g)) \overset{\sim}{\to} \bigoplus_{x \in M^i_{n-i}} \mathbb{C}x\ ,$$

where M^i_{n-i} is the set of i-admissible monomials of order n-i.

4. Cyclic homology of twisted group rings.

4.1. Let k be a commutative ring with unit, A an algebra over k, G a discrete group, $\alpha : G \to \operatorname{Aut} A$ a group homomorphism: $g \mapsto \alpha_g$ (we will write a^g for $\alpha_g(a)$). Consider the algebra $A_\alpha[G]$ such that $A_\alpha[G] = k[G] \otimes A$ as a k-module and $(g_o,a_o).(g_1,a_1) = (g_og_1,\ a_o^{g_1}.a_1)$.

Let $\hat{G}$ be the set of conjugacy classes of elements of G and if $g \in G$ denote by $[g] \in \hat{G}$ the class containing g. Let N_g be the centralizer of g in G and (g) the central subgroup generated by g in N_g.

Theorem 4.1.1. The groups $HC_\cdot(A_\alpha[G])$ admit a natural decomposition into the direct sum

$$HC_\cdot(A_\alpha[G]) \overset{\sim}{\leftarrow} \bigoplus_{[g] \in \hat{G}} HC_\cdot(A_\alpha[G])_{[g]}\ ;$$

where $HC_\cdot(A_\alpha[G])_{[g]}$ are abutments of certain sequences of the form

$$E^2_{ij} = H_i(N_g/(g),\ HC_j(A\ ;\ \text{ord } g,\ \alpha_g)) \Rightarrow HC_{i+j}(A_\alpha[G])_{[g]} \qquad (4.0)$$

Proof. First describe the cyclic k-module $A_\alpha[G]^\natural$. We claim that it is isomorphic to another cyclic k-module $A_\alpha[G]^{\natural\natural}$ where $A_\alpha[G]^{\natural\natural}_n = A^{\otimes(n+1)} \otimes k[G]^{\otimes(n+1)}$; and if $x = (a_o,\dots,a_n) \otimes (g_o,\dots,g_n)$, $a_i \in A$, $g_i \in G$, then

$$xd_i = (a_o,\dots,a_{i-1}a_i,\dots,a_n) \otimes (g_o,\dots,g_{i-1}g_i,\dots,g_n),\ 1 \leqslant i < n\ ;$$

$$xd_n = (a_o,\dots,a_{n-1}a_n) \otimes (g_o,\dots,g_{n-2},g_n)\ ;$$

$$xd_o = (a_1^{g_o^{-1}},\dots,(a_n a_o^{g_n})^{g_o^{-1}}) \otimes (g_o,\dots,g_{n-1},g_o^{-1}g_n g_o)\ ;$$

$$xt = (a_1^{g_o^{-1}},\dots,a_n^{g_o^{-1}},(a_o^{g_n})^{g_o^{-1}}) \otimes (g_1,\dots,(g_o\cdots g_{n-1})^{-1}g_n,g_o^{-1}g_n g_o)\ ;$$

$$xs_i = (a_o,\dots,a_{i-1},1,a_i,\dots,a_n) \otimes (g_o,\dots,g_{i-1},1,g_i,\dots,g_n)\ .$$

An isomorphism $A_\alpha[g]^\natural \tilde{\to} A_\alpha[G]^{\natural\natural}$ is determined by the following formula:

$$((a_o,g_o),\dots,(a_n,g_n)) \mapsto (a_o,a_1^{g_o},\dots,a_n^{g_o\cdots g_{n-1}}) \otimes (g_o,\dots,g_{n-1},g_o\cdots g_n)\ .$$

Notice that the operators d_i, s_i, t in $A_\alpha[G]^{\natural\natural}$ preserve the conjugacy class of g_n. Therefore we have a decomposition

$$A_\alpha[G]^{\natural\natural} = \bigoplus_{[g]\in\hat{G}} M_{[g]}\ ; \tag{4.1}$$

where $M_{[g]}$ is generated by elements $(a_o,\dots,a_n) \otimes (g_o,\dots,g_n)$ such that $g_n \in [g]$.

For a simplicial k-module $M_{[g]}$ we have the following mappings

$$\begin{aligned} M_{[g]} &\overset{a}{\rightleftarrows} P(G) \boxtimes_{K[G]} \Big(\bigoplus_{x\in[g]} A^{\natural,\mathrm{ord}\ g}_{\alpha_x}\Big) \overset{b}{\rightleftarrows} \\ &\overset{b}{\rightleftarrows} P(G) \boxtimes_{K[G]} (k[G] \otimes_{k[N_g]} A^{\natural,\mathrm{ord}\ g}_{\alpha_g}) \overset{c}{\hookleftarrow} \\ &\overset{c}{\hookleftarrow} P(N_g) \boxtimes_{k[N_g]} A^{\natural,\mathrm{ord}\ g}_{\alpha_g} \end{aligned} \tag{4.2}$$

where a is determined by the formula

$$(a_o,\dots,a_n) \otimes (g_o,\dots,g_n) \mapsto (g_o,\dots,g_n) \boxtimes (a_o,\dots,a_n)$$

and b by the formula

$$b = 1_{P(G)} \boxtimes \theta,\ \theta(x;(a_o,\dots,a_n)) = y \otimes (a_o^{y^{-1}},\dots,a_n^{y^{-1}})\ ,$$

where $ygy^{-1} = x$.

The maps a, b are isomorphisms. The map c induces the group isomorphism for $\mathbb{L}_\cdot \varinjlim_{\Lambda^o} (\cdot)$ (by Shapiro's lemma). Applying 1.2.3 and comparing spectral sequences we see that it induces the group isomorphism for $\mathbb{L}_\cdot \varinjlim_{\Lambda^o_{\mathrm{ord}\ g}} (\cdot)$.

To deduce the statement of Theorem 4.1.1 from the obtained direct decomposition of $A_\alpha[G]^{\natural\natural}$ we introduce several notations and definitions.

Denote by $\mathcal{A}$ the category of k-modules. If C is a category, M and M' are functors $C \to \mathcal{A}$ then denote by $M \boxtimes M'$ the functor whose values on the objects and morphisms are tensor products of values of M and M'.

Let N be a group and $g \in$ Cent N. Consider the category C_N with the single object $*$ and the monoid of morphisms $\hom_{C_N}(*, *) = N^o$. Consider the Cartesian product $\Lambda_\infty \times C_N$. Introduce the additional relation

$$(t_n^{n+1}, g^{-1}) = id_{([n],\, *)} \quad .$$

Denote the obtained category by $\Lambda_{N,g}$. Notice that Δ is a subcategory in $\Lambda_{N,g}$ such that Ob $\Lambda_{N,g}$ = Ob Δ and if $f \in \hom_{\Lambda_{N,g}}([n], [n'])$ then there exists a unique factorization

$$f = g\cdot\sigma,\ g \in \hom_\Delta([n],[n']),\ \sigma \in \mathrm{Aut}_{\Lambda_{N,g}}([n]) \quad .$$

There is an exact sequence of groups

$$1 \to Z \xrightarrow{p} N^o \times Z \to \mathrm{Aut}_{\Lambda_{N,g}}([n]) \to 1 \ ;$$

$$p(l) = (g^{-l},\ l(n+1)),\ l \in Z \ .$$

Notice that $A_{\alpha_g}^{\natural,\mathrm{ord}\ g}$, $P(N_g, g)$ are actually contravariant functors $\Lambda_{N_g,g} \to \mathcal{A}$. The N_g-action is determined by the formulae

$$h(a_o,\dots,a_n) = (a_o^h,\dots,a_n^h),\ a_i \in A,\ h \in N_g \ ;$$

$$h(h_o,\dots,h_n) = (hh_o,h_1,\dots,h_n),\ h,\ h_i \in N_g \ .$$

It follows from (4.1) and (4.2) that

$$HC_\cdot(A_\alpha[G]) \cong \bigoplus_{[g]\in G} L_\cdot \varinjlim_{\Lambda^o} ((A_{\alpha_g}^{\natural,\mathrm{ord}\ g} \boxtimes P(N_g,g)) \otimes_{k[N_g]} k) \qquad (4.3)$$

Let S be a category such that Ob S = $\{[n] : n \geq 0\}$, Mor S = $\{id_{[n]} : n \geq 0\}$.

We have the following commuting diagrams of categories and functors

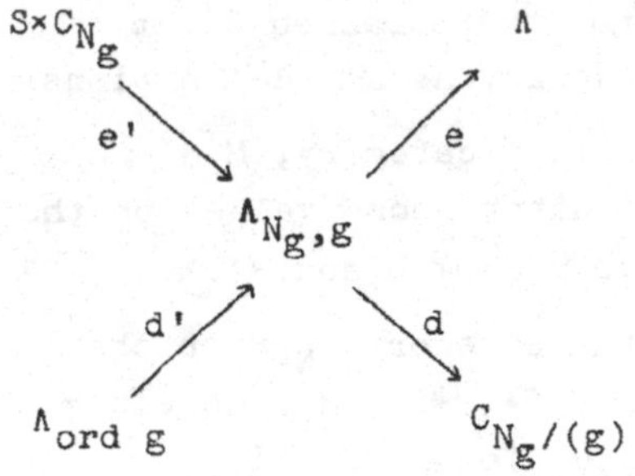

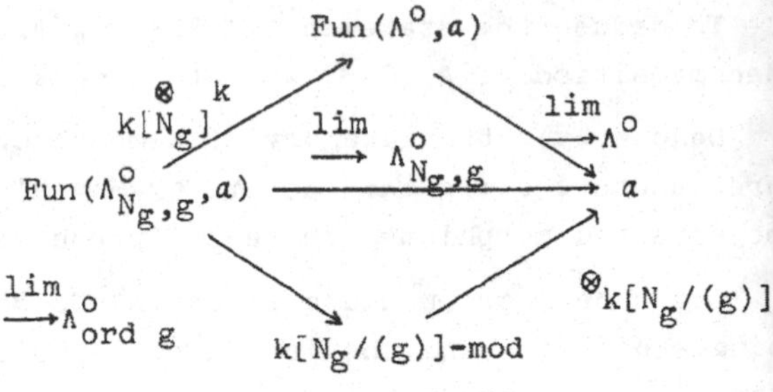

Here the functors d, e are the obvious embeddings of categories;

$$e'([n], *) = * \; ; \; e'(f,h) = h, \; f \in \mathrm{Mor}\, \Lambda_\infty, \; h \in N_g \; ;$$

h is the projection of $h \in N_g$ onto $N_g/(g)$ and

$$d'([n],*) = [n] \; ; \; d'(f,h) = f \; ;$$

$$f \in \mathrm{Mor}\, \Delta, \; h \in N_g \; ; \; d'(t,h) = t, \; h \in N_g \; .$$

The functors $\otimes_{k[N_g]} k$ and $\underset{\longrightarrow \Lambda^o_{\text{ord } g}}{\lim}$ admit right exact duals (restriction functors) and therefore send projective objects to projective ones.

The projection $A^{\natural, \text{ord } g}_{\alpha_g} \blacksquare P(N_g, g) \to A^{\natural, \text{ord } g}_{\alpha_g}$ induces a group isomorphism $\mathbb{L}_\cdot \underset{\longrightarrow \Lambda^o}{\lim} (\cdot)$. This easily implies that it induces the group isomorphism $\mathbb{L}_\cdot \underset{\longrightarrow \Lambda^o_{N_g,g}}{\lim} (\cdot)$. On the other hand, the higher derived functors of $\otimes_{k[N_g]} k$ vanish for $A^{\natural, \text{ord } g}_{\alpha_g} \blacksquare P(N_g, g)$. Therefore from (4.3) we get the isomorphism

$$HC_\cdot(A_\alpha[G]) \simeq \bigoplus_{[g] \in \hat{G}} \mathbb{L}_\cdot \underset{\longrightarrow \Lambda^o_{N_g,g}}{\lim} A^{\natural, \text{ord } g}_{\alpha_g} \; .$$

Now Theorem 4.1.1 follows from considering the spectral sequence of the composition

$$\underset{\longrightarrow \Lambda^o_{N_g,g}}{\lim} = (\otimes_{k[N_g/(g)]} k) \circ \underset{\longrightarrow \Lambda^o_{\text{ord } g}}{\lim} \; .$$

Remark 4.1.2. It follows from 1.2.3 that for a ∞-cyclic object M of a we have

$$\mathbb{L}_\cdot \underset{\longrightarrow \Lambda^o}{\lim} (M) \simeq \mathbb{L}_\cdot \underset{\longrightarrow \Lambda^o_\infty}{\lim} (M)$$

(s_o is a contracting homotopy for b). Therefore if ord g = ∞ then

$$HC_\cdot(A; \alpha_g, \infty) \simeq H_\cdot(A, A_{\alpha_g}) = \mathrm{Tor}^{A \otimes A^o}_\cdot (A, A_{\alpha_g}) \; .$$

Here A_{α_g} is a bimodule over A which is A as a k-module and the A-action on A_{α_g} is determined by the formula

$$a.b.c = abc^g .$$

It follows from the proof of Theorem 4.1.1 that $HC_\bullet(A_\alpha[G])_{[1]} \cong \mathbb{H}_\bullet(G, C(A))$ is the hyperhomology of G in the complex of Λ-modules. The G-action on $C_\bullet(A)$ is determined by the formula $g(a_0,\dots,a_n) = (a_0^g,\dots,a_n^g)$.

Corollary 4.1.2. There are functorial embeddings as direct summands (see [3, 10]):

$$\mathbb{H}_\bullet(G,C_\bullet(A)) \hookrightarrow HC_\bullet(A_\alpha[G]) ;$$

$$H_\bullet(G, k) \oplus H_{\bullet-2}(G, k) \oplus \dots \hookrightarrow HC_\bullet(k[G]) .$$

Applying this statement we get the Chern character of [9], [5], [11].

Consider the composition

$$H_i(GL_k(A)) \hookrightarrow HC_{i+2n}(\mathbb{Z}[GL_k(A)]) \to HC_{i+2n}(M_k(A)) .$$

It is known from [5] that $HC_\bullet(M_k(A)) \cong HC_\bullet(A)$. Passing to the direct limit we get a homomorphism $H_i(GL(A)) \to HC_{i+2n}(A)$ and the composition with the Hurewicz homomorphism gets us the Chern character

$$ch_{n,i} : K_i(A) \to HC_{i+2n}(A) .$$

For $g \neq 1$ the problem of constructing the spectral sequence (4.0) is more delicate.

The groups $HC_\bullet(\mathbb{Z}[G])_{[g]}$ can be interpreted as homology of certain topological spaces (3). The situation is simpler for zero characteristic.

Let $k \supseteq \mathbb{Q}$ and M be an r-cyclic object of the category of k-modules, $r < \infty$. As in 1.1 consider the complex $(M_\bullet/\mathrm{im}(1-\tau); \delta)$. It follows from 1.2.3 that the homology of this complex is isomorphic to $\mathbb{L}_\bullet \varinjlim_{\Lambda_r^o} M$. If $M = A_\alpha^{\natural,r}$ we denote this complex by $C_\bullet^{cycl.}(A; \alpha, r)$. By a standard argument of homological algebra we get the following

Proposition 4.1.4. Let $k \supseteq \mathbb{Q}$ and $\mathrm{ord}\, g < \infty$; then

$$HC_\bullet(A_\alpha[G])_{[g]} \cong H_\bullet(N_g/(g), C_\bullet^{cycl}(A; \alpha_g, \mathrm{ord}\, g)).$$

We conclude this paper by an example concerning $G = \mathbb{Z}$. Consider the algebra J(A) of matrices $(a_{ij})_{i,j\in\mathbb{Z}}$ over A such that $a_{ij} = 0$ for $|i - j| > N$ for some N (generalized Jacobian matrices). There is an isomorphism of k-modules

$$J(A) \cong \bigoplus_{i\in\mathbb{Z}} \left(\prod_{j\in\mathbb{Z}} A \right) ;$$

on $\prod_{j\in\mathbb{Z}} A$ determine the Tikhonov product topology (assuming the topology on A is discrete). Then J(A) with the direct sum topology is a topological k-algebra.

Generalizing the main definitions of 1.1 it is not difficult to determine continuous Hochschild homology $\hat{H}_\cdot(B, B)$ and cyclic homology $\hat{HC}_\cdot(B)$ for a large class of topological algebras containing J(A) (cf. [11]). Let $B = \prod_{j\in\mathbb{Z}} A$; $\alpha \in \mathrm{Aut}\, B$; $\alpha(b_i)_{i\in\mathbb{Z}} = (b_{i+1})_{i\in\mathbb{Z}}$. Then $J(A) \cong B_\alpha(\mathbb{Z})$. For $\hat{HC}_\cdot$ an analogue of Theorem 4.1.1 holds. We have

$$\hat{HC}_\cdot(B) \cong \prod_{j\in\mathbb{Z}} HC_\cdot(A) ; \quad H_0(\mathbb{Z}, \hat{HC}_\cdot(B)) = 0 ;$$

$$H_1(\mathbb{Z}, HC_\cdot(B)) \cong HC_\cdot(A) ;$$

and if $\mathbb{Q} \subseteq k$ then $\hat{HC}_\cdot(B_\alpha[\mathbb{Z}])_{[g]} = 0$, $g \neq 1$, implying

Proposition 4.1.5. Let $\mathbb{Q} \subseteq k$ then

$$HC_\cdot(A) \cong \hat{HC}_{\cdot-1}(J(A)) .$$

Bibliography.

1. Beilinson A.A. Higher regulators and values of L-functions. In: Modern problems of mathematics, Vol. 24, VINITI, Moscow.

2. Burghelea D., Fiedorowicz Z. Cyclic homology of group rings. - Preprint, 1984.

3. Cartan A., Eilenberg S. Homological algebra, Princeton Univ. Press, Princeton, 1956.

4. Connes A. Non commutative differential geometry. - Prepr. IHES, Oct. 1982, mars 1983; Publ. Math. IHES, Vol. 62, 1985.

5. Connes A. Cohomologie cyclique et foncteurs Ext_n. - C.R.A.S. Paris.

6. Drinfeld V.G. On quadratic commutation relations in quasi-classical case. In: Mathematical Physics, Funct. Anal.,Naukova Dumka, Kiev, 1986, 25-34 (Russian).

7. Daletsky Yu.L., Tsygan B.L. Hamiltonian operators and Hochschild homology. Funct. Anal. Appl, Vol. 19, No. 4, 1985, 82-83 (Russian)

8. Feigin B.L., Tsygan, B.L. Homology of Lie algebras of generalized Jacobi matrices. Funct. Anal. Appl., Vol. 17, No. 2, 1983, 86-87 (Russian).

9. Feigin B.L., Tsygan B.L. Additive K-theory and crystalline cohomology. - Funct. Anal. Appl., Vol. 19, No. 2, 1985, 52-62 (Russian

10. Feigin B.L., Tsygan B.L. Additive K-theory. This volume.

11. Karoubi M. Homologie cyclique et K-théorie I, II. - Preprint Paris VII, 1985.

12. Loday J.L., Quillen D. Cyclic homology and the Lie algebras of matrices. - Comm. Math. Helv. 59, No. 4, 1984, 559-594.

13. May P. Simplicial objects in algebraic topology. - Van Nostrand, Princeton, 1967.

14. Priddy S. Koszul resolutions. - Trans. Amer. Math. Soc.

15. Quillen D. On the cyclic homology of algebras. - Prepr. Oxford, 1984.

16. Quillen D. Homotopical algebra. - Springer Lect. Notes, Vol. 43, 1967.

17. Tsygan B.L. Homology of Lie algebras over rings and Hochschild homology. Russian Mathematical Surveys, Vol. 38, No. 2, 1983, p. 217-218 (Russian).

18. Waldhausen F. Algebraic K-theory of generalized free products, I, II. - Ann. Math., II.Ser., Vol. 108, 1978, No. 1, p.135-204; No. 2, p. 205-256.

ON HOMOTOPY LIMIT OF HOMOTOPY ALGEBRAS

V.A. Hinich, V.V. Schechtman

Contents

Introduction

Let $F: C \to Cdg(k)$ be a functor from a small category C into the category of commutative DG algebras over a commutative ring k. Then the cohomology $H^*(C,F)$ has a canonical structure of a graded commutative k-algebra. This structure is induced by a structure of commutative algebra in the homotopy category $K^{\cdot}(k)$ on cochain complex $C^*(X,F)$.

In the present paper we show that $C^*(X,F)$ has a finer structure which must be thought of as "a commutative algebra up to higher homotopies". More explicitly, we show that $C^*(X,F)$ is an algebra over an acyclic operad in the sense of May (see [May 1]; the definition from this paper works directly in the category $C^{\cdot}(k)$ of complexes instead of the category of topological spaces). We call such algebras May algebras.

The structure of May algebra is more rich then that of commutative algebra in homotopy category. For example, in the cohomology of a May algebra over F_p act Steenrod powers, see (1.4.2).

More generally, we introduce the notion of "a functor up to higher homotopies" into the category of May algebras which is a simultaneous generalization of the notions of pseudo-functor from [HS] and of May algebra. This notion can also be considered as a variant - in the spirit of May - of the definitions of Boardman-Vogt, [BV]. Our main result says that homotopy inverse limit of such a functor has a structure of a May algebra.

In §1 we introduce basic definitions and formulate the main theorem (see (1.6)). In §2 we define a natural structure of May algebra on the normalization of a cosimplicial algebra. A beautiful construction of the corresponding operad presented in (2.4)-(2.5) is due to A. Beilinson. §3 contains the proof of (1.6). In §4 we show, using a generalization of Sullivan construction, that if $k \supset \mathbb{Q}$ then for a functor $F: C \to Cdg(k)$ there exists a commutative DG algebra $Th(F)$ such that

$$H^*(C,F) = H^*(Th(F)).$$

The Appendix is logically independent from the main part of the paper and may be read immediately after reading the definitions of C-operad and pseudo-functor.

We present here some examples of free acyclic operads and coefficient systems. All of them have a kind of a universal property similar to the universal property of a free left resolution of a module among all its left resolutions.

They key ingredient in the constructions is the formula of Sugawara (cf. [BV, 1.15]). In (A2) we apply these ideas to the construction of "higher octahedrons" in $C(k)$ (cf. (A2.2)); and in (A3) - to the construction of a free resolution of an arbitrary associative ring. In (A4) is presented a variant of W-construction of Boardman-Vogt, [BV].

The starting point of our paper was the construction of multiplication in D-cohomology due to A. Beilinson, [B], where appears the family of multiplications (parametrized by k) on homotopy fiber product of Cdg-algebras.

The notion of May algebra also appears (without this name) in the paper of V. Smirnov, [S], where a variant of (2.3) is proved by the other method.

We wish to express our gratitude to A. Beilinson whose ideas enable us to simplify original proofs. We also thank V. Drinfeld for stimulating correspondence.

Notations

Throughout this paper we fix a commutative ring with unity k. (k-mod) is the category of k-modules;

Δ is the category of finite ordered sets;

$\mathcal{E}ns$ is the category of sets.

If C is a category then $\Delta^{o}C$ (resp., ΔC) denotes the category of simplicial (resp., cosimplicial) objects in C.

If A is an additive category, then $C^{\cdot}(A)$ will denote the category of complexes over A (with deg d = +1). A is identified with the full subcategory of $C^{\cdot}(A)$ of complexes concentrated in degree zero. For $X \in C^{\cdot}(A)$, $i \in \mathbb{Z}$, $\tau_{\leq i}X$ is the complex

$$(\ldots \to X^{i-2} \xrightarrow{d^{i-2}} X^{i-1} \to \ker d^{i-1} \to 0 \to \ldots).$$

Cdg = Cdg(k) will denote the category of commutative differential graded k-algebras, i.e. commutative algebras in $C^{\cdot}(k)$.

For $a,b \in \mathbb{Z}$ $C^{\geq a}(A)$, $C^{\leq b}(A)$, $C^{[a,b]}(A)$ are the full subcategories of C(A) with objects

$$\mathrm{Ob}\, C^{\geq a}(A) = \{X^{\cdot} \in \mathrm{Ob}\, C^{\cdot}(A) \mid X^{i}=0 \quad \text{for} \quad i < a\};$$

$$\mathrm{Ob}\, C^{\leq b}(A) = \{X^{\cdot} \quad \mathrm{Ob}\, C^{\cdot}(A) \mid X^{i}=0 \quad \text{for} \quad i > b\};$$

$$\mathrm{Ob}\, C^{[a,b]}(A) = \mathrm{OB}\, C^{\geq a} \cap \mathrm{Ob}\, C^{\leq b}.$$

For brevity we shall write C(k), Δ(k), etc. instead of $C^{\cdot}$(k-mod), Δ(k-mod), etc.; also we shall often write $X \in C$ instead of $X \in \mathrm{Ob}\, C$.

We shall identify the set $\mathbb{N} = \{0, 1, 2, \ldots\}$ with Ob Δ.

For a morphism we'll denote by $s\varphi$ (resp., $t\varphi$) its source (resp., its target).

§1. May algebras

(1.0) Let C be a category. For $x \in C$ we'll denote by C/x the category of objects over x. For $\varphi = (\varphi_1, \ldots, \varphi_n) \in \mathrm{Ob}\,(C/x)^n$, $\varphi_i\colon y_i \dashrightarrow x$, we put $t\varphi := x$.

(1.1) Definition. An operad $\mathcal{O}$ over C, or C-operad, consists of

a) a set of complexes $\mathcal{O}(\varphi) = \mathcal{O}(\varphi_1, \ldots, \varphi_n) \in C(k)$ given for every $n \in \mathbb{N}$, $\varphi = (\varphi_1, \ldots, \varphi_n) \in (C/x)^n$, $x \in \mathrm{Ob}\, C$;

b) morphisms of multiplication: for $\varphi = (\varphi_1, \ldots, \varphi_n) \in (C/x)^n$, $\psi_i = (\psi_{i1}, \ldots, \psi_{im_i}) \in (C/s\varphi_i)^{m_i}$, $i = 1, \ldots, n$,

$$\gamma_{\varphi,\psi} : \mathcal{O}(\varphi) \otimes \mathcal{O}(\psi) \longrightarrow \mathcal{O}(\varphi\psi)$$

here $\varphi\psi := (\varphi_1\psi_{11}, \ldots, \varphi_1\psi_{1m_1}, \ldots, \varphi_n\psi_{nm_n})$,

$\mathcal{O}(\psi) := \mathcal{O}(\psi_1) \otimes \ldots \otimes \mathcal{O}(\psi_n)$ (see fig. 1).

c) unities: elements $e_x \in \mathcal{O}(\mathrm{id}(x))^0$ given for every $x \in \mathrm{Ob}\, C$.

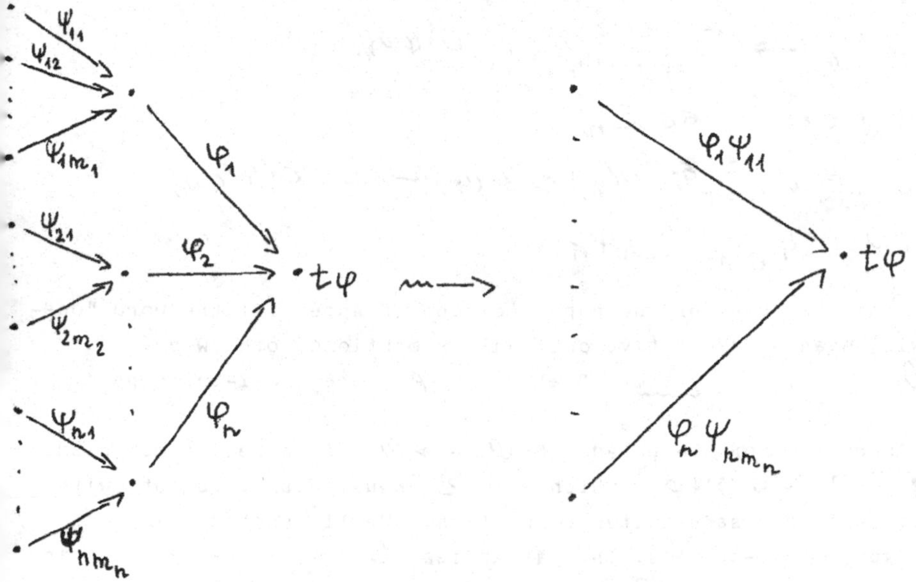

Fig. 1. Multiplication in a C-operad.

The following axioms must hold.

Ass) For composable $\varphi = (\varphi_i)$, $\psi = (\psi_{ij})$, $\chi = (\chi_{ijk})$

$$\gamma_{\varphi\psi,\chi} \cdot \gamma_{\varphi,\psi} = \gamma_{\varphi,\psi\chi} \circ \gamma_{\psi,\chi} : \mathcal{O}(\varphi) \otimes \mathcal{O}(\psi) \otimes \mathcal{O}(\chi) \longrightarrow \mathcal{O}(\varphi\psi\chi).$$

Un) For every $\varphi = (\varphi_1, \ldots, \varphi_n) \in \mathrm{Ob}\,(C/x)^n$

$$\mathrm{id}(\mathcal{O}(\varphi)) = \gamma_{\mathrm{id}(x),\varphi}\big|_{e_x \otimes \mathcal{O}(\varphi)} = \gamma_{\varphi,\mathrm{id}(s\varphi)}\big|_{\mathcal{O}(\varphi) \otimes e_{s\varphi}}$$

here $\mathrm{id}(s\varphi) = (\mathrm{id}(s\varphi_1), \ldots, \mathrm{id}(s\varphi_n))$, $e_s = e_{s\varphi_1} \otimes \ldots \otimes e_{s\varphi_n}$.

$\mathcal{O}$ is called commutative if, for every $\varphi \in (C/x)^n$ the (left) action f symmetric group Σ_n on n letters on $\mathcal{O}(\varphi)$ is given such that

Comm) for composable $\varphi = (\varphi_1, \ldots, \varphi_n)$, $\psi = (\psi_1, \ldots, \psi_n)$,

$\psi_i = (\psi_{i1}, \ldots, \psi_{im_i})$

(a) $\gamma_{\varphi,\psi}$ is $\Sigma_{m_1, \ldots, m_n} := \Sigma_{m_1} \times \ldots \times \Sigma_{m_m}$ -equivariant, where $\Sigma_{m_1, \ldots, m_n}$ acts trivially on $\mathcal{O}(\varphi)$, through the action of Σ_{m_i} on $\mathcal{O}(\psi_i)$, and through the inclusion

$\Sigma_{m_1, \ldots, m_n} \to \Sigma_{m_1+\ldots+m_n}$ on $\mathcal{O}(\varphi\psi)$.

(b) for every $\sigma \in \Sigma_n$

$$\sigma \cdot \gamma_{\varphi,\psi} = \gamma_{\varphi,\sigma\psi} \cdot \sigma : \mathcal{O}(\varphi) \otimes \mathcal{O}(\psi) \to \mathcal{O}(\sigma(\varphi\psi)),$$

where $\sigma\psi = (\psi_{\sigma(1)}, \ldots, \psi_{\sigma(n)})$.

In the main body of the paper (except of Appendix) the word "operad" will mean a commutative one, unless mentioned otherwise.

$\mathcal{O}$ is called <u>acyclic</u> if all $\mathcal{O}(\varphi)$ are quasi-isomorphic to k.

A morphism of C-operads $f: \mathcal{O} \to \mathcal{O}'$ is a set of morphisms $f(\varphi): \mathcal{O}(\varphi) \to \mathcal{O}'(\varphi)$ which are Σ_n-equivariant, commute with multiplication and send unites to unities. We'll denote by $\mathcal{O}p(C)$ the category of C-operads. The categories $\mathcal{O}p(C)$, C varying, form the category of operads $\mathcal{O}p$, fibered over Cat (the category of small categories). By $\mathcal{A}cop$ we'll denote its full fibered subcategory of acyclic operads.

If C = * (the final object in Cat) then C-operads will be called simply operads, cf. [May 1].

(1.2) <u>Examples</u>. 1. The trivial operad $I_C(\varphi) = k$ for all φ. A C-operad $\mathcal{O}$ endowed with a morphism $\varepsilon: \mathcal{O} \to I_C$ is called <u>augmented</u>.

2. Suppose one has a map A: Ob C $\to$ Ob C(k). For

$\varphi = (\varphi_1, \ldots, \varphi_n) \in \mathrm{Ob}\ (C/x)^n$ put

$$\mathrm{Op}(A)(\varphi) = \underline{\mathrm{Hom}}(A(s\varphi), A(x)),$$

where $A(s\varphi) := A(s\varphi_1) \otimes \ldots \otimes A(s\varphi_n)$. With the evident action of Σ_n's and amorphisms of multiplication, these complexes form the C-operad Op(A).

(1.3) Definition. 1. An algebra A over C-operad $\mathcal{O}$ is map A : Ob C $\longrightarrow$ Ob C(k) together with a morphism of C-operads $\mathcal{O} \to Op(A)$. In other words, to define on A a structure of -algebra, one must define morphisms

$$\mu_\varphi : A(s\varphi) \otimes \mathcal{O}(\varphi) \longrightarrow A(t\varphi),$$

for all $\varphi = (\varphi_1, \dots, \varphi_n) \in Ob\,(C/x)^n$. The maps μ_φ must satisfy certain evident properties.

A morphism of $\mathcal{O}$-algebras $f : A \to A'$ is a set of morphisms $f(x) : A(x) \to A'(x)$, $x \in Ob\,C$, which commute with μ_φ's. $\mathcal{O}$-alg(C) will denote the category of $\mathcal{O}$-algebras.

2. A May algebra over C is a pair $(\mathcal{O}, A)$, where $\mathcal{O}$ is an acyclic augmented C-operad, and A is an $\mathcal{O}$-algebra. A May algebra over * is called simply May algebra. May algebras form a category $\mathcal{M}ay$ which is fibered over $\mathcal{A}cop$ and Cat with fibers May(C) for C $\in$ Cat.

(1.4) Examples. Steenrod powers.

(1.4.1) An algebra over I_C just a function from C to the category Cdg.

(1.4.2) Let us consider the case C = *. If $A \in C(k)$ is a May algebra then it becomes a commutative algebra in the derived category D(k), cf. [May 1]. In particular, $H^*(A) \in$ Cdg. But the structure of May algebra gives more.

Suppose for example that k is a field of characteristics $p > 0$. Then one can introduce the action of "Steenrod powers" on $H^*(A)$ which measure the deviation of A from being strictly commutative. Namely, let a $H^q(A)$ be represented by a cycle $u \in Z^q(A)$. Consider an element $u^{\otimes p} = u \otimes \dots \otimes u$ (p times) $\in Z^{pq}(A^{\otimes p})$. It is a $\mathbb{Z}/p$-equivariant cocycle, and therefore represents an element in $H^{pq}(\mathbb{Z}/p, A^{\otimes p})$.

Augmentation $\varepsilon : \mathcal{O} \longrightarrow I$ induces isomorphism

$$\varepsilon_* : H^*(\mathbb{Z}/p, \mathcal{O}(p) \otimes A^{\otimes p}) \text{ --- } H^*(\mathbb{Z}/p, A^{\otimes p}).$$

Consider the image of $\varepsilon_*^{-1}(u^{\otimes p})$ under the morphism

$$\mu_* : H^*(\mathbb{Z}/p, \mathcal{O}(p) \otimes A^{\otimes p}) \longrightarrow H^*(\mathbb{Z}/p, A).$$

As the action of $\mathbb{Z}/p$ on A is trivial, one has by Künneth

$$H^*(\mathbb{Z}/p, A) = H^*(\mathbb{Z}/p) \otimes H^*(A) = \sum_{n \in \mathbb{Z}} H^{*-2n}(A).$$

The non-zero components of $\mu_* \mathcal{E}_*^{-1}(u^{\otimes p})$ in groups $H^{pq-2n}(A)$, $n \in \mathbb{Z}$, are just Steenrod powers of a and their "compositions with Bokstein". They satisfy the Adem relations (cf. [May 2]).

(1.5) Pseudo-functors (cf. [HS]).

(1.5.1) Definition. A coefficient system E over C is a set of complexes $E(\varphi)$ given for every $\varphi \in \mathrm{Mor}\, C$, unities $e_x \in E(id(x))^0$ given for every $x \in \mathrm{Ob}\, C$, and morphisms

$$\gamma_{\varphi,\psi} : E(\varphi) \times E(\psi) \longrightarrow E(\varphi\psi) \quad .$$

given for composable pairs (φ,ψ), which satisfy the axioms

(Ass) For composable (φ, ψ, χ)

$$\gamma_{\varphi\psi,\chi} \cdot \gamma_{\varphi,\psi} = \gamma_{\varphi,\psi\chi} \circ \gamma_{\psi,\chi}$$

(Un) For every $\varphi \in \mathrm{Mor}\, C$

$$\gamma_{id(s\varphi),\varphi}\big|_{e_s \otimes E(\varphi)} = \gamma_{\varphi,id(t\varphi)}\big|_{E(\varphi) \otimes e_{t\varphi}} = id(E(\varphi)).$$

By analogy with (1.1) one defines categories of coefficient systems $Cs(C)$, Cs. One has evident functor

$$\# : \mathcal{O}_p \longrightarrow Cs$$

A pseudofunctor $F : C \rightsquigarrow C(k)$ over a c.s. E is a map $F : \mathrm{Ob}(C) \to \mathrm{Ob}(C(k))$ together with a morphism of coefficient systems $E \to (\mathrm{Op}(F))^{\#}$.

By analogy with (1.1) one defines categories of pseudofunctors $Pf(E,C)$, $Pf(C)$, Pf.

We'll denote also by $\#$ the evident functor

$$May(\mathcal{O}, C) \longrightarrow Pf(\mathcal{O}, C).$$

(1.5.2) Let $\mathcal{N}C$ denote the category of simplices of NC. Its objects are simplices of NC, or, which is the same, functors $\sigma : \{n\} \to C$, where $\{n\}$ is the category $(0 \to 1 \to \ldots \to n)$, $n \in \mathbb{N}$. Morphism from $\sigma : \{n\} \to C$ to $\tau : \{m\} \to C$ is a functor $f : \{n\} \to \{m\}$ such that $\sigma = \tau \circ f$.

Let $E \in Cs(C)$ and suppose E to be augmented, i.e. endowed with a morphism $\mathcal{E} : E \longrightarrow (I_C)^{\#}$.

Every $F \in Pf(E,C)$ defines the functor

$$\mathcal{N}F : \mathcal{N}C \longrightarrow C(k)$$

by the formula

$$\mathcal{N}F(\sigma) = \underline{\mathrm{Hom}}_k(E_\sigma, F(t\sigma)),$$

where for $\sigma:(x_0 \xrightarrow[f_1]{} \dots \xrightarrow[f_n]{} x_n)$ $t\sigma := x_n$,

$E_\sigma := E_{f_n} \otimes \dots \otimes E_{f_1}$. The action of $\mathcal{N}F$ on morphisms is induced in the usual way by multiplication in E, action of E on F, unities in E and augmentation of E (cf. [HS]).

Let $NF \in \Delta C(k)$ denote the "cochain complex" of $\mathcal{N}F$,

$$(NF)^n = \prod_{\dim \sigma = n} \mathcal{N}F(\sigma).$$

1.5.3) Definition. Let $\mathrm{Tot} : \Delta C(k) \longrightarrow C(k)$ denote the functor which sends $X \in \Delta C(k)$ to the simple complex associated with the double complex obtained from X by the normalization in the cosimplicial direction.

We put

$$\underleftarrow{\mathrm{holim}}\, F = \mathrm{Tot}(NF).$$

1.5.4) Example. Suppose that $E = (I_C)^\#$. Then F is just the functor $C \rightarrow C(k)$ and $\underleftarrow{\mathrm{holim}}\, F$ is the complex of normalized cochains of F.

1.5.5) If E is acyclic, i.e. $\varepsilon : E \rightarrow (I_C)^\#$ is a quasi-isomorphism, then one has the spectral sequence

$$E_2^{pq} = H^p(C, H^q(F)) \Longrightarrow H^{p+q}(\underleftarrow{\mathrm{holim}}\, F)$$

(note that $H^*(F)$ form a strict functor $C \rightarrow$ Ab.groups).

Now we can formulate our main result.

1.6) Theorem. Let (A, ω) be a May algebra over a connected category C. Then $\underleftarrow{\mathrm{holim}}\, A$ admits a structure of May algebra.

For a proof, see §3.

§2. Eilenberg-Zilber operad

2.1) Definition. Let $F : C \longrightarrow C(k)$ be a functor, $C \in \mathrm{Cat}$.

An operad of endomorphisms of F, End F, is the final object in the category of operads[1] $\mathcal{O}$ with natural (with respect to morphisms in C) action on all $F(x)$, $x \in \mathrm{Ob}\, C$.

More explicitly, End $F(n)$ is the complex whose p-component consists of natural transformations of degree p

$$gr_\bullet(F^{\otimes n}) \text{ --- } gr_\bullet F[p], \quad \text{where}$$

$gr : C(k)$ --- (Graded k-modules) is the forgetful functor, and $F^{\otimes n} : C \longrightarrow C(k)$ is defined by $F^{\otimes n}(x) = F(x) \otimes \ldots \otimes F(x)$ (n times).

(2.2) Definition. Let $C^* : (\Delta^\circ \mathcal{E}nsf)^\circ \longrightarrow C(k)$ be the functor of normalized k-valued cochains on the category of finite simplicial sets. The Eilenberg-Zilber operad $\mathcal{Z}$ (over k) is the operad of endomorphisms of C^*.

Thus, $\mathcal{Z}(n)$ may be called "the complex of n-ary cochain operations".

(2.3) Theorem. $\mathcal{Z}$ is acyclic. For every cosimplicial k-algebra A its normalization $\mathrm{Norm}(A) \in C^{\geq 0}(k)$ has a natural structure of $\mathcal{Z}$-algebra.

The proof will occupy the rest of the section.

(2.4) Dold-Puppe correspondence.

Let $\mathrm{Norm} : \Delta(k) \longrightarrow C^{\geq 0}(k)$

(resp., $\mathrm{Norm}^0 : \Delta^\circ(k) \longrightarrow C^{\leq 0}(k)$) denote functors of normalization. Let Y be a simplicial cosimplicial set with $Y^n_m = \mathrm{Hom}\,(m,n)$, $kY \in \Delta^\circ\Delta(k)$ be the free k-module on Y and put finally $Z = \mathrm{Norm}^0 Y \in C^{\leq 0}\Delta(k)$ (normalization of Y in simplicial direction).

(2.4.1) Proposition. For every $A \in \Delta(k)$

$$\mathrm{Norm}\, A = \mathrm{Hom}_{\Delta(k)}(Z,A).$$

Here Z is considered as the complex $\longrightarrow Z^i \longrightarrow Z^{i+1} \longrightarrow \ldots$ in $\Delta(k)$, and $\mathrm{Hom}_{\Delta(k)}(Z,A)$ denotes the complex

$$\longrightarrow \mathrm{Hom}_{\Delta(k)}(Z^{-i+1},A) \longrightarrow \mathrm{Hom}_{\Delta(k)}(Z^{-i},A) \longrightarrow \ldots$$

Proof. Straightforward. ▲

---

1) Not C-operads!

2.4.2) Exercise. Prove that for $B \in C^{\leq 0}(k)$ $\mathrm{Hom}_{C(k)}(Z,B) = PB$ here $P : C^{\leq 0}(k) \longrightarrow \Delta^0(k)$ is the Dold-Puppe inverse of Norm^0.

2.5) Construction of $\mathcal{Z}(n)$.

The category $\Delta(k)$ is endowed with tensor product $\boxtimes$: for $A,B \in \Delta(k)$ we put $(A \boxtimes B)^n = A^n \otimes B^n$, cofaces and codegeneracies acting diagonally. So, considering Z as a complex in $\Delta(k)$, we can form its n-th tensor power $Z^{\boxtimes n} \in C^{\leq 0}\,\Delta(k)$, and the complex of homomorphisms

$$\mathrm{Hom}_{\Delta(k)}(Z, Z^{\boxtimes n}) \in C(k).$$

These complexes form an operad: the multiplication is defined as composition

$$\mathrm{Hom}(Z,Z^{\boxtimes n}) \otimes \mathrm{Hom}(Z,Z^{\boxtimes k_1}) \otimes \ldots \otimes \mathrm{Hom}(Z,Z^{\boxtimes k_n}) \longrightarrow$$

$$\longrightarrow \mathrm{Hom}(Z,Z^{\boxtimes n}) \otimes \mathrm{Hom}(Z^{\boxtimes n}, Z^{\boxtimes \Sigma k_i}) \longrightarrow \mathrm{Hom}(Z,Z^{\boxtimes \Sigma k_i}),$$

Σ_n acts on $\mathrm{Hom}(Z,Z^{\boxtimes n})$ through the permutation of factors in $Z^{\boxtimes n}$, $e = \mathrm{id} \in \mathrm{Hom}(Z,Z)^0$.

2.5.1) Proposition. The above operad is isomorphic to the Eilenberg-Zilber operad $\mathcal{Z}$.

Proof. Left to the reader. (Hint: Pass to dual complexes $Z^* \in C^{\geq 0}\,\Delta^0(k)$ and use the equality

$$\mathrm{Hom}_{\Delta(k)}(Z,Z^{\boxtimes n}) = \mathrm{Hom}_{\Delta^0(k)}(Z^{*\boxtimes n}, Z^*)).\ \blacktriangle$$

Now let A be a cosimplicial k-algebra. We have natural morphisms

$$\mathrm{Hom}(Z,Z^{\boxtimes n}) \otimes (\mathrm{Norm}\ A)^{\otimes n} = \mathrm{Hom}(Z,Z^{\boxtimes n}) \otimes \mathrm{Hom}(Z,A)^{\otimes n} \longrightarrow$$

$$\longrightarrow \mathrm{Hom}(Z,Z^{\boxtimes n}) \otimes \mathrm{Hom}(Z^{\boxtimes n}, A^{\boxtimes n}) \longrightarrow \mathrm{Hom}(Z,A^{\boxtimes n}) \longrightarrow$$

$$\longrightarrow \mathrm{Hom}(Z,A) = \mathrm{Norm}\ A.$$

They define the structure of $\mathcal{Z}$-algebra on Norm A.

2.6) It remains to prove the acyclicity of $\mathcal{Z}(n)$. The morphisms $n \longrightarrow 0$ in Δ induce objectwise quasi-isomorphism $\varepsilon: Z^{\boxtimes n} \longrightarrow k$, where k is considered as the trivial cosimplicial complex. Passing to the corresponding morphism of spectral sequences one obtains the required property.

Theorem (2.3) is proved. $\blacktriangle$

(2.7) Remark: The Alexander-Whitney product.
Formulas of Alexander-Whitney define a morphism of non-commutative operads $AW : I \longrightarrow \mathcal{Z}$ which provides Norm A ($A \in \Delta(k\text{-alg})$) with the structure of strictly associative algebra.

§3. Proof of main theorem

(3.1) Some reformulations of Tot; functors $\overleftarrow{\mathrm{hom}}$, $\overrightarrow{\mathrm{hom}}$, $\overleftarrow{\otimes}$, $\overrightarrow{\otimes}$.

(3.1.1) Definition. Let $C \in Cat$. A category $\mathcal{M}or_C$ is defined as follows:

$$\mathrm{Ob}\, \mathcal{M}or_C = \mathrm{Mor}\, C;$$

$$\mathrm{Hom}(f,g) = \left\{ \text{the commutative squares } \begin{array}{ccc} \bullet & \xrightarrow{f} & \bullet \\ \uparrow & & \downarrow \\ \bullet & \xrightarrow{g} & \bullet \end{array} \text{ in } C \right.$$

(pay attention to the direction of arrows!).

(3.1.2) Let A be an abelian category which is closed in the sense of MacLane, i.e. is endowed with a pair of adjoint functors $\otimes : A \times A \longrightarrow A$ and $\mathrm{hom}: A^0 \times A \longrightarrow A$ such that $\otimes$ turns A into a monoidal category.

For $X, Y : C \longrightarrow A$ define a functor

$$\mathrm{hom}(X,Y) : \mathcal{M}or_C \longrightarrow A$$

by formula $\mathrm{hom}(X,Y)(f) = \mathrm{hom}(X(sf), Y(tf))$. Analogously, for $X : C^0 \longrightarrow A$ and $Y : C \longrightarrow A$ define a functor

$$X \otimes Y : \mathcal{M}or_C \longrightarrow A$$

by $(X \otimes Y)(f) = X(sf) \otimes Y(tf)$.

Definition. For $X, Y : C \longrightarrow A$ put

$$\overleftarrow{\mathrm{hom}}(X,Y) = \overleftarrow{\lim}\, \mathrm{hom}(X,Y)$$

$$\overrightarrow{\mathrm{hom}}(X,Y) = \overrightarrow{\lim}\, \mathrm{hom}(X,Y).$$

Similarly, for $X : C^0 \longrightarrow A$, $Y : C \longrightarrow A$ we define $X \overleftarrow{\otimes} Y$ and $X \overrightarrow{\otimes} Y$.

(3.1.3) Proposition. For $X \in \Delta C(k)$

$$\mathrm{Tot}(X) = \overleftarrow{\mathrm{hom}}(Z, X) = Z^* \overleftarrow{\otimes} X,$$

where $Z^* = \mathrm{Hom}_k(Z,k) \in \Delta^0 C^{\geq 0}(k)$ (cf. (1.5.3)). Here we consider X and Z as functors $\Delta \longrightarrow C(k)$. ▲

(3.2) Proof of (1.6), the first step.

Here we define an operad ω over $\mathcal{N}C$ (cf. (1.5.2)) such that $\mathcal{N}A : \mathcal{N}C \longrightarrow C(k)$ is an ω-algebra.

Let $\alpha_i : \sigma_i \longrightarrow \sigma$, $i = 1, \ldots, r$, be a collection of arrows in $\mathcal{N}C$. We put

(3.2.1) $$\omega(\alpha_1, \ldots, \alpha_r) = \underline{\mathrm{Hom}}_k(\mathcal{O}(\sigma), \bigotimes_i \mathcal{O}(\sigma_i) \otimes$$
$$\otimes \mathcal{O}(\mathrm{target}(\alpha_1), \ldots, \mathrm{target}(\alpha_r))),$$

where target : $\mathcal{N}C \longrightarrow C$ sends σ to $t\sigma$, and so target(α_i) is a morphism from $t\sigma_i$ to $t\sigma$.

$\mathcal{O}(\sigma)$ is, as in the definition of $\mathcal{N}A$,(1.5.2), a tensor product of $\mathcal{O}(\varphi_i)$, φ_i runs through the morphisms belonging to σ.

The associative multiplication, the symmetric group action and unities $e_\sigma \in \omega(id)$ are defined in the obvious way.

Remark. ω is not augmented.

$\mathcal{N}A : \mathcal{N}C \longrightarrow C(k)$ is an ω-algebra: to define multiplication

$$\mu : \omega(\alpha_1, \ldots, \alpha_r) \otimes \mathcal{N}A(\sigma_1) \ldots \otimes \mathcal{N}A(\sigma_r) \longrightarrow \mathcal{N}A(\sigma)$$

we put $\mu(o \otimes f_1 \otimes \ldots \otimes f_r)$ to be the composition

$$\mathcal{O}(\sigma) \xrightarrow{o} \bigotimes_i \mathcal{O}(\sigma_i) \otimes \mathcal{O}(\mathrm{target}(\alpha_1), \ldots, \mathrm{target}(\alpha_r)) \xrightarrow{f_1 \otimes \ldots \otimes f_r}$$

$$\rightarrow \mathcal{O}(\mathrm{target}(\alpha_1), \ldots, \mathrm{target}(\alpha_r)) \otimes A_{t\sigma_1} \otimes \ldots \otimes A_{t\sigma_r} \xrightarrow{\mu} A_{t\sigma}.$$

(3.2.2) Remark. One could ask why $\mathcal{N}A$, being an ω-algebra, and so, via $\#$, a pseudo-functor, is in fact a genuine functor. This is because there is a morphism $I^{\#} \longrightarrow \omega^{\#}$ of the coefficient systems. This morphism is given by the maps

$$\mathcal{O}(\sigma) \longrightarrow \mathcal{O}(\sigma_1) \otimes \mathcal{O}(\mathrm{target}(\alpha)), \alpha : \sigma_1 \longrightarrow \sigma \in \mathcal{N}C,$$

which are obtained by a composition and tensor product of multiplications μ and unities e_x.

Definition. An operad endowed with a quasi-isomorphism $I^{\#} \longrightarrow \omega^{\#}$, will be called special operad.

(3.3) Reduction to cosimplicial objects.

(3.3.1.) Put for $(\varphi_1, \ldots, \varphi_r) \in \mathrm{Mor}\Delta$, $t\varphi_i = n$,

$$\Omega(\varphi_1, \ldots, \varphi_r) = \prod_{\sigma \in \mathcal{N}_n C} \omega((\sigma, \varphi_1), \ldots, (\sigma, \varphi_r)).$$

Ω is a (non-acyclic) operad over Δ, and NA (cf. (1.5.2)) is an Ω-algebra.

(3.3.2) One has $H(\Omega(\varphi_1,\dots,\varphi_r)) = \prod_{\sigma\in N_n C} H(\omega((\sigma,\varphi_1),\dots,(\sigma,\varphi_r))) =$

$= \prod k\cdot\langle\sigma\rangle$, $\langle\sigma\rangle$ being a free generator of degree 0.

So, $H(\Omega(\varphi_1,\dots,\varphi_r)) = C^n(C,k)$, a k-module of n-cochains of C with coefficients in k, $n = t\varphi$.

(3.4) Now we have the following data:

(a) a special Δ-operad Ω;

(b) an Ω-algebra $X = NA \in C(k)$.

We shall build an operad which turns Tot(X) into a May algebra.

(3.4.1) For $n, r \in \mathbb{N}$ we denote by $\Delta_r(n)$ the category $\underbrace{(\Delta/n)^o \times \dots \times (\Delta/n)^o}_{r \text{ times}}$.

Every special Δ-operad Ω defines a functor

$$\Omega(r)^n : \Delta_r(n) \longrightarrow C(k),$$

$\Omega(r)(\varphi_1,\dots,\varphi_r) := \Omega(\varphi_1,\dots,\varphi_r) \in C(k)$,

for $\alpha = (\alpha_1,\dots,\alpha_r) : \varphi \to \psi$ $\Omega(r)(\alpha_1,\dots,\alpha_r)$ being a composition

$$\Omega(\varphi_1,\dots,\varphi_r) \to \Omega(\varphi_1,\dots,\varphi_r) \otimes \Omega(\alpha_1) \otimes \dots \otimes \Omega(\alpha_r) \longrightarrow$$

$\longrightarrow \Omega(\psi_1,\dots,\psi_r)$ (the first morphism is induced by the section $I \longrightarrow \Omega$).

Consider now $Z^{\otimes r}$ as a functor from $(\Delta/n)^r = \Delta_r(n)^o$ to $C(k)$, $Z^{\otimes r}(\varphi_1,\dots,\varphi_r) = Z(s\varphi_1) \otimes \dots \otimes Z(s\varphi_r)$, and define $\Omega(r)^n \underset{\rightarrow}{\otimes} Z^{\otimes r}$ as in (3.1.2). We claim that

(a) $\Omega(r)^n \otimes Z^{\otimes r}$ form a cosimplicial complex, n running over Δ;

(b) The formula

$$E(r) = \hom(Z, \Omega(r) \underset{\rightarrow}{\otimes} Z^{\otimes r})$$

gives the required operad.

The assertion (a) is evident.

(3.4.2) In order to prove (b) we need the following evident

<u>Fact</u>. The Ω-algebra structure on $X \in \Delta C(k)$ is given by a morphism

$$\Omega(r) \underset{\rightarrow}{\otimes} X^{\otimes r} \longrightarrow X \quad \text{in} \quad \Delta C(k)$$

ith obvious associativity properties, actions of symmetric groups, tc.

Now the action of $E(r)$ on $\mathrm{Tot}(X)^{\otimes r}$ is defined to be the composition

3.4.3)

$$\mathrm{hom}(Z,\Omega(r)\underset{\rightarrow}{\otimes} Z^{\otimes r})\otimes \mathrm{Tot}(X)^{\otimes r} = \underset{\leftarrow}{\mathrm{hom}}(Z,\Omega(r)\underset{\rightarrow}{\otimes} Z^{\otimes r})\otimes \mathrm{hom}(Z,X)^{\otimes r} \longrightarrow$$

$$\to \underset{\leftarrow}{\mathrm{hom}}(Z,\Omega(r)\underset{\rightarrow}{\otimes} Z^{\otimes r})\otimes \mathrm{hom}_{\Delta^r}(Z^{\otimes r},X^{\otimes r})\longrightarrow \underset{\leftarrow}{\mathrm{hom}}(Z,\Omega(r)\underset{\rightarrow}{\otimes} Z^{\otimes r})\otimes$$

$$\underset{\leftarrow}{\mathrm{hom}}(\Omega(r)\underset{\rightarrow}{\otimes} Z^{\otimes r},\ \Omega(r)\underset{\rightarrow}{\otimes} X^{\otimes r})\longrightarrow \underset{\leftarrow}{\mathrm{hom}}(Z,\Omega(r)\underset{\rightarrow}{\otimes} X^{\otimes r})\longrightarrow \mathrm{hom}(Z,X) =$$

$\mathrm{Tot}(X)$.

The operad multiplication on $E(r)$'s is defined similarly.

3.4.4) It remains to prove the acyclicity of E. For this we need n explicit computation of $\Omega(r)\underset{\rightarrow}{\otimes} Z^{\otimes r}$.

3.4.4.1) <u>Proposition</u>. $n \in \mathrm{Ob}\,\Delta = \mathbb{N}$ being fixed, $(\Omega(r)\underset{\rightarrow}{\otimes} Z^{\otimes r})_n$ s equal to

$$\bigoplus_{\substack{\varphi = (\varphi_1,\dots,\varphi_r)\\ t\varphi_i = n\\ \varphi_i \text{ are injective}}} \Omega(\varphi_1,\dots,\varphi_r)[-|\varphi|],\quad |\varphi| = \sum_i s\varphi_i .$$

3.4.4.2) <u>Corollary</u>. $I(r)\underset{\rightarrow}{\otimes} Z^{\otimes r} = Z^{\otimes r}$.

3.4.4.3) <u>Corollary</u>. If $\mathcal{O} \longrightarrow \mathcal{O}'$ is an objectwise quasi-isomorphism f functors $\Delta^{o} \longrightarrow C(k)$ then

$$\mathcal{O}\underset{\rightarrow}{\otimes} Z^{\otimes r} \longrightarrow \mathcal{O}'\underset{\rightarrow}{\otimes} Z^{\otimes r}$$

s also a quasi-isomorphism.

<u>Proof</u>. It suffices to consider a spectral sequence associated ith the filtration of

$$\bigoplus_{\varphi} \mathcal{O}(\varphi_1,\dots,\varphi_r)[-|\varphi|]$$

y $|\varphi|$. (This filtration is finite on every diagonal.)

3.4.3.3.) <u>Corollary</u>. $\Omega(r)\underset{\rightarrow}{\otimes} Z^{\otimes r}$ is quasi-isomorphic to the cosimplicial complex of cochains of C with coefficients in k.

<u>Proof</u>. Consider an operad $\tau_{\leq 0}\Omega = \{\tau_{\leq 0}\Omega(r)\}$. One has the ollowing pair of quasi-isomorphisms

$$\begin{array}{ccc} \tau_{\leq 0}(\Omega(r)) & \longrightarrow & \Omega(r) \\ \downarrow & & \\ H(\tau_{\leq 0}(\Omega(r))) = H(\Omega(r)) = C^*(C,k). & & \end{array}$$

So $\Omega(r) \underset{\rightarrow}{\otimes} Z^{\otimes r}$ is quasi-isomorphic to $C^*(C,k) \otimes (I(r) \underset{\rightarrow}{\otimes} Z^{\otimes r}) =$
$= C^*(C,k) \otimes Z^{\otimes r} \xrightarrow{\simeq} C^*(C,k)$. ▲

Returning to the conditions of (1.6), we see that $\underleftarrow{\mathrm{holim}}\, A^{\#}$ is an algebra over E and, therefore, over $\tau_{\leq 0}E$. But by (3.4.3.3) $\tau_{\leq 0}E$ is acyclic if NC is connected. This completes the proof of (1.6). ▲

§4. Thom-Sullivan cochains

Throughout this section we suppose $k \supset \mathbb{Q}$ (the field of rational numbers).

(4.1) Theorem. Let $A: C \longrightarrow Cdg(k)$ be a functor from a small category C to commutative DG algebras over k. Then there exists an algebra $Th(A) \in Cdg(k)$ and a natural quasi-isomorphism

$$\int : Th(A) \longrightarrow \underleftarrow{\mathrm{holim}}\, A,$$

compatible with the algebra structures on $Th(A)$ and $\underleftarrow{\mathrm{holim}}\, A$ in the derived category $D(k)$.

(4.2) Definition. Let $\Omega : \Delta^{o} \longrightarrow Cdg(k)$ be defined as follows. $\Omega(p)$ = the algebra of polynomial k-valued differential forms on the standard p-simplex Δ_p (cf. [BG]).

Then if $A : C \longrightarrow Cdg(k)$ is a functor, we put

$$Th(A) = \Omega \underset{\leftarrow}{\otimes} NA.$$

$Th(A)$ is obviously a commutative algebra as inverse limit of commutative algebras.

(4.3) We point out two properties of $\underset{\leftarrow}{\otimes}$ which imply (4.1).

(4.3.1) Let $f : U \longrightarrow V : C^{o} \longrightarrow C(k)$ be a homotopy equivalence, i.e. $f_n : U_n \longrightarrow V_n$ are homotopy equivalences and respective homotopy inverses and chain homotopies are functorial. Then for $X : C \longrightarrow C(k)$ $f_* : U \underset{\leftarrow}{\otimes} X \longrightarrow V \underset{\leftarrow}{\otimes} X$ is also a homotopy equivalence.

Proof. This is because $\lim$ commutes with tensoring by the "interval" $I = (k \longrightarrow k \oplus k)$.

(4.3.2) Let $\mathcal{O}$ be an operad and U be a simplicial $\mathcal{O}$-algebra. Then, if $X \in Cdg(k)$. Then $U \underset{\leftarrow}{\otimes} X$ has a natural structure of $\mathcal{O}$-algebra.

Proof. Straightforward. ▲

4.4) Proof of (4.1).

4.4.1) Morphism $\int : \Omega \longrightarrow Z^*$.

Recall that $Z^* \in \Delta^o C^{\geq 0}(k)$ consists of normalized cochains com-
lexes of standard simplices (cf. (3.1.3)).

Let $\omega \in \Omega_n$. We define $\int\omega \in Z^*_n$ as follows.
or $\alpha : p \longrightarrow n$, $\alpha \in X^n_p$, we put $(\int\omega)\alpha = \int_{\Delta_p} \alpha^*\omega$, where
$\alpha^* : \Omega_n \longrightarrow \Omega_p$ is induced by simplicial structure of Ω. The stan-
ard arguments (cf. [BG, §2]) show this is well-defined morphism in
$^oC(k)$. It is a homotopy equivalence (loc. cit), hense, by (4.3.1), it
nduces a homotopy equivalence

$$\int : Th(A) \longrightarrow Tot(N(A)) = \underleftarrow{holim}\, A.$$

4.4.2) Compatibility of $\int$ with multiplication follows from the
ore general

Proposition. There exists a non-commutative acyclic operad Ch
ver $C = (0 \longrightarrow 1)$ such that $Ch_0 = Ch_1 = I$, and a Ch-algebra B
uch that $B|_0 = Th(A)$, $B|_1 = \underleftarrow{holim}\, A$ with Alexander-Whitney multi-
lication, (2.7).

Proof. See (A1). ▲

Theorem (4.1) is proved. ▲

4.4.3) Remark. It is likely that there exists a May algebra over
$\longrightarrow 1$, which, restricted to 0, is Th(A), and to 1, is
$\underleftarrow{olim}$ A over $\mathbb{Z}$.

Appendix: "Universal" operads.

A0. Cubic sets.

A0.1) Let $\square$ denote the subcategory of the category of topological
paces whose objects are cubes

$$\square^n = \{(x_1,\dots,x_n) \in \mathbb{R}^n \mid 0 \leq x_i \leq 1 \text{ for all } i\},\quad n \in \mathbb{N},$$

nd morphisms are compositions of maps ("cofaces")

$$\partial'^i, \partial''^i : \square^{n-1} \longrightarrow \square^n,$$

$$\partial'^i(x_1,\dots,x_{n-1}) = (x_1,\dots,x_{i-1},0,x_i,\dots,x_n),$$

$$\partial''^i(x_1,\dots,x_{n-1}) = (x_1,\dots,x_{i-1},1,x_i,\dots,x_n),$$

$1 \leqslant i \leqslant n$.

Definition. A cubic object in a category C is a contravariant functor from $\square$ to C.

Thus, a cubic object

$$X : X_0 \leftleftarrows X_1 \overset{\longleftarrow}{\underset{\longleftarrow}{\leftleftarrows}} X_2$$

is a set of objects $\{X_n\}$, $n \in \mathbb{N}$, endowed with morphisms

$$d_i', d_i'' : X_n \longrightarrow X_{n-1}, \quad i = 1,\ldots,n,$$

which satisfy the relations

(A0.1.1) $\quad d_i^k d_j^l = d_{j-1}^l d_i^k$ for $i < j$, $k, l = '$ or $''$.

We denote the category of cubic objects in C by $\square^\circ C$.

(A0.1.2) Remark. To define the "right" notion of cubic object one has to add to the category the codegeneracies

$$\sigma^i : \square^n \longrightarrow \square^{n-1}, \; (x_1,\ldots,x_n) \longrightarrow (x_1,\ldots,\hat{x}_i,\ldots,x_n),$$

$i = 1,\ldots,n$. But we shall not need degeracies in the sequel.

(A0.1.3) In $\square^\circ \mathcal{E}ns$ one has standard cubes $\square^n$-functors represented by $\square^n$, $n \in \mathbb{N}$.

For $X \in \square^\circ \mathcal{E}ns$ its geometric realization is the cellular complex

$$|X| = \coprod_n X_n \times \square^n / \sim$$

where $X_n \times \square^n$ is the disjoint sum of $\#X_n$ standard cubes $\square^n$ and the equivalence relation $\sim$ is generated by $(f^*x, y) \sim (x, f_*y)$ for $x \in X_n$, $y \in \square^m$, $f : \square^m \longrightarrow \square^n$.

(A0.1.4) For $X, Y \in \square^\circ \mathcal{E}ns$ we define their product $X \times Y$ by

$$(X \times Y)_n = \coprod_{p+q=n} X_p \times Y_p,$$

$$d_i^k(x,y) = \begin{cases} (d_i^k x, y) & \text{for } i \leq p, \\ (x, d_{i-p}^k y) & \text{for } i > p, \end{cases}$$

where, as above, $k = '$ or $''$.

One has canonically $|X \times Y| \cong |X| \times |Y|$.

AO.2) Let A be an additive category. The functor of chains

$$Ch : \square^{\circ}A \longrightarrow C^{\leqslant 0}(A)$$

s defined as follows.

For $X \in \square^{\circ}A$ we put $Ch(X)_n = Ch(X)^{-n} = X_n$,

$$d = \sum_{i=1}^{n} (-1)^i (d_i' - d_i'') : Ch(X)_n \longrightarrow Ch(X)_{n-1}.$$

he relations (AO.1.1) imply $d^2 = 0$.

Let $X \in \square^{\circ}\mathcal{E}ns$. The complex of chains of X with coefficients n k is the complex $C_*(X,k) = Ch(kX)$, where $kX \in \square^{\circ}(k\text{-mod})$ is he free k-module on X.

One has canonically

$$C_*(X \times Y,k) \cong C_*(X,k) \otimes C_*(Y,k).$$

A1. Sugawara operad and Chen integrals

A.1.1) Definition (cf. [BG, (3.3)]). Let A,B be DG k-algebras. he Sugawara morphism $A \rightsquigarrow B$ is a sequence of morphisms of graded k-modules

$$\varphi_p : A^{\otimes p} \longrightarrow B[-p+1]$$

efined for all integers $p \geqslant 1$, satisfying the relations

$$(*) \quad D\varphi_p(x_1 \otimes \dots \otimes x_p) = \sum_{j=1}^{p-1} (-1)^{j+1} (\varphi_{p-1}(x_1 \otimes \dots \otimes x_j x_{j+1} \otimes$$

$$\otimes \dots \otimes x_p) - \varphi_j(x_1 \otimes \dots \otimes x_j)\, \varphi_{p-j}(x_{j+1} \otimes \dots \otimes x_p)),$$

here

$$D\varphi_p := d_B \varphi_p + (-1)^p \varphi_p d_A \quad (\text{we put } \varphi_0 = 0).$$

Thus, $\varphi_1 : A \longrightarrow B$ is a morphism of complexes, $\varphi_2 : A^{\otimes 2} \longrightarrow B$ a homotopy which connects $\varphi_1 \circ \mu_A$ with $\mu_B \cdot \varphi_1$, where μ_A and are multiplications in A and B, and so on.

1.2) Remark. Suppose that A,B are augmented over k and $\overline{A}$, $\overline{B}$ e their augmentation ideals. Then to give a system of morphisms

$$\varphi_p : \overline{A}^{\otimes p} \longrightarrow \overline{B}$$

tisfying the relations (*) is the same thing as to give a morphism

$\varphi: B(A) \longrightarrow B(B)$ of DG k-coalgebras, where $B(A)$, $B(B)$ are two-sided bar-constructions $B(k,A,k)$ and $B(k,B,k)$.

(A1.3) Now we shall define an operad S over $C = (0 \xrightarrow[f]{} 1)$ such that

(a) $S|_0 = S|_1 = I$;

(b) an S-algebra A is just a Sugawara morphism from $A(0)$ to $A(1)$, cf. [BV, 1.15].

For $(f_1,\ldots,f_n) \in (C/1)^n$ we put

$$S(f_1,\ldots,f_n) = C_*(\square^{p-1},k),$$

if exactly p of the f_i are equal to f (and others to $id(1)$).

Proposition. There is a unique collection of maps $S(f) \otimes S(g) \longrightarrow S(fg)$ which turns S into an operad, if it is required that an S-algebra A is uniquely determined by maps

$$\varphi_p = \mu_p(e_p, \_) : A(0)^{\otimes p} \longrightarrow A(1)[-p+1],$$

satisfying (*), where

$$\mu_p : S(\underbrace{f,\ldots,f}_{p \text{ times}}) \otimes A(0)^{\otimes p} \longrightarrow A(1)[-p+1], \quad \text{and}$$

$e_p \in C_{p-1}(\square^{p-1},k)$ is the class of the highest dimensional cube.

Proof left to reader. The action of S is illustrated on Fig. A1.

$$\varphi_1(xy) \xrightarrow{\varphi_2} \varphi_1(x)\varphi_1(y)$$

$$\begin{array}{ccc} \varphi_1(xyz) & \xrightarrow{\varphi_2} & \varphi_1(x)\varphi_1(yz) \\ \downarrow{\scriptstyle\varphi_2} & \varphi_3 & \downarrow{\scriptstyle\varphi_2} \\ \varphi_1(xy)\varphi_1(z) & \longrightarrow & \varphi_1(x)\varphi_1(y)\varphi_1(z) \end{array}$$

Fig. A1.

(A1.4) Proposition. The Sugawara operad has the following universal property. Given any acyclic operad $\mathcal{O}$ over C with $\mathcal{O}|_0 = \mathcal{O}|_1 = I$, there exists a morphism of operads $S \longrightarrow \mathcal{O}$ (compare with (A2.5), (A4.3)). ▲

(A1.5) Example. For $X \in \Delta^\circ\mathcal{E}ns$ let $A(0)$ be the DeRham-Thom-Sullivan complex of X with coefficients in k, $A(1) = C^*(X,k)$, $\varphi_1 = \int : A(0) \longrightarrow A(1)$.

These data can be extended to an S-algebra, see [BG, prop. 3.3]. Maps $\varphi_p|_{(A(0)^1)^{\otimes p}}$ are just Chen iterated integrals.

A2. "Right" simplices in C(k) and "universal" coefficient system

A2.0) NB ! In this section we adopt the following convention. he maps in C(k) are written to the right from the objects. We do o to make formulas below more readable.

Thus, fg means the composition usually denoted by gf. Also f $X,Y \in C(k)$ and $f : X \longrightarrow Y[q]$ is a map of graded k-modules of egree $|f| = q$, we put

$$D(f) = d_X f + (-1)^{|f|+1} f d_Y .$$

ith this convention we have the usual Leibnitz formula

$$D(fg) = (Df)g + (-1)^{|f|} f\, D(g).$$

A2.1) Definition. A Sugawara n-simplex in C(k) consists of

a) a set of complexes $A_0, \ldots, A_n$ C(k);

b) a set of maps of graded k-modules

$$f(i_0, \ldots, i_k) : A_{i_0} \longrightarrow A_{i_k}[-k+1]$$

iven for every k, $1 \leqslant k \leqslant n$, and every $(k+1)$-tuple $(i_0, \ldots, i_k)$ uch that $0 \leqslant i_0 < \ldots < i_k \leqslant n$.

These maps must satisfy the following conditions

C0) $Df(i_0, i_1) = 0$ for all $i_0 < i_1$;

C1) if $k > 1$ then

$$Df(i_0, \ldots, i_k) = \sum_{j=1}^{k-1} (-1)^j \big(f(i_0, \ldots, \hat{i_j}, \ldots, i_k) - f(i_0, \ldots, i_j) f(i_j, \ldots, i_k)\big).$$

Thus, a Sugawara 0-simplex is an object $A \in C(k)$, 1-simplex s a map $f : A \longrightarrow B \in \mathrm{Mor}\, C(k)$, 2-simplex is a triangle

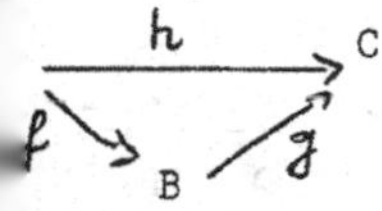

in C(k) with

fixed homotopy $h \simeq fg$, etc.

A2.2) The cone construction.

Every map $f : A \longrightarrow B$ defines the cone $C(f) \in \mathrm{Ob}\, C(k)$; every riangle (f,g,h) with a homotopy $h \simeq fg$ defines a map of cones

$Cf \longrightarrow Ch$. More generally, one has the following proposition which indicates that Sugawara simplices are "right" simplices in $C(k)$.

Proposition. Let $S = \{A_i,\ 0 \leq i \leq n,\ f(i_1,\dots,i_k)\}$ be a Sugawara n-simplex. Put for $1 \leq i \leq n$ $B_i = \text{Cone } f(0,i)$, and for $1 \leq i_1 < \dots < i_k \leq n$

$$g(i_1,\dots,i_k) : B_{i_1} \longrightarrow B_{i_k}[-k+2],$$

$$g(i_1,i_2)(a_0,a_1) = (a_0,\ a_0 f(0,i_1,i_2) + a_1 f(i_1,i_2));$$

$$g(i_1,\dots,i_k)(a_0,a_1) = (0,\ a_0 f(0,i_1,\dots,i_k) + a_1 f(i_1,\dots,i_k))$$

for $k > 2$, where $(a_0,a_1) \in B^p_{i_1} = A_0^{p+1} \oplus A^p_{i_1}$.

Then $\{B_i,\ g(i_1,\dots,i_k)\}$ form a Sugawara $(n-1)$-simplex (cf. Fig. A2).

Proof. Straightforward checking. ▲

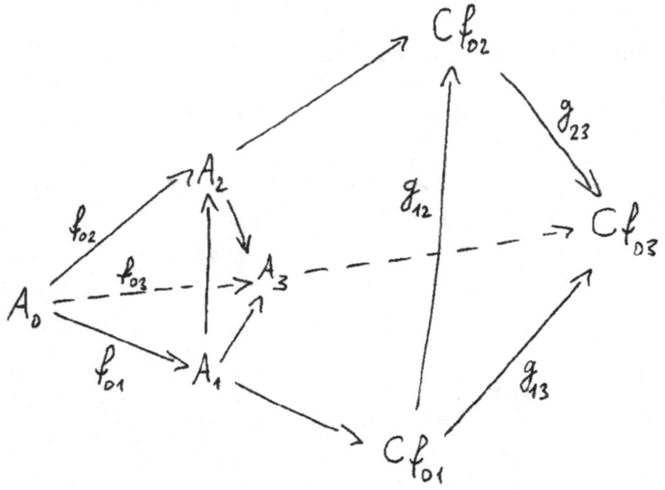

Fig. A2. A cone of a Sugawara 3-simplex.

(A2.3) Definition. Let $RNC(k)$ denote the simplicial set whose set of n-simplices $RN_nC(k)$ is a set of Sugawara n-simplices in $C(k)$, with evident faces and degeneracies. A Sugawara functor, or S-functor $F : C \rightsquigarrow C(k)$ from a small category C to $C(k)$ is a map in $\Delta^\circ \mathcal{E}ns$

$$NC \longrightarrow RNC(k).$$

(A2.4) Let C be a small category. For $\varphi \in \text{Mor } C$ define a cubic set $Q(\varphi)$ as follows.

Its p-cubes are n-simplices

$$\sigma: x_0 \xrightarrow{\varphi_1} \dots \xrightarrow{\varphi_{n-1}} x_n$$

with (n-1-p) distinguished internal vertices

$x_{i_1}, \dots, x_{i_{n-1-p}}$, $1 \leq i_1 < i_2 < \dots < i_{n-1-p} \leq n-1$,

and such that $\varphi = \varphi_1 \varphi_2 \dots \varphi_n$.

For $\sigma \in Q_p(\varphi)$ d'_j (resp., d''_j) is obtained from by omitting its j'th (from left to right) undistinguished vertex (resp., by making this vertex distinguished).

The concatenation induces maps

$$Q(\varphi) \times Q(\psi) \longrightarrow Q(\varphi\psi)$$

for composable (φ, ψ).

Put $R(\varphi) = C_*(Q(\varphi), k)$.

So we obtain a coefficient system R over C.

Let $\langle\varphi\rangle \in R_0(\varphi)$ denote the class of the trivial 1-simplex $\xrightarrow{\varphi}$.

A2.5) Proposition. (a) All $Q(\varphi)$ are contractible, and so R is acyclic.

(b) Given any acyclic c.s. E over C and a set of cycles $e_\varphi \in E_0(\varphi)$ representing the unique non-trivial cohomology classes, the map

$$\langle\varphi\rangle \longrightarrow e$$

may be extended to a map of c.s's $R \longrightarrow E$.

(c) Pseudo-functors $F : C \rightsquigarrow C(k)$ over R are in one-to-one correspondance with S-functors $C \rightsquigarrow C(k)$.

Proof. Only (a) is non-trivial. A more general fact will be proved in (A4.3a). ▲

A3. Free resolution of an associative algebra

A3.1) Let R be an associative k-algebra which is free as a k-module. Let us define a free associative graded k-algebra $\widehat{R}$ as follows. Put

$$\widehat{R} = \bigotimes_{n \geq 0} T(R^{\otimes n+1}),$$

where $R^{\otimes i} = R \otimes \ldots \otimes R$ (i times), $T(R^{\otimes i})$ is the tensor algebra (over k) of a k-module $R^{\otimes i}$.

We put for $x = x_0 \otimes \ldots \otimes x_n \in T^1(R^{\otimes n+1})$ deg $x = n$.

Now define the differential $d : \hat{R}_n \longrightarrow \hat{R}_{n-1}$ by the rule: for $x \in \hat{R}$ as above

$$dx = \sum_{i=0}^{n-1} (-1)^{i-1}(x_0 \otimes \ldots \otimes x_{i-1} \otimes x_i x_{i+1} \otimes \ldots \otimes x_n -$$

$$- (x_0 \otimes \ldots \otimes x_i)(x_{i+1} \otimes \ldots \otimes x_n)).$$

The associativity of R implies $d^2 = 0$ and $\hat{R}$ becomes a free DG k-algebra. We have also a natural augmentation $\varepsilon : \hat{R}_0 \longrightarrow R$, $x \longmapsto x$.

(A3.2) <u>Proposition</u>. ε induces isomorphism $H_0(\hat{R}) \longrightarrow R$. For $i > 0$ $H_i(\hat{R}) = 0$.

<u>Proof.</u> Note that if $R = k\,M$ where M is a monoid, then this is equivalent to (A2.5) (a) with $C = M$. The general case follows from considering the spectral sequence associated to filtration $\{F_i\hat{R}\}$,

$$F_i\hat{R} = \sum_{\sum m_k \leq i} (x_{11} \otimes \ldots \otimes x_{1m_1}) \cdot \ldots \cdot (x_{n1} \otimes \ldots \otimes x_{nm_n}). \blacktriangle$$

A4. <u>The "universal" C-operad.</u>

(A4.1) <u>Trees.</u>

(A4.1.1) Let us call <u>a tree</u> a finite poset T with the unique maximal element, tT, and such that for all $x, y \in T$ if $x < y$ then the poset $(x,y] = \{z \in T \mid x < z \leq y\}$ has the unique minimal element.

Put sT = the set of minimal elements of T, int $T = T - sT - tT$, ht(T) (height of T) = length of the longest chain in T.

<u>A pointed tree</u> is a triple (T - a tree, $P(T) \subset$ int T, σ - total order on int T). We do not claim any compatibility of σ with the partial order on int T induced by the structure of a tree.

(A4.1.2) Let $(T, P(T), \sigma)$ be a pointed tree, card $P(T) = n+1$. Using σ we shall numerate the elements of $P(T)$: $P(T) = \{v_0, \ldots, v_n\}$.

For $0 \leq i \leq n$ we define pointed trees $d_i'T$, $d_i''T$ as follows

$$d_i'T = (T, P(T) - \{v_i\}, \sigma),$$

$$d_i''T = (T - \{v_i\}, P(T) - \{v_i\}, \sigma).$$

A4.2) The construction of operad.

Let C be a category.

A4.2.1) A tree in C is a pair

A pointed tree $(T,P(T),\sigma)$; a functor $\tau : T \longrightarrow C$) (we regard poset T as a category).

The support of τ, supp τ, is a tree of height 1 in C obtained from τ by omitting all its internal vertices; thus, supp τ = $\{\varphi_1,\dots,\varphi_m\}$, $\varphi_i : \tau(x_i) \longrightarrow \tau(tT)$, $\{x_i\} = sT$.

A4.2.2) For $x \in \mathrm{Ob}\, C$, $\varphi = (\varphi_1,\dots,\varphi_n) \in (C/x)^n$ put

$$U_m(\varphi) = \{\text{a set of trees } ((T,P(T),\sigma),\ \tau : T \to C) \text{ with } \operatorname{supp}\tau = \varphi,\ \operatorname{card} P(T) = m\}.$$

he operators d_i', d_i'' from (A2.1.2) induce maps

$$d_i',\ d_i'' : U_m(\varphi) \longrightarrow U_{m-1}(\varphi),\quad 1 \leqslant i \leqslant m,$$

hich turn $U.(\varphi) = \{U_m(\varphi)\}_{m\geqslant 0}$ into a cubic set.

Further, a concatenation of trees induces maps

$$\mu : U(\varphi) \times U(\psi_1) \times \dots \times U(\psi_m) \longrightarrow U(\varphi\psi)$$

or composable (φ,ψ); also in every $U_0(id(x))$, $x \in \mathrm{Ob}\, C$, one has distinguished element $e_x = (x \xrightarrow[id]{} x)$.

Put $W(\varphi) = C_*(U(\varphi), k)$.

he arguments above imply these complexes form an operad $W = W(C)$.

A4.3) Proposition. (a) All of the $|U(\varphi)|$ are contractible; therefor, W is acyclic.

b) Given any acyclic C-operad $\mathcal{O}$ and a collection of cycles $z_\varphi \in \mathcal{O}(\varphi)^0$ which represent $1 \in H^0(\mathcal{O}(\varphi))$, there exists a map of C-operads $W \longrightarrow \mathcal{O}$ which sends the unique tree of height one $\langle\varphi\rangle \in U_0(\varphi)$ to z_φ.

Proof. (a) Put $V_m(\varphi) = \{x \in U_m(\varphi) \mid \langle\varphi\rangle \text{ is the vertex of } x\}$. hus, $x \in V_m(\varphi)$ iff $P(x) = \mathrm{int}\, x$. d_i'' induce operators $d_i'' : V_m(\varphi) \longrightarrow V_{m-1}(\varphi)$, $i = 1,\dots,m$. Further, for any $x \in U(\varphi)$, ut $\bar{x} = x$ as a tree, and $P(\bar{x}) = \mathrm{int}\, x$. Then $\bar{x} \in V.(\varphi)$ and every $x' \in V.(\varphi)$ such that x is a face of x', contains $\bar{x}$ as a face.

This easily implies that

$$(*)\quad |U.(\varphi)| = \coprod_m V_m(\varphi) \times \square^m / \{(d_i''x,y) \sim (x, \ _i'y)\}.$$

But the right hand side of (*) is evidently contractible.

b) Left to reader. ▲

References

[B] Beilinson, A. Higher regulators and values of L-functions (in Russian). In: Modern problems of mathematics, VINITI series, Moscow, v.24, p. 181-238.

[BG] Bousfield A.K., Gugenheim V.K.A.M. On PL DeRham theory and rational homotopy type, Memoirs AMS, t. 8, n°179, 1976.

[BV] Boardman, Vogt. Homotopy invariant structures on topological spaces. Lecture Notes in Math., 347, Springer, 1973.

[HS] Hinich V.A., Schechtman V.V. Geometry of a category of complexes and algebraic K-theory, to appear.

[May 1] May J. The geometry of iterated loop spaces. Lecture Notes in Math., 271, Springer, 1972.

[May 2] May J. A general algebraic approach to Steenrod operations, In: Steenrod algebra and its applications. Lecture Notes in Math., 168, Springer, 1970, p. 153-231.

[S] Smirnov V. On a cochain complex of topological spaces (in Russian). Mat. Sbornik, v. 115, n°1, 1981, p. 146-158.

On the delooping of Chern character and Adams operations

V.V. Schechtman

Abstract

The polysimplicial schemes representing the spectrum of algebrai K-theory are constructed. Their cohomology and K-theory is partially computed, with application to the construction of the delooping of Chern character and Adams operations. The interpretation of $ch|_{K_i}$ as a cohomology operation of higher order with respect to $ch|_{K_{i-1}}$ is given.

Table of Contents

Introduction

1. In [HS] we advocated the following point of view on algebraic

K-groups, which goes back to Waldhausen's definition using S-construction, [L], [G].

For an exact category $\mathcal{E}$

$K_0(\mathcal{E}) = \{$objects of $\mathcal{E}\} / \{$short exact sequences$\}$, or $\{$complexes of $\mathcal{E}\} / \{$exact triangles$\}$;

$K_1(\mathcal{E}) = \{$closed configurations of short exact sequences$\}$ / $\{$objects with three stage filtrations$\}$, or $\{$closed configurations of exact triangles$\}$ / $\{$exact octahedrons$\}$,

and so on.

In the present paper I wish to explain how characteristic classes enter in this picture. The heuristic answer is: "an additive characteristic class ch on K_i appears as an operation of the higher order with respect to the corresponding class on K_{i-1}". For example, $ch|_{K_1}$ is a kind of obstruction to the additivity of $ch|_{K_0}$ "on the level of cochains" (see (2.16)).

More precisely, we construct additive characteristic classes in terms of Waldhausen's definition of K-theory. In such a way we get the delooping of corresponding classes defined using +-construction, [G], [S].

We perform the construction in two steps.

At first, we construct in §1 a sequence of polysimplicial schemes*) $N\mathcal{B}^m P$, $m \geq 1$, together with maps φ^m: $\Sigma N\mathcal{B}^m P \longrightarrow N\mathcal{B}^{m+1} P$ (Σ denotes suspension) which may be called "the algebraic classifying spectrum of K-theory" in view of the following property.

For every local scheme X, the polysimplicial set of X-points of $N\mathcal{B}^m P$, $N\mathcal{B}^m P(X)$, is the m-fold delooping of the classifying space of K-theory of X, $K(X) = \Omega BQP(X)$. The delooping maps are just $\varphi^m(X)$. If X is not local then one must replace X by its Chech resolution X which is a pro-simplicial scheme over X consisting of the nerves of Zarissky coverings of X. The polysimplicial topological spaces of $\mathbb{C}$-points, $\mathcal{B}^m P(\mathbb{C})$ (resp., $\mathbb{R}$-points, $\mathcal{B}^m P(\mathbb{R})$) form the spectrum of (-1)-connected complex (resp., real) topological K-theory.

This result (Thm. (1.9)) is the main point of §1.

Secondly, we compute partially in §2 cohomology of $N\mathcal{B}^m P$'s.

Namely, we prove (Thm. (2.8)) that for every cohomology theory $H^*(\cdot, Z(*))$ satisfying the axioms of A. Beilinson, [B], (in particular, etale cohomology, Chow cohomology, Beilinson's absolute Hodge co-

*) Here the word "scheme" means "a kind of ringed space", and not "a kind of combinatorial object".

homology, etc.) the cohomology ring of $N\mathcal{B}^1 P$ is the exterior algebra in variables $c^1_{2i+1} \in H^{2i+1}(N\mathcal{B}^1 P, Z(i))$ over the cohomology of a base scheme.

In Thm (2.10) we compute $H^*(N\mathcal{B}^m P)$ for $m > 1$ modulo torsion. The cohomology classes of $N\mathcal{B}^\bullet P$ give by standart procedure the required deloopings of additive characteristic classes.

In §3 we give a sketch of the similar computation of algebraic K-theory of $N\mathcal{B}^\bullet P$. This provides the construction of m-fold delooping of additive operations in K-theory, modulo torsion for $m > 1$.

?. The schemes $N\mathcal{B}^m P$ are the nerves of the objects $\mathcal{B}^m P$. For $m = 1$ $\mathcal{B}^1 P$ is a bisimplicial scheme. However, for $m > 1$, because of non-commutavity of Whitney sum, $\mathcal{B}^m P$ is only "pseudo-functor" (in the sense of [HS]) from the m-th power of the category of finite ordered sets into (Simplicial schemes).

In Appendix we recall the required technique from [loc. cit.], in a slightly more general setting. Namely, we introduce the notion of weak functor which generalizes the notion of pseudo-functor in the same spirit as G. Segal's notion of Γ-space generalize May's algebra over an acyclic operad. I hope that this notion, more simple than that of pseudo-functor, may be of independent interest.

Secondly, we introduce the notion of weak m-functor from an m-category. The aim of this generalization, which is more or less straightforward but occupies much place, is to prove the existence of spectral sequences (2.9.1) which is a consequence of (A8.1), and the existence of comultiplication on it-which is the key point in proving the degeneration (see (2.7.1) -which is a consequence of (A8.2). The reader who is ready to believe these facts may omit reading n°'s (A5-A8).

Part B of the Appendix is devoted to the construction of the spectrum of a permutative category which is a base for constructing $\mathcal{B}^\bullet P$. The reader who wants to ignore completely the infinite loop subtleties may consider Whitney sum as commutative and to treat $\mathcal{B}^m P$ simply as (m+1)-simplicial schemes. In fact, the cohomological computations are such as it were the case. This reader can omit reading the Appendix.

. This work arose from thinking over some ideas contained in the papers of E. Friedlander, [F], and A. Beilinson, [B], I'm especially gratefull to A. Beilinson for his interest to the work, and a lot of inspiring discussions.

§0. Notations and some simplicial constructions*)

(0.1) Categories

(0.1.1) $\mathcal{E}ns$ = category of sets

Top = category of topological spaces

Sch = category od schemes

Cat = category of small categories

If C is a category, $x \in C$ means $x \in \mathrm{Ob}\, C$

For $C, D \in \mathrm{Cat}$

CD, or D^C = category of functors $C \longrightarrow D$

(0.1.2) Let C be a category.

C^o = opposed category

C^k = $C\times\ldots\times C$ (k-fold cartesian product).

For $x \in C$

C/x (resp., $C\backslash x$) = category of objects over x (resp., under x).

C is called pointed if it contains an object $*$ which is at the same time final and initial.

If C possesses the final object $*$ then $C^{\bullet} = C/*$; $C^{\bullet}$ is pointed.

(0.2) Simplicial notions

(0.2.1) $N = 0,1,2,\ldots$, $N^+ = N \cup \{-1\}$

For $n \in N$

$[n]$ = finite ordered set $\{0,1,\ldots,n\}$,

$[-1] = \emptyset$

Let $0 < i < n$.

∂_i^n (or simply ∂_i): $[n-1] \longrightarrow [n]$ = increasing inclusion such that $i \notin \mathrm{Im}\, \partial_i$.

$\hat{s}_i^n$ (or simply $\hat{s}_i$): $[n] \longrightarrow [n+1]$ = non decreasing surjection such that $\hat{s}_i(i) = \hat{s}_i(i+1)$.

Δ (resp., Δ^+) = the category whose objects are $[n]$, $n \in \mathbb{N}$ (resp., $n \in \mathbb{N}^+$) and morphisms non-decreasing maps.

(0.2.2) Let C be a category, $k \in \mathbb{N}$.

k-simplicial object in C = object of $(\Delta^o)^k C$;

Augmented k-simplicial object in C = object of $(\Delta^{+o})^k C$.

*) Details about simplicial notions can be found in [GZ], [Q2].

;implicial object = 1-simplicial object.

.et $X \in (\Delta^\circ)^k C$. For $(i_1,\dots,i_k) \in \mathbb{N}^k$

$$X_{i_1\dots i_k} = X([i_1],\dots,[i_k]) \in C.$$

'or $1 \leqslant r \leqslant k$, $0 \leqslant i \leqslant i_r$

$${}^r d_i = {}^r d_i^{i_1\dots i_k} \equiv X(id_{[i_1]},\dots,id_{[i_{r-1}]}, \partial_i^{i_r}, id_{[i_{r+1}]},\dots$$

$$\dots, id_{[i_k]}) : X_{i_1,\dots,i_r-1,\dots,i_k}.$$

$${}^r s_i = {}^r s_i^{i_1\dots i_k} = X(id_{[i_1]},\dots,id_{[i_{r-1}]}, s_i^{i_r}, id_{[i_{r+1}]},\dots,id_{[i_k]}):$$

$$X_{i_1,\dots,i_k} \longrightarrow X_{i_1,\dots,i_r+1,\dots,i_k}.$$

'orphisms ${}^r d_i$ (resp., ${}^r s_i$) are called faces (resp., degeneracies).

$$d_i = {}^1 d_i;\quad s_i = {}^1 s_i$$

:-cosimplicial object in C = object of $\Delta^k C$.

By duality one defines (augmented) (k-)cosimplicial objects, co-
'aces and codegeneracies.

0.2.3) The categories $(\Delta^\circ)^k$, $k \in \mathbb{N}$, themselves form the augmented
implicial category

$$* \longleftarrow \Delta^\circ \rightleftarrows (\Delta^\circ)^2 \dots$$

'hose degeneracies (resp., faces) are induced by diagonal inclusions
$^\circ \longrightarrow \Delta^\circ \times \Delta^\circ$ (resp., by projections).

Therefore, for every $C \in Cat$ we obtain the augmented cosimplici-
l category*)

$$C \longrightarrow \Delta^\circ C \rightrightarrows (\Delta^\circ)^2 C \dots$$

hose cofaces (resp., codegeneracies) we'll denote by

$$\delta^i : (\Delta^\circ)^n C \longrightarrow (\Delta^\circ)^{n+1} C$$

---

*) The idea of considering it is due to A. Beilinson.

(resp., $\overset{\delta}{o}^{i} : (\Delta^\circ)^{n+1}C \longrightarrow (\Delta^\circ)^n C$).

Also

$$\delta : (\Delta^\circ)^n C \longrightarrow \Delta^\circ C$$

(resp., $\sigma : C \longrightarrow (\Delta^\circ)^n C$)

will denote functor corresponding to the unique morphism $[n] \longrightarrow [0]$ (resp., $[-1] \longrightarrow [n]$).

(0.2.4) For $p \in \mathbb{N}$

$\Delta[p] \in \Delta^\circ \mathcal{E}ns$ = the standart p-simplex.

$N : Cat \longrightarrow \Delta^\circ \mathcal{E}ns$ = nerve.

For $\sigma = (x_o \longrightarrow \dots \longrightarrow x_n) \in N_n C$ we put $s\sigma := x_o$, $t\sigma = x_n$.

$|\cdot| : \Delta^\circ \mathcal{E}ns \longrightarrow Top$ = geometric realisation.

For brevity we'll denote also by the same letter compositions

$$(\Delta^\circ)^n \mathcal{E}ns \xrightarrow{\delta} \Delta^\circ \mathcal{E}ns \longrightarrow Top,$$

and

$$(\Delta^\circ)^n Cat \xrightarrow{N} (\Delta^\circ)^{n+1} \mathcal{E}ns \longrightarrow Top.$$

$\simeq$ = homotopy equivalence

$\Sigma, \Omega : Top \longrightarrow Top$ = functors of suspension and loops

$\mathcal{H}o = \mathcal{H}o(Top)$, $\mathcal{H}o^\bullet = \mathcal{H}o(Top^\bullet)$ = homotopy categories

(0.3) <u>Some constructions in categories with limits.</u>

Let C be a category.

(0.3.1) Assume that C has coproducts.

(a) For $X \in C$, $Y \in \mathcal{E}ns$

$X \otimes Y \in C$ = object representing the functor $Z \mapsto Hom(Y, Hom(X,Z))$. From this one obtains functors

$$\otimes : C \times \Delta^\circ \mathcal{E}ns \longrightarrow \Delta^\circ C,$$

$$\otimes : \Delta^\circ C \times \Delta^\circ \mathcal{E}ns \longrightarrow \Delta^{\circ 2} C \longrightarrow \Delta^\circ C.$$

Note that for $D \in Cat$, $X \in \Delta^\circ \mathcal{E}ns$

$$N(D \otimes X) \cong ND \otimes X.$$

b) <u>Simplicial suspension</u>.

Assume C to be pointed. For $X \in C$

$$\Sigma X = X \otimes \Delta[1]/X \otimes 0 \cup X \otimes 1 \in \Delta^\circ C.$$

In other words, ΣX represents the functor

$$Y \in \Delta^\circ C \longmapsto \{ f \in \mathrm{Hom}(X, Y_1) \mid d_0 f = d_1 f = * \}.$$

Note that for $D \in \mathrm{Cat}$

$$\delta N \Sigma D \cong \Sigma ND.$$

c) For $X, Y \in \Delta^\circ C$

$\underline{\mathrm{Hom}}(X,Y) = \underline{\mathrm{Hom}}_{\Delta^\circ C}(X,Y) \in \Delta^\circ \mathcal{E}ns$ = object representing the functor

$$S \longmapsto \mathrm{Hom}_{\Delta^\circ C}(X \times S, Y).$$

For example, if $X \in C$, $Y \in \Delta^\circ C$ then

(0.3.1.1) $\underline{\mathrm{Hom}}(\sigma X, Y) = \mathrm{Hom}_C(X,Y)$,

that is, $\underline{\mathrm{Hom}}(\sigma X, Y)_n = \mathrm{Hom}(X, Y_n)$ and faces and degeneracies in this object are induced by those of Y.

(0.3.2) <u>Skeleton and coskeleton</u>. For $n \in \mathbb{N}$

$\Delta_{\leqslant n}$ = the full subcategory of Δ with objects $[p]$, $0 < p < n$.

$i_n : \Delta_{\leqslant n} \longrightarrow \Delta$ = the canonical inclusion.

i_n induces $i_n^* : \Delta^\circ C \longrightarrow \Delta^\circ_{\leqslant n} C$;

if C has finite limits (resp., colimits)

then i_n^* admits the right adjoint cosk_n

(resp., the left adjoint sk_n).

We'll also write cosk_n (resp., sk_n) for $\mathrm{cosk}_n \circ i_n^*$

(resp., $\mathrm{sk}_n \circ i_n^*$).

(0.3.3) <u>The classifying space of a monoid</u>.

Assume C to have finite products.

Let X be an associative monoid with unit e or more generally a categorical object in C. Then $BX \in \Delta^\circ C$ is defined to be its nerve.

(0.4) If A is abelian category then

$D(A)$ = derived category of A, D^+, D^-, D^b have the standart meaning.

$D^{\geqslant 0}(A) \subset D(A)$ = full subcategory of complexes with zero negative cohomology.

If A is a commutative ring then

A-mod = category of A-modules

Ab = $\mathbb{Z}$-mod

For $A \in Ab$ $A_{\mathbb{Q}} = A \otimes \mathbb{Q}$.

§1. Spectrum of algebraic K-theory

The aim of this section is to construct the algebraic "classifying spectrum" of algebraic K-theory.

A. Waldhausen's S-construction

For $n \in \mathbb{N}$ let Cat(n) be the category associated with the ordered set [n]; let M_n be the category of morphisms of Cat(n).

In other words, $\mathrm{Ob}\, M_n = \{(i,j) \in \mathbb{N} \times \mathbb{N} \mid 0 \leqslant i \leqslant j \leqslant n\}$, $\mathrm{Hom}((i,j),(k,l))$ consists of the unique element if $i \leqslant k$ and $j \leqslant l$ and is empty otherwise.

Categories M_n, $n \geqslant 0$, form the cosimplicial category M.

(1.1) Definition (cf. [L]).

Let $\mathcal{E}$ be a small exact category in the sense of [Q2]. For $n \in \mathbb{N}$

$\underline{\underline{S}}_n(\mathcal{E})$ = full subcategory of $M_n\mathcal{E}$ whose objects are functors $(i,j) \rightsquigarrow E_{i,j}$ such that

a) $E_{ii} = 0$ for all i;

b) for $i \leqslant j \leqslant k$ $E_{ij} \longrightarrow E_{ik} \longrightarrow E_{jk}$ is an exact sequence in $\mathcal{E}$.

Categories $\underline{\underline{S}}_n(\mathcal{E})$, $n \in \mathbb{N}$, form the simplicial exact category $\underline{\underline{S}}.(\mathcal{E})$.

$S.(\mathcal{E}) = \mathrm{Ob}\, \underline{\underline{S}}.(\mathcal{E}) \in \Delta^\circ \mathcal{E}ns$

$\underline{\underline{S}}.^{Is}(\mathcal{E})$ = simplicial subcategory of $\underline{\underline{S}}.(\mathcal{E})$ whose category of n-simplices $\underline{\underline{S}}_n^{Is}(\mathcal{E})$ is the subcategory of $\underline{\underline{S}}_n(\mathcal{E})$ with the same objects but having only isomorphisms as morphisms.

(1.2) Proposition. There exist natural homotopy equivalences

$$|Q\mathcal{E}| \simeq |S.(\mathcal{E})| \simeq |\underline{\underline{S}}.^{Is}(\mathcal{E})|$$

where Q is the Quillen Q-construction, [Q2].

This result is formulated in [L]. The first homotopy equivalence is proven in [G, 6.3].

For completeness let us prove the second. We'll construct a homotopy equivalence in $\Delta^\circ \mathcal{E}ns^{\cdot}$ between $S.(\mathcal{E})$ and $\delta N\underline{\underline{S}}.^{Is}(\mathcal{E})$ (cf. [HS, 3.6]). An n-simplex of $\delta N\underline{\underline{S}}.^{Is}(\mathcal{E})$ is a set of objects of $\mathcal{E}$

$\{P^k_{ij}\}$, $0 \leq k \leq n$, $0 \leq i \leq j \leq n$, and morphisms

$P^k_{ij} \longrightarrow P^k_{i'j'}$, $i \leq i'$, $j \leq j'$;

$P^k_{ij} \xrightarrow{\sim} P^{k'}_{ij}$, $k \leq k'$, satisfying the obvious compatibility condition.

Define inclusion $f: S.\mathcal{E} \longrightarrow \delta\underline{\underline{NS}}^{Is}(\mathcal{E})$ by the rule $f(\{P_{ij}\}) = \{Q^k_{ij}\}$, where $Q^k_{ij} = P_{ij}$ for all k, maps $Q^k_{ij} \longrightarrow Q^{k'}_{ij}$ are equal to $Id(P_{ij})$, and Q^k_{ij} — $Q^k_{i'j'}$ to $P_{ij} \longrightarrow P_{i'j'}$.

Conversely, define retraction $r: \delta\underline{\underline{NS}}^{Is}(\mathcal{E}) \longrightarrow S.(\mathcal{E})$ by the rule $r(\{Q^k_{ij}\}) = \{P_{ij}\}$, where $P_{ij} = Q^i_{ij}$, $P_{ij} \longrightarrow P_{i'j'}$ is the composition $Q^i_{ij} \xrightarrow{\sim} Q^{i'}_{ij} \longrightarrow Q^{i'}_{i'j'}$.

It is clear that $r \circ f = id$. Define homotopy

$$H : f \circ r \simeq id ; \delta\underline{\underline{NS}}^{Is}(\mathcal{E}) \times \Delta[1] \longrightarrow \delta\underline{\underline{NS}}^{Is}(\mathcal{E}).$$

Namely, for $Q = \{Q^k_{ij}\} \in \delta\underline{\underline{NS}}^{Is}(\mathcal{E})_n$, $x = \langle 00\ldots01\ldots1\rangle \in \Delta[1]_n$ ($p+1$ zeroes, $-1 \leq p \leq n$), put $H(Q,x) = \{R^k_{ij}\}$, where $R^k_{ij} = Q^k_{ij}$ for $i \leq p$, and Q^k_{ij} for $i > p$.

(1.2.1) <u>Corollary</u>. $\pi_{i+1}(|S.(\mathcal{E})|,0) = \pi_{i+1}(|\underline{\underline{S}}.^{Is}(\mathcal{E})|,0) = K_i(\mathcal{E})$ for $i \in \mathbb{N}$ (we put $K_{-1}(\mathcal{E}) = 0$).

(1.3.0) For $m \geq 1$ put by induction

$$\underline{\underline{S}}^m_{\cdot}(\mathcal{E}) = \underline{\underline{S}}.(\underline{\underline{S}}^{m-1}(\mathcal{E})) \in (\Delta^\circ)^m \, Cat;$$

$$\underline{\underline{S}}^{m,Is}_{\cdot}(\mathcal{E}) = \underline{\underline{S}}^{Is}_{\cdot}(\underline{\underline{S}}^{m-1}_{\cdot}(\mathcal{E})) \in (\Delta^\circ)^m \, Cat;$$

$$S^m_{\cdot}(\mathcal{E}) = Ob\, \underline{\underline{S}}^m_{\cdot}(\mathcal{E}) \in (\Delta^\circ)^m \mathcal{E}ns.$$

(1.3) <u>Proposition</u> (cf. [G, §7]).

There exist natural arrows in $(\Delta^\circ)^{m+1}$ Cat

$$\varphi_m \; : \; \Sigma \underline{\underline{S}}^{m,Is}(\mathcal{E}) \longrightarrow \underline{\underline{S}}^{m+1,Is}(\mathcal{E})$$

(cf. (0.3.1) b)), inducing homotopy equivalences

$$|\underline{\underline{S}}^{m,Is}(\mathcal{E})| \xrightarrow{\sim} \Omega\, |\underline{\underline{S}}^{m+1,Is}(\mathcal{E})|,$$

$$|S^m(\mathcal{E})| \longrightarrow |\Omega \;\; S^{m+1}(\mathcal{E})|.$$

Proof. Let $\underline{\underline{R}}.(\mathcal{E})$ be the simplicial exact category with

$$\underline{\underline{R}}_n(\mathcal{E}) = \underline{\underline{S}}_{n+1}(\mathcal{E}); \quad d_{i,R} = d^{n+1}_{i+1,S}, \; s^n_{i,R} = s^{n+1}_{i+1,S}.$$

One has the sequence in $(\Delta^\circ)^2\mathcal{E}ns$

(1.3.1) $$\sigma_1 S.(\mathcal{E}) \xrightarrow{\;\varphi\;} S.\underline{\underline{R}}(\mathcal{E}) \xrightarrow{\;\psi\;} S^2(\mathcal{E})$$

where φ is induced by inclusions $\mathcal{E} \longrightarrow \underline{\underline{R}}_n(\mathcal{E})$,

$$x \longmapsto \{x_{ij}\}, \quad x_{ij} = \begin{cases} x & \text{if } (i,j) = (n, n+1), \\ 0 & \text{otherwise}; \end{cases}$$

ψ is induced by $\partial_{n+1} : M_n \longrightarrow M_{n+1}$.

From Quillen's exact sequence theorem [Q2, Thm 2] and (1.2) follows that $|S.\underline{\underline{R}}_n(\mathcal{E})| \simeq |S.(\mathcal{E})|^{n+1}$, $|S.\underline{\underline{S}}^n(\mathcal{E})| \simeq |S.(\mathcal{E})|^n$ and the map ψ can be identified with the natural projection with fibre $S.(\mathcal{E})$. Therefore, for $n \in \mathbb{N}$ $|(1.3.1)_n|$ is fibration in $Top^\bullet$. Therefore, (1.3.1) is fibration.

Moreover, the superfluous degeneracy in $\underline{\underline{R}}.(\mathcal{E})$ gives the contracting homotopy

$$H : S.\underline{\underline{R}}.(\mathcal{E}) \otimes \Delta[1] \longrightarrow S.R.(\mathcal{E})$$

and therefore map

$$\varphi_1 : \Sigma S.(\mathcal{E}) \longrightarrow S^2(\mathcal{E}),$$

inducing homotopy equivalence

$$|S.(\mathcal{E})| \xrightarrow{\sim} \Omega.\, |S^2(\mathcal{E})|.$$

This proves the case $m = 1$ of proposition. The general case is obtained by application of the proved assertion to $\underline{\underline{S}}^{m-1}(\mathcal{E})$. ▲

B. Representability of functors $\underline{\underline{S}}^{Is}_{l_1,\dots,l_m}$.

(1.4.0) Let X be a scheme; $U = \{U_i\}$ - an open Zarissky covering of X.

Put

$\mathcal{P}(X)$ = exact category of finite dimensional vector bundles over X;
for $n \in \mathbb{N}$ $\mathcal{P}_n(X)$ = subcategory of $\mathcal{P}(X)$ of rank n vector bundles and isomorphisms;

$\mathcal{P}(X,U)$ = full subcategory of $\mathcal{P}(X)$ whose objects are bundles admitting trivialization over U;

$\mathcal{P}_n(X,U) = \mathcal{P}_n(X) \cap \mathcal{P}(X,U)$.

Let $NU = \mathrm{Cosk}_o(\coprod U_i \longrightarrow X) \in \Delta^\circ \, Sch/X$

be the nerve of U.

(1.4) Proposition (cf. [F])

For all $n \in \mathbb{N}$ there is natural homotopy equivalence

$$|\mathcal{P}_n(X,U)| \simeq |\underline{\mathrm{Hom}}(NU, BGL(n))|$$

where $\underline{\mathrm{Hom}} = \underline{\mathrm{Hom}}_{\Delta^\circ Sch/X}$, see (0.3.1.c).

Proof. $\mathcal{P}_n(X,U)$ is equivalent to the category whose objects are $\mathcal{O}_{N_1U}$ - linear isomorphisms $\alpha : \mathcal{O}^n_{N_1U} \xrightarrow{\sim} \mathcal{O}^n_{N_1U}$ such that $d_0^{2*}\alpha \cdot d_2^{2*}\alpha = d_1^{2*}\alpha$ (where $d_i^2 : N_2U \to N_1U$) and morphisms between α and α' are isomorphisms $\varphi : \mathcal{O}^n_{N_oU} \xrightarrow{\sim} \mathcal{O}^n_{N_oU}$ such that $\alpha \cdot d_o^{1*}\varphi = d_1^{1*}\varphi \circ \alpha$. But the nerve of the last category is isomorphic to $\underline{\mathrm{Hom}}(NU, BGL(n))$. ▲

(1.5) Let $(l_1,\ldots,l_m) \in \mathbb{N}^m$,

$\underline{n} = \{n_{j_1,\ldots,j_m}\}_{1 \leq i_1 \leq l_1, \ldots, 1 \leq i_m < l_m}$ be a set of elements of $\mathbb{N}$ indexed by $\{1,\ldots,l_1\} \times \ldots \times \{1,\ldots,l_m\}$.

Put $|\underline{n}| = m$, $\dim \underline{n} = (l_1,\ldots,l_m)$.

(1.5.1) Definition. Let us define for a ring A by induction on m the standart object $\mathcal{L}(\underline{n},A) \in S^m_{l_1,\ldots,l_m}(\mathrm{Spec}\ A)$.

For $m = 1$, $n = n_1,\ldots,n_l$ let $\mathcal{L}(\underline{n},A)_{ol}$ be the free A-module of rank $n = n_1 + \ldots + n_l$ with the distinguished base $e_1,\ldots,e_n$. For $(i,j) : 0 \leq i \leq j \leq l$ put $\mathcal{L}(\underline{n},A)_{ij} \subset \mathcal{L}(n,A)_{ol}$ to be the A-submodule generated by $e_{n_1+\ldots n_{i+1}},\ldots, e_{n_1+\ldots+n_j}$.

For $i \leq i'$, $j \leq j'$ let $\mathcal{L}(n,A)_{ij} \longrightarrow \mathcal{L}(n,A)_{i'j'}$ be the map sending e_q to e_q if $n_1 + \dots + n_{i'} + 1 \leqslant q \leqslant n_1 + \dots + n_{j'}$ and to 0 otherwise.

Suppose now $m > 1$, $|\underline{n}| = m$. For $1 \leqslant i \leqslant l_m$ put $p_i(\underline{n}) = \{n_{j_1,\dots,i_{m-1},i}\}_{i_1,\dots,i_{m-1}}$. By induction $(p_i(\underline{n}),A)$ are already defined.

For $0 \leqslant i \leqslant j \leqslant l_m$ put $\mathcal{L}(\underline{n},A)_{ij} = \coprod_{q=i+1}^{j} \mathcal{L}(p_q(\underline{n}),A) \in$

$\in S^{m-1}_{l_1,\dots,l_{m-1}}$ (Spec A); a map $\mathcal{L}(\underline{n},A)_{ij} \longrightarrow \mathcal{L}(n,A)_{i'j'}$ is identical on $\mathcal{L}(p_q(\underline{n}),A)$ if $n_1 + \dots + n_{j'} + 1 \leqslant q \leqslant n_1 + \dots + n_{j'}$ and zero otherwise.

This defines $\mathcal{L}(\underline{n},A) \in S_{l_m}(\underline{\underline{S}}^{m-1}_{l_1,\dots,l_{m-1}}) = S^m_{l_1,\dots,l_m}(\text{Spec } A)$.

(1.5.2) <u>Lemma</u> $\mathcal{L}(n,A)$ is projective object of $\underline{\underline{S}}^m_{l_1,\dots,l_m}$ (Spec A).

<u>Proof</u>. Call $\mathcal{L}(\underline{n},A)$ elementary if all $n_{i_1,\dots,i_m}$ are equal to zero, except the one, which is equal to 1. In this case for $x \in S^m_{l_1,\dots,l_m}$ $\text{Hom}_{\underline{\underline{S}}^m}(x, \mathcal{L}(n,A)) = x_{op_1,\dots,op_m}$ if $n_{p_1,\dots,p_m} = 1$. Therefore, such $\mathcal{L}(\underline{n},A)$ is projective. But arbitrary $\mathcal{L}(n,A)$ is a direct sum of elementary ones. ▲

(1.5.3) <u>Corollary</u>. Under conditions of (1.4.0) suppose X to be connected, and U affine (i.e. all U_i affine).

Then every object of $\underline{\underline{S}}^m_{l_1,\dots,l_m} \mathcal{P}(X,U)$ becomes under restriction on N_oU isomorphic to some $\mathcal{L}(\underline{n},A)$, where $A = \Gamma(N_oU, \mathcal{O}_{N_oU})$.

<u>Proof</u>. Induction on m. The case $m = 1$ is obvious, and the inductive step follows from (1.5.2). ▲

(1.5.4) <u>Definition</u>.

a) Let $\underline{n},A$ be as in (1.5.1).
$P(\underline{n},A)$ is the group of automorphisms of $\mathcal{L}(\underline{n},A)$ in $S^m_{l_1,\dots,l_m}$ (Spec A).

By definition we have

$$\mathcal{L}(\underline{n},A)_{\underline{\underline{ol}}} := \mathcal{L}(\underline{n},A)_{ol_1,\dots,ol_m} = \sum_{i_m=1}^{l_m} \dots \sum_{i_1=1}^{l_1} A^{n_{i_1\dots i_m}}$$

$\mathcal{L}(\underline{n},A)_{\underline{\underline{ol}}}$ is equipped with m filtrations $F^1, \dots, F^m$ with

$$F_p^q \, \mathcal{L}(\underline{n},A)_{\underline{ol}} = \sum_{i_m=1}^{l_m} \cdots \sum_{i_q=1}^{p} \cdots \sum_{i_1=1}^{l_1} A^{n_{i_1 \ldots i_m}}, \quad 1 \leqslant q \leqslant n,$$

$\leqslant p \leqslant l_m$, and $P(\underline{n},A)$ is the group of A - linear automorphisms f $\mathcal{L}(\underline{n},A)$, respecting these filtrations.

Let $L(\underline{n},A) \subset P(\underline{n},A)$ be the subgroup of automorphisms respect- ng the natural k-grading on $L(n,A)$.

Denote by $L(n)$, $P(n)$ the corresponding algebraic groups.

xample. For $m = 1$, $\underline{n} = \{n_1,\ldots,n_l\}$, $P(n) = P(n_1,\ldots,n_l)$ is tandart parabolic subgroup of $GL(n_1 + \ldots + n_l, A)$ consisting of mat- ices of type

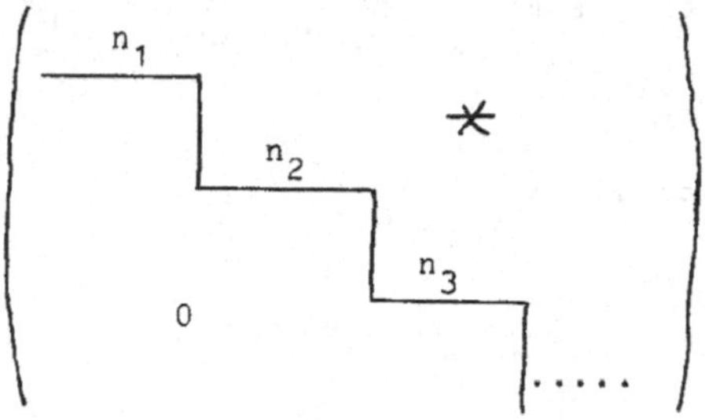
;

$L(\underline{n}) \subset P(\underline{n})$ is the standart Levi subgroup.

b) For $(l_1,\ldots,l_m) \in \mathbb{N}^m$ put

$$P_{l_1,\ldots,l_m} = \coprod_{\underline{n}:\dim \underline{n} = (l_1,\ldots,l_m)} P(\underline{n});$$

$$BP_{l_1,\ldots,l_m} = \coprod BP(\underline{n}) \in \Delta^\circ Sch,$$

nd in the same way define $L_{l_1,\ldots,l_m}$ and $BL_{l_1,\ldots,l_m}$.

The following generalization of (1.4) is a consequence of (1.5.3).

.6) Proposition. Under conditions of (1.5.3) there exist natural omotopy equivalences in $\mathcal{H}o^{\bullet}$:

$$|S^{m,Is}_{l_1,\ldots,l_m} \mathcal{P}(X,U)| \simeq |\underline{Hom}(NU, BP_{l_1,\ldots,l_m})|. \blacktriangle$$

C. Representability of $\underline{\underline{S}}^{m,Is}$.

w we'll build from all $BP_{l_1,\ldots,l_m}$'s an algebraic "spectrum" which represents the spectrum constructed in n°A.

(1.7) Construction of $\mathcal{B}^{\bullet}L$. Let us denote by L the simplicial scheme

$$\coprod_{n\in\mathbb{N}} BGL(n).$$

L has natural structure of a permutative category in $\Delta^{\circ}Sch$ (cf. (A9)). Namely,
we put $Ob\,L = \mathbb{N}$ (discrete scheme),

$$Hom_L(n,m) = \begin{cases} BGL(n) & \text{if } n = m \\ \emptyset & \text{otherwise;} \end{cases}$$

0 is BGL(0);
$+ : L \times L \longrightarrow L$ is induced by Whitney sum

$$BGL(n) \times BGL(m) \longrightarrow BGL(m+n)$$

The natural transformation of commutativity

$$\varphi : + \circ s \longrightarrow +$$

consists of morphisms

$$\varphi_{m,n} : BGL(m+n) \longrightarrow BGL(m+n)$$

induced by conjugation by the matrix $\sigma_{m,n} \in GL(m+n)$,

$$\sigma_{m,n}(e_i) = \begin{cases} e_{i+m}, & \text{if } 1 \leqslant i \leqslant n \\ e_{i-n}, & \text{if } n+1 \leqslant i \leq n+m \end{cases}$$

Thus, we can apply to L the construction of (App., B)., and obtain the spectrum $\mathcal{B}^{\bullet}L$ which consists of m-pseudofunctors

$$\mathcal{B}^{m}L : \Delta^{\circ}Sch \longrightarrow \Delta^{\circ}Sch$$

(cf. (A10.4)).

For example, $\mathcal{B}^{1}L$ is the strict functor

$$\Delta^{\circ} \longrightarrow \Delta^{\circ}Sch$$

with

$$(\mathcal{B}^{1}L)_l = \coprod_{(n_1,\dots,n_l)\in\mathbb{N}^l} BGL(n_1) \times\dots\times BGL(n_l).$$

(1.8) Construction of $\mathcal{B}^{\cdot}P$.

More generally, for $l = (l_1, \ldots, l_m) \in \mathbb{N}^m = \mathrm{Ob}\ \Delta^{\boxtimes m}$,

$$(\mathcal{B}^m L)_{l_1, \ldots, l_m} = BL_{l_1, \ldots, l_m}$$

cf. (1.5.4.b)). Put

$$(\mathcal{B}^m P)_{l_1, \ldots, l_m} = BP_{l_1, \ldots, l_m}.$$

(1.8.1) Lemma-definition. The above construction can be naturally complemented to an m-pseudo-functor

$$\mathcal{B}^m P : \Delta^{\circ \boxtimes m} \sim\longrightarrow \Delta^\circ Sch$$

over the same m-c.s. as $\mathcal{B}^m L$ such that inclusions

$$(\mathcal{B}^m L)_{\underline{l}} \longrightarrow (\mathcal{B}^m P)_{\underline{l}}, \qquad \underline{l} \in \mathrm{Ob}\ \Delta^{\boxtimes m},$$

form the morphism of m-p.f.'s.

Proof. We'll treat the case $m = 1$ which contains the essential difference between $\mathcal{B}L$ and $\mathcal{B}P$, leaving the general case to the reader. Namely, we have

$$(\mathcal{B}^1 P)_l = \coprod_{(n_1, \ldots, n_l) \in \mathbb{N}^l} BP(n_1, \ldots, n_l).$$

$\mathcal{B}^1 P$ is the strict simplicial object in $\Delta^\circ Sch$. Its faces

$$d_i : (\mathcal{B}^1 P)_l \text{———} (\mathcal{B}^1 P)_{l-1}$$

are induced by

- projections $P(n_1, \ldots, n_l) \longrightarrow P(n_2, \ldots, n_l)$ (resp., $P(n_1, \ldots, n_l) \longrightarrow P(n_1, \ldots, n_{l-1})$) when $i = 0$ (resp., $i = l$);
- inclusions $P(n_1, \ldots, n_l) \longrightarrow P(n_1, \ldots, n_{i-1}, n_i + n_{i+1}, n_{i+2}, \ldots, n_l)$ when $0 < i < l$.

Degeneracies

$$s_i : (\mathcal{B}^1 P)_l \longrightarrow (\mathcal{B}^1 P)_{l+1}, \qquad 0 \leqslant i \leqslant l,$$

are induced by isomorphisms

$$P(n_1, \ldots, n_l) \xrightarrow{\sim} P(n_1, \ldots, n_i, 0, n_{i+1}, \ldots, n_l).$$

We have natural morphisms

(1.8.2) $\Sigma \mathcal{B}^m P \longrightarrow \mathcal{B}^{m+1} P, \quad m \geq 1.$

Now we came to the main point of this section.

(1.9) Theorem. Under the conditions of (1.5.3), for every $m \geq 1$, there exist natural homotopy equivalences

$$|\underline{\underline{S}}^{m,Is} \mathcal{P}(X,U)| \simeq |\mathrm{Hom}(NU, \mathcal{B}^m P)|$$

Therefore, if we put

$$\underline{\mathrm{Hom}}\,(X^V, \mathcal{B}^m P) = \varinjlim \underline{\mathrm{Hom}}\,(NU, \mathcal{B}^m P)$$

($\varinjlim$ taken over all Zarissky coverings of X), then

$$|\underline{\underline{S}}^{m,Is} \mathcal{P}(X)| \simeq |\underline{\mathrm{Hom}}\,(X^V, \mathcal{B}^m P)|.$$

All the squares

$$\begin{array}{ccc} |\Sigma \underline{\underline{S}}^{m,Is} \mathcal{P}(X)| & \longrightarrow & |\Sigma \underline{\mathrm{Hom}}\,(X^V, \mathcal{B}^m P)| \\ {\scriptstyle (1.3)}\downarrow & & \downarrow{\scriptstyle (1.8.2)} \\ |\underline{\underline{S}}^{m+1,Is} \mathcal{P}(X)| & \longrightarrow & |\mathrm{Hom}\,(X^V, \mathcal{B}^{m+1} P)| \end{array}$$

commute in $\mathcal{H}o^{\bullet}$.

Proof. One has only to gather all homotopy equivalences of (1.6). They do not form the morphism of m - p.-f's but can be completed to an $(m+1)$-p.f. over $\Delta^{\circ \boxtimes m} \times \mathcal{J}$, $\mathcal{J} = (0 \longrightarrow 1)$, and thus, by finality theorem (A7.2), induce the unique homotopy class of homotopy equivalences

$$|\underline{\underline{S}}^{m,Is} \mathcal{P}(X,U)| \longrightarrow |\underline{\mathrm{Hom}}\,(NU, \mathcal{B}^m P)|$$

(cf. the argument in (A8.2) or [HS]). ▲

(1.10) Remarks. 1. Suppose that the covering $U = \{U_i\}$ of X has the property that over each U_i every finitely generated projective module is free. Then one has an equivalence

$$\mathcal{P}(X,U) \simeq \mathcal{P}(X),$$

and therefore,

$$|\underline{\underline{S}}^{m,Is} \mathcal{P}(X)| \simeq |\underline{\mathrm{Hom}}\,(NU, \mathcal{B}^m P)|.$$

Example. X local, $U = (X \xrightarrow{id} X)$,

$$\underline{\mathrm{Hom}}\,(NU, \mathcal{B}^m P) = \mathrm{Hom}\,(X, \mathcal{B}^m P), \quad \text{cf. } (0.3.1.1.).$$

Thus, for local X simply X-points of the spectrum $\mathcal{B}^{\cdot}P$ give the classifying spectrum $\mathbb{K}(X)$ of K-theory of X.

2. In general one has always

$$\pi_i \,|\, \underline{\underline{S}}^{m,Is}\mathcal{P}(X,U) | \cong \pi_i \,|\, \underline{\underline{S}}^{m,Is}\mathcal{P}(X) |$$

for $i > m$, i.e.

$$\pi_{i+m} \,|\, \mathrm{Hom}\,(X, \mathcal{B}^m P) | \cong K_i(X)$$

for $i > 0$, if X is affine.

3. The space of $\mathbb{C}$-points, $\mathcal{B}^m P(\mathbb{C})$ is just the m-fold delooping of $\mathbb{Z} \times BGL^{top}(\mathbb{C})$. This fact probably helps to understand the cohomological computations of the next section.

§2. Cohomology of $\mathcal{B}^{\cdot}P$ and Chern character

In this section we partially compute the cohomology of schemes $\mathcal{B}^m P$ cf. Thm's (2.8), and (2.10). In n°C deduce from this result the construction of the delooping of Chern character.

Notation

2.1) At first we introduce the appropriate axioms for our cohomology theory. These axioms are taken from [B].

2.1.1) Fix a regular base scheme S.

Let $\mathcal{V}$ be a full subcategory of Sch/S satisfying the following requirements

a) $\mathcal{V}$ contains all schemes which are smooth over S.

b) $\mathcal{V}$ is closed under disjoint unions and fibre products.

c) If $X \in \mathcal{V}$, $U \subset X$ - Zarissky open, then $U \in \mathcal{V}$

d) If $X \in \mathcal{V}$, $E \in \mathcal{P}(X)$, then the corresponding affine $\mathbb{A}(E)$, and projective, $\mathbb{P}(E)$, spaces lie in $\mathcal{V}$.

Supply $\mathcal{V}$ with Zarissky topology.

Put

$\mathcal{V}^{\sim}$ = category of abelian sheaves over $\mathcal{V}$.

For $C \in \mathrm{Cat}$, $X \in \mathcal{V}^C$, put

$X^{\sim}$ = category of abelian sheaves over X.

An object of $X^{\sim}$ is a set of Zarissky sheaves $F(c)$ over $X(c)$ for all $c \in C$, and morphisms $F(c') \to F(c)$ over $X(c) \to X(c') \in \mathrm{Mor}\,\mathcal{V}$ for each $c \to c' \in \mathrm{Mor}\,C$, satisfying the standart compatibility condition.

One has evident functors

$$\mathcal{V}^{\sim} \xrightarrow{(\cdot)_X} X^{\sim} \xrightarrow{\Gamma} \mathcal{Ab}^{C} \xrightarrow{\varprojlim} \mathcal{Ab}.$$

Let $\Gamma(X,\cdot)$ denote their composition. Deriving, one gets

$$R\Gamma(X,\cdot) : \mathbb{D}^{+}(\mathcal{V}^{\sim}) \longrightarrow \mathbb{D}^{+}(\mathcal{Ab}).$$

For $\mathcal{F} \in \mathbb{D}^{+}(\mathcal{V}^{\sim})$ put

$$H^*(X,\mathcal{F}) = H^*R\Gamma(X,\mathcal{F}).$$

Thus we have defined the cohomology of every small diagram in $\mathcal{V}$ with coefficients in $\mathcal{F} \in \mathbb{D}^{+}(\mathcal{V}^{\sim})$. In particular, the cohomology of (poly-) simplicial objects are defined.

(2.1.1.1) Cohomology of weak diagramms

Let $\mathcal{F} \in \mathbb{D}^{+}(\mathcal{V})$, $C \in \mathrm{Cat}^{(m)}$,

$X : C \sim\!\longrightarrow \Delta^{\circ}\mathcal{V}$ - an m-pseudo-functor (see Appendix)

Then we define $H^*(X,\mathcal{F})$ to be the cohomology of

$\underrightarrow{\mathrm{holim}}\, X \in (\Delta^{\circ})^{m+1}\mathcal{V}$ (A6.3.1).

Equivalent definition: taking the composition of X with $R\Gamma(\cdot,\mathcal{F})$ we obtain the weak m-functor

$$R\Gamma(X,\mathcal{F}) : C^{\circ} \sim\!\longrightarrow \Delta^{\circ}C^{\bullet}(\mathcal{Ab}) \xrightarrow{\mathrm{Tot}} C^{\bullet}(\mathcal{Ab})$$

and then take the cohomology of $NR\Gamma(X,\ ) \in \Delta^{\circ m}C^{\bullet}(\mathcal{Ab})$.

(2.1.2) Let $\mathbb{Z} \in \mathcal{V}^{\sim}$ denote the constant sheaf with fibre $\mathbb{Z}$, and $\mathcal{O}^* \in \mathcal{V}^{\sim}$ - the sheaf of units of the structure sheaf $\mathcal{O}$.

Definition. A cohomology theory on $\mathcal{V}$ consists of

a) The set of complexes

$$Z(i) \in \mathbb{D}^{b}(\mathcal{V}^{\sim}),\quad i \in \mathbb{Z},$$

such that for all $X \in \mathcal{V}$, $Z(i)_X \in \mathbb{D}^{b}(X^{\sim})$;

b) the morphisms

$$c_0 : \mathbb{Z} \longrightarrow Z(0),$$

and

$$c_1 : \mathcal{O}^*[1] \longrightarrow Z(1)$$

in $\mathbb{D}^{b}(\mathcal{V}^{\sim})$;

c) the pairings

$$\cup : Z(i) \overset{L}{\otimes} Z(j) \longrightarrow Z(i+j).$$

These data must satisfy the following axioms.

Axiom (A). (Algebra structure). The pairings c) are commutative and associative in $\mathbb{D}^{b}(\mathcal{V})$,

c_0 is the unit for this multiplication.

Axiom (DT). (Dold-Thom isomorphisms). Let $n \in \mathbb{N}$, $\mathbb{P}^n = \mathbb{P}^n_S$ be the n-dimensional projective space over S; $\xi = c_1(H) \in H^2(\mathbb{P}^n, Z(1))$, where $H = cl(\mathcal{O}(1)) \in H^1(P^n, \mathcal{O}^*)$.
Let $X \in \mathcal{V}$, $\pi_{\mathbb{P}^n}, \pi_X : \mathbb{P}^n \times X \longrightarrow \mathbb{P}^n, X$ be the projections

Then the map

$$\coprod_i \pi^*_{\mathbb{P}^n}(\xi^i) \cup \pi^*_X : \coprod_{i=0} Z(j-i)_X[-2i] \longrightarrow R\pi_* Z(j)_{\mathbb{P}^n \times X}$$

is an isomorphism in $\mathbb{D}^{b}(X^{\sim})$ for all $j \in \mathbb{Z}$.

Axiom (H). (Homotopy invariance). Let $\mathbb{A}^1$ be the affine line over S, $X \in \mathcal{V}$, X smooth, $\pi : X \times \mathbb{A}^1 \longrightarrow X$ projection. The map

$$\pi^* : Z(j)_X \longrightarrow R\pi_* Z(j)_{X \times \mathbb{A}^1}$$

is an isomorphism in $\mathbb{D}^{b}(X^{\sim})$ for all $j \in \mathbb{Z}$ *).

(2.1.3) It is known that, with the appropriate definition of $Z(j)$ and $\mathcal{V}$, the most cohomology theories satisfy the above axioms, for example:

a) etale cohomology;

b) DeRham cohomology in characteristic zero;

c) Chow cohomology;

d) Singular cohomology of varieties over $\mathbb{C}$;

e) Beilinson's absolute Hodge cohomology.

(2.1.4) From now on, we fix in this section a cohomology theory. We'll also use the following notations.

---

*) Note that (DT) (at any rate for smooth X) is a consequence of (H) and a certain weak form of Poincare duality. We shall not use the Poincare duality in the sequel.

$$H^*(X,Z(*)) = \coprod_{i,j} H^i(X,Z(j));$$

$$A = H^*(S,Z(*)); \quad Q(j) = Z(j) \times Q.$$

We'll write simply BGL, $\mathcal{B}^m P$, etc. instead of BGL x S, $\mathcal{B}^m P \times S$, etc.

(2.1.5) NB ! Let $f : B \longrightarrow C$ be an inclusion of graded commutative algebras, $e_i \in C$, $i \in I$, - a set of homogenious elements.

The words "there is an isomorphism $B[e_i]_{i \in I} \xrightarrow{\sim} C$" will mean "the homomorphism $B[T_i]_{i \in I} \longrightarrow C$ from the free graded commutative algebra over B generated by T_i, $\deg T_i = \deg e_i$, which on B coincides with f and sends T_i to e_i, is an isomorphism".

A. BGL(n)

(2.2) Proposition. On $\Delta^\circ \mathcal{V}$ there exists a unique theory of Chern classes which to every $X \in \Delta^\circ \mathcal{V}$ and $E \in \mathcal{P}(X)$ associates the element

$$c(E) = \prod c_i(E) \in \prod_{i \geq 0} H^{2i}(X,Z(i))$$

such that

a) c_0 and c_1 (for line bundles) are induced by, respectively, c_0 and c_1 from (2.1.2.b);

b) $f : X \longrightarrow Y$ Mor $\Delta^\circ \mathcal{V}$, $E \in \mathcal{P}(Y)$, then $f^*c(E) = c(f^*E)$.

c) (Whitney sum formula). If $0 \longrightarrow E' \longrightarrow E \longrightarrow E'' \longrightarrow 0$ is an exact sequence in $\mathcal{P}(X)$, then $c(E) = c(E') \cdot c(E'')$.

Sketch Proof. Use the method of [J]. One has to define the class of a divizor on a simplicial scheme - this is done using c_1 and to check for it a particular case of projection formula which follows directly from definition. ▲ Put $T(1) = GL(1)$, $T(n) = T(1)^n$ for $n \in \mathbb{N}$.

(2.3) Lemma.

Let $a \in H^2(BT(1),Z(1))$ be the first Chern class of the canonical line bundle e_1 over $BT(1)$;

$$a_i = pr_i^* a \in H^2(BT(n), Z(1)),$$

where $pr_i : BT(n) \longrightarrow BT(1)$, $1 \leq i \leq n$, - projections.

Then one has an isomorphism

$$H^*(BT(n), Z(*)) = A[a_1, \ldots, a_n].$$

Proof. It's sufficient to treat the case $n = 1$; the general case follows by induction.

For $m \geqslant 1$ consider the open covering $U = \{U_o, \ldots, U_m\}$ of $\mathbb{P}^m$ by affine charts $U_i = \{(x_o : \ldots : x_m) \mid x_i \neq 0\}$. The standart trivialization of $\mathcal{O}(1)$ over U defines a map $\varphi_m : NU \longrightarrow BT(1)$. We want to show that φ_m is "a homotopy equivalence" on nondegenerate parts up to dimension $m-1$.

More precisely, let $B \subset \Delta$ be the subcategory with the same objects and with monotone inclusions as morphisms, $i : B \longrightarrow \Delta$ the inclusion. The evident functor $i_* : \Delta^o Sch/_S \longrightarrow B^o Sch/_S$ has the left adjoint, $i^* : B^o Sch/_S \longrightarrow \Delta^o Sch/_S$ (the formal adjoining of degeneracies). For every $X \in \Delta^o Sch/_S$ the adjunction morphism $i^* i_* X \longrightarrow X$ induces an isomorphism in cohomology $H^*(X, Z(*)) \xrightarrow{\sim} H^*(i^* i_* X, Z(*))$.

Further, note that $0 \leqslant i_o < \ldots \leqslant i_p \leqslant m$ one has an isomorphism $U_{i_o} \cap \ldots \cap U_{i_p} = T(p) \times \mathbb{A}^{m-p}$. From this follows easily, that φ_m is the composition of the morphism $\varphi'_m : NU \longrightarrow Sk_m(i^* i_* BT(1) \times \Delta[m])$, whose fibres in each dimension are affine spaces, and the natural map $Sk_m(i^* i_* BT(1) \times \Delta[m]) \longrightarrow i^* i_* BT(1) \times \Delta[m] \longrightarrow BT(1) \times \Delta[m] \longrightarrow BT(1)$. By axiom (H), φ'_m is cohomology isomorphism. In the other hand, $H^*(NU, Z(*)) = H^*(\mathbb{P}^m, Z(*))$, because $Z(*)$ are complexes of Zarissky sheaves, and by construction $\varphi^*_m(a) = c_1(\mathcal{O}(1))$. From this follows that, given $p \in \mathbb{N}$ and $j \in \mathbb{Z}$ there exists $m_o \in \mathbb{N}$ such that for $m \geqslant m_o$ $H^p(\varphi_m, Z(j))$ is an isomorphism. Taking m tending to infinity, and using (DT), one deduces the assertion of the Lemma. ▲

(2.4) Proposition. a) Let $n \in \mathbb{N}$,

$$c_i = c_i(e_n) \in H^{2i}(BGL(n), Z(i)), \quad 1 \leqslant i \leqslant n,$$

be the Chern classes of the canonical n-dimensional vector bundle over $BGL(n)$. Then one has an isomorphism

$$H^*(BGL(n), Z(*)) = A[c_1, \ldots, c_n].$$

(cf. (2.1.5)).

Moreover, for every $p \in \mathbb{N}$, $(n_1, \ldots, n_p) \in \mathbb{N}^p$

$$H^*(\prod_{k=1}^{p} BGL(n_k),\ Z(*)) = \bigotimes_{k=1}^{p} H^*(BGL(n_k),\ Z(*)).$$

b) The Whitney sum

$$+ : BGL(n) \times BGL(m) \longrightarrow BGL(n+m)$$

induces the map in cohomology

$$\mu : A[c_1,\dots,c_{n+m}] \longrightarrow A[c'_1,\dots,c'_n;\ c''_1,\dots,c''_m]$$

given by formula

$$\mu(c_i) = \sum_{p+q=i} c'_p \times c''_q$$

(we put as usually $c_o = c'_o = c''_o = 1$).

Proof. a) Again it's sufficient to treat the case $p = 1$. Consider the inclusions

$$BT(n) \xrightarrow{f} BB(n) \xrightarrow{g} BGL(n)$$

where $B(n) \subset GL(n)$ is the standart Borel subgroup. By (H), f is cohomology isomorphism. The map g may be included in commutative triangle

$$\begin{array}{ccc} & BB(n) & \\ \nearrow & & \searrow^{g} \\ NU & \xrightarrow{\quad h \quad} & BGL(n) \end{array}$$

where U is an appropriate Zarissky covering of the complete flag space of e_n. By (DT), $h^* := H^*(h)$ is mono, therefore g^* is mono.

On the other hand, $\mathrm{Im}(gf)^* \subset H^*(BT(n), Z(*))^{\Sigma_n}$, and by Whitney sum formula, $(gf)^* c_i = \sigma_i(a_1,\dots,a_n)$, where σ_i is the i-th elementary symmetric function. Therefore, $\mathrm{Im}(gf)^* = A[\sigma_1,\dots,\sigma_n]$.

b) Follows from Whitney sum formula. ▲

B. $\mathcal{B}^1 P$

Let us turn now to the cohomology of $\mathcal{B}P = \mathcal{B}^1 P$.

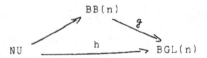

(2.5) By axiom (H), $H^*(\mathcal{B}P, Z(*)) = H^*(\mathcal{B}L, Z(*))$. Consider $\mathcal{B}L$ as simplicial object in $\Delta^o\mathcal{V}$. To its cohomology converges the spectral sequence

(2.5.1) $\mathcal{E}(\mathcal{B}L) : E^{pq} = H^p(\mathcal{B}L_q, Z(*)) \Longrightarrow H^{p+q}(\mathcal{B}L, Z(*))$,

whose E_1 term is the complex of (graded A-modules corresponding to the cosimplicial A-module.

(2.5.2) $A = H^*(\mathcal{B}L_o, Z(*)) \rightrightarrows H^*(\mathcal{B}L_1, Z(*)) \rightrightarrows H^*(\mathcal{B}L_2, Z(*)) \ldots$

Put $C = H^*(\mathcal{B}L_1, Z(*))$. $\mathcal{B}L_1 = \coprod_{n \in \mathbb{N}} BGL(n)$,

so by (2.4)

$$C = \prod_{n \geqslant 0} A[c_{1n}, \ldots, c_{nn}],$$

where $c_{in} \in H^{2i}(BGL(n), Z(i))$ are the universal Chern classes. Moreover,

$$H^*(\mathcal{B}L_k, Z(*)) \cong C^{\widehat{\otimes} k}$$

where $\widehat{\otimes}$ is the completed tensor product over A (i.e., if $C = \varprojlim C_{\leqslant i}$, where $C_{\leqslant i} = \prod_{0 \leqslant n \leqslant i} H^*(BGL(n), Z(*))$, then

$$C^{\widehat{\otimes} k} = \varprojlim_{(i_1, \ldots, i_k)} C_{\leqslant i_1} \otimes_A \cdots \otimes_A C_{\leqslant i_k}.$$

The Whitney sum $\mathcal{B}L_1 \times \mathcal{B}L_1 \longrightarrow \mathcal{B}L_1$ induces comultiplication $\mu : C \longrightarrow C \widehat{\otimes} C$ given by the formula

$$\mu(c_{in}) = \sum_{\substack{p+q=i \\ r+s=n}} c_{pr} \otimes c_{qs}$$

(cf. (2.4.b)),

which supplies C with the structure of commutative graded Hopf algebra over A. The term $E_1(\mathcal{B}L)$ is the completed cobar construction of C.

Let us compute $E_2(\mathcal{B}L) = H^*(E_1)$. It's clear that $H^0(E_1) = A$, and $H^2(E_1)$ is the space of primitive elements of C.

(2.5.2) <u>Lemma</u>. $H^1(E_1)$ is a free A-module admitting as a base the set $\{s_k\}_{k \in \mathbb{N}}$, where $s_k = (s_{ko}, s_{k1}, \ldots) \in C$, $s_{ki} \in A[c_{1i}, \ldots, c_{ii}]$ is the k-th Newton polynomial in c_{ij} (the expression of the sum of the k-th powers through elementary symmetric functions).

<u>Proof</u>. Put $C_i = A[c_{1i}, \ldots, c_{ii}]$; we can assume that $A = \mathbb{Z}$. Let $f = (f_o, f_1, \ldots)$, $f_i = f_i(c_{1i}, \ldots, c_{ii}) \in C_i$, be a primitive element. Supply C_i with a grading, assuming $\deg c_{ji}$ to be $2i$. Components

of comultiplication $\mu_{pq} : C_{p+q} \longrightarrow C_p \otimes C_q$ are homogenious, so we can assume deg(f) to be k for all p. The condition of primitivity is

$$\mu_{pq}(f_{p+q}) = p_1(f_p) + p_2(f_q),$$

where $p_1 : C_p \longrightarrow C_p \otimes C_q$ (resp., $C_q \longrightarrow C_p \otimes C_q$) are induced by $A \longrightarrow C_q$ (resp., $A \longrightarrow C_q$).

In other words,

$$(*) \quad f_{p+q}(c_{op}d_{1q} + c_{1p}d_{oq}, \ldots, c_{pp}d_{qq}) =$$

$$f_p(c_{1p}d_{oq}, \ldots, c_{pp}d_{oq}) + f_q(c_{op}d_{1q}, \ldots, c_{op}d_{qq}),$$

where $C_p \otimes C_q = A[c_{1p}, \ldots, c_{pp}; d_{1q}, \ldots, d_{qq}]$, $c_{op} = d_{oq} = 1$.

If $k = 0$, we obtain $f_p + f_q = f_{p+q}$, i.e. $f_p = pf_1 = s_{po} f_1$, $f_1 \in A$, for all p.

Suppose that $k > 0$. Substituting in $(*)$ $c_{1p} = \ldots = c_{pp} = 0$, we obtain $f_{p+q}(d_{1q}, \ldots, d_{qq}, 0, \ldots, 0) = f_q(d_{1q}, \ldots, d_{qq})$. On the other hand, substituting $c_{ip} = \ldots = c_{pp} = d_{ip} = \ldots = d_{pp} = 0$, $i \geqslant 2$, one deduces, by induction on i that

$$f_i(\sigma_1(x_1, \ldots, x_i), \ldots, \sigma_i(x_1, \ldots, x_i)) = f_1(x_1) + \ldots + f_1(x_i).$$

Therefore, $f_i = \mathrm{const} \cdot s_{k_i}$. ▲

To compute the whole ring $H^*(E_1)$ let us pass to the dual Hopf algebra. More precisely, put

$$C^\vee = \coprod_{i \geqslant 0} C_i^\vee,$$

where $C_i^\vee = \coprod_{p \geqslant 0} \mathrm{Hom}_A(C_i^p, A)$, C_i^p is the homogenious component of C_i of degree p.

(2.5.4) <u>Lemma</u> $C^\vee \cong A[b_o, b_1, \ldots]$ (as an algebra), where $b_i = c_{11}^{i*}$; comultiplication in C^V is given by the formula

$$b_i \longmapsto \sum_{p+q=i} b_p \otimes b_q.$$

Proof. This is essentially the known fact that

$$H_*(\coprod_{n>0} BGL^{top}(n,\mathbb{C}), \mathbb{Z}) \cong \mathbb{Z}[b_0, b_1, \ldots], \quad \text{where}$$

$$b_i = c_1^{i*}, \quad c_1 \in H^2(BGL^{top}(1, \mathbb{C}), \mathbb{Z}).$$

cf. [Q1 , Example 1, p. 198]. For completeness, let us sketch its proof.

Include C_i as the subalgebra of symmetric functions into the polynomial ring $B_i = A[t_{1i}, \ldots, t_{ii}]$, deg $t_{ji} = 2$. There exist an evident (non-commutative) comultiplication on $B = \prod B_i$ such that $C \hookrightarrow B$ is the morphism of Hopf algebras. The dual algebra B^V is isomorphic to the polymonial ring in non-commuting variables $a_i = t_{11}^{i*}$, $i \geq 0$.

Therefore, $C^V = B^V / \Sigma_\infty = A[b_0, b_1, \ldots]$ ▲

The dual compex E_1 is thus the bar-construction of B:

$$A \leftarrow B \leftarrow B^{\otimes 2} \longleftarrow \ldots,$$

whence $H_*(E_1^\vee) = \mathrm{Tor}_*^B(A,A)$. The last algebra is known to be the exterior algebra of the space $Q = I/I^2$ over A, $I = (b_0, b_1, \ldots)$. The coalgebra structure on $\mathrm{Tor}_*^B(A,A)$ is such that the space $Q = \mathrm{Tor}_1^B(A,A)$ consists of primitive elements. Passing again to dual objects, we get

(2.6) Proposition. The algebra $E_2(\mathcal{B}L)$ is isomorphic to the exterior algebra

$$\Lambda_A^*[H^1(E_1)] = A[s_0, s_1, \ldots],$$

where deg $s_i = 2i + 1$, s_i are defined in (2.5.3).

(2.7) Proposition. The differentials d_r in $\mathcal{E}(\mathcal{B}L)$ are zero when $r \geq 2$.

Proof. Let us denote by a the inclusion of $I = (0 \longrightarrow 1) \in \mathrm{Cat}$ in Δ° as $[1] \xrightarrow{d_1} [2]$. Then $(\mathrm{Id} \times a)^* \mathcal{B}^2 L$ is the 2-p.f. $\Delta^\circ \boxtimes I \longrightarrow \Delta^\circ \mathrm{Sch}$, whose restriction to $\Delta^\circ \times 0$ (resp., $\Delta^\circ \times 1$) is equal to $\mathcal{B}L$ (resp., $(\mathcal{B}L)^2$)). Thus, by (A8.2) we get the map of spectral sequences

(*) $\mu : \mathcal{E}(\mathcal{B}L) \longrightarrow \mathcal{E}(\mathcal{B}L \times \mathcal{B}L).$

From (2.6) we know that $E_2(\mathcal{B}L)$ is free over A and consequently, by Künneth

$$E_2(\mathcal{B}L \times \mathcal{B}L) = E_2(\mathcal{B}L) \otimes_A E_2(\mathcal{B}L).$$

Thus, (*) induces the structure of Hopf algebra on $E_2(\mathcal{B}L)$, and all s_i are primitive. But then all $d_2(s_i)$ are primitive and hence are equal to zero by the reason of degree. Therefore, $d_2 = 0$. Hence, $E_3(\mathcal{B}L) = E_2(\mathcal{B}L)$. Applying the same argument to E_3, we conclude that $d_3 = 0$, hence $E_4 = E_3$, and so on. ▲

(2.8) Theorem. There exist elements

$$c_{2i+1} \in H^{2i+1}(\mathcal{B}P, Z(i)), \quad i \geqslant 1,$$

such that

a) one has an isomorphism

(2.8.1) $H^*(\mathcal{B}P, Z(*)) \cong A[c_1^1, c_3^1, \ldots]$,

(cf. (2.1.5);

b) c^1_{2i+1} goes to $s_i \in H^{2i}(\mathcal{B}P_1, Z(i))$ under edge homomorphism

(2.8.2) $H^{2i+1}(\mathcal{B}P, Z(i)) \longrightarrow H^{2i}(\mathcal{B}P_1, Z(i))$.

Proof. By (2.7) there is a filtration $F^\bullet$ on $H^*(\mathcal{B}P, Z(*))$ such that

(2.8.3) $Gr_F H^*(\mathcal{B}P, Z(*)) = A[s_0, s_1, \ldots]$.

More exactly,

$$H^n = F^o H^n = F^1 H^n \supset F^2 H^n \supset \ldots \supset F^{n+1} H^n = 0,$$

and

$$s_i \in F^1 H^{2i+1}(\mathcal{B}P, Z(i))/F^2.$$

Let

$$c^1_{2i+1} \in H^{2i+1}(\mathcal{B}P, Z(I)).$$

be any liftings of s_i. Then b) holds by definition, and (2.8.1) follows from (2.8.3).

C. $\mathcal{B}^m P$, $m > 1$

(2.9) By (A8.1) and (A10.1 b) we have natural spectral sequences

(2.9.1) $\mathcal{E}(\mathcal{B}^m P) : E_1^{pq} = H^p((\mathcal{B}^{m-1}P)^q; Z(*)) \Rightarrow H^{p+q}(\mathcal{B}^m P, Z(*))$.

(2.10) Theorem. There exist elements

$$c_{2i+m}^m \in H^{2i+m}(\mathcal{B}^m P, Q(i)), \quad i \geq 1, \quad m \geq 1,$$

such that

a) one has an isomorphism

$$H^*(\mathcal{B}^m P, Q(*)) = A_Q[c_m^m, c_{m+2}^m, \ldots],$$

cf. (2.1.5));

b) c_{2i+m}^m goes to c_{2i+m-1}^{m-1} under the edge homomorphism

$$H^{2i+m}(\mathcal{B}^m P, Q(i)) \longrightarrow H^{2i+m-1}(\mathcal{B}^{m-1} P, Q(i)),$$

oming from $(2.9.1)_Q$.

roof. The theorem is proved by induction on m. The case $m = 1$ is ontained in (2.8). The implication $(m = k) \Rightarrow (m = k+1)$ can be prov- d by the same method as in the proof of (2.8).

At first one computes $E_2(\mathcal{B}^k P)$, using the fact that the cohomo- ogy of the polynomial algebra in characteristic zero is the polynomi- l algebra with the degrees of generators shifted by one.

Secondly, one proves that $\mathcal{E}(\mathcal{B}^k P)$ degenerates from E_2 on by he same argument as in (2.7).

Finally, one concludes using

2.10.1) Lemma. Let C be a connected graded commutative Hopf algeb- a over a ring R, which is isomorphic as an algebra to a polynomial ing $R[e_i]_{i \in I}$. If $R \supset Q$ or degrees of all e_i's are odd, then here exists a homogeneous change of variables of the form

$$e_i' = e_i + \text{(decomposable element)}$$

uch that all e_i' are primitive.

roof. Correct the e_i's in consecutive order, by lifting on their egree. ▲

The theorem is proved. ▲

D. Chern character

In our treatment of the theory of Chern classes we partially fol- ow [B].

(2.11) First recall some constructions from homological algebra.

For a (simplicial) scheme X, and $\mathcal{F} \in X^{\sim}$ let $K(X,\mathcal{F};i) \in \Delta^{\circ}X^{\sim}$ denote the simplicial sheaf corresponding by Dold-Puppe to the complex $\tau_{\leq 0}\mathcal{F}[i]$, $i \in \mathbb{Z}$.

Let

$$\Gamma : \mathcal{H}o^{\cdot}(\Delta^{\circ}X) \longrightarrow \mathcal{H}o^{\cdot}$$

be (the right derived of) the functor of global sections (recall that $\mathcal{H}o$ denotes the homotopy category, $\mathcal{H}o = \mathcal{H}o(\mathrm{Top})$). One has

(2.11.1) $\pi_j \Gamma K(X,\mathcal{F};i) = H^{i-j}(X,\mathcal{F})$, $j \geqslant i$.

(2.11.2) Lemma. For $E \in \mathcal{E}ns$, $i \geqslant 0$,

$$H^i(X \otimes E, \mathcal{F}) \cong \mathrm{Hom}_{\mathcal{H}o^{\cdot}}(E, \Gamma K(X,\mathcal{F};\ i))$$

(cf. [S, lemma (1.5.4)]).▲

(2.12) Return to the situation of (1.9). We have tautological morphisms

$$\varphi^m : \check{X} \otimes \underline{\mathrm{Hom}}\ (\check{X}, \mathcal{B}^m P) \longrightarrow \mathcal{B}^m P$$

(where we put for brevity $\check{X} = \text{"}\varprojlim\text{"}\ NU \in \text{Pro-Sch}$). Therefore, universal classes c^m_{2i+m} constructed in (2.8) and (2.10), define elements

$$s^1_i = \varphi^{1*} c^1_{2i+1} \in H^{2i+1}(\check{X} \otimes \underline{\mathrm{Hom}}\ (\check{X}, \mathcal{B}P), Z(i))$$

(resp.,

$$s^m_i = \varphi^{m*} c^m_{2i+m} \in H^{2i+m}(\check{X} \times \underline{\mathrm{Hom}}\ (\check{X}, \mathcal{B}^m P), Q(i)),\quad m > 1),$$

or by (2.11.2) and (1.9), morphisms in

(2.12.1) $s^1_i : \underline{\underline{S}}^{Is}\mathcal{P}(X) \longrightarrow \Gamma K(X,Z(i);\ 2i+1)$

(2.12.2) $s^m : \underline{\underline{S}}^{m,Is}\mathcal{P}(X) \longrightarrow \Gamma K(X,Q(i);\ 2i+m)$, $m > 1$

(note that because $Z(i)$ are complexes of Zarissky sheaves,

$$\Gamma K(\check{X},\ Z(i);\ *) = \Gamma K(X,\ Z(i);\ *)).$$

From (2.10.b) follows

(2.13) Theorem. The squares

$$\begin{array}{ccc} \Sigma \underline{\underline{S}}^{m,Is}\mathcal{P}(X) & \xrightarrow{\Sigma s_i^m} & \Sigma K(X, Q(i); 2i+m) \\ \downarrow & & \downarrow \\ \underline{\underline{S}}^{m+1,Is}\mathcal{P}(X) & \xrightarrow{s_i^{m+1}} & K(X, Q(i); 2i+m+1) \end{array}$$

m ≥ 1, commute in $\mathcal{H}o^{\cdot}$.

In particular,

$$\P_{j+m}(s_i^m) = \P_{j+m+1}(s_i^{m+1}) : K_j(X) \longrightarrow H^{2i-j}(X, Q(i))$$

(2.14) Now let us compare s_i^1 with Chern classes from [G], [S].

Let $BGL = \varinjlim BGL(n)$. By (2.4)

$$H^*(BGL, Z(*)) \cong A[c_2^o, c_4^o, \dots],$$

$$c_{2i}^o \in H^{2i}(BGL, Z(i)).$$

(2.14.1) At first let X be affine, $X = \mathrm{Spec}\ C$.

As in (2.12) one has the adjunction morphism

$$\varphi: X \otimes BGL(C) \text{ —— } BGL$$

Therefore, for a homogeneous $f \in \mathbb{Z}[T_1, T_2, \dots]$ of degree i (we put $\deg T_i = 2i$) we obtain classes

$$c(f) = \varphi^* f(c_2^o, c_4^o, \dots) \in H^{2i}(X \times BGL(C), Z(i)) \cong$$

$$\mathrm{Hom}_{\mathcal{H}o^{\cdot}}(BGL(C), K(X, Z(i); 2i)) =$$

$$\mathrm{Hom}_{\mathcal{H}o^{\cdot}}(\mathbb{Z}_\infty BGL(C), K(X, Z(i); 2i)),$$

where $\mathbb{Z}_\infty$ denotes the Bousfield-Kan completion, and hence

$$\P_j c(f) : K_j(C) \longrightarrow H^{2i-j}(X, Z(i)), \quad j \geq 1.$$

(2.14.2) Now let $X = \{X_p\}_{p \geq 0}$ be a simplicial affine scheme.

Note that the results of (2.11) may be generalized to a category of C-diagramms for every small category C, $\mathcal{H}o$ being replaced by homotopy category of pointed C-simplisial sets.

Therefore, applying the same construction as in (2.14.1), we obtain morphisms of cosimplicial topological spaces (defined up to homotopy).

$$\widetilde{c(f)} : \quad \{\mathbb{Z}_\infty BGL(X_p)\}_p \longrightarrow \{K(X_p, Z(i); 2i\}_p$$

whence

$$c(f) := \underset{\overleftarrow{p}}{\text{holim}}\ c(f) : \mathcal{K}(X)_{<0} := \underset{\overleftarrow{p}}{\text{holim}}\ \mathbb{Z}_{\infty} BGL(X_p) \longrightarrow$$

$$\underset{\overleftarrow{p}}{\text{holim}}\ \Gamma K(X_p, Z(i); 2i) \cong \Gamma K(X, Z(i); 2i).$$

(2.14.3) Now, for $X \in Sch$ put

$$\mathcal{K}(X)_{<0} = \mathcal{K}(\check{X})_{<0} := \underset{\overleftarrow{U}}{\lim}\ \mathcal{K}(NU)_{<0},$$

where U runs over affine coverings of X. We have

$$\mathcal{K}(X)_{<0} \cong \Omega_0\ \underline{\text{Hom}}(X, \mathcal{B}P) \cong \Omega_0 |Q\mathcal{P}(X)|,$$

where Ω_0 denotes the connected component of identity in the space of loops.

Morphisms $c(f)(x)$ are just the classes from [G], [S]. More precisely, in notations of [S]

(2.14.4) $\quad c_{ji} = \P_j\, c\, (T_i)$

<u>Remark</u>. This method enables one to define Chern classes for every simplicial scheme X.

(2.14.5) <u>Lemma</u>. Let $f = ax_i +$ (decomposable element). Then for $j > 0$

$$\P_j\, c\,(f) = a\, c_{ji}.$$

<u>Proof</u>. This follows for instance from [S, (2.3.2)]. ▲

Let us introduce the notation:

$$s_i^o = c(s_i),$$

where $s_i(T_i,\ldots,T_i)$ is the i-th Newton polynomial. From (2.14.5) and Newton formulas follows that for $j > 0$

(2.14.6) $\quad \P_j(s_i^o) = (-1)^{i-1} i c_{ji}.$

(2.15) <u>Theorem</u>. Classes s_i^1 are deloopings of s_i^o, i.e., the squares

$$\begin{array}{ccc} \Sigma\,\mathcal{K}(X)_{<0} & \xrightarrow{\ s_i^o\ } & K(X, Z(i); 2i) \\ \downarrow & & \downarrow \\ \underline{\underline{S}}^{Is}\mathcal{P}(X) & \xrightarrow{\ s_i^1\ } & K(X, Z(i); 2i+1) \end{array}$$

commute in $\mathcal{H}o^{\cdot}$.

Proof. This follows from the definition of s^1_i, (2.5.3).

(2.16) Let us look at s^1_i more attentively.

Suppose that $Z(i)$ consists of injective sheaves, and put $D^{\cdot} = \Gamma(X, Z(i))$, i.e., $HD^{\cdot} = H^*(X, Z(i))$. More generally, for every $E \in \Delta^{\circ}\mathcal{E}ns^{\cdot}$

$$H^*(X \otimes E, Z(i)) = H^*(\mathbb{H}om(E, D^{\cdot})),$$

where $\mathbb{H}om(E, D^{\cdot})$ is the simple complex associated with the double complex

$$Hom(E, D^{\cdot})^{pq} = Hom_{\mathcal{E}ns}(E_p, D^q) = Hom_{\mathbb{Z}}(\mathbb{Z}E_p, D^q)$$

the p-differential being induced by the differential in the complex $\mathbb{Z}E$ of integer chains of E, the q-differential - by differential in $D^{\cdot}$.

Let us apply this to $E = S.\mathcal{P}(X)$, and consider the element

$$s^1_i \in H^{2i+1}(X \otimes S.\mathcal{P}(X), Z(i)) = H^{2i+1}(X \otimes \underline{\underline{S}}.^{Is}\mathcal{P}(X), Z(i))$$

(cf. (1.2)). Let us represent s^1_i by a cocycle

$$z \in Z^{2i+1}\,\mathbb{H}om(S.\mathcal{P}(X), D^{\cdot})$$

(see Fig. 2.1).

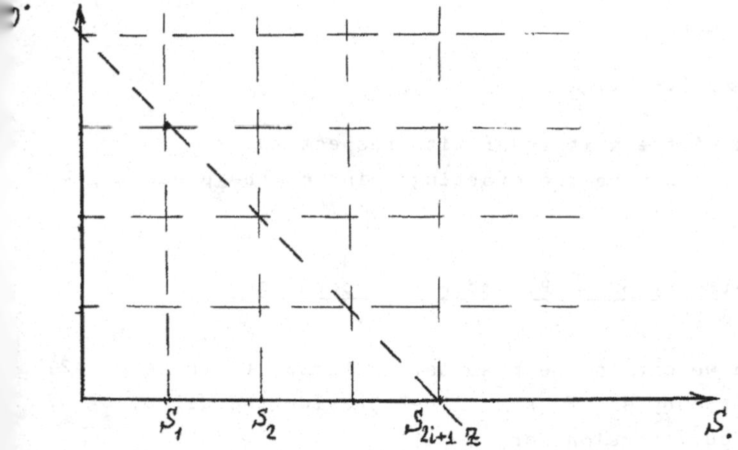

Fig. 2.1 $\mathbb{H}om(S.\mathcal{P}(X), D^{\cdot})$

z is the set of elements

$$\mathfrak{z}(x^n) \in D^{2i+1-n}$$

given for every $x^n = \{x_{ij}\}_{0\leq i\leq j\leq n} \in S_n\mathcal{P}(X)$, $n > 0$,

and satisfying the compatibility condition:

$$(*) \qquad d_D\, \mathfrak{z}(x^n) = \pm \sum_{\alpha=0}^{n} (-1)^{\alpha}\, \mathfrak{z}(d_{S,\alpha}\, x^n).$$

In particular, for $x \in S_1\mathcal{P}(X) = \mathrm{Ob}\,\mathcal{P}(X)$,

$\mathfrak{z}(x) \in Z^{2i}(D)$, and the cohomology class of $\mathfrak{z}(x)$ is $s_i([x])$ where $[x] \in K_0(X)$ is the class of x. Further, let $y \in K_1(X)$, and $\gamma(y) \in H_1S.\mathcal{P}(X)$ be the image of y under the Hurewitz homomorphism. $\gamma(y)$ is represented by a linear combination

$$\sum n_q y^q, \quad y^q = (y^q_{01} \rightarrowtail y^q_{02} \twoheadrightarrow y^q_{12}) \in S_2\mathcal{P}(X), \quad n_q \in \mathbb{Z},$$

and

$$(**) \qquad \sum_q n_q(y^q_{01} - y^q_{02} + y^q_{12}) = 0 \quad \text{in} \quad \mathbb{Z}S_1P(X).$$

From (*) and (**) follows that

$$\mathfrak{z}(y) := \sum n_q\, \mathfrak{z}(y^q) \in Z^{2i-1}(D^{\cdot})$$

The class of $\mathfrak{z}(y)$ in $H^{2i-1}(D^{\cdot}) = H^{2i-1}(X, Z(i))$ is just $s_{1i}(y)$.

In other words, one can say that s_{1i} is the obstruction to additivity of s_{0i} on the level of cocycles. Analogously, s_{ji} passes through the Hurewitz map

$$K_j(X) \longrightarrow H_jS.\mathcal{P}(X),$$

and is an operation of the next order with respect to $s_{j-1,i}$.

These considerations were the starting point of the present paper.

§3. K-theory of $\mathcal{B}^{\cdot}P$, and Adams operations

In this section we obtain the results, analogous to those of §2, for the algebraic K-theory of $\mathcal{B}^mP$'s. They provide the delooping of Adams operations (modulo torsion for $m > 1$).

(3.0) In this section the word "scheme" means "the scheme of finite Krull dimension". The schemes $\mathcal{B}^mP$ are again considered over $\mathrm{Spec}\,\mathbb{Z}$.

Put $A = K_*(\mathbb{Z})$.

Let us say a few words about the K-theory of simplicial schemes. Call a simplicial scheme X finite dimensional, if it coincides with its m-skeleton, $Sk_m X$, for some $m \in \mathbb{N}$. For example the nerve of finite Zarissky covering is finite dimensional.

For such a scheme define its K-theory in the following way. Choose an integer $q > m$ and form the cosimplicial space $\{|S^q P(X_i)|\}_{i>0}$. Let $S^q P(X)$ be the associated total space; we put

$$K_i(X) = \pi_{i+q} S^q P(X)$$

With this definition (which is independent of the choise of $q > m$), X can have a priori non zero negative K-groups $K_i(X)$ for $i \geq -m$.

One has a usual <u>converging</u> spectral sequence

$$E_2^{pq} = H^p(n \mapsto K_{-q}(X_n)) \Rightarrow K_{-p-q}(X),$$

which is located in the shaded region:

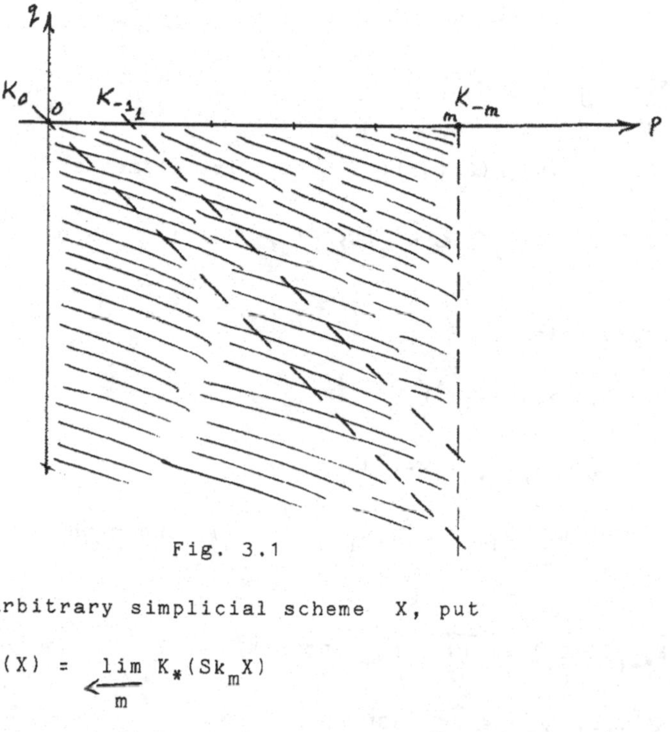

Fig. 3.1

For an arbitrary simplicial scheme X, put

$$\widehat{K}_*(X) = \varprojlim_m K_*(Sk_m X)$$

The analogous definitions we'll have in mind for polysimplicial schemes.

By the K-theory of an m-pseudo-functor $F : C \sim\longrightarrow \Delta^{\circ}Sch$ we'll mean the K-theory of its nerve $NF \in \Delta^{\circ m+1}Sch$.

With these definitions the following results are proved by the same arguments as in §2.

(3.1) Lemma. In the notations of (2.3) one has canonical isomorphisms

$$\widehat{K}_*(BT(n)) \cong A[[a_1, \ldots, a_n]],$$

where $a_i = pr_i^* a$, $a = cl(e_1) - [1] \in K_o(BT(1))$. ▲

(3.2) Proposition. In the notations of (2.4) let $\gamma^i \in \widehat{K}_0(BGL(n))$ denote the i-th Grothendieck operation of $cl(e_n) - [n] \in K_o(BGL(n))$. Then one has canonical isomorphism

$$\widehat{K}_*(BGL(n)) \cong A[[\gamma^1, \ldots, \gamma^n]].$$

Moreover, for every $p > 0$ one has Kunneth isomorphism

$$\widehat{K}_*\left(\prod_{i=1}^{p} BGL(n_i)\right) \cong \widehat{\bigotimes_{n=1}^{p}} \widehat{K}_*(BGL(n_i)).$$

Further,

$$\oplus^* \gamma^i = \sum_{p+q=i} \gamma'^p \otimes \gamma''^q.$$

(3.3) Theorem. (cf. (2.8), (2.10)). There exist elements

$$\beta^1_{2i+1} \in \widehat{K}_{-1}(\mathcal{B}P), \quad \beta^m_{2i+m} \in \widehat{K}_{-m}(\mathcal{B}^m P)_Q, \quad m > 1, \quad i \geqslant 0,$$

such that

a) one has isomorphisms

$$\widehat{K}_*(\mathcal{B}P) = A[[\beta^1_1, \beta^1_3, \ldots]];$$

$$K_*(\mathcal{B}^m P)_Q = A_Q[[\beta^m_m, \beta^m_{m+2}, \ldots]].$$

b) β^1_{2i+1} goes to $s_i = (s_{i0}, s_{i1}, \ldots, s_{in})$ under the edge homomorphism

$$\widehat{K}_*(\mathcal{B}P) \longrightarrow \widehat{K}_{*+1}(\mathcal{B}P_1) = \prod_{n>0} \widehat{K}_{*+1}(BGL(n))$$

where $s_{in} = s_i(\gamma^1, \ldots, \gamma^n) \in \widehat{K}_o(BGL(n))$.

β^{m+1}_{2i+m+1} goes to β^{m}_{2i+m} under the edge homomorphism

$$\widehat{K}_*(\mathcal{B}^{m+1}P)_Q \longrightarrow \widehat{K}_{*+1}(\mathcal{B}^{m}P)_Q. \blacktriangle$$

(3.4) Corollary. (cf. §2, D). there exist elements $\Psi^1_k \in \widehat{K}_{-1}(\mathcal{B}^1 P)$, $\Psi^m_k \in \widehat{K}_{-m}(\mathcal{B}^m P)_Q$, $m > 1$, which are the deloopings of k-th Adams operation, that is, for every $X \in Sch$ maps

$$\mathcal{B}P(\check{X}) \longrightarrow \mathcal{B}P(\check{X}),$$

$$\mathcal{B}^m P(X)_Q \longrightarrow \mathcal{B}^m P(X)_Q$$

induced by Ψ^m_k (cf. (2.12)) are m-fold deloopings of Adams operations defined in terms of BGL^+-construction (cf. for example, [So]).▲

Appendix

Weak functions and pseudo-punctors

A. Generalities

(A.1) Weak functors

(A 1.1) Definition. Let $C \in Cat$, $X : C^o \longrightarrow \mathcal{E}ns$ be a functor. By $Cat(X)$ we'll denote the fibered category over C corresponding to X, with discrete fibers $X(c)$, $c \in C$.

In other words,

$$Ob\,Cat(X) = \coprod_{c \in C} X(c),$$

and for $x \in X(c)$, $y \in X(d)$

$$Hom_{Cat(X)}(x,y) = \{ f \in Hom_C(c,d) \mid x = X(f)y \}.$$

In particular, we get functors

$$Cat : \Delta^{o m}\mathcal{E}ns \longrightarrow Cat, \quad m \geqslant 0.$$

(A1.2) Definition. A category with weak equivalences is a pair (C,W) where C is a category, and $W \subset Mor\,C$ is a class of morphisms (called weak equivalences) such that

a) W contains all isomorphisms;

b) if $\cdot \xrightarrow{f} \cdot \xrightarrow{g} \cdot \in N_2C$, and two elements of f, g, gf lie in W then so does the third.

A morphism $F : (C,W) \to (C', W')$ is a functor $F : C \to C'$ such that $F(W) \subset W'$. Categories with weak equivalences form category Catw.

(A1.3) Definition a) Let $(D,W) \in$ Catw, $C \in$ Cat. A category of weak functors $Wf(C,(D,W))$ or simply $Wf(C,D)$, is a full subcategory of $Hom_{Cat}(Cat(NC)^o,D)$, consisting of functors

$$F : Cat(NC)^o \longrightarrow D$$

such that

(W) for every $n \in \mathbb{N}$, $\sigma \in N_nC$

$$F(d_n^n) : F(\sigma) \longrightarrow F(d_n^n\sigma) \text{ lies in } W \text{ or, equivalently,}$$

$(F(\sigma) \longrightarrow F(s\sigma)) \in W$.

We'll denote weak functors by $F : C \dashrightarrow D$.

b) One has evident maps

(b1) $\underline{Hom}(C,C') \longrightarrow \underline{Hom}(Wf(C',D), Wf(C,D))$,

and

(b2) $\underline{Hom}_{Catw}(D,D') \longrightarrow \underline{Hom}(Wf(C,D), Wf(C,D'))$,

which make from $Wf(C,D)$'s a category Wf fibered over Cat, and cofibered over Catw.

(A1.4) Example. One has inclusions

$$i : Hom_{Cat}(C,D) \longrightarrow Wf(C,D),$$

sending a functor $F : C \to D$ to the weak functor $i(F)$ for which $i(F)(\sigma) = F(s\sigma)$, and $i(F)(s_i) = i(F)(d_i)$ (for $i>0$) $= id(F(s\sigma))$, and $i(F)(d_o) = F(f)$ for $\sigma = (s\sigma \xrightarrow{f} \cdot \to \ldots \to \cdot) \in N_nC$.

Weak functors of the form $i(F)$ we'll identify with the usual functors, and sometimes call strict functors.

(A1.5) Definition. Let $F \in Wf(C,D)$.
Suppose that D has coproducts. The nerve $NF \in \Delta^oD$ is defined in

the following way:

$$N_nF = \coprod_{\sigma \in N_nC} F(\sigma);$$

for $f \in Hom_\Delta([n],[m])$ $f^* : N_mF \longrightarrow N_nF$ is induced by $F(f) : F(\sigma) \longrightarrow F(f^*\sigma)$.

(A2) Weak functors from pseudo-functors.

Pseudo-functors from [HS] lead to non-trivial examples of weak functors.

(A2.1) Definition. a) Let $C \in$ Cat. A coefficient system (c.s.) E over C consists of

1) the simplicial sets E_α given for every $\alpha \in$ Mor C, all $E_{id(x)}$, $x \in$ Ob C being equipped with a distinguished vertice e_x.

2) the morphisms

$$\mu_{\alpha\beta} : E_\alpha \times E_\beta \longrightarrow E_{\beta\alpha}$$

given for every $(\cdot \xrightarrow{\alpha} \cdot \xrightarrow{\beta} \cdot) \in N_2C$,

such that

3) for every $\alpha : x \longrightarrow y \in$ Mor C

$$\mu_{id(x),\alpha}|_{e_x \times E} = id_E = \mu_{\alpha,id(y)}|_{E_\alpha \times e_y}$$

4) for every $\cdot \xrightarrow{\alpha} \cdot \xrightarrow{\beta} \cdot \xrightarrow{\gamma} \cdot \in N_3C$ the square

$$\begin{array}{ccc} E_\alpha \times E_\beta \times E_\gamma & \longrightarrow & E_\alpha \times E_{\gamma\beta} \\ \downarrow & & \downarrow \\ E_{\beta\alpha} \times E_\gamma & \longrightarrow & E_{\gamma\beta\alpha} \end{array}$$

commutes.

A c.s. $E = \{E_\alpha\}$ is called contractible if all E_α are contractible.

A morphism

$$\varphi : E \longrightarrow E'$$

of c.s.'s over C is a set of morphisms

$$\varphi_\alpha : E_\alpha \longrightarrow E'_\alpha, \qquad \alpha \in \text{Mor } C,$$

commuting with μ's and respecting e_x's.

C.S's over C form a category $C\lambda(C)$. For a functor

$$f : C \longrightarrow C'$$

one has evident functor

$$f^* : Cs(C') \longrightarrow Cs(C).$$

Thus, all $Cs(C)$, $C \in Cat$, form a fibered category Cs over Cat.

b) Let D be a category which is a (right) module over $\Delta^\circ \mathcal{E}ns$, that is, equipped with a functor

$$\times : D \times \Delta^\circ \mathcal{E}ns \longrightarrow D$$

and natural isomorphisms

$$(X \times A) \times B \xrightarrow{\sim} X \times (A \times B), \quad X \times * \xrightarrow{\sim} X,$$

satisfying the obvious coherence conditions.
In applications we'll need only the categories of type $\Delta^\circ C$ where C has coproducts and $\times$ is defined in (0.3.1).

A pseudo-functor (p.f.)

$$F : C \sim\longrightarrow D$$

consists of

1) a contractible c.s. $E = \{E_\alpha\}$ over C;
2) a set of objects $F_x \in D$ defined for all $x \in Ob\ C$;
3) a set of morphisms

$$\nu_\alpha : F_x \times E_\alpha \longrightarrow F_y$$

defined for all $\alpha : x \longrightarrow y \in Mor\ C$, such that

4) for all $x \in Ob\ C$

$$id(x)|_{F_x \times e_x} = id(F_x);$$

5) for all $x \xrightarrow{\alpha} y \xrightarrow{\beta} z \in N_2 C$ the square

$$\begin{array}{ccc} F_x \times E_\alpha \times E_\beta & \xrightarrow{\nu} & F_y \times E \\ \downarrow \mu & & \downarrow \nu \\ F_x \times E_{\beta\alpha} & \longrightarrow & F_z \end{array}$$

commutes.

A morphism of p.f.'s

$$f : F \longrightarrow F'$$

over the same c.s. E is a set of morphisms

$$f_x : F_x \longrightarrow F_x'$$

commuting with ν_α.

The evident inverse image functors, as in a) define a category

Pf(C,D) of pseudo-functors over C, fibered over Cs(C), and also Pf(•,D) fibered over Cs (and over Cat).

Varying D (in the category of right $\Delta^{\circ}\mathcal{E}ns$-modules $Cat_{\Delta^{\circ}\mathcal{E}ns}$) we obtain the category Pf fibered over Cat and cofibered over $Cat_{\Delta^{\circ}\mathcal{E}ns}$.

(A2.2) Definition. a) Let $(D,W) \in Catw$, and D be a right $\Delta^{\circ}\mathcal{E}ns$-module. Assume that the following condition holds:

(*) for every $x \in D$ functor

$$x \times \cdot \ : \Delta^{\circ}\mathcal{E}ns \longrightarrow D$$

sends weak equivalences to weak equivalences, were "weak equivalence" in $\Delta^{\circ}\mathcal{E}ns$ means weak homotopy equivalence.

Let $F \in Pf(C,D)$ be a p.f. over a c.s. E .

A weak functor $W(F) \in \mathcal{W}f(C,D)$ is defined in the following way. For $\sigma \in NC$

$$W(F) = F_{s\sigma} \times E_{\sigma} ,$$

where we use the notation $E_{\sigma} = E_{\alpha_1} \times \dots \times E_{\alpha_n}$ for $\sigma = x_0 \xrightarrow{\alpha_1} \dots \xrightarrow{\alpha_n} x_n$.

We have evident maps

$$W(F)(\sigma) \longrightarrow W(F)(d_0\sigma) \quad \text{induced by } \nu ,$$

$$W(F)(\sigma) \longrightarrow W(F)(d_i\sigma), \quad i > 0, \quad \text{induced by } \mu,$$

and

$$W(F)(\sigma) \text{ —— } W(F)(s_i\sigma) \quad \text{induced by } * \longrightarrow E_{id(x_i)}.$$

These maps define the structure of a weak fucntor on W(F). Condition (W) follows from the contractibility of E, and (*).

Thus, we have defined a functor

$$W : Pf \longrightarrow \mathcal{W}f.$$

b) In addition to the conditions of a), suppose that admits a right adjoint, i.e. that there exists a functor

$$Hom: \ (\Delta^{\circ}\mathcal{E}ns)^{\circ} \times D \longrightarrow D$$

such that there is functorial isomorphism

$$Hom(X, Hom(A,Y)) \cong Hom(X \times A, Y), \quad X,Y \in D, \quad A \in \Delta^{\circ}\mathcal{E}ns.$$

Suppose also that

$(*)^{V}$ for every $X \in D$

$$\underline{\mathrm{Hom}}\,(\cdot, X) : (\Delta^{\circ}\mathcal{E}ns)^{o} \longrightarrow D$$

sends weak equivalences to weak equivalences.
For $F \in Pf(C,D)$, $\sigma \in N_nC$ put

$$W^{V}(F)(\sigma) = \underline{\mathrm{Hom}}\,(E\ , F_t\ ),$$

and for $f \in \mathrm{Hom}_{\Delta}([m],[n])$ define maps

$$W^{V}(F)(f^*\sigma) \longrightarrow W^{V}(F)(\sigma)$$

in the way, dual to that of a).

We obtain a weak functor $W^{V}(F) \in \mathcal{W}f(C,D^{o})$.

(A2.3) Definition. In conditions of (2.2), for $F \in \mathcal{P}f(C,D)$

$$\underset{\longrightarrow}{\mathrm{holim}}\ F := NW(F) \in \Delta^{\circ}D;$$

$$\underset{\longleftarrow}{\mathrm{holim}}\ F := NW^{V}(F) \in \Delta^{\circ}D^{o} = \Delta D$$

(we assume of course that D has coproducts or products).
This definition essentially coincides with [HS, (2.2.2)].

(A3) The spectral sequence

Let $C(\mathcal{A})$ be a category of complexes over some abelian category $\mathcal{A}$.
From now on, weak equivalence in $C(\mathcal{A})$ will mean quasiisomorphism.
Let $F \in \mathcal{W}f(D,C^{\cdot}(\mathcal{A}))$. Then we have $|NF| \in C^{\cdot}(\mathcal{A})$ were $|\cdot|$ denotes the composition

$$\Delta^{\circ}C(\mathcal{A}) \cong C^{\leq 0}C(\mathcal{A}) \xrightarrow{\mathrm{Tot}} C^{\cdot}(\mathcal{A}).$$

From the other hand, the cohomology groups $H^{i}(F)$ obviously form strict functors $D \longrightarrow \mathcal{A}$ and one has a natural cohomological spectral sequence

$$E^{pq} = H_{-p}(D, H^{q}(F)) = H^{p}(D^{o}, H^{q}(F)^{o}) \Longrightarrow H^{p+q}(|NF|)$$

($H^{*}(F)^{o}$ denotes $H^{*}(F)$ considered as a contravariant functor $D^{o} \longrightarrow \mathcal{A}$).

(A4) Finality theorem

Let $f : C \longrightarrow D \in \mathrm{Mor}\ \mathcal{C}at$, $F \in \mathcal{W}f(D, C^{\cdot}(\mathcal{A}))$.

Suppose that for every $y \in D$ fiber $y \setminus f$[1] is contractible. Then the natural map

$$f_* : |Nf^*F| \longrightarrow |NF|$$

is quasi-isomorphism.

Sketch proof. Form the bisimplicial complex $N(f,F)$ with

$$N(f,F)_{pq} = \coprod F(y_o \rightarrow \ldots \rightarrow y_q \rightarrow f(x_o) \rightarrow \ldots \rightarrow f(x_p)),$$

the sum being taken over all triples

$$(\sigma = (x_o \rightarrow \ldots \rightarrow x_p) \in N_pC; \tau = (y_o \rightarrow \ldots \rightarrow y_q) \in N_qD; \varphi: t\tau \rightarrow sf(\sigma)).$$

Then proceed as in the proof of [HS, (2.4)], or [Q2, Thm, A]. (cf. also (A7.2). below).▲

In applications we'll need a certain generalization of the previous constructions and results. We develop it in nn°5-8.

Fix an integer $m > 1$.

(A5) Weak m-functors

(A5.1) m-categories

(A5.1.1) Definition a) Category of small m-categories, $Cat^{(m)}$, is defined inductively: $Cat^{(1)} = Cat$, $Cat^{(m)}$ is category of categorical objects in $Cat^{(m-1)}$.

b) We have a functor of nerve

$$Cat^{(m)} \longrightarrow \Delta^\circ Cat^{(m-1)},$$

and iterating it we obtain

$$N : Cat^{(m)} \longrightarrow \Delta^{\circ m}\mathcal{E}ns$$

(A5.1.2) Let $X \in Cat^{(m)}$. It is convenient to imagine X geometrically in the following way.

Let $\square^m = \{(x_1,\ldots,x_m) \in \mathbb{R}^m \mid \forall_i \ 0 \leqslant x_i \leqslant 1\}$ be the standart m-dimensional cube. Let $\mathcal{F}aces(\square^m)$ denote the following category. Its objects are faces of $\square^m$ i.e. subsets $F_{A,B} \subset \square^m$ indexed by pairs of non-intersecting subsets $A,B \subset \{1,\ldots,m\}$ such that $F_{A,B} := \{(x_1,\ldots,x_m) \mid x_i = 0$ if $i \in A$ and $x_i = 1$ if $i \in B.\}$

1) Recall that it is a category with objects - pairs $(x \in C, \varphi : y \rightarrow f(x))$, and maps from (x,φ) to (x',φ') such $\psi: x \rightarrow x'$ that $\varphi' \cdot \psi = \varphi$.

Maps in $\mathcal{F}aces\,(\square^m)$ are compositions of inclusions $F_{A,B} \hookrightarrow F_{A',B'}$ and projections $F_{A',B'} \longrightarrow F_{A,B}$, $A \supset A'$, $B \supset B'$.

Let $\mathcal{T}ypes\,(m)$ be the full subcategory of $\mathcal{F}aces\,(\square^m)$ with objects of the form $F_{A,\phi}$. We shall identify $\mathrm{Ob}\,\mathcal{T}ypes\,(m)$ with $\{0,1\}^m$ using the correspondence $(\alpha = \{\alpha_1,\ldots,\alpha_m\},\ \alpha_i = 0 \text{ or } 1) \longleftrightarrow (F(\alpha) = F_{A,\phi}$ where $i \in A \Longleftrightarrow \alpha_i = 0)$, and denote this set simply $\mathcal{T}ypes$.

For $\alpha \in \mathcal{T}ypes$ call its <u>dimension</u>, $\dim\alpha$, the dimension of the corresponding face, i.e. the number of ones among α_i's. We'll denote by $\mathbb{1}(\alpha) = \{\alpha_{i_1},\ldots,\alpha_{i_k}\}$, $i_1 < \ldots < i_k$, the set of ones in α.

Put $\mathcal{T}ypes^k = \{\alpha \in \mathcal{T}ypes \mid \dim\alpha = k\}$. One has evidently $\mathcal{T}ypes = \coprod_{k=0}^{m} \mathcal{T}ypes^k$. For $\alpha \in \mathcal{T}ypes^k$, $1 \leqslant p \leqslant k$, $1 \leqslant q \leqslant m-k$, let $\partial_p\alpha \in \mathcal{T}ypes^{k-1}$ (resp., $\delta_q\alpha \in \mathcal{T}ypes^{k+1}$) denote the type obtained by replacing p-th 1 (from left to right) in α by 0 (resp., q-th by 1).

The following data form a description of an m-category X equivalent to (5.1.1).

a) A functor

$$\mathrm{Mor} = \mathrm{Mor}\,X : \mathcal{T}ypes\,(m)^{\circ} \longrightarrow \mathcal{E}ns,$$

$$\alpha \longmapsto \mathrm{Mor}^{\alpha}(X).$$

We put $\mathrm{Mor}^k(X) = \coprod_{\alpha\,:\,\dim\alpha = k} \mathrm{Mor}^{\alpha}(X)$.

In the language of (5.1.1) Mor^{α} is the set $\mathcal{O}_1\mathcal{O}_2\ldots\mathcal{O}_m(X)$, where $\mathcal{O}_i = \mathrm{Ob}$ if $\alpha_i = 0$ and $\mathcal{O}_i = \mathrm{Mor}$ if $\alpha_i = 1$. Elements of Mor (X) are called <u>morphisms of type α</u> and are pictured as k-dimensional cubes, $k=\dim\alpha$. Morphisms of type (0,0,...,0) are called <u>objects</u>. Abusing the notations, we put $\mathrm{Mor}\,(X) = \coprod_{\alpha} \mathrm{Mor}^{\alpha}(X)$.

For $\alpha \in \mathcal{T}ypes^k$ one has 2k maps

$$s_i,\ t_i : \mathrm{Mor}^{\alpha}(X) \rightrightarrows \mathrm{Mor}^{\partial_i\alpha}(X),\quad 1 \leqslant i \leqslant k, -$$

k "sources", or front faces, and k "targets", or back faces. Also every k-morphism has source and target which are objects:

$$s,\ t : \mathrm{Mor}^k \rightrightarrows \mathrm{Mor}^{\circ},\ s = s_{i_1}\ldots s_{i_k},\quad t = t_{i_1}\ldots t_{i_k}$$

(see Fig. A.1)

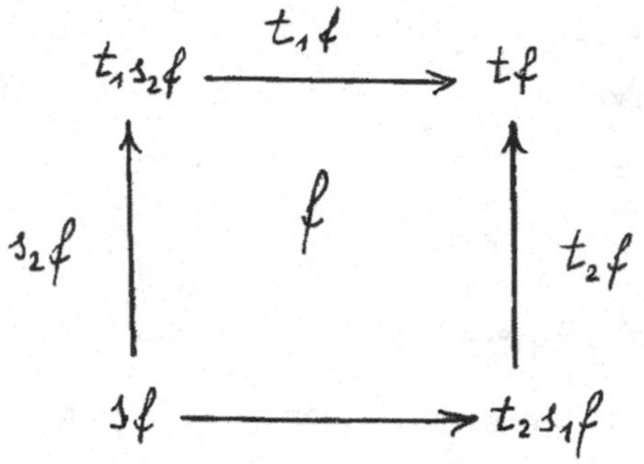

Fig. A.1. 2-morphism and its faces.

One has also m-k "identities"

$$id_j : Mor^{\alpha} \longrightarrow Mor^{\sigma_j \alpha}, \quad j = 1,\dots,m-k.$$

b) Further, one has operations of composition, which correspond to concatenation of cubes (see Fig. A.2).

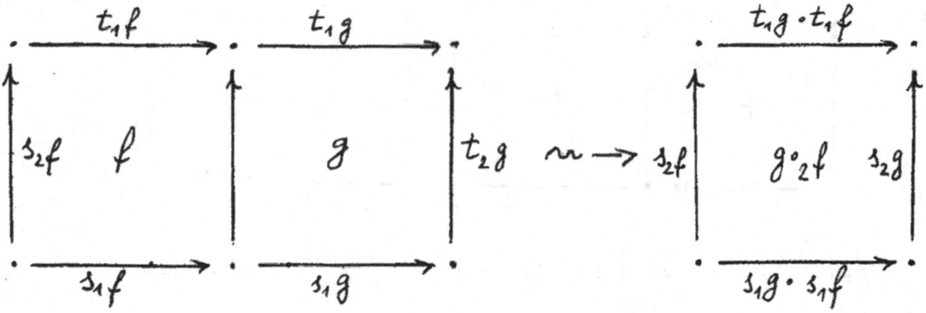

Fig. A. 2. Composition of 2-morphisms

More exactly, for every $k \geqslant 1$, $\alpha \in \mathcal{T}ypes^k$, $1 \leqslant i \leqslant k$ one has map

$$\circ_i : \{(f, g) \in Mor^{\alpha} \times Mor^{\alpha} \mid s_i g = t_i f\} \longrightarrow Mor^{\alpha},$$

$$(f, g) \longmapsto g \circ_i f, \quad \text{or simply} \quad g \circ f.$$

These maps must satisfy the following properties.

Ax. o) For $j \neq i$ $\quad s_j(g \circ_i f) = s_j(g) \circ_{i'} s_j(f)$;

$t_j(g \circ_i f) = t_j(g) \circ_{i'} t_j(f)$, $i' = i$ if $i < j$, $i' = i - 1$ if $i > j$.

$s_i(g \circ_i f) = s_i(f)$; $t_i(g \circ_i f) = t_i(g)$.

Ax. 1) They must be associative.

(Ax2) For all $\alpha \in \mathcal{T}ypes^k$, $f \in Mor^{\alpha}(X)$, $1 \leqslant i \leqslant k$,

$$f \circ_i id_j(s_i f) = id_j(t_i f) \circ_i f = f,$$

where j is such that $\alpha = \sigma_j \partial_i \alpha$.

Ax. 3) If $1 \leqslant i < j \leqslant k$, and f, g, h, e are such that

$$t_i f = s_i g, \quad t_i f = s_j h, \quad t_i h = s_i e, \quad t_j g = s_j e$$

(See Fig. A.3),

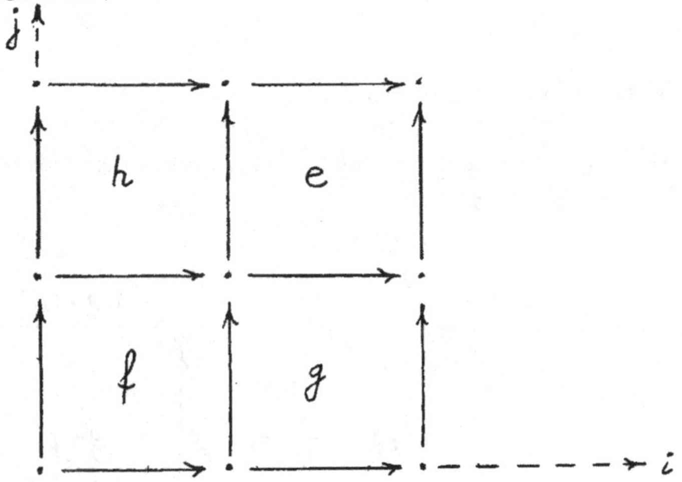

Fig. A.3. to Axiom 3: Two ways of making composition

then

$$(e \circ_i h) \circ_j (g \circ_i f) = (e \circ_j g) \circ_i (h \circ_j f).$$

(A5.1.3) Let us describe in these terms the nerve.

If $(n_1, \ldots, n_m) \in \mathbb{N}^m$, and exactly $n_{i_1}, \ldots, n_{i_{m-k}}$ are equal to zero, then an $(n_1, \ldots, n_m)$ - simplex of NC is a k-dimensional parallelogramm divided into $n_{j_1} \ldots n_{j_k}$ cubes, where $\{j_1, \ldots, j_k\} = \{1, \ldots, m\} - \{i_1, \ldots, i_{m-k}\}$ (see Fig. A.4).

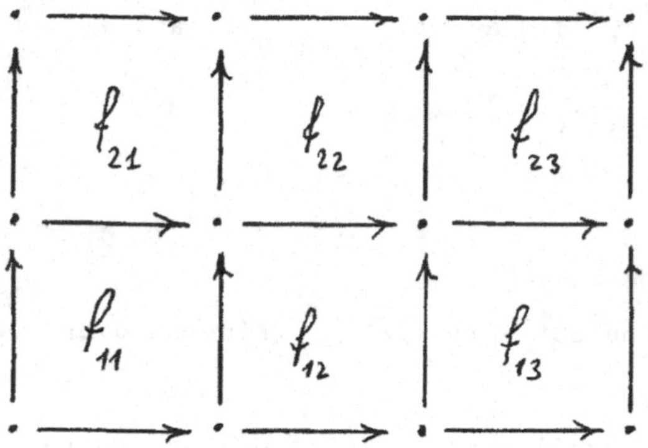

Fig. A.4. A (3,2)-simplex in 2-category

t is convenient to imagine it as k-dimensional matrix of k-morphisms:

$$\sigma = \{f_{p_1,\dots,p_k}\}, \quad 1 \leq p_i \leq n_{j_i} \;; \quad s_i f_{p_1,\dots,p_k} =$$

$$= t_i f_{p_1,\dots,p_i-1,\dots,p_k}.$$

ut

$$s\sigma = sf_{1,1,\dots,1}; \quad t\sigma = tf_{n_{j_1},\dots,n_{j_k}}.$$

A5.2) Example. Let $C_1, \dots, C_m \in$ Cat. Their external product

$$C_1 \boxtimes \dots \boxtimes C_m \in \mathrm{Cat}^{(m)}$$

s defined by

$$\mathrm{Mor}^{\alpha}(C_1 \boxtimes \dots \boxtimes C_m) = X_1 \times \dots \times X_m,$$

here for $\alpha = (i_1,\dots,i_m)$ $X_j = \mathrm{Ob}\, C_j$ if $i_j = 0$, and $\mathrm{Mor}\, C_j$ if $i_j = 1$. One has

$$N(C_1 \boxtimes \dots \boxtimes C_m)_{n_1,\dots,n_m} = N_{n_1}C_1 \times \dots \times N_{n_m}C_m.$$

In fact, we shall not need other examples of m-categories in applications.

A5.3) Definition. Let $C \in \mathrm{Cat}^{(m)}$, $(D,W) \in$ Catw. category of weak m-functors $\mathcal{W}f^{(m)}(C,D)$ is a full subcategory of $\mathrm{Hom}_{\mathrm{Cat}}$ (Cat NC,D) consisting of such functors

$$F : \mathrm{Cat}\, NC \longrightarrow D$$

hat

$(W)_m$ for all $\sigma \in NC$, $\dim \sigma = (n_1,\dots,n_m)$, and all $i = 1,\dots, m$,

$$F({}^{i}d^{n_i}_{n_i}) : F(\sigma) \longrightarrow F({}^{i}d^{n_i *}_{n_i}\sigma) \in W,$$

or, equivalently,

$$F(\sigma) \longrightarrow F(s\sigma) \in W$$

Weak m-functors form the category $\mathcal{W}f^{(m)}$ fibered over $Cat^{(m)}$ and cofibered over Catw.

(A.5.4) <u>Definition</u>. Let $F \in \mathcal{W}f^{(m)}(C,D)$. Assume D to have coproducts. <u>The nerve</u> $NF \in \Delta^{\circ m}D$ is defined by

$$(NF)_{n_1,\dots,n_m} = \coprod_{\substack{\sigma \in NC: \\ \dim\sigma = (n_1,\dots,n_m)}} F(\sigma).$$

(A6) <u>m-pseudo-functors</u>

(A6.1) <u>Definition</u>. Let $C \in Cat^{(m)}$. An <u>m-coefficient system</u> over C is an m-categorical object E in $\Delta^{\circ}\mathcal{E}ns$ such that

a) Ob E = Ob C (discrete simplicial set)

b) $\pi_o E = C$, where $\pi_o : Cat^{(m)}(\Delta^{\circ}\mathcal{E}ns) \longrightarrow Cat^{(m)}$

is the functor of connected components.

E is called <u>contractible</u> if for every $\alpha \in \mathcal{T}ypes$ every connected component of $Mor^{\alpha}(E)$ is contractible.

(A6.1.1) Thus, to define a coefficient system over C one has:

a) for every $\alpha \in \mathcal{T}ypes$, $\dim \alpha > 0$, $f \in Mor^{\alpha} C$ to give a simplicial set E_f (it will be the connected component of $Mor^{\alpha}(E)$ corresponding to f);

b) for every $\varphi : \beta \longrightarrow \alpha$ $Mor\,\mathcal{T}ypes$, $g = Mor(E)(\beta)(f)$ to give a morphism $\varphi_* : E_g \longrightarrow E_f$;

c) for composable $f,g \in Mor^{\alpha} C$, $t_i f = s_i g$ for some i, to give morphism

$$\mu_{f,g} : E_f \times E_g \longrightarrow E_{gf}.$$

These data must satisfy certain compatibility conditions which follow from (A6.1).

(A6.2) Let D be a category satisfying the conditions of (A2.1.b).

(A6.2.1) Let us attach to D a certain m-category $\widehat{D}$ in $\Delta^{\circ}\mathcal{E}ns$.

$$\mathrm{Ob}\,\hat{D} = \mathrm{Ob}\,D \quad \text{(discrete simplicial set).}$$

Let $\alpha \in \mathcal{T}ypes^k$. The symplicial set $\mathrm{Mor}^\alpha(\hat{D})$ is defined as follows. Its p-symplex $f \in \mathrm{Mor}^\alpha(\hat{D})_p$ consists of

a) a set of objects $f_\beta \in \mathrm{Ob}\,D$ indexed by those $\beta \in \mathcal{T}ypes$ which are faces of α i.e. such that $\mathbb{I}(\beta) \subset \mathbb{I}(\alpha)$. One must imagine f_β as standing in the vertices of $\alpha = \square^k$.

b) a set of maps $f_{\beta,\beta'} : f_\beta \times \Delta[p] \longrightarrow f_{\beta'}$, given for every pair (β,β') such that $\mathbb{I}(\beta) \subset \mathbb{I}(\beta') \subset \mathbb{I}(\alpha)$.

c) If $\mathbb{I}(\beta) \subset \mathbb{I}(\beta') \subset \mathbb{I}(\beta'') \subset \mathbb{I}(\alpha)$ then the composition

$$f_\beta \times \Delta[p] \xrightarrow{\text{diag}} f_\beta \times \Delta[p] \times \Delta[p] \xrightarrow{f_{\beta,\beta'}} f_{\beta'} \times \Delta[p] \xrightarrow{f_{\beta',\beta''}} f_{\beta''}$$

must to equal to $f_{\beta,\beta''}$.

If $m = 1$ then the last condition is vacuous.

The simplicial structure on $\{\mathrm{Mor}^\alpha(\hat{D})_p\}_{p\geqslant 0}$ is induced by the standart cosimplicial structure on $\{\Delta[p]\}_{p\geqslant 0}$.

If $\varphi : \alpha \longrightarrow \beta \in \mathrm{Mor}\,\mathcal{T}ypes(M)$ is a face then one has obvious map $\varphi^* : \mathrm{Mor}^\beta(\hat{D}) \longrightarrow \mathrm{Mor}^\alpha(\hat{D})$. If φ is a projection, $f \in \mathrm{Mor}^\beta(\hat{D})_p$ then $\varphi^* f$ is obtained from f by adding the projections $f_\gamma \times \Delta[p] \longrightarrow f_\gamma$ in the directions of degeneracy.

Thus, we have defined functor

$$\mathrm{Mor}\,(\hat{D}) : \mathcal{T}ypes\,(m)^o \longrightarrow \Delta^o \mathcal{E}ns.$$

It remains to define the composition. Let $\alpha \in \mathcal{T}ypes^k$,

$f, g \in \mathrm{Mor}^\alpha(\hat{D})_p$, $\quad 1 \leqslant i \leqslant k$, $\quad s_i g = t_i f$.

We have to define $\quad g \cdot_i f = gf \in \mathrm{Mor}^\alpha(\hat{D})_p$.

Let $\mathbb{I}(\alpha) = \{n_1, \dots, n_k\}$, $\beta \in \mathcal{T}ypes$, $\mathbb{I}(\beta) \subset \mathbb{I}(\alpha)$.

We put $(gf)_\beta = f_\beta$ if $\beta_{n_i} = 0$, and g_β if $\beta_{n_i} = 1$.

Finally, by property c) it suffices to define $(gf)_{\beta,\beta'}$ only for such (β,β') that $\dim \beta' = \dim + 1$.

We put $(gf)_{\beta,\beta'} = f_{\beta,\beta'}$ if $\beta_{n_i} = \beta'_{n_i} = 0$,

$(gf)_{\beta,\beta'} = g_{\beta,\beta'}$ if $\beta_{n_i} = \beta'_{n_i} = 1$, and to be the composition

$$f_\beta \times \Delta[p] \xrightarrow{\text{diag}} f_\beta \times \Delta[p] \times \Delta[p] \xrightarrow{f_{\beta,\beta'}} f_{\beta'} \times \Delta[p] = g \times \Delta[p] \xrightarrow{g_{\beta,\beta'}} g_{\beta'}$$

if $\beta_{n_i} = 0$ and $\beta'_{n_i} = 1$.

The required checkings are straightforward.

(A6.2.2) Definition. An m-pseudo-functor $F : C \sim\!\!\longrightarrow D$ consists of

a) a contractible m-c.s. E over C;

b) a morphism

$$F : E \longrightarrow \widehat{D}$$

in $Cat^{(m)}(\Delta^{\circ}\mathcal{E}ns)$.

Thus, to define an m-p.-f. over an m-c.s. $E = \{E_f\}$ one must define a map $F : Ob\, C \longrightarrow Ob\, D$, and for every $f \in Mor\, C$ a map

$$F_f : F(sf) \times E_f \longrightarrow F(tf).$$

These maps must satisfy certain compatibilities which we leave the reader to write down.

(A.6.3) Definition. Let $(D,W) \in Catw$ be as in (A2.2a), $C \in Cat^{(m)}$. Let us construct the functor

$$W : \mathcal{P}f(C,D) \longrightarrow \mathcal{W}f(C,D).$$

Namely, let $F \in \mathcal{P}f^{(m)}(C,D)$ be a p.f. over a c.s. E.

Let $\underline{n} = (n_1,\dots,n_m) \in \mathbb{N}^m$. Let $n_{p_1},\dots,n_{p_k}$, $p_1 < \dots < p_k$, be the positive entries of n. Suppose that σ consists of the morphisms $f_{i_1,\dots,i_k}$, $1 \leq i_1 \leq n_{p_1},\dots,$ $1 \leq i_k \leq n_{p_k}$ (see. (A5.1.3)).

Put

$$E = \prod_{(i_1,\dots,i_k)} E_{f_{i_1,\dots,i_k}},$$

where the set of indices $[1, n_{p_1}] \times \dots \times [1, n_{p_k}]$ is ordered lexicographically. Finally, put

$$W(F)(\sigma) = F_{s\sigma} \times E_{\sigma}.$$

We leave to the redder the definition of structure maps

$$W(F)(\sigma) \longrightarrow W(F)(g^{*}\sigma), \quad g \in Mor(\Delta^m).$$

(cf. A2.2 a) for $m = 1$).

Dually, under conditions of (A2.2b) we define

$$W^{\vee} : \mathcal{P}f^{(m)}(C,D) \longrightarrow \mathcal{W}f(C,D^{\circ}).$$

We put

(A6.3.1) $\quad \underrightarrow{\mathrm{holim}}\, F = NW(F) \in \Delta^{\circ m}D$

(A6.3.2) $\quad \underleftarrow{\mathrm{holim}}\, F = NW^{\vee}(F) \in \Delta^{m}D$

(A7) Finality theorem.

Let $X \in \mathrm{Cat}^{(m)}$. For every $(p_1,\ldots,p_m) \in \mathbb{N}^m$, $1 \le i \le m$, we have evident maps of i-th source and target (or front and back face)

$$s_i, t_i : (NX)_{p_1,\ldots,p_m} \rightrightarrows (NX)_{p_1,\ldots,p_{i-1},0,p_{i+1},\ldots,p_m}$$

(A7.1) Definition. a) A category $\mathrm{Cat}\,(X;\, p_1,\ \ldots,\ p_{m-1})$ is defined by

$$\mathrm{Ob}\ \mathrm{Cat}\,(X;\ p_1,\ldots,p_{m-1}) = (NX)_{p_1,\ldots,p_{m-1},0};$$

$$\mathrm{Hom}(\tau,\tau') = \{\sigma \in (NX)_{p_1,\ldots,p_{m-1}} \mid s_m\sigma = \tau,\ t_m\sigma = \tau'\},$$

composition is defined in the evident way.

b) Let $f : X \longrightarrow Y \in \mathrm{Mor}\ \mathrm{Cat}^{(m)}$, $\tau \in (NX)_{p_1,\ldots,p_{m-1},0}$.

A category $\tau \setminus f$ is by definition

$$\tau \setminus \mathrm{Cat}\ (f;\ p_1,\ldots,p_{m-1}), \qquad \text{(cf. A4) where}$$

$$\mathrm{Cat}\ (f;p_1,\ldots,p_{m-1}) : \mathrm{Cat}\ (X;p_1,\ldots,p_{m-1}) \longrightarrow \mathrm{Cat}(Y;p_1,\ldots,p_{m-1}).$$

c) A functor f is called final (in direction m) if for every $(p_1,\ldots,p_{m-1}) \in \mathbb{N}^{m-1}$, $\tau \in (NY)_{p_1,\ldots,p_{m-1},0}$ $\tau\setminus f$ is contractible.

(A7.2) Theorem. Let $f : C \longrightarrow D$ Mor $\mathrm{Cat}^{(m)}$, $F \in Wf^{(m)}(D, C^{\cdot}(\mathcal{A}))$, cf. (A3). Suppose that f is final. Then

$$f_* : |Nf^* F| \longrightarrow |NF|$$

is quasiisomorphism, where $|\cdot|$ denotes the composition

$$\Delta^{\circ m} C^{\cdot}(\mathcal{A}) \longrightarrow (C^{\cdot})^{m+1}(\mathcal{A}) \xrightarrow{\ \mathrm{Tot}\ } C^{\cdot}(\mathcal{A}).$$

Proof. Let us consider the $(m+1)$-simplicial object $N(f,F)$ of $C^{\cdot}(A)$ with

$$N(f,F)_{p_1,\ldots,p_{m-1},p,q} =$$

$$= N(\mathrm{Cat}\ (f;p_1,\ldots,p_{m-1}),\ \mathrm{Cat}\ (F;p_1,\ldots,p_{m-1}))_{p,q}$$

(cf. (A4)), where $\mathrm{Cat}(F;p_1,\ldots,p_{m-1}) : \mathrm{Cat}(D;p_1,\ldots,p_{m-1}) \to C^{\cdot}(\mathcal{A})$ is induced by F. In other words, $(p_1,\ldots,p_{m-1},p,q)$-simplices of

$N(f,F)$ are of the form

$$F(\tau_0 \to \dots \to \tau_q \to f(\sigma_0) \to \dots \to f(\sigma_p)),$$

$$\tau_i \in Cat(D;\, p_1,\dots,p_{m-1}), \quad \sigma_j \in Cat(C;\, p_1,\dots,p_{m-1}).$$

When we fix $\tau = (\tau_0 \to \dots \to \tau_q) \in NY_{p_1,\dots,p_{m-1},q}$ and vary $\sigma = (\sigma_0 \to \dots \to \sigma_p)$, the realization of the corresponding simplicial complex is quasiisomorphic to

$$F(\tau_0 \to \dots \to \tau_q) \times N(\tau_q \setminus f) \simeq F(\tau_0 \to \dots \to \tau_q)$$

by the condition of finality. Thus, the natural map $|N(f,F)| \to |NF|$ is quasiisomorphism. From the other hand, when σ is fixed, then

$$|N(f,F)|_{(\sigma \text{ fixed, } \tau \text{ varying})} \simeq F(f(\sigma))$$

(cf. [HS, Lemma (2.4.1.1)]), so

$$|N(f,F)| \to |Nf^*F|.$$

The result now follows from commutative diagramm

$$\begin{array}{ccccc} |NF| & \xleftarrow{\sim} & |N(f,F)| & \xrightarrow{\sim} & |Nf^*F| \\ \| & & \downarrow f_* & & \downarrow f_* \\ NF & \xleftarrow{\sim} & N(id(D),F) & \xrightarrow{\sim} & NF \end{array}$$

(A8) <u>Spectral sequences</u>.

(A8.1) Let $F : C \dashrightarrow C^{\cdot}(\mathcal{A})^{o}$ be a weak m-functor. Then NF is an m-cosimplicial complex, or (m+1)-complex in $\mathcal{A}$ to which correspond the standart spectral sequences.

For example let $C = C_1 \boxtimes \dots \boxtimes C_m$, $C_i \in Cat$.
When we fix $x \in Ob\, C_m$, we obtain by restriction weak (m-1)-functor

$$F_x : C_1 \boxtimes \dots \boxtimes C_{m-1} \boxtimes x \to C^{\cdot}(\mathcal{A})^{o},$$

and there is a natural spectral sequence

$$E(F) : E_2^{pq} = H^p(C_m,\, x \mapsto H^q(|NF_x|)) \Rightarrow H^{p+q}(|NF|)$$

(A8.2) <u>Functoriality</u>. Suppose we have a weak (m+1)-functor

$$H : C_1 \boxtimes \dots \boxtimes C_m \boxtimes I \to C^{\cdot}(\mathcal{A}),$$

where $I = (0 \longrightarrow 1)$.

Then H induces the unique in the filtered derived category map

(*) $|NH_o| \longrightarrow |NH_1|$ and thus, map of spectral sequences

(A8.2.1) $E(H_o) \longrightarrow E(H_1)$

In fact, by (A7.2), the natural inclusion

$$|NH_1| \longrightarrow |NH|$$

is a quasiisomorphism, which respects the filtrations as is seen from the proof of (A7.2), and (*) is induced by

$$|NH_o| \longrightarrow |NH|.$$

B. Spectrum of a permutative category

(A.9) Recall that a permutative category is a category M equipped with a functor

$$+ : M \times M \longrightarrow M,$$

a distinguished object $0 \in \mathrm{Ob}\, M$, and a natural isomorphism

$$\varphi : + \circ s \longrightarrow +,$$

where $s : M \times M \longrightarrow M \times M$ is the transposition, which satisfy the following conditions:

For every $a, b, c \in \mathrm{Ob}\, M$

1) $a + o = o + a = a$;

$a + (b+c) = (a+b) + c$;

2) $\varphi(o,a) = \varphi(a,o) = id(a)$;

3) $\varphi(a,b)\, \varphi(b,a) = id(a+b)$

4) triangle

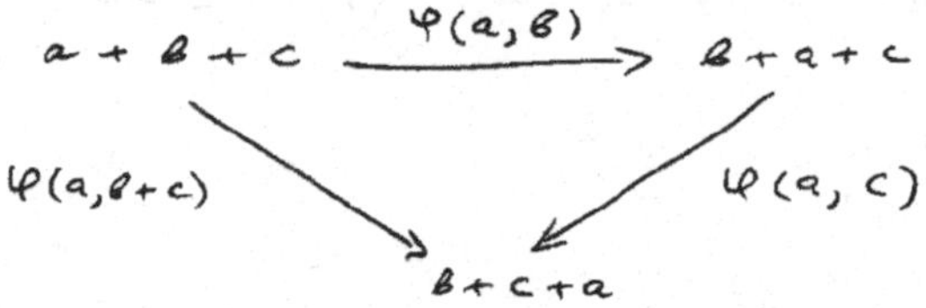

commutes.

(A9.1) Coherence theorem. Every diagramm in M involving only the maps $\varphi(a,b)$, commutes.

Proof. Exercise. Hint: use the presentation of symmetric groups as Coxeter groups by generators and relations. ▲

(A10). Let M be a small permutative category.

(A10.1) In this n^o we shall construct a spectrum $\mathcal{B}M$. It will be set $\{\mathcal{B}^m M\}$, $m \in \mathbb{N}$, $m \geq 1$, where

$$\mathcal{B}^m M : \Delta^{o \otimes m} \dashrightarrow \mathrm{Cat}$$

is an m-pseudo-functor from $\Delta^{o \otimes m} := \underbrace{\Delta^o \otimes \ldots \otimes \Delta^o}_{m}$ to Cat[1]) such that

a) $\mathcal{B}^1 M$ is a strict functor $BM : \Delta^o \longrightarrow \mathrm{Cat}$, the classifying space of M considered as a monoidal object in Cat (cf. (0.3.2))

b) For all $k \in \mathbb{N}$

$$\mathcal{B}^m M|_{\Delta^{o \otimes m-1} \times [k]} = (\mathcal{B}^{m-1} M)^k.$$

In particular,

$$\mathcal{B}^m M|_{\Delta^{o \otimes m-1} \times [o]} = *,$$

and

$$\mathcal{B}^m M|_{\Delta^{o \otimes m-1} \times [1]} = \mathcal{B}^{m-1} M,$$

so we obtain canonical maps

$$\Sigma \mathcal{B}^{m-1} M \longrightarrow \mathcal{B}^m M.$$

c) Moreover, for any i: $1 \leq i \leq m$, and $k \in \mathbb{N}$

$\mathcal{B}^m M|_{\Delta^{o \otimes i-1} \times [k] \times \Delta^{o \otimes m-i}}$ is "pseudo-isomorphic" to $(\mathcal{B}^{m-1} M)^k$

in the sense of [HS, (2.4.3)].

(A10.2) Now we shall construct $\mathcal{B}^m M$.

First of all define inductively for $(n_1, \ldots, n_m) \in \mathbb{N}^m$ the permutative category

$$M^{n_1 \cdot n_2 \cdot \ldots \cdot n_m}.$$

Namely, put

$$M^n = M \times \ldots \times M \qquad (n \text{ factors}),$$

$$M^{n_1 \ldots n_m} = M^{n_1 \ldots n_{m-1}} \times \ldots \times M^{n_1 \ldots n_{m-1}} \qquad (n_m \text{ factors}).$$

---

1) Cat has the evident structure of $\Delta^o \mathcal{E}ns$-module such that $N(X \times Y) = NX \times Y$ for $X \in \mathrm{Cat}$, $Y \in \Delta^o \mathcal{E}ns$.

Here is important the order of factors;
Strictly speaking, $M^{n_1 \cdots n_k}$ is $M^{\prod n_i}$ with the fixed structure of brackets.
For example,

$$M^{2\cdot 3} = (M \times M) \times (M \times M) \times (M \times M),$$

$$M^{3\cdot 2} = (M \times M \times M) \times (M \times M \times M).$$

Now put

$$\mathcal{B}^m M([n_1],\dots,[n_m]) = M^{n_1 \cdots n_m},$$

and define $\mathcal{B}^m M$ on 1-morphisms in such a way that when $n_1,\dots, n_{i-1}, n_{i+1},\dots,n_m$ are fixed,

$$\mathcal{B}^m M\big|_{[n_1]\times\dots\times[n_{i-1}]\times\Delta^\circ\times[n_{i+1}]\times\dots\times[n_m]}$$

is the simplicial category

$$(B(M^{n_1\cdots n_{i-1}}))^{n_{i+1}\cdots n_m}$$

If M were strictly commutative, this would provide the strict functor $\Delta^{\circ m} \longrightarrow \mathrm{Cat}$. In general we can construct an m-pseudo-functor.

For clarity let us treat the case $m = 2$. We have to define 2-c.s. E, and $\mathcal{B}^2M$ on 2-morphisms. Of course, we put $E_f = *$ for $f \in \mathrm{Mor}^1(\Delta^{\circ \boxtimes 2})$.

Call the 2-morphism of $\Delta^\circ \times \Delta^\circ$ elementary if it has the form $f_1 \boxtimes f_2$, where $f_1 = s_i$ or d_j, and $f_2 = s_p$ or d_q.
Every 2-morphism f is the composition of elementary ones; every decomposition $\sigma: f = (f^1 \boxtimes g^1)\dots(f^i \boxtimes g^i)$ into elementary ones defines a functor

$$\mathcal{B}^2M(\sigma) : \mathcal{B}^2M(sf) \longrightarrow \mathcal{B}^2M(tf),$$

and for two decompositions σ, σ' of the same f the maps φ induce the unique (by Coherence theorem) natural isomorphism

$$\mathcal{B}^2M(\sigma) \longrightarrow \mathcal{B}^2M(\sigma').$$

So, if we define the category $\widetilde{E}_f$ with objects - decompositions of f into elementary 2-morphisms and with the unique isomorphism between every two objects, we obtain the maps

$$\nu_f : \mathcal{B}^2M(sf) \times \widetilde{E}_f \longrightarrow \mathcal{B}^2M$$

These maps define the 2-pseudo-functor $\mathcal{B}^2M$ over the c.s. $\{N\tilde{E}_f\}$ (cf. [HS]).

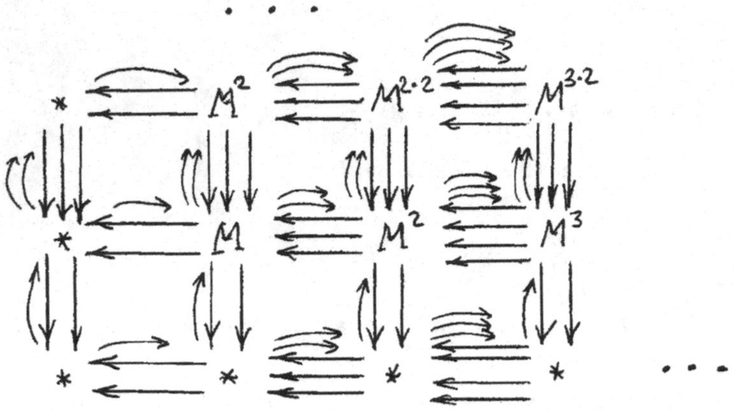

Fig. A.6. $\mathcal{B}^2M$. Squares commute up to a natural isomorphism.

The case $m > 2$ is treated analogously.
The properties a) - c) are obvious.

(A10.3) By taking the diagonal we obtain 1-pseudo-functors

$$\Delta\mathcal{B}^m M : (\Delta^\circ)^m \sim\longrightarrow \mathrm{Cat}$$

Also we have compositions

$$N\mathcal{B}^m M : \Delta^{\circ\times m} \longrightarrow \mathrm{Cat} \xrightarrow{N} \Delta\mathcal{E}ns$$

(A10.4) Important remark. Suppose that M is a monoidal category in some category $\mathcal{D}$ with products and coproducts. Then by the same construction we obtain m-pseudo-functors

$$N\mathcal{B}^m M : \Delta^{\circ\times m} \longrightarrow \Delta^\circ\mathcal{D}$$

We'll apply this remark to the case $\mathcal{D}$ = Sch.

References

[B] Beilinson A.A. Higher regulators and values of L-functions (in Russian). In: "Modern problems of mathematics (VINITI series), v.24, p. 181-238.

[F] Friedlander E. Etale K-theory, II. Ann. Sci. E.N.S.

[G] Gillet H. Riemann-Roch theorems in higher K-theory. Adv. Math., 40, 1981, 203-289.

[GZ] Gabriel P., Zisman M. Calculus of fractions and homotopy theory. Ergebnisse, Bd. 35, Springer, 1967.

[HS] Hinich V.A., Schechtman V.V. Geometry of a category of complexes and algebraic K-theory. Preprint, Moscow, 1983. Cf. also: J. Func. Anal. Appl., 1984, N 2, p. 83-84 (in Russian).

[J] Jouanolou P. Cohomologie des quelques schemas classiques et theorie cohomologique des classes de Chern. SGA5, exp. VII. Lect. Notes Math., 589.

[L] Loday J.-L. Homotopie des espaces des concordances. (d'après F. Waldhausen). Sem. Bourbaki, N 516, 1977/78. Lect. Notes Math.

[M] Maclane S. Categories for working mathematician GTM 5, Springer, 1971.

[Q1] Quillen D. Characteristic classes of representations. Lect. Notes Math., 551, 189-216.

[Q2] Quillen D. Higher K-theory I, Lect. Notes Math., 341, 85-147.

[S] Schechtman V.V. Algebraic K-theory and characteristic classes (in Russian). Trudy MMO (= Proc. Moscow Math. Soc.), 45, 1983, 237-264.

[So] Soule C. Operations dans K-theorie algébrique. Preprint CNRS, 1983.

NONCOMMUTATIVE RESIDUE

Chapter I. Fundamentals

by

Mariusz Wodzicki*

Mathematical Institute, University of Oxford, 24-29 St. Giles, Oxford OX1 3LB, England**.

Mathematical Institute, Polish Academy of Sciences, ul. Śniadeckich 8, P.O. Box 137, 00-950 Warsaw, Poland.

* Supported by the Royal Society Grant.

** Current address: The Institute for Advanced Study, Princeton, NJ 08540, U.S.A.

Introduction

Noncommutative residue was discovered originally about 1978 in the special case of one-dimensional symbols by (working independently) M. Adler (cf. [2]) and Yu. I. Manin (cf. [10]). They considered the half of the algebra of one-dimensional symbols whose elements are formal Laurent series:

$$a = \sum_{i=-\infty}^{m} a_i \partial^i \qquad (m = m(a)) \tag{1}$$

(a_i belong to a K-algebra A with differentiation $\partial : A \to A$; K is a field of characteristic zero). Two such series are composed as:

$$a \circ b = \sum_i \Big(\sum_{j+k=i} \sum_{\ell=0}^{\infty} \frac{(j+\ell)\dots(j+1)}{\ell!} a_{j+\ell} b_k^{(\ell)} \Big) \partial^i \tag{2}$$

($b^{(\ell)} \equiv \partial^\ell b$). Let $\bar{A}$ denote the K-vector space $A/([A,A] \cup \mathrm{Im}\,\partial)$. It was observed that assigning to a series (1) the class of its coefficient a_{-1} in $\bar{A}$ defines a trace functional on the algebra of series (1) with the composition law (2). This functional was given a name of *noncommutative* residue in analogy to Cauchy residue on ordinary (commutative) Laurent series. Noncommutative residue was discovered in the context of completely integrable systems and since then it was recognized as one of standard tools in the theory of such systems. It allows, e.g. to view linear functionals on the space of differential operators $A[\partial]$ as being represented by symbols of negative order (in terminology of Yu.I. Manin: formal Volterra operators) by means of the correspondence

$$A[[\partial^{-1}]]\partial^{-1} \ni q \mapsto \phi_q \in (A[\partial])', \qquad \phi_q(a) := \chi(\mathrm{res}(q \circ a))$$

($\chi \in \bar{A}'$; $\bar{A}$ is usually a one-dimensional space which is canonically identified with K so that there is no need in choosing χ).

In applications A is generally an algebra of functions on the real line (e.g. fast decaying, periodic etc.) with $\partial = d/dx$, and such that the projection $A \to \overline{A}$ has the form of integration:

$$A \ni f \mapsto \overline{f} = \int f dx \ .$$

Noncommutative residue has in this situation the extra feature: the form $a_{-1}dx$ is functorial with respect to orientation preserving automorphisms of $\mathbb{R}^1$ which also preserve A.

All mentioned facts seemed for a long time to be fairly peculiar to dimension one, probably because of their relation to the invariant decomposition of the algebra of one-dimensional symbols into the direct sum of its positive and negative parts (i.e. differential and formal Volterra operators). No such decomposition exists in higher dimensions indeed. So it was rather surprising when the author, when thinking about the residues at negative integers of the zeta function of an elliptic *pseudodifferential* operator (ψDO), discovered in Spring 1983 that $\mathrm{Res}_{s=-1}\zeta(s;A)$ behaved much like a trace functional. This suggested to put for an *arbitrary* ψDO Q:

$$\mathrm{res}\ Q := \frac{d}{dt}\left(\mathrm{ord}\ A\ \mathrm{Res}_{s=-1}\zeta(s;A+tQ)\right)\Big|_{t=0} \tag{3}$$

where A was an elliptic ψDO admitting complex powers and of the order strictly bigger than that of Q. Since, as can be shown, (3) does not depend on A, it is not difficult to deduce that (3), in fact defines a trace functional. In early Summer 1983 another discovery was made - that this trace on the algebra of symbols is not only exotic - it is also unique. Since then it was established that non-commutative residue was inherent in all known local invariants of ψDOs (e.g. heat kernel expansion coefficients, local formulae for the variation of $\log\det A$). In dimension one this new trace, defined in the framework of spectral geometry, reduces to the one introduced by M. Adler and Yu.I. Manin.

The progress which followed the discovery of higher-dimensional noncommutative residue allowed to change a set up from rather subtle analytical methods of the zeta-function theory to more general and more flexible methods of homogeneous symplectic geometry and homological algebra. There is still no satisfactory complete theory however. Despite this, the author prompted by richness of the material available decided to publish first few "chapters" of the theory of noncommutative residue to come. These will include:

Chapter I. Fundamentals.

Chapter II. The commutator structure of pseudodifferential operators [18].

Chapter III. Bi-multiplicative index formula and exotic log det [19].

Chapter IV. Homology of the algebra of differential operators and the algebra of symbols [20].

The contents of Chapter I is, very briefly, as follows. In Section 1 we introduce an elementary formalism of the symplectic residue in a fairly general set up. Basic for all further applications and constructions is the morphism of chain complexes

$$\underline{\text{Res}}: \ C_{\bullet}(P^{\bullet}(Y);\text{ad}) \to \Omega_{\bullet}(Z)$$

where $P^{\bullet}(Y) = \bigoplus_{\ell\in\mathbb{Z}} P^{\ell}(Y)$ denotes the graded Poisson algebra of a symplectic cone Y^{2n} with a base Z^{2n-1}, and $\Omega_q(Z) \equiv \Omega^{2n-1-q}(Z)$ (see 1.17 and 1.28 below).

The core of Chapter I is the next section. There we further develop the technique of Section 1 in the special case: $Y = T_0^*U$, $Z = S^*U$ (U is a domain).

This allows us to define for an arbitrary ψDO A in a domain a certain functorial matric density $\text{res}_u A$ called its noncommutative residue (cf. 2.13-20). Then we prove (cf. Proposition 2.28) that

$$\operatorname{tr}\operatorname{res}_u AB - \operatorname{tr}\operatorname{res}_u BA = d\rho_u(A,B) \tag{4}$$

holds for a certain explicit form $\rho_u(A,B) \equiv -\rho_u(B,A)$. The form ρ_u is functorial with respect to gauge transformations (Proposition 2.31) but fails in general to be functorial with respect to open embeddings (cf. Example 2.44). In Proposition 2.38 we establish that its "non-functoriality" is measured by another explicit exact form $d\sigma_u$. One should dwell, at this point, on a special feature of the considered situation: whenever some equality holds only up to certain exact terms (and this happens more often than the point-wise equality does) it appears always possible to write down a suitable local primitive form.

In Section 3 we consider global ψDOs acting on sections of general vector bundles on arbitrary manifolds. We introduce a canonical vector-valued density $\operatorname{res}_x A$ (see 3.1) and deduce its basic properties from results of previous section (see Proposition 3.2). We prove that (4) still holds but now the analogue of form ρ_u is canonical only when defined up to an exact form (Proposition 3.7). The corresponding class modulo exact forms turns outto be a cyclic 1-cocycle on the algebra of ψDOs (Proposition 3.9).

Still another invariant quantity related to the noncommutative residue can be associated with an arbitrary ψDO. It is called a *subresidue*, and is a higher-dimensional analogue of a quadratic differential (see 3.14-17).

In 3.18-21 we rewrite R.T. Seeley's formulae for the residues and values of the zeta-function of an elliptic ψDO purely in terms of the noncommutative residue formalism. This includes a new proof of the invariance under gauge transformations and changes of local coordinates Our proof is intrinsic and purely algebraic. We believe that our reformulation of Seeley's results clarifies many issues. For example, it turns out that the zeta-function of a ψDO is not so much different from that of an endomorphism of a finite-dimensional space. The difference is comparable, using a free parallel, to that between rational elliptic curves of rank one and rank zero (if Birch-Swinnerton-Dyer conjecture has to hold). For both ψDOs and matrices there are similar formulae for coefficients at leading terms in power expansions of zeta-function at certain points. In the case of ψDOs the *residues* of $\zeta(s;A)$ are given by

$$\operatorname{Res}_{s=\sigma}\zeta(s;A) = \frac{\operatorname{res} A^{-\sigma}}{\operatorname{ord} A}$$

(see (8) of Section 3; one is reminded that res is the only trace available in this situation), whereas for matrices the *values* of $\zeta(s;A)$ are given by

$$\zeta(\sigma;A) = \operatorname{tr} A^{-\sigma}.$$

Both formulae are virtually the same if to neglect the standard in ψDO theory appearance of the order of an operator.

Next turn to the behaviour of $\zeta(s)$ at the origin. In the former case $\zeta(s;A)$ has *a priori* residue at that point but this turns out to be *equal* to zero. For a matrix the *value* $\zeta(0;A)$ also does not depend on A: it is equal to the dimension of the linear space (in both cases we assume for brevity that A is invertible). Thus the leading terms in both cases can be fairly ignored, and one may examine the next terms. For matrices this turns out to be the *derivative*

$$\zeta'(0;A) = -\log \det A \equiv -\operatorname{tr} \log A,$$

for ψDOs - the value:

$$\zeta(0;A) = \frac{Z(A)}{\operatorname{ord} A}$$

where $Z(A) \equiv \int_X \operatorname{tr} Z_{\theta,x}(A)$, and $Z_{\theta,x}(A)$ is defined in Lemma 3.10.

(In [17] we proved that $\int_X \operatorname{tr} Z_{\theta,x}(A)$ does not depend on the choice of a cut in the spectral plane, so, we omit the subscript θ). We shall prove, indeed, in Chapter III [19] that the functional $Z(A)$ extends to a certain exotic "-log det" on the group of invertible elliptic ψDOs and that it is even a *homomorphism* from this group into the additive group of complex numbers.

The noncommutative residue approach yields equally simple formulas for more general zeta-functions (see 3.22). It gives also a fresh insight into the standard high-temperature (or small-time) expansion

$$\operatorname{tr} e^{-t\Box_E}(x,x) - \operatorname{tr} e^{-t\Box_F}(x,x) \sim \sum_{j=0}^{\infty} \tilde{\alpha}_j(x) t^{(j-n)/2m} + \sum_{q=1}^{\infty} \tilde{\beta}_q(x) t^q \log t$$

($t \searrow 0$; for notation cf. 3.23 and 3.28). As is well known, all coefficients of this expansion except $\tilde{\alpha}_n(x)$ integrate to zero. We deduce from the noncommutative residue formalism (and its further extension called the Λ-twisted noncommutative residue which is sketched in 3.26-27) that the coefficients mentioned can be expressed as differentials of explicit forms which depend *locally* on the symbol of an operator and metrics involved (see Prop. 3.24 and the further discussion). From an entirely different point of view, an analogous statement was proved (by P.B. Gilkey and, perhaps, also by others) for certain *differential* operators arising in Riemannian geometry (cf. e.g. [4] and [7]). The two approaches seem to provide different answers, the difference being measured by certain cohomology classes (see. 3.31) Finally, local cancellation phenomena which are widely known in physics and Riemannian geometry are responsible in a number of cases for vanishing of $\tilde{\alpha}_j(x)$'s ($j < n \equiv \dim X$) point-wise. Whenever such a case occurs, our approach produces a certain secondary cohomology class which lives in $H^{n-1}(X,\varepsilon_X)$ where ε_X denotes the orientation local system (see 3.28). These secondary classes call for further investigation (cf., however, the related example in 3.29).

Chapter I was designated to contain basic constructions so that it could serve (for a while) as a standard reference source for the subsequent chapters' and other works. (When the higher-dimensional residue was originally discovered and some of its fundamental properties announced in the final section of [17] the complete exposition was intended to appear in Functional Analysis and its Applications. This soon turned out to be unrealistic since the bulk of the available material overgrew limits of an article permissible for that journal.) The first draft of the first two chapters was published (in September 1984) as a part of the author's Steklov Habilitation Thesis [21]. In the different context some results were reported in [16]. Even in the present long overdue publication one fundamental topic is still absent: the formalism of "super-residue". As in a purely even case its original discovery was made in the different context of super-integrable systems, the underlying space having dimension $1|1$ (actually, the "superisation" used there is fairly nontrivial, e.g. the corresponding time-direction is odd, for details the reader is referred to the article by Yu.I. Manin and A.O. Radul [12]). The author introduced general $n|m$-dimensional super-residue and proved it to be a supertrace as well as being essentially unique. The notion of super-residue might also deserve more attention because it seems to be well suited to heat kernel methods which are current in physics and Riemannian geometry.

One final remark must be made. Although the name of Alain Connes is not mentioned in the present paper it is needless to say that noncommutative residue is well suited to his general picture of noncommutative integration. The careful reader will find in 1.25 an indication of a more direct link existing between noncommutative residue and his approach based on relations derived from the failure of trace property (of the ordinary trace). We hope to trace this link in subsequent papers. It should be also clear from the exposition of noncommutative residue which is presented below that all ways lead us to homological algebra. Despite this we consciously decided to remain in the introductory Chapter I on an elementary level of exposition.

During my long stay in Moscow I was fortunate to be surrounded by a unique community consisting of outstanding mathematicians as well as remarkable people. To all of them my warmest thanks. I owe special gratitude to Yuri Ivanovich Manin (among other things for his invaluable support) and to Sasha Beilinson. I wish also to thank Mikhail Alexandrovich Shubin for numerous discussions on the subject of the present paper.

These notes were finally prepared during my appointment at Oxford with Sir Michal F. Atiyah, whose support I greatly acknowledge.

0. Notation

0.1. Manifolds are usually denoted by X, Y, Z etc. Domains in $\mathbb{R}^n$ are denoted U, V etc. The space of vector fields on X is denoted by $\mathcal{T}X$, the cosphere bundle by S^*X.

0.2. Vector bundles are denoted by E, F, G, H etc. The canonical trivial bundle of rank k is denoted θ^k, sometimes θ^k_X, to indicate its base space.

0.3. Triples $(X;E,F)$ consisting of a manifold and two vector bundles on it form the category whose morphisms are triples $\phi \equiv (f;r,s) : (X';E',F') \to (X;E,F)$ such that

1) $f : X' \to X$ is an open embedding,
2) $r \in \mathrm{Hom}(E',f^{\cdot}E)$ and $s \in \mathrm{Hom}(f^{\cdot}F,F')$

0.4. Pairs $(X;E)$ form a subcategory whose morphisms $\psi = (f;s)$ correspond to triples $(f;s^{-1},s)$. In particular s should be an isomorphism.

0.5. For every linear map $A : C^\infty_{comp}(X,E) \to C^\infty(X,F)$ a morphism $\phi : (X';E',F) \to (X;E,F)$ defines the induced map

$\phi^{\#} A : C^{\infty}_{comp}(X',E') \to C(X',F')$ so that the diagram

$$\begin{array}{ccc} C^{\infty}_{comp}(X,E) & \xrightarrow{A} & C^{\infty}(X,F) \\ \phi_! \uparrow & & \downarrow \phi^* \\ C^{\infty}_{comp}(X',E') & \xrightarrow{\phi^{\#} A} & C(X',F') \end{array}$$

ommute. If $E' = f^{\cdot}E$, $F' = f^{\cdot}F$ and r, s are identical maps we shall se also notation $f^{\#} A$. If $Y \subset X$ and f is the natural embedding e shall write also $A|_Y$ instead of $f^{\#}A$.

.6. By a *local coordinate patch* we mean a morphism $= (j;r,s) : (U;\theta^k,\theta^{\ell}) \to (X;E,F)$ where $k = \text{rank } E$, $\ell = \text{rank } F$ and oth r and s are isomorphisms.

.7. All pseudodifferential operators (ψDOs) considered are supposed o be of *classical type*. Recall that $A : C^{\infty}_{comp}(X,E) \to C^{\infty}(X,F)$ is a DO of classical type if for every local coordinate patch ϕ the nduced operator $B := \phi^{\#} A$ can be expressed as

$$Bf(u) = \int_{\mathbb{R}^n_{\xi}} \int_{\mathbb{R}^n_{v}} e^{i(u-v)\cdot\xi} \alpha(u,\xi) f(v) đ\xi dv + Tf(u)$$

here $đ\xi \equiv (2\pi)^{-n} d\xi$, $n = \dim X$ and T is an operator with smooth ernel, and $\alpha \in C^{\infty}(T^*U) \otimes \text{Mat}_{\ell,k}(\mathbb{C})$ admits the asymptotic expansion

$$\alpha \sim \sum_{j=0}^{\infty} a_{m-j} \quad \text{as} \quad |\xi| \to \infty \tag{1}$$

$$(a_{m-j}(u,t\xi) \equiv t^{m-j} a_{m-j}(u,\xi) \;, \quad j = 0,1,2,\ldots) \;.$$

.8. As opposed to the hitherto existing practice we shall mean by he *symbol* of B the formal sum on the right-hand side of (1). When e refer to α itself we shall call it the *smoothed* symbol. Usually t is not uniquely defined unlike the formal sum in (1).

.9. The number m in (1) is called the *order* of B (it can be any omplex number). An operator A (as in 0.7) is called of order m

(or 'less') if $\phi^{\#}A$ is such for every local coordinate patch. The space of such ψDOs is denoted by $CL^m(X;E,F)$. If $E = F$ we shall also write $CL^m(X;E)$.

Usually m will be an integer.

0.10. A ψDO is called *proper* if the support of its Schwartz kernel is a proper neighbourhood of the diagonal $\Delta \subset X \times X$. A proper ψDO can be composed with other ψDOs. In particular, the linear space $CL^{\cdot}_{prop}(X,E) = \bigcup_{m\in\mathbf{Z}} CL^m_{prop}(X,E)$ forms an associative algebra.

0.11. By $L^{-\infty}(X;E,F)$ we denote the space of operators with smooth kernels. It is a subspace in $CL^m(X;E,F)$ and the factor-space $CL^m(X;E,F)/L^{-\infty}(X;E,F)$ is denoted by $CS^m(X;E,F)$ and called the space of (global) symbols of order m. It is isomorphic locally to the space of formal series as in (1).

0.12. Composition formula. The composition of ψDOs defines the composition law on symbols:

$$a\circ b = \sum_{\alpha}\frac{1}{\alpha!}\ \partial_\xi^\alpha a D_u^\alpha b$$

(the sum over all multi-indices α; $D_u = -\sqrt{-1}\partial_u$) .

0.13. Base change formula. Let $f : V \to U$ be an open embedding, and (u,ξ) and (v,η) be canonical coordinates in T^*U and T^*V respectively. We shall denote by T^*f the map

$$T^*V \to T^*U\ , \quad (v,\eta) \to (f(v),(f_v^*)^{-1}\eta)\ .$$

Let $a(u,\xi)$ be the symbol of $A \in CL^m(U;\theta^k,\theta^\ell)$ and $a^f(v,\eta)$ be the symbol of $f^{\#}A \in CL^m(V;\theta^k,\theta^\ell)$. The base change formula expresses a^f in terms of a and f :

$$a^f = (T^*f)^*\{\sum_\alpha a^{(\alpha)}\psi_\alpha\} \tag{2}$$

where $a^{(\alpha)}(u,\xi) \equiv \partial_\xi^\alpha\, a(u,\xi)$, and $\psi_\alpha \equiv \psi_\alpha(u,\xi)$ are universal function depending only on f, and polynomial with respect to ξ's of degree not exceeding $|\alpha|/2$.

Description of ψ'_αs. Let $g : f(V) \to V$ denote the inverse mapping, i.e. $gf \equiv id_V$, considered as a function $f(V) \to \mathbb{R}^n$. Its linear part at a point $u \in f(V)$ will be denoted by $j^1_u(g)$. Then $\bar{g}_u := g - j^1_u(g)$ vanishes at u up to second order. We put

$$\psi_\alpha(u,\xi) = \frac{1}{\alpha!} D^\alpha_z \, e^{\sqrt{-1}<\bar{g}_u(z), f^*_v\xi>} \Big|_{\substack{z = u \\ v = f(u)}} \tag{3}$$

($<\cdot,\cdot>$ denotes the ordinary scalar product in $\mathbb{R}^n$).

Formula (2) can be found in a slightly different form in two standard reference works on ψDOs: [9] (cf. 2.1.14-16), and [15] (cf. Thm. 4.2).

We shall need, however, still another form of the base change formula.

.14. Base change formula (Second form).

$$a^f = (T^*f)^*\{(1 + \sum_{j=1}^{n} \partial_{\xi_j} \circ \Psi_j)a\} \tag{4}$$

here $\Psi_j \equiv \Psi^f_j(u,\xi,\partial_\xi)$ are certain differential operators on T^*U of nfinite order (such that $\Psi_j a$ has sense). Their explicit form can e deduced easily from (2) and (3).

.15. The complex of integral forms. Let $o : \tilde{X} \to X$ be the canon- cal orientation double cover of X. Then the complex $o_*\Omega^\cdot\tilde{X}$ is cted by the deck involution. The anti-invariants of this action form he complex of *integral forms* denoted by $A^\cdot(X)$ (N. Bourbaki uses ifferent terminology: $T(X)$-twisted differential forms (formes ifférentielles $T(X)$-tordues), cf. [3], 10.4.1). It coincides, of ourse, with the complex of *smooth* currents. When choosing its name he author was guided by two reasons: integral volume forms (known also under the name of 1-densities) can be canonically integrated and, hence, the corresponding sheaf is a dualizing sheaf in C^∞-category (and not Ω^{vol}). An analogous situation is well-known in super-geometry where the corresponding dualizing sheaf is called Berezinian and the complex generated by it - the complex of integral forms (cf. e.g. [11], 4.5.4. or [13]). In the classical (even) complex geometry the complex of integral forms is canonically isomorphic to de Rham complex.

In this connection we want to draw attention to the fact that local invariants of ψDOs (as e.g. local residues and values of the zeta function) are usually canonically integrable. When they happen to be 'exact' or when one looks on higher relations existing between them one immediately evokes the whole complex of integral forms.

<u>0.16</u>. The properties of integral forms are fairly dual to those of differential forms. In particular, an integral form of codimension q (the space of such forms will be denoted by A_q) can be integrated over co-oriented chains of the same codimension (co-orientation = orientation of the normal bundle).

<u>0.17</u>. For any submersion $\tau : Z \to X$ there is defined a push-out morphism $A_.(Z) \to A_.(X)$ dual to the pull-back of differential forms. This is the composition of the canonical projection $A_.(Z) \to A_.(X, \tau_. A_{Z/X,0})$ with the integral

$$\int : \tau_. A_{Z/X,0} \to \mathcal{O}_X^\infty$$ ($A_{Z/X,0}$ denotes the sheaf of relative densities while $\mathcal{O}_X^\infty$ is the sheaf of (complex valued) smooth functions on X).

<u>0.18</u>. Integral forms can be pulled back under open embeddings. Pull backs commute with push-outs:

$$\tau_*^! g^* = f^* \tau_* \tag{5}$$

if in the corresponding commutative diagram

$$\begin{array}{ccc} Z' & \overset{g}{\hookrightarrow} & Z \\ \tau' \downarrow & & \tau \downarrow \\ X' & \overset{f}{\hookrightarrow} & X \end{array}$$

g induces diffeomorphisms between fibres of projections τ and τ' ((5) is nothing but change of variables under the sign of integral).

<u>0.19</u>. $|\ | : \Omega^{\cdot}(X) \tilde{\to} A^{\cdot}(X)$ denotes the identification induced by an orientation (if X is oriented).

<u>0.20</u>. For a general chain complex $C_{\cdot}[k]$ is the standard notation for its k-shift:

$$(C[k])_q \equiv C_{q-k}$$

(boundary maps are those in $C_{\cdot}$ but multiplied by $(-1)^k$).

§1. <u>Symplectic residue</u>

Our aim in this section will be to introduce and briefly discuss in a general setting of homogeneous symplectic geometry an elementary formalism of symplectic residue which we expect to become one of 'standards' in the global ψDO-theory.

1.1. Let Y^{2n} be a symplectic manifold and ω be the corresponding non-degenerated 2-form. Recall that any function $f \in C^{\infty}(Y)$ determines the associated Hamiltonian vector field H_f such that

$$i_{H_f}\omega = -df .$$

(For brevity reasons i_{H_f} will be later denoted as i_f and similarly Lie derivative L_{H_f} as L_f) .

The Poisson bracket

$$\{f,g\} := L_f g = \omega(H_f, H_g) = i_g i_f \omega$$

can be alternatively defined in terms of volume forms:

<u>1.2.</u> $$\{f,g\} = \frac{n df \wedge dg \wedge \omega^{n-1}}{\omega^n} = \frac{d(g i_f \omega^n)}{\omega^n} .$$

Indeed, we have:

$$0 = i_f i_g \omega^{n+1} = -(n+1) i_f (dg \wedge \omega^n) = -(n+1)[dg(H_f)\omega^n + n dg \wedge df \wedge \omega^{n-1}]$$

$$= (n+1)[n df \wedge dg \wedge \omega^{n-1} - \{f,g\}\omega^n].$$

<u>1.3.</u> From now on Y is assumed to be acted by the multiplicative group of positive real numbers and the action (denote it by $\chi : \mathbb{R}_+^\times \to \mathrm{Diff}(Y)$) is required to be a conformal one:

$$\chi_t^* \omega = t\omega \qquad (t \in \mathbb{R}_+^\times) . \tag{1}$$

The infinitesimal action $\chi_* : \mathrm{Lie}(\mathbb{R}_+^\times) \to TY$ defines the Euler vector field $\Xi = \chi_*(r \frac{d}{dr})$ on Y. Put:

$$\alpha = i_\Xi \omega \qquad \text{and} \qquad \mu = \alpha \wedge (d\alpha)^{n-1}$$

(α is the standard "canonical" 1-form of classical mechanics).

Then the following identities:

<u>1.4.</u> $\chi_t^* \alpha = t\alpha, \quad \chi_t^* \mu = t^n \mu;$

<u>1.5.</u> $d\alpha = L_\Xi \omega = \omega, \qquad L_\Xi \alpha = \alpha , \qquad d\mu = \omega^n, \qquad L_\Xi \mu = n\mu .$

can be easily deduced from (1) using the fact that Ξ generates the flow $\phi_t = \chi_{\exp t}$. The next identity:

<u>1.6.</u> $$\{f,g\}\mu = d(g i_f \mu) - (n-1)\, g df \wedge \omega^{n-1} - L_\Xi(g df) \wedge \omega^{n-1}$$

follows from:

$$\{f,g\}\mu = \frac{1}{n}\, i_{\Xi}\{f,g\}\omega^n \overset{1.2}{=} \frac{1}{n}\, i_{\Xi} d(g i_f \omega^n) = \frac{1}{n} L_{\Xi}(g i_f \omega^n) - \frac{1}{n}\, d(g i_{\Xi} i_f \omega^n)$$

$$\overset{(1)}{=} -L_{\Xi}(gdf) \wedge \omega^{n-1} - (n-1) gdf \wedge \omega^{n-1} + d(g i_f \mu) \ .$$

Starting from this point we shall deal exclusively with functions on Y which are homogeneous. For a function f its homogeneity order will be denoted by ℓ_f .

The following identities will be used frequently:

1.7. $[\Xi, H_f] = (\ell_f - 1) H_f$.

1.8. $L_{f\Xi}(g\mu) = (n + \ell_g) fg\mu$.

1.9. $i_f \alpha = \ell_f f, \quad i_f \mu = \ell_f f \omega^{n-1} + (n-1)\alpha \wedge df \wedge \omega^{n-2}$.

1.10. $L_f \alpha = (\ell_f - 1) df, \quad L_f \mu = (\ell_f - 1) df \wedge \omega^{n-1}$.

1.11. $\{f,g\}\mu = d(g i_f \mu) - (\ell_f + \ell_g - 1 + n)\, gdf \wedge \omega^{n-1}$.

Identity 1.7 follows from:

$$d_{[\Xi, H_f]}\omega = [L_{\Xi}, i_f]\omega = -L_{\Xi}(df) - i_f L_{\Xi}\omega \overset{1.5}{=} (-\ell_f + 1) df = i_{(\ell_f - 1)f}\omega$$

(we used the Euler identity $L_{\Xi} f = \ell_f f$). The last four identities follow from 1.5-6 and the Euler identity.

We finish this somewhat lengthy register with the following items:

1.12. $L_{\Xi}(f_0 i_{f_1} \dots i_{f_q} \omega^n) = (\ell_{f_0} + \dots + \ell_{f_q} - q + n) f_0 i_{f_1} \dots i_{f_q} \omega^n$.

1.13. $L_{\Xi}(f_0 i_{f_1} \dots i_{f_q} \mu) = (\ell_{f_0} + \dots + \ell_{f_q} - q + n) f_0 i_{f_1} \dots i_{f_q} \mu$.

1.14. $X_t^*(f_0 i_{f_1} \dots i_{f_q} \mu) = t^{(\ell_{f_0} + \dots + \ell_{f_q} - q + n)} f_0 i_{f_1} \dots i_{f_q} \mu$.

The former two identities can be verified by a recurrent use of 1.7. and of the standard identity $[L_\eta, i_\zeta] = i_{[\eta,\zeta]}$. The last one is global version of 1.13.

1.15. Hereafter we shall assume Y to be a *connected* symplectic cone with a smooth base Z^{2n-1}, i.e. Y has to be a principal $\mathbb{R}_+^\times$-bundle with $Y/\mathbb{R}_+^\times = Z$. The natural projection $Y \to Z$ will be denoted by ρ.

Consider the differential form $f_0 i_{f_1} \cdots i_{f_q} \mu$. Identity 1.14 says that it is $\mathbb{R}_+^\times$-invariant precisely when

$$\sum_{i=0}^{q} \ell_{f_i} = q - n .$$

Since it is always horizontal (i.e. is annihilated by i_Ξ), it must exist a unique $(2n - q - 1)$-form $\mu_{f_0;f_1,\ldots,f_q}$ on Z such that

$$\rho^* \mu_{f_0;f_1,\ldots,f_q} = f_0 i_{f_1} \cdots i_{f_q} \mu .$$

Its differential can be calculated by a recurrent use of Cartan identity and the mentioned identity $[L_\eta, i_\zeta] = i_{[\eta,\zeta]}$:

$$d(f_0 i_{f_1} \cdots i_{f_q} \mu) = \frac{(-1)^q}{n} d(i_\Xi f_0 i_{f_1} \cdots i_{f_q} \omega^n)$$

$$= \frac{(-1)^q}{n} L_\Xi (f_0 i_{f_1} \cdots i_{f_q} \omega^n) + \frac{(-1)^{q-1}}{n} i_\Xi d(i_{f_1} \cdots i_{f_q} f_0 \omega^n)$$

$$= \frac{(-1)^{q-1}}{n} i_\Xi (L_{f_1} i_{f_2} \cdots i_{f_q} f_0 \omega^n - i_{f_1} d(i_{f_2} \cdots i_{f_q} f_0 \omega^n)) \qquad (1.12)$$

$$= \frac{(-1)^{q-1}}{n} i_\Xi [(\sum_{1<j\le q} i_{f_2} \cdots i_{\{f_1,f_j\}} \cdots i_{f_q} f_0 \omega^n + i_{f_2} \cdots i_{f_q} \{f_1, f_0\} \omega^n)$$

$$- i_{f_1} d(i_{f_2} \cdots i_{f_q} f_0 \omega^n)]$$

$$= \frac{(-1)^{q-1}}{n} i_\Xi [\sum_{h=1}^{q} (-1)^{h-1} (\sum_{j=h+1}^{q} i_{f_1} \cdots \hat{i}_{f_h} \cdots i_{\{f_h,f_j\}} \cdots i_{f_q} f_0 \omega^n$$

$$+ i_{f_1} \cdots \hat{i}_{f_h} \cdots i_{f_q} \{f_h, f_0\} \omega^n) + (-1)^q i_{f_1} \cdots i_{f_q} d(f_0 \omega^n)]$$

$$= \sum_{1\le h<j\le q} (-1)^{h-1} i_{f_1} \cdots \hat{i}_{f_h} \cdots i_{\{f_h,f_j\}} \cdots i_{f_q} f_0 \mu$$

$$+ \sum_{h=1}^{q} (-1)^{h-1} i_{f_1} \cdots \hat{i}_{f_h} \cdots i_{f_q} \{f_h, f_0\} \mu$$

$$= \sum_{1\le h<j\le q} (-1)^{h+j-1} f_0 i_{\{f_h,f_j\}} i_{f_1} \cdots \hat{i}_{f_h} \cdots \hat{i}_{f_j} \cdots i_{f_q} \mu$$

$$+ \sum_{h=1}^{q} (-1)^{h-1} \{f_h,f_0\} i_{f_1} \cdots \hat{i}_{f_h} \cdots i_{f_q} \mu . \qquad (2)$$

Since $\ell_{\{f_h,f_j\}} = \ell_{f_h} + \ell_{f_j} - 1$, each term on the right hand side of (2) must be a corresponding $(2n-q)$-form on Z lifted to Y. We obtain thus:

1.16.

$$\partial(\mu_{f_0;f_1,\ldots,f_q}) = \sum_{1\le h<j\le q} (-1)^{h+j-1} \mu_{f_0;\{f_h,f_j\},f_1,\ldots,\hat{f}_h,\ldots,\hat{f}_j,\ldots,f_q}$$

$$+ \sum_{h=1}^{q} (-1)^{h-1} \mu_{\{f_h,f_0\};f_1,\ldots,\hat{f}_h,\ldots,f_q} .$$

1.17. Graded Poisson algebra. The right hand side of 1.16 is nothing but a common expression for the boundary operator in the standard chain complex of a Lie algebra with coefficients in its adjoint representation. More precisely, let us introduce the graded Lie algebra

$$P^{\cdot}(Y) = \bigoplus_{\ell\in\mathbb{Z}} P^{\ell}(Y) \qquad (P^{\ell}(Y) = \{f \in C^{\infty}(Y) \mid L_{\Xi} f = \ell f\})$$

with ordinary Poisson bracket as commutator. This will be called the graded Poisson algebra of a symplectic cone Y (we shall often leave out in notation its dependence on Y where it does not lead to confusion).

Recall the definition of the standard chain complex $C_{\cdot}(P^{\cdot};\mathrm{ad})$:

$$C_q(P^{\cdot};\mathrm{ad}) = P^{\cdot} \otimes \Lambda^q P^{\cdot} ,$$

$$\partial(f_0 \otimes f_1 \wedge \cdots \wedge f_q) = \sum_{1\le h<j\le q} (-1)^{h+j-1} f_0 \otimes \{f_h,f_j\} \wedge f_1 \wedge \cdots \wedge \hat{f}_h \wedge \cdots \wedge \hat{f}_j \wedge \cdots \wedge f_q$$

$$+ \sum_{h=1}^{q} (-1)^{h-1} \{f_h,f_0\} \otimes f_1 \wedge \cdots \wedge \hat{f}_h \wedge \cdots \wedge f_q .$$

Since $P^{\cdot}$ is graded:

$$\{P^h, P^j\} \subset P^{h+j-1} ,$$

the complex $C_{\bullet}(P^{\bullet};\mathrm{ad})$ splits into the direct sum of its subcomplexes $C^{(k)}_{\bullet}(P^{\bullet};\mathrm{ad})$ which are defined by

$$C^{(k)}_q(P^{\bullet};\mathrm{ad}) = \bigoplus_{\sum_{i=0}^{q} \ell_i = k+q} \mathbb{C} f_0 \otimes f_1 \wedge \ldots \wedge f_q .$$

Put

$$\mathrm{Res}_q(f_0 \otimes f_1 \ldots f_q) = \begin{cases} \mu_{f_0; f_1, \ldots, f_q} & \text{if } f_0 \otimes f_1 \wedge \ldots \wedge f_q \in C^{(-n)}_q \\ 0 & , \text{if otherwise.} \end{cases} \quad (3)$$

Now, 1.16 simply says that Res_q's $(q \in \mathbb{Z}_+)$ define the morphism of chain complexes

$$\underline{\mathrm{Res}}: \; C_{\bullet}(P^{\bullet};\mathrm{ad}) \to \Omega_{\bullet}(Z) \quad (4)$$

which, hereafter, will be called the (*total*) *symplectic residue* morphism ($\Omega_{\bullet}(Z)$ denotes de Rham *chain* complex of Z, i.e. $\Omega_q(Z) \equiv \Omega^{2n-1-q}(Z)$) .

<u>1.18</u>. The morphism <u>Res</u> is surjective.

<u>Proof</u>. Recall that there is a canonical $C^\infty(Z)$-isomorphism

$$\sigma : \Omega^{\mathrm{vol}}(Z) \otimes_{C^\infty(Z)} (\Lambda^{\bullet}_{C^\infty(Z)} TZ) \xrightarrow{\sim} \Omega_{\bullet}(Z) , \quad (5)$$

$$\nu \otimes \eta_1 \wedge \ldots \wedge \eta_q \mapsto i_{\eta_1} \ldots i_{\eta_q} \nu$$

which is also an isomorphism of TZ-modules.

On the other hand any vector field $\eta \in TZ$ clearly can be expressed as

$$\eta = \sum_{\alpha} g_\alpha \cdot \rho_*(H_{f_\alpha}) ,$$

for some $g_\alpha \in C^\infty(Z)$ and $f_\alpha \in P^1$. Moreover, any volume form ν on Z lifts to $f_0\mu$ for some $f_0 \in P^{-n}$. Therefore all forms

$i_{\eta_1}\cdots i_{\eta_q}\nu$ lie in a space spanned by $\mu_{f_0;f_1,\dots,f_q}$'s (where $f_0 \in P^{-nq}$ and $f_i \in P^1$, $i = 1,\dots,q$). □

1.19. Corollary. The total symplectic residue induces the epimorphisms:

$$\mathrm{Res}_q : H_q(P^{\cdot}(Y);\mathrm{ad}) \longrightarrow H_{DR}^{2n-1-q}(Z) \qquad (q = 0,1). \tag{6}$$

Proof: This follows from the fact that in the commuting diagram:

$$\begin{array}{ccc} C_1^{(-n)}(P^{\cdot};\mathrm{ad}) & \xrightarrow{\partial} & C_0^{(-n)}(P^{\cdot};\mathrm{ad}) \\ \mathrm{Res}_1 \big\downarrow & & \big\downarrow \mathrm{Res}_0 \\ \Omega^{2n-2}(Z) & \xrightarrow{d} & \Omega^{2n-1}(Z) \end{array}$$

both vertical arrows are surjective whereas the right one is even bijective. □

1.20. Remark. We shall determine later the commutator structure of the Lie algebra $P^{\cdot}(Y)$ in the case of closed Z (see [18], Sect. 3). In particular, it will be shown that (6), for $q = 0$, is an isomorphism (at least when Z is closed).

1.21. The base of a symplectic cone has a canonical orientation determined by any volume form μ_g corresponding to a positive $g \in P^{-n}$. Hence for $f \in P^{\cdot}$ with compact support the integration of $\mathrm{Res}_0(f)$ yields the number

$$\mathrm{Res}\, f := \int_Z \mathrm{Res}_0(f) \tag{7}$$

which will be called the (*symplectic*) *residue* of f. (This definition was introduced first by V.W. Guillemin who also proved 1.22 below in the special case $\ell = 1$ by a different method, cf. [8]).

Notice that (7) can be calculated by using any "section" $Z' \subset Y$ of the projection $\rho : Y \to Z$

$$\text{Res } f = \int_{Z'} \mu_f \ .$$

The following assertion is a corollary of identity 1.11 and (3).

1.22. Let $f \in P^\ell$ and $g \in P^m$ ($\ell, m \in \mathbb{C}$ and $\ell + m \in \mathbb{Z}$), and $\rho(\text{supp } f \cap \text{supp } g)$ be compact. Then $\text{Res}\{f, g\}$ is defined and equal to zero. □

1.23. Corollary. The map $f \mapsto \text{Res } f$ defines a trace functional on the Lie algebra $P^\cdot_{\text{comp}}(Y) = \{f \in P^\cdot(Y) \mid \rho(\text{supp } f) \text{ is compact}\}$.

Later it will be demonstrated that when Y is connected and Z is compact this trace is actually unique (see [18], 3.4).

1.24. Remark. Assume that $\dim Y = 2$ and we are given two functions f_0 and f_1 such that $\ell_{f_0} + \ell_{f_1} = 0$ (which is the actual value of $1 - n$). Then:

$$\rho^* \mu_{f_0; f_1} = f_0 i_{f_1} \mu = -i_\Xi (f_0 i_{f_1} \omega) = f_0 df(\Xi) = f_0 L_\Xi f_1 = \ell_{f_1} f_0 f_1 \ .$$

In particular, $\text{Res}_1(f_0 \otimes f_1)$ is seen to be skew-symmetric in f_0 and f_1. Since the whole Res-morphism reduces in this case to Res_0 and Res_1 alone we infer that Res factorises through the 'anti-symmetrisation' map $\nu : C_\cdot(P^\cdot; \text{ad}) \to C_\cdot(P^\cdot)[-1]$

$$(f_0 \otimes f_1 \wedge \cdots \wedge f_q \mapsto f_0 \wedge f_1 \wedge \cdots \wedge f_q)$$

$$\begin{array}{ccc}
C_\cdot(P^\cdot; \text{ad}) & \xrightarrow{\underline{\text{Res}}} & \Omega_\cdot(Z) \\
\nu \searrow & & \nearrow \underline{R} \\
& C_\cdot(P^\cdot)[-1] &
\end{array}$$

where $R_0(f_0) := \text{Res}_0(f_0)$ and $R_1(f_0 \wedge f_1) := \text{Res}_1(f_0 \otimes f_1)$.

Although this factorisation is peculiar to dimension 2, its analogue exists in higher dimensions for symplectic cones of special type (e.g. for $Y = T_0^* X$ and $Z = S^* X$, cf. 2.8).

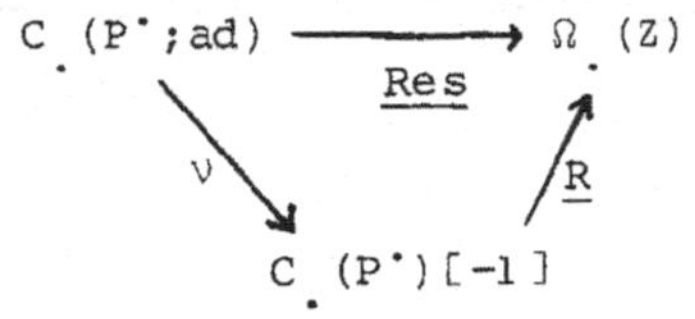

1.25. Trace morphisms. Consider the following hierarchy of "trace" morphisms:

$$C_\bullet(TX;\Omega^{vol}(X)) \to \Omega_\bullet(X), \qquad (8a)$$

$$C_\bullet(T_\nu X;C^\infty(X)) \to \Omega_\bullet(X), \qquad (8b)$$

$$C_\bullet(Ham(X);C^\infty(X)) \to \Omega_\bullet(X), \qquad (8c)$$

$$C_\bullet(Poiss(X);ad) \to \Omega_\bullet(X). \qquad (8d)$$

$T_\nu X$ denotes the Lie algebra of vector fields on X which preserve a fixed volume form ν, $Ham(X)$ is the algebra of Hamiltonian vector fields and $Poiss(X)$ is the Poisson algebra. In the last two cases X is assumed to be a symplectic manifold.

Arrow (8a) is defined as the composition of the "brute" $C^\infty(X)$-linearisation map

$$\Omega^{vol}(X) \underset{\mathbb{C}}{\otimes} \Lambda^\bullet_{\mathbb{C}} TX \to \Omega^{vol}(X) \underset{C^\infty(X)}{\otimes} (\Lambda^\bullet_{C^\infty(X)} TX)$$

with isomorphism (5). A direct computation shows that so defined map is actually a morphism of chain complexes.

The next trace (8b) is obtained from the previous one by composing it with the obvious morphism $C_\bullet(T_\nu X;C^\infty(X)) \to C_\bullet(TX;\Omega^{vol}(X))$ induced by the inclusion $T_\nu X \subset TX$ and the map $C^\infty(X) \to \Omega^{vol}(X)$, $f \to f\nu$.

Third trace is induced by the second one and the inclusion:

$$Ham(X) \subset T_\nu X \qquad (\nu = \omega^n).$$

Finally, the last trace is induced by the previous one using the canonical homomorphism $Poiss(X) \to Ham(X)$.

Assume now that the underlying manifold is a symplectic cone Y with a base Z (as in 1.15). Its graded Poisson algebra $P^\bullet = P^\bullet(Y)$ sits inside $Poiss(Y)$ and the restriction of (8d) to $P^\bullet(Y)$ induces the commutative diagram

$$\begin{array}{ccc} C_\bullet(\mathrm{Poiss}(Y);\mathrm{ad}) & \longrightarrow & \Omega_\bullet(Y) \\ \cup & & \cup \\ C_\bullet(P^\bullet(Y);\mathrm{ad}) & \longrightarrow & \Omega_{\bullet\bullet}(Y) \end{array} \tag{9}$$

where $\Omega_{\bullet\bullet}(Y)$ is the corresponding graded de Rham chain complex:

$$\Omega_{\bullet\bullet}(Y) := \bigoplus_{\ell \in \mathbb{Z}} \Omega_{\bullet\ell}(Y) , \qquad \Omega_{\bullet\ell}(Y) := \{\psi \in \Omega_\bullet(Y) \mid L_\Xi \psi = \ell\psi\} .$$

Identity 1.12 says that the lower arrow in (9) sends the k-th subcomplex $C^{(k)}_\bullet(P^\bullet;\mathrm{ad})$ (cf. 1.17) to $\Omega_{\bullet,k+n}(Y)$ ($n = \dim Y/2$). In particular, up to this shift, it is a graded morphism (called, hereafter, the *graded Poisson trace*).

Replace Y by $Y^c := Y \times_{\mathbb{R}_+^\times} \mathbb{C}^\times$. This is a $\mathbb{C}^\times$-bundle over Z associated to $\rho : Y \to Z$. Then $\Omega_{\bullet\bullet}(Y)$ embeds canonically as a subcomplex into $\Omega^{hol}_\bullet(Y^c)$ - the complex of differential forms on Y^c which are holomorphic along fibres. The ordinary fibre-wise Cauchy residue induces a canonical push-out morphism

$$\Omega^{hol}_\bullet(Y) \to \Omega_\bullet(Z) . \tag{10}$$

Notice that its composition with the graded Poisson trace that was introduced above is exactly our symplectic residue morphism (4) (this may serve as a justification of its name).

Projection (10), however, reflects precisely one half of the homology of $\Omega^{hol}_\bullet(Y^c)$. The other half, which is equally important, is carried by the subcomplex $(\rho^c)^*\Omega_\bullet(Z)[1]$. We may represent this by means of the diagram:

$$\begin{array}{ccccc} & & C_\bullet(P^\bullet(Y);\mathrm{ad}) & & \\ & \swarrow\!\!\dashrightarrow & \downarrow \mathrm{ptr} & \searrow \mathrm{Res} & \\ \Omega_\bullet(Z)[1] & \xrightarrow{(\rho^c)^*} & \Omega^{hol}_\bullet(Y^c) & \xrightarrow[\text{Residue}]{\text{Cauchy}} & \Omega_\bullet(Z) \end{array} \tag{11}$$

where ptr denotes the (graded) Poisson trace.

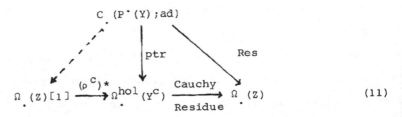

This suggests exactly two ways of constructing invariants of the graded Poisson algebra. One of them uses the canonical projection provided by Cauchy residue while no canonical projection exists in the other case. Choosing this missing projection is equivalent essentially to fixing the differential form $d\log r$ where r is an everywhere positive function on Y of homogeneity 1. Apart from this difference the two approaches are roughly speaking equivalent.

We proved in 1.18 that <u>Res</u> is a surjective map. The following lemma shows that this holds also for the graded Poisson trace.

1.26. <u>Lemma</u>. The map $ptr: C_{\cdot}(P^{\cdot}(Y);ad) \to \Omega_{\cdot\cdot}(Y)$ is surjective.

<u>Proof</u>. We have to prove that ptr induces for every $k \in \mathbb{Z}$ a surjective map $C_{\cdot}^{(k)}(P^{\cdot}(Y);ad) \to \Omega_{\cdot,k+n}(Y)$. Obviously it suffices to do this for just one k, say $k = -n$. We already know that the composition

$$C_{\cdot}^{(-n)}(P^{\cdot}(Y);ad) \to \Omega_{\cdot,0}(Y) \subset \Omega_{\cdot}^{hol}(Y) \underset{\text{residue}}{\overset{\text{Cauchy}}{\to}} \Omega_{\cdot}(Z)$$

is surjective (see 1.18). In 1.18, in fact, we demonstrated also that $\pi^*\Omega_{\cdot}(Z)[1]$, which is the kernel of the projection $\Omega_{\cdot,0}(Y) \to \Omega_{\cdot}(Z)$, is spanned by forms $f_0 i_{\Xi} i_{f_1} \dots i_{f_q} \omega^n$ where $f_0,\dots,f_q$ are homogeneous of the total order $q - n$. What remains, thus, is to show that Euler field Ξ can be represented in the form

$$\Xi = \sum_{\alpha} h_{\alpha} H_{g_{\alpha}} \tag{12}$$

where $g_{\alpha}, h_{\alpha} \in P^{\cdot}$ and $\ell_{g_{\alpha}} + \ell_{h_{\alpha}} = 1$. Actually (12) is equivalent to the equality:

$$\alpha = -\sum_{\alpha} h_{\alpha} dg_{\alpha} \tag{13}$$

where α is the canonical 1-form introduced in 1.3. In view of 1.5 α is of weight 1 and thus it can be written as $h dr + r\phi$ where r is everywhere non-zero and homogeneous of order 1, $h \in C^{\infty}(Z)$ and

$\phi \in \Omega^1(Z)$. Since obviously both hdr and $r\phi$ have required form (13), the proof is complete. (In fact, $h = (i_{\Xi}\alpha)r^{-1} = 0$, in view of definition of α). □

Lemma 1.26 implies the following refinement of Corollary 1.19.

1.27. Corollary. The graded Poisson trace induces an epimorphism

$$H_1(C_{\cdot}(P^{\cdot}(Y));ad) \to H_1(\Omega_{\cdot\cdot}(Y)) \simeq H^{2n-2}(Z) \oplus H^{2n-1}(Z). \quad \square$$

1.28. Remark. The graded Poisson trace (9) (and its variants) play an important role in computing Hochschild homology of the associative algebras of differential operators and pseudodifferential symbols (cf. [20]). The point is that the space $\Omega_{\cdot\cdot}(Y)$ is equal to the Hochschild homology of the suitable *commutative* algebra of homogeneous functions on Y. For more details the reader is referred to [20].

For completeness sake we make a note of the variant 'with coefficients' of the symplectic residue construction.

1.29. Symplectic residue with coefficients. Fix a vector bundle H on Z and let

$$P_H^{\ell} = \{s \in C^{\infty}(Y,\rho^{\cdot}H) \mid s(ty) = t^{\ell}s(y); \quad t \in \mathbb{R}_+^{\times},\ y \in Y\},$$

and $P_H^{\cdot} = \bigoplus_{\ell \in \mathbb{Z}} P_H^{\ell}$. Choose a connection ∇ on H, then

$$C_q(P^{\cdot};P_H^{\cdot}) = P_H^{\cdot} \otimes \Lambda^q P^{\cdot}$$

and

$$\partial^{\nabla}(s \otimes f_1 \wedge \dots f_q) = \sum_{1 \le h < j \le q} (-1)^{h+j-1} s \otimes \{f_h, f_j\} \wedge f_1 \wedge \dots \wedge \hat{f}_h \wedge \dots \hat{f}_j \wedge \dots \wedge f_q$$
$$+ \sum_{1 \le h \le q} (-1)^{h-1} \tilde{\nabla}_{f_h} s \otimes f_1 \wedge \dots \wedge \hat{f}_h \wedge \dots \wedge f_q$$

define the variant 'with coefficients' of the complex $C_{\cdot}(P^{\cdot};ad)$ (here $\tilde{\nabla}_f \equiv (\rho^*\nabla)_{H_f}$). Of course, $C_{\cdot}(P^{\cdot};P_H^{\cdot})$ is a chain complex in the traditional sense of the word only when ∇ is integrable.

Another instance of a "complex with curvature" is supplied by the standard de Rham complex with coefficients in H:

$$\Omega_q(Z;H) = C^\infty(Z, H\otimes\Lambda^{2n-1-q}T^*), \quad d^\nabla = \mathrm{id}_H \otimes d + \nabla \wedge \mathrm{id}_{\Omega^\bullet} \ .$$

We want to connect these two generalised complexes by a morphism (i.e. a graded map respecting boundary homomorphisms) similar to (4). The definition of forms $\mu_{s;f_1,\ldots,f_q}$ reproduces almost literally the definition from 1.15, so that the construction of the corresponding maps

$$\mathrm{Res}_q : C_q(P^\bullet;P^\bullet_H) \to \Omega_q(Z;H)$$

goes as before.

.30. <u>Proposition</u>. The mappings $\mathrm{Res}_q (q = 0,1,\ldots)$ form a morphism of chain complexes 'with curvature'

$$\underline{\mathrm{Res}} : (C_\bullet(P^\bullet;P^\bullet_H), \partial^\nabla) \to (\Omega_\bullet(Z;H), d^\nabla) \ .$$

<u>Proof</u>. Notice that the two standard identities still hold for Lie derivative "with coefficients" $L^\nabla_\eta := \nabla_\eta \otimes \mathrm{id}_{\Omega^\bullet} + \mathrm{id}_H \otimes L_\eta$, namely:

$$L^\nabla_\eta = i_\eta d^\nabla + d^\nabla i_\eta \quad \text{and} \quad [L^\nabla_\eta, i_\zeta] = i_{[\eta,\zeta]} \ .$$

Hence the proof reduces to reproducing computation (2) with L_η replaced by L^∇_η.

.31. <u>Final remark</u>. In the course of work with symbols of ψDOs it will be very convenient to extend <u>Res</u> to a slightly bigger complex $_\bullet(P^\bullet;P^\bullet_H)$ defined as:

$$_q(P^\bullet;P^\bullet_H) = \{\sum_{\nu=1}^{\infty} s^\nu \otimes f_1^\nu \wedge\ldots\wedge f_q^\nu \mid \mathrm{ord}\, s^\nu + \mathrm{ord}\, f_1^\nu + \ldots + \mathrm{ord}\, f_q^\nu \to -\infty \ \text{ as } \ \nu\to\infty\} \ .$$

§2. Noncommutative residue (local theory)

We will define for an arbitrary pseudodifferential operator of classical type a certain global density which will be called its noncommutative residue. Throughout this section, however, we are occupied only with local considerations. The general case is postponed to the next section.

2.1. Fix a domain $U \subset \mathbb{R}^n$ and let $(u^1,\ldots,u^n;\xi_1,\ldots,\xi_n)$ be the corresponding coordinates in $T^*U = U \times \mathbb{R}^n$.

Constructions of the previous section will be applied to the following special case

$$Y = T_0^*U = T^*U\backslash U \qquad \text{and} \qquad Z = S^*U$$

($R_+^\times$ acts as a homothety in ξ-space), and

$$\omega = \sum_{i=1}^n d\xi_i \wedge du^i, \qquad \Xi = r\frac{\partial}{\partial r} \quad (\text{where } r = |\xi|),$$

$$\alpha = \sum_{i=1}^n \xi_i\, du^i,$$

$$\omega^n = (-1)^{\frac{n(n-1)}{2}} n!d\xi_1\wedge\ldots\wedge d\xi_n\wedge du^1\wedge\ldots\wedge du^n \equiv (-1)^{\frac{n(n-1)}{2}} n!d\xi du,$$

$$\mu = (-1)^{\frac{n(n-1)}{2}} (n-1)!(*rdr)\wedge du^1\wedge\ldots\wedge du^n$$

$$= (-1)^{\frac{n(n-1)}{2}} (n-1)!\left\{\sum_{i=1}^n (-1)^{i-1}\xi_i d\xi_1\wedge\ldots\wedge\widehat{d\xi_i}\ldots d\xi_n\right\}\wedge du^1\wedge\ldots\wedge du^n$$

$$= \beta_n^{-1}\, r đ\xi'_{(r)}du \tag{1}$$

where $đ\xi'_{(r)}$ is the volume form divided by $(2\pi)^n$, of the sphere of radius r in $\mathbb{R}^n_\xi$ and, β_n will hereafter denote the constant

$$\beta_n = \frac{(-1)^{\frac{n(n-1)}{2}}}{(n-1)!\,(2\pi)^n} \; .$$

The described situation possesses one extra feature which we will take advantage of: Y is a Lagrangean fibre bundle over U with its projection fitting into the triangle

$$\begin{array}{ccc} Y = T_0^*U & & \\ & \overset{\rho}{\searrow} & \\ \pi \big\downarrow & & Z = S^*U \; . \\ & \underset{\tau}{\swarrow} & \\ U & & \end{array}$$

The following lemma will be often used. Assume we are given a vector bundle G on U and a homogeneous section s of $\pi^{\cdot}G$ of order $1-n$.

2.2. Lemma. One has

$$\mathrm{Res}_0\left(\frac{\partial}{\partial \xi_j}\, s\right) = d_{\mathrm{fibre}}\, \gamma_j \wedge du^1 \wedge \ldots \wedge du^n \qquad (j = 1,\ldots,n)$$

where $\gamma_j \in C^\infty(U, G \otimes \tau_{\cdot}\Omega^{n-2} Z/U)$ is a relative $(n-2)$-form determined by the equality

$$\rho^*\gamma_j = (-1)^{\frac{n(n-1)}{2}+j+1}(n-1)!\, s \otimes \left\{ \sum_{i=1}^{j-1} (-1)^i \xi_i\, d\xi_1 \wedge \ldots \wedge \widehat{d\xi_i} \wedge \ldots \wedge \widehat{d\xi_j} \wedge \ldots \wedge d\xi_n \right.$$

$$\left. - \sum_{i=j+1}^{n} (-1)^i \xi_i\, d\xi_1 \wedge \ldots \wedge \widehat{d\xi_j} \wedge \ldots \wedge \widehat{d\xi_i} \wedge \ldots \wedge d\xi_n \right\} ,$$

and d_{fibre} denotes de Rham differential along fibres of the projection τ.

Proof. Take any connection ∇^G on G and put $H = \tau^{\cdot}G$, $\nabla = \tau^{\cdot}\nabla^G$. According to 1.30 the boundary homomorphism

$$\partial^\nabla : C_1(P^{\cdot}; P_H^{\cdot}) \to C_0(P^{\cdot}; P_H^{\cdot})$$

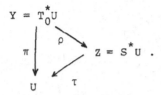

is related to de Rham differential via the equality $\mathrm{Res}_0 \partial^\nabla = d^\nabla \mathrm{Res}_1$.
Since the connection ∇ is trivial along fibres of τ and
$H_{u^j} = -\partial/\partial_{\xi_j}$, one obviously has

$$\partial^\nabla (s \otimes u^j) = -\partial_{\xi_j} s .$$

Hence

$$\mathrm{Res}_0(\partial_{\xi_j} s) = -d^\nabla \mathrm{Res}_1(s \otimes u^j) = d^\nabla (s \otimes i_{\partial/\partial_{\xi_j}} \mu)$$

$$= (-1)^{\frac{n(n-1)}{2}} (n-1)!\; d_\xi (s \otimes i_{\partial/\partial_{\xi_j}} (*rdr)) \wedge du^1 \wedge \ldots \wedge du^n$$

and the assertion follows. □

<u>2.3.</u> <u>Corollary</u>. $\tau_* \left| \mathrm{Res}_0 \left(\frac{\partial}{\partial_{\xi_j}} \right) s \right| = 0 \qquad (j = 1,\ldots,n)$. □

Here τ_* denotes the push-out map $A_\cdot(Z;\tau^\cdot G) \to A_\cdot(U;G)$ and $|\ |$ denotes the canonical identification $\Omega_\cdot(Z;\tau^\cdot G) \tilde{\to} A_\cdot(Z;\tau^\cdot G)$, (cf. 0.17 and 0.19).

Return to the scalar case. We make two elementary observations:

<u>2.4.</u> $\mathrm{Res}_q(f_0 \otimes f_1 f_1' \wedge f_2 \wedge \ldots \wedge f_q) = \mathrm{Res}_q(f_0 f_1 \otimes f_1' \wedge f_2 \wedge \ldots \wedge f_q) +$

$\mathrm{Res}_q(f_0 f_1' \otimes f_1 \wedge f_2 \wedge \ldots \wedge f_q)$.

<u>2.5.</u> $\mathrm{Res}_q(f_0 \otimes f_1 f_1' \wedge f_2 \wedge \ldots \wedge f_q) = \mathrm{Res}_q(f_0 \otimes f_1 \wedge f_1' f_2 \wedge f_3 \wedge \ldots \wedge f_q) +$

$\mathrm{Res}_q(f_0 \otimes f_1' \wedge f_1 f_2 \wedge f_3 \wedge \ldots \wedge f_q)$

which are valid for a general symplectic cone Y and $f_i, f_i' \in P^\cdot(Y)$.

Identity 2.4 follows from:

$$i_{ff'} = f i_{f'} + f' i_f \quad ,$$

and 2.5 also from:

$$i_{f_1 f_1'} i_{f_2} = i_{f_1} i_{f_1' f_2} + i_{f_1'} i_{f_1 f_2} .$$

The former identity implies, in particular, that:

$$\mathrm{Res}_q(f_0 \otimes f_1 \wedge f_2 \wedge \ldots \wedge f_q) + \mathrm{Res}_q(f_1 \otimes f_0 \wedge f_2 \wedge \ldots \wedge f_q) =$$

$$\mathrm{Res}_q(1 \otimes f_0 f_1 \wedge f_2 \wedge \ldots \wedge f_q) .$$

2.6. Lemma. For any set of homogeneous functions $f_1, \ldots, f_q$ on T_0^*U one has

$$\tau_* |\mathrm{Res}_q(1 \otimes f_1 \wedge \ldots \wedge f_q)| = 0 .$$

Therefore $\tau_* |\mathrm{Res}_q(f_0 \otimes f_1 \wedge \ldots \wedge f_q)|$ is skew-symmetric in all entries.

Proof. Introduce an abbreviated notation:

$$i_j = i_{\partial/\partial u^j} , \quad \partial_j f = \partial_{\xi_j} f .$$

It follows directly from the definition of Res_q that:

$$\rho^* \mathrm{Res}_q(1 \otimes f_1 \wedge \ldots \wedge f_q) = i_{f_1} \cdots i_{f_q} \mu = \sum_{j_1, \ldots, j_q = 1}^{n} i_{j_1} \cdots i_{j_q} (\partial_{j_1} f_1) \ldots (\partial_{j_q} f_q) \mu$$

$$+ \sum_{k=1}^{n} i_{\partial/\partial \xi_k} \omega_k$$

where ω_k are certain horizontal forms (i.e. $i_\Xi \omega_k = 0$) on $Y = T_0^*U$ of weight 1.

The forms $r^{-1}\omega_k$ $(r = |\xi|)$ descend to uniquely defined forms $\tilde{\omega}_k$ on $Z = S^*U$, and the vector fields $r\partial/\partial\xi_k$ possess well-defined projections $\Xi_k := \rho_*(r\partial/\partial\xi_k)$, hence the equalities

$$i_{\partial/\partial\xi_k} \omega_k = \rho^*(i_{\Xi_k} \tilde{\omega}_k) \qquad (k = 1, \ldots, n)$$

hold.

Next, let us notice that, because $i_k i_\ell (\partial^2_{k\ell} f_{q-1}) f_q$ is (k,ℓ)-skew-symmetric, we must have:

$$\sum_{j_{q-1}=1}^{n} \sum_{j_q=1}^{n} i_{j_{q-1}} i_{j_q} (\partial_{j_{q-1}} f_{q-1})(\partial_{j_q} f_q) =$$

$$= \sum_{j_{q-1}=1}^{n} \sum_{j_q=1}^{n} i_{j_{q-1}} i_{j_q} \partial_{j_{q-1}} (f_{q-1} \partial_{j_q} f_q) .$$

An inductive application of this argument yields the identity:

$$\sum_{j_1,\dots,j_q=1}^{n} i_{j_q} \cdots i_{j_q} (\partial_{j_1} f_1) \cdots (\partial_{j_q} f_q)\mu$$

$$= \sum_{j_1,\dots,j_q=1}^{n} i_{j_1} \cdots i_{j_q} \partial_{j_1} (f_1 \partial_{j_2} (f_2 \partial_{j_3} \cdots \partial_{j_{q-1}} (f_{q-1} \partial_{j_q} f_q) \dots))\mu$$

$$\equiv \sum_{j_1,\dots,j_q=1}^{n} i_{j_1} \cdots i_{j_q} (\partial_{j_1} \phi_{j_2 \dots j_q})\mu$$

$$= \sum_{j_1,\dots,j_q=1}^{n} i_{j_1} \cdots i_{j_q} \rho^* \mathrm{Res}_0 \left(\frac{\partial}{\partial \xi_{j_1}} \phi_{j_2 \dots j_q} \right) .$$

Thus we obtain

$$\mathrm{Res}_q (1 \otimes f_1 \wedge \dots \wedge f_q) = \sum_{j_1,\dots,j_1=1}^{n} i_{j_1} \cdots i_{j_q} \mathrm{Res}_0 \left(\frac{\partial}{\partial \xi_{j_1}} \phi_{j_2 \dots j_q} \right) + \sum_{k=1}^{n} i_{\Xi_k} \omega_k .$$

We will apply to both sides the operation $\tau_* |\cdots|$. Recall that τ_* is a composition of two operations - the projection:

$$A_{\cdot}(Z) \to A_{\cdot}(U; \tau_{\cdot} A_{0,\mathrm{rel}}), \qquad (3)$$

and the integration:

$$\int : \tau_{\cdot} A_{0,\mathrm{rel}} \to \mathcal{O}_U^{\infty} .$$

(A_{rel} denotes the sheaf of relative integral forms $A_{Z/U}$). Since the fields Ξ_k are tangent to fibres of τ we infer the application of (3) alone kills the second sum in (2). The first sum vanishes after the integration in view of Lemma 2.2 and the observation that τ_* commutes with i_j's. □

2.7. Remark. Lemma 2.2 allows us to write the image of $\mathrm{Res}_q(1\otimes f_1\wedge\ldots\wedge f_q)$ under (3) as an explicit fibre-wise exact form.

2.8. Lemma 2.6 implies, in particular, commutativity of the following diagram of morphisms of chain complexes:

$$\begin{array}{ccc} C_{\cdot}(P^{\cdot}(Y);\mathrm{ad}) & \xrightarrow[\underline{\mathrm{Res}}]{} & \Omega_{\cdot}(Z) \\ \downarrow \nu & & \downarrow \tau_*|\ldots| \\ C_{\cdot}(P^{\cdot}(Y))[-1] & \xrightarrow[\underline{R}]{} & \Omega_{\cdot}(U) \end{array} \qquad (4)$$

where $\underline{R}$ is defined as:

$$R_q(f_0\wedge\ldots\wedge f_q) = \tau_*|\mathrm{Res}_q(f_0\otimes f_1\wedge\ldots\wedge f_q)| .$$

Notice that (4) is precisely a higher-dimensional version of actorisation 1.24. Another important property of $\underline{R}$-morphism is its $^{\infty}(U)$-linearity:

.9. $\quad R_q(gf_0\wedge f_1\wedge\ldots\wedge f_q) = gR_q(f_0\wedge f_1\wedge\ldots\wedge f_q) \qquad (g\in C^{\infty}(U))$.

Indeed, by definition we have $\mathrm{es}_q(gf_0\otimes f_1\wedge\ldots\wedge f_q) = g\,\mathrm{Res}_q(f_0\otimes f_1\wedge\ldots\wedge f_q)$ for an arbitrary $\in C^{\infty}(Z)$, while the multiplication by $g\in C^{\infty}(U)$ commutes with $_*|\ldots|$.

Lemmas 2.6. and 2.9. together imply:

.10. $\quad R_q(f_0\wedge\ldots\wedge f_q) = 0$ if at least one f_i belongs to $C^{\infty}(U)$.

.11. $\quad \sum_{i=0}^{q} R_q(f_0\wedge\ldots\wedge\frac{\partial f_i}{\partial\xi_j}\wedge\ldots\wedge f_q) = 0 \qquad (j\in\{1,\ldots,n\})$.

Only the latter identity, perhaps, requires any explanation. It ollows easily from:

$$\sum_{i=0}^{q} f_0\wedge\ldots\wedge\frac{\partial f_i}{\partial\xi_j}\wedge\ldots\wedge f_q = u^j\wedge\underline{\partial}(f_0\wedge\ldots\wedge f_q) - \underline{\partial}(u^j\wedge f_0\wedge\ldots\wedge f_q), \qquad (5)$$

identity 2.10 and the equality $R_q\underline{\partial} = -dR_{q+1}$. (To avoid confusion with partial derivatives or with other boundary maps the boundary in the complex $C_\bullet(P^\bullet)$ is denoted in (5) and in the rest of this section by $\underline{\partial}$).

Armed with the morphism $\underline{R}$ we proceed to the definition of the residue for pseudodifferential symbols.

2.12. Consider an arbitrary ψDO with matric coefficients $A : C^\infty_{comp}(U,\theta^k) \to C^\infty(U,\theta^\ell)$ and of order m. Its complete symbol is a formal series

$$a(u,\xi) = \sum_{j=0}^{\infty} a_{m-j}(u,\xi)$$

such that each component $a_{m-j} \in C^\infty(T_0^*U, \underline{Hom}(\theta^k,\theta^\ell))$ is homogeneous of order $m - j$.

2.13. Definition. $\mathrm{res}_u A = \left(\int_{|\xi|=1} a_{-n}(u,\xi)\,|đ\xi'|\right)|du|$

($đ\xi' \equiv đ\xi'_{(1)}$ denotes the normalised volume form on the unit sphere).

If $\mathrm{ord}\,A \notin \mathbb{Z}$ we put $\mathrm{res}_u A \equiv 0$.

By definition $\mathrm{res}_u A$ is a matric-valued density on U. An alternative expression for $\mathrm{res}_u A$ is

$$\mathrm{res}_u A = \beta_n \tau_* |\mathrm{Res}_0 a| \qquad (\tau : S^*U \to U)$$

where Res_0 is the residue map of 1.31 in the case:

$$H = \underline{Hom}(\theta^k,\theta^\ell) \text{ and } \nabla = \text{canonical flat connection on } H.$$

The density $\mathrm{res}_u A$ will be called the *noncommutative residue* of A.

2.14. Remark. For $n = 1$ a slightly different density will be neede[d]. In this case S^*U consists of two copies of U and it can be canonically identified with the orientation double cover of U. The volume form $đ\xi'$ reduces to

$$đ\xi' = \frac{1}{2\pi}\frac{|du|}{du} ,$$

and $(2\pi)^{-1} a_{-1}(u,\xi)|du| = a_{-1}(u,\xi)đ\xi' du$ when restricted to $\{\xi = \pm 1\}$ can be viewed as a density on U with coefficients in

$$\underline{\mathrm{Hom}}(\theta^k, \theta^\ell) \otimes S$$

($S = \tau_{\cdot}\theta^1_{S^*U}$ is a rank 2 vector bundle on U, $\tau_{\cdot}$ denotes the direct image).

This density will be denoted by $\underline{\mathrm{res}}_u A$. Choosing orientation identifies S with θ^2_U and $\underline{\mathrm{res}}_u A$ with a 2-component density:

$$\begin{pmatrix} \mathrm{res}_{+,u}A \\ \\ \mathrm{res}_{-,u}A \end{pmatrix} .$$

2.15. Let $f : V \to U$ be an open embedding of another domain $V \subset \mathbb{R}^n$ (possibly $V = U$, f being a diffeomorphism). Then the induced operator $f^{\#} A: C^\infty_{comp}(V,\theta^k) \to C^\infty(V,\theta^\ell)$ is defined (cf. 0.5).

The following lemma asserts the functoriality of $\mathrm{res}_u A$ with respect to open embeddings of domains.

2.16. Lemma. $\mathrm{res}_v f^{\#} A = f^* \mathrm{res}_u A$.

Proof. Denote the complete symbol of $f^{\#} A$ by a^f. According to the 2nd form of the base change formula (cf. 0.14) we have

$$a^f = (T^*f)^*\left\{(1 + \sum_{j=1}^{n} \partial_{\xi_j} \circ \Psi_j)a\right\}$$

where $\Psi_j = \Psi^f_j(u,\xi,\partial_\xi)$ are some differential operators of infinite order (depending on f).

On the other hand, in virtue of the identities

$$(T^*f)^*\left(\sum_{i=1}^{n} d\xi_i \wedge du^i\right) = \sum_{i=1}^{n} d\eta_i \wedge dv^i, \quad (T^*f)_* \Xi^V = \Xi^U$$

where Ξ^U, Ξ^V are the Euler fields on T_0^*U and T_0^*V respectively, it follows that

$$\mathrm{Res}_0^V (T^*f)^* = (S^*f)^* \mathrm{Res}_0^U .$$

Therefore

$$\mathrm{Res}_0(a^f) = (S^*f)^* \mathrm{Res}_0\left((1 + \sum_{j=1}^{n} \partial_{\xi_j} \circ \Psi_j)a\right) ,$$

and

$$\mathrm{res}_V f^{\#} A = \beta_n \tau_*^V |\mathrm{Res}_0(a^f)|$$

$$= \beta_n\left(\tau_*^V (S^*f)^*\right) \left|\mathrm{Res}_0\left((1 + \sum_{j=1}^{n} \partial_{\xi_j} \circ \Psi_j)a\right)\right|$$

$$= \beta_n(f^*\tau_*^U) \left|\mathrm{Res}_0\left((1 + \sum_{j=1}^{n} \partial_{\xi_j} \circ \Psi_j)a\right)\right|$$

$$= f^*\mathrm{res}_u A + \beta_n f^*\tau_*^U \left|\mathrm{Res}_0 \sum_{j=1}^{n} \partial_{\xi_j}(\Psi_j a)\right| \qquad (6)$$

(we used here the formula for change of variables under the sign of integral: $\tau_*^V(S^*f)^* = f^*\tau_*^U$ (cf. 0.18)). According to Lemma 2.2 the second term in (6) vanishes since $\mathrm{Res}_0 \sum_{j=1}^{n} \partial_{\xi_j}(\Psi_j a)$ belongs to $\Omega^n(U,\tau_* d_{\mathrm{fibre}}\Omega^{n-2}_{S^*U/U})$, i.e. is fibre-wise exact. □

2.17. Let $r \in C^\infty(U, \underline{\mathrm{Hom}}(\theta^{k'},\theta^k))$ and $s \in C^\infty(U,\underline{\mathrm{Hom}}(\theta^\ell,\theta^{\ell'}))$ be arbitrary matric functions. In particular, for $A : C^\infty_{\mathrm{comp}}(U,\theta^k) \to C^\infty(U,\theta^\ell)$ the operator $s \circ A \circ r : C^\infty_{\mathrm{comp}}(U,\theta^{k'}) \to C^\infty(U,\theta^{\ell'})$ is well defined.

2.18. Lemma. $\mathrm{res}_u(s \circ A \circ r) = s(\mathrm{res}_u A)r$.

($\mathrm{res}_u A$ is a density with coefficients in $\underline{\mathrm{Hom}}(\theta^k,\theta^\ell)$ so that its composition with s on the left and r on the right makes sense).

Proof. Let $a^{(r,s)}$ be the complete symbol of $s\circ A\circ r$, then by the composition formula (cf. 0.12) $a^{(r,s)}$ can be written as

$$a^{(r,s)} = sar + s \sum_{j=1}^{n} \partial_{\xi_j} (\Phi_j a)$$

where $\Phi_j = \Phi_j^r(u,\partial_\xi)$ are differential operators of infinite order (depending on r) defined from:

$$\sum_{j=1}^{n} \partial_{\xi_j} \circ \Phi_j \equiv \sum_{|\alpha| \geq 1} \frac{1}{\alpha!} \partial_\xi^\alpha a \, D_u^\alpha r \, .$$

Another application of Lemma 2.2 yields:

$$\text{res}_u(s\circ A\circ r) - s(\text{res}_u A)r = \rho_n s\tau_* \Big|\text{Res}_0 \sum_{j=1}^{n} \partial_{\xi_j} (\Phi_j a)\Big| = 0 \, . \qquad \square$$

2.19. Finally, let $\phi = (f;r,s)$ be a general morphism between two triples $(V;\theta^{k'},\theta^{\ell'})$ and $(U;\theta^k,\theta^\ell)$ (cf. 0.3) .

Lemmas 2.16 and 2.18 together imply the functoriality of the residue density under such morphisms.

2.20. Proposition. $\text{res}_v(\phi^\# A) = \phi^* \text{res}_u A$. $\square$

2.21. Remark. In fact the proofs of both Lemma 2.13 and 2.15 give more than just identity 2.20. They produce an explicit relative (n-2)-form γ on V (depending on ϕ) such that

$$\text{Res}_0 a^\phi - (S^* f)^* (\text{res}_0 a) = d_{\text{fibre}}\gamma \wedge dv^1 \wedge \ldots \wedge dv^n \, .$$

Here a^ϕ denotes the complete symbol of $\phi^\# A$).

2.22. The noncommutative residue via the amplitudes. Recall that a ΨDO $A : C^\infty_{\text{comp}}(U,\theta^k) \to C^\infty(U,\theta^\ell)$ possesses many different representations in the form:

$$Af(u) = \int_{R_\xi^n} \int_U e^{i(u-v)\xi} \beta(u,v,\xi) f(v) dv đ\xi + Tf(u) \qquad (7)$$

with property that the *amplitude* β admits the standard quasi-classic expansion:

$$\beta(u,v,\xi) \sim b(u,v,\xi) = \sum_{j=0}^{\infty} b_{m-j}(u,v,\xi) \tag{8}$$

and T is an operator with smooth kernel. If the dependence on (u,v) is through the linear combination $(1-t)u + tv$ (t being fixed real number):

$$\beta(u,v,\xi) \equiv \alpha((1-t)u + tv,\xi)$$

then $\alpha(u,\xi)$ is called a (smoothed) t-*symbol* of A. For $\tau = 0$ we obtain an ordinary (smoothed) symbol.

For completeness sake we should mention that even if the smoothing operator T can also be made an oscillatory integral like the first term in (7) its amplitude may not be reducible to a 'symbol-like' one.

When appealing to amplitudes or symbols we shall mean, however, (if not stated otherwise) the *formal* amplitudes and *formal* symbols, which are formal series with homogeneous components as in (8). If so, then a standard assertion from the ψDO theory (cf. e.g. [15], IV.23.3-5) says that a t-symbol exists and is unique (for each t), and that the *ordinary* symbol $a(u,\xi)$ is related to an *arbitrary* amplitude $b(u,v,\xi)$ by the identity

$$a = (e^{-L_v} b)_{|T_0^*\Delta} \tag{9}$$

where L denotes the second order differential operator on T^*U

$$L = \sqrt{-1} \sum_{i=1}^{n} \frac{\partial^2}{\partial\xi_i \partial u^i} \tag{10}$$

(the letter v indicates that in (9) L acts on the variables (v,ξ)) and $T_0^*\Delta \subset U \times T_0^*U$ consists of points (u,u,ξ) with $\xi \neq 0$; the exponent e^{-L} in (9) should be meant as a differential operator of infinite order).

Moreover, t-symbols $a^{(t)}$ for two different t's are related to each other by the identity

$$a^{(t_1)} = e^{(t_1-t_2)L} a^{(t_2)} \tag{11}$$

which is a particular case of (9).

In applications apart from the ordinary (left) symbol appear usually only *Weyl* symbol $(t = \frac{1}{2})$ and the *right* symbol $(t = 1)$.

Assume we are given some amplitude b. Then we can apply Res_0 to its restriction to $T_0^*\Delta$. This turns out to give another method of calculating the noncommutative residue of the operator.

2.23. Lemma. a) For an arbitrary amplitude b of a ψDO A one has

$$res_u A = \beta_n \tau_* \left| Res_0 (b_{|T_0^*\Delta}) \right| .$$

b) Similarly, for an arbitrary t-symbol $a^{(t)}$ of A one has:

$$res_u A = \beta_n \tau_* \left| Res_0 a^{(t)} \right| . \tag{12}$$

In particular, the right-hand side of (12) does not depend on t.

Proof. One has

$$e^{-L} - 1 = \sum_{i=1}^{n} \partial_{\xi_i} \cdot K^{(i)}$$

where

$$K^{(i)} := D_i T(L) \qquad (D_i \equiv \frac{1}{\sqrt{-1}} \partial_{u^i}) ,$$

and

$$T(L) := \frac{1-e^{-L}}{L}$$

is a differential operator on T^*U of infinite order (notice that $T(X)$ is precisely the Todd series).

In particular, the difference between the ordinary symbol of A and $b_{|T_0^*\Delta}$ equals (cf. (9)):

$$a - (b_{|T_0^*\Delta}) = (e^{-L_v} - 1) b_{|T_0^*\Delta} = \sum_{i=1}^{n} \frac{\partial}{\partial \xi_i} (K_v^{(i)} b_{|T_0^*\Delta}) .$$

Appealing to Lemma 2.2 finishes the proof of part a). Part b) is proved similarly but with (11) used instead of (9). □

2.24. Let $B : C^{\infty}_{comp}(U,\theta^{\ell}) \to C^{\infty}(U,\theta^{k})$ be another ψDO and assume either A or B to be a *proper* ψDO (cf. 0.10). Then the compositions AB and BA are both defined. In particular, there are two densities:

$$\mathrm{res}_u(AB) \in A_0(U;\underline{\mathrm{End}}\ \theta^{\ell}) \quad \text{and} \quad \mathrm{res}_u(BA) \in A_0(U;\underline{\mathrm{End}}\ \theta^{k}) \ .$$

Our aim is to demonstrate that the *scalar* densities $\mathrm{tr}\ \mathrm{res}_u(AB)$ and $\mathrm{tr}\ \mathrm{res}_u(BA)$ actually coincide up to a certain explicit exact form. We consider first the case when both A and B are scalar operators. The matric case will follow easily from the scalar one.

Make some auxiliary remarks before we state the result. It will be convenient to extend $\underline{R}$-morphism $C_{\bullet}(P^{\bullet}(Y))[-1] \to A_0(U)$ to the formally completed complex $\hat{C}_{\bullet}(P^{\bullet}(Y))[-1]$ much as we did in the case of symplectic residue morphism (cf. 2.31).

Recall that

$$\hat{C}_q = \{ \sum_{\nu=1}^{\infty} f_1^{\nu}\wedge\ldots\wedge f_q^{\nu} \mid \mathrm{ord}\ f_q^{\nu}+\ldots+\mathrm{ord}\ f_q^{\nu} \to -\infty \quad \text{as} \quad \nu \to \infty\} \ .$$

For basically aesthetic reasons, we shall assume orders of all operators considered to be integers (cf. however, Remark 2.26).

2.25. Lemma. Let A and B be scalar operators. Then one has:

$$\mathrm{res}_u[A,B] = d\rho_u(A,B)$$

where $\rho_u(A,B) \in A_1(U)$ is defined as

$$\rho_u(A,B) = \sqrt{-1}\beta_n R_1 \left(\sum_{i,j=0}^{\infty} \frac{1}{(1+i+j)!} L^i a \wedge L^j b \right) \ . \tag{13}$$

(a and b denote the symbols of A and B respectively, and L is the second order operator (10) on T^*U).

2.26. Remark. We deal only with the case of integer order operators but since the differential forms $\mu_{f_0;f_1,\dots,f_q}$ of 1.15 are defined under the sole condition that $\ell_{f_0} + \dots + \ell_{f_q}$ equals $-n$, and crucial identity 1.16. still holds in that situation, one can easily extend both Lemma 2.25 and all considerations below to operators with any complex order. Of course, $\rho_u(A,B)$ can be non-zero only when $\operatorname{ord} A + \operatorname{ord} B \in \mathbb{Z}$.

Proof. We shall need the completed tensor square $P^{\cdot} \hat{\otimes} P^{\cdot}$ which consists of series $\sum_{v=1}^{\infty} f_1^{\nu} \otimes f_2^{\nu}$ satisfying the condition $\ell_{f_1^{\nu}} + \ell_{f_2^{\nu}} \to -\infty$ as $\nu \to \infty$ and the similarly completed Poisson algebra $\hat{P}^{\cdot}$; the usual commutative multiplication $f \otimes g \to fg$ defines then a map $m : P^{\cdot} \hat{\otimes} P^{\cdot} \to \hat{P}^{\cdot}$. The simple identity

$$m \circ (1 \otimes L - L \otimes 1) = \frac{1}{\sqrt{-1}} \{.,.\} + \frac{1}{\sqrt{-1}} \sum_{i=1}^{n} \partial_{\xi_i} \cdot m \cdot (\partial_{u^i} \otimes 1 - 1 \otimes \partial_{u^i})$$

and Lemma 2.2 imply together commutativity of the diagram:

$$\begin{array}{ccccc} P^{\cdot}\hat{\otimes}P^{\cdot} & \xrightarrow{\frac{1}{\sqrt{-1}}\{,\}} & \hat{P}^{\cdot} & & \\ & & & \searrow^{R_0} & \\ {\scriptstyle 1\otimes L - L\otimes 1}\downarrow & & & & \Omega_{\cdot}(U) \\ & & & \nearrow_{R_0} & \\ P^{\cdot}\hat{\otimes}P^{\cdot} & \xrightarrow{\ m\ } & \hat{P}^{\cdot} & & \end{array} \qquad (14)$$

(Here $\{,\}$ is the Poisson bracket).

The operator AB possesses one especially nice amplitude:

$$c(u,v,\xi) := a(u,\xi) b^{(1)}(v,\xi) \qquad (15)$$

where $b^{(1)}$ is the *right* symbol (cf. 2.19) of B. Similarly, $AB-BA$ possesses the amplitude:

$$\tilde{c}(u,v,\xi) = a(u,\xi) b^{(1)}(v,\xi) - b(u,\xi) a^{(1)}(v,\xi) .$$

Restrict it to $T^*_U \Lambda$ and then apply formula (11) to obtain

$$\tilde{c}_{|T_0^*\Delta} = ab^{(1)} - a^{(1)}b = m(1 \otimes e^L - e^L \otimes 1)(a \otimes b).$$

In the ring of formal power series $\mathbb{C}[L] \hat{\otimes} \mathbb{C}[L]$ the element $1 \otimes e^L - e^L \otimes 1$ decomposes as

$$1 \otimes e^L - e^L \otimes 1 = (1 \otimes L - L \otimes 1) \sum_{i,j=0}^{\infty} \frac{1}{(1+i+j)!} L^i \otimes L^j .$$

Hence we have

$$\begin{aligned}
\mathrm{res}_u AB - \mathrm{res}_u BA &\overset{(2.23)}{=} \beta_n R_0(\tilde{c}_{|T_0^*\Delta}) \\
&= \beta_n R_0 \left| m(1 \otimes L - L \otimes 1) \left(\sum_{i,j=0}^{\infty} \frac{1}{(1+i+j)!} L^i a \otimes L^j b \right) \right| \\
&\overset{(14)}{=} \beta_n R_0 \left| \frac{1}{\sqrt{-1}} \{,\} \circ \left(\sum_{i,j=0}^{\infty} \frac{1}{(1+i+j)!} L^i a \otimes L^j b \right) \right| \\
&= \frac{\beta_n}{\sqrt{-1}} (R_0 \partial) \left(\sum_{i,j=0}^{\infty} \frac{1}{(1+i+j)!} L^i a \wedge L^j b \right) \\
&= \sqrt{-1}\, \beta_n d R_1 \left(\sum_{i,j=0}^{\infty} \frac{1}{(1+i+j)!} L^i a \wedge L^j b \right)
\end{aligned}$$

as required (recall that in $\hat{C}_\cdot(P^\cdot)[-1]$ all boundary morphisms have opposite sign). □

2.27. Let us extend the definition of the integral form (13) to general matric ψDOs:

$$A = (A^{\lambda\kappa}), \qquad B = (B^{\kappa\lambda}) \qquad (\kappa = 1,\dots,k; \lambda = 1,\dots,\ell)$$

by putting

$$\rho_u(A,B) = \sqrt{-1}\, \beta_n R_1 \left(\sum_{\kappa=1}^{k} \sum_{\lambda=1}^{\ell} \sum_{i,j=0}^{\infty} \frac{1}{(1+i+j)!} L^i a^{\lambda\kappa} \wedge L^j b^{\kappa\lambda} \right) .$$

Lemma 2.25 implies then immediately

2.28. Proposition. $\mathrm{tr}\, \mathrm{res}_u AB - \mathrm{tr}\, \mathrm{res}_u BA = d\rho_u(A,B)$.

2.29. Remark. Let $F(X,Y) \in \mathbb{C}[[X,Y]]$ denote the formal series:

$$F(X,Y) = \frac{e^X - e^Y}{X-Y} . \tag{16}$$

Make substitutions $X = 1 \otimes L$ and $Y = L \otimes 1$ and regard $\underline{F} := F(1 \otimes L, L \otimes 1)$ as a linear map $P^{\cdot} \hat{\otimes} P^{\cdot} \to P^{\cdot} \hat{\otimes} P^{\cdot}$. Then (13) reads as

$$\rho_u(A,B) = \sqrt{-1}\ \beta_n(R_1 \circ \underline{F})(a \otimes b) . \tag{17}$$

Since

$$F(X,Y) \equiv \hat{A}(Z)e^{T/2} \qquad (Z := X - Y,\ T := X + Y)$$

and

$$e^{\frac{1}{2}(1\otimes L + L\otimes 1)}(a \otimes b) = a^W \otimes b^W$$

where a^W and b^W are Weyl symbols of operators A and B respectively (cf. (11)) and $\hat{A}(Z)$ is the standard $\hat{A}$-series

$$\hat{A}(Z) = \frac{\sinh(Z/2)}{(Z/2)} ,$$

one can rewrite (17) also in the form:

$$\rho_u(A,B) = \sqrt{-1}\ \beta_n(R_1 \circ \underline{\hat{A}})(a^W \otimes b^W)$$

($\underline{\hat{A}} \equiv \hat{A}(1 \otimes L - L \otimes 1)$). In particular, Lemma 2.25 asserts nothing more than commutativity of the following diagram:

$$\begin{array}{ccc} P^{\cdot}\hat{\otimes}P^{\cdot} & \xrightarrow{[\,,\,]_{Weyl}} & \hat{P}^{\cdot} \\ \sqrt{-1}\ R_1 \circ \underline{\hat{A}} \downarrow & & \downarrow R_0 \\ A_1(U) & \xrightarrow{\ d\ } & A_0(U) \end{array} \tag{18}$$

Here $[\,,\,]_{Weyl}$ denotes commutator defined by the composition law of Weyl symbols

$$a \circ_{\text{Weyl}} b \equiv \sum_{\alpha,\beta} \frac{(-1)^{|\beta|}}{\alpha!\beta!2^{|\alpha+\beta|}} (\partial_\xi^\alpha D_u^\beta a)(\partial_\xi^\beta D_u^\alpha b)$$

(cf. [15], IV (23.8)). Recall that we have a similar commutative diagram but with $[\,,]_{\text{Weyl}}$ replaced by Poisson bracket:

$$\begin{array}{ccc} P^{\cdot}\hat{\otimes}P^{\cdot} & \xrightarrow{\{,\}} & \hat{P}^{\cdot} \\ R_1 \big\downarrow & & \big\downarrow R_0 \\ A_1(U) & \longrightarrow & A_0(U) \end{array} \qquad (19)$$

If one interprets vertical arrows in (18) and (19) as "non-commutative integrations" along fibres of the projection $S^*U \times S^1 \to U$ (integration over the circle corresponds to taking Cauchy residue, cf. diagram (11) of previous section) which are defined either in terms of the full commutator of symbols or of only its principal part (i.e. in terms Poisson bracket) then one reaches the conclusion that the difference between the two is roughly speaking, the "$\hat{A}$-genus".

In light of still another evidence (referring to Hochschild homology of differential operators and pseudodifferential symbols, cf.[20]) it is inspiring to regard the observed phenomenon as a manifestation of the "noncommutative" Riemann-Roch.

2.30. Functorial properties of the form ρ_u. Assume there are given two matric operators A and B and two matric functions s and r such that the compositions $s\circ A$, $A\circ r$, $r\circ B$ and $B\circ s$ are all well defined. In other words:

$$s \in \text{Mat}_{\ell_2,\ell_1}(C^\infty(U)), \quad r \in \text{Mat}_{k_1,k_2}(C^\infty(U)), \qquad (k_1,k_2,\ell_1,\ell_2 \in \mathbb{Z}_+)$$

and

$$A : C^\infty_{\text{comp}}(U,\theta^{k_1}) \to C^\infty(U,\theta^{\ell_1}), \qquad B : C^\infty_{\text{comp}}(U,\theta^{\ell_2}) \to C^\infty(U,\theta^{k_2})$$

(as before we assume either A or B to be proper).

2.31. Proposition. $\rho_u(s \circ A \circ r, B) = \rho_u(A, r \circ B \circ s)$.

Proof. It is clearly enough to prove the statement in the scalar case (i.e. when $k_1 = k_2 = \ell_1 = \ell_2 = 1$). Because for scalar operators $\rho_u(A,B)$ is skew-symmetric in A and B and its value at every point depends only on finite jets of their symbols at that point all we need is, actually, to prove 2.31 for $r \equiv 1$ and $s = s(u)$ - a polynomial function of u's. This last statement is an immediate consequence of the following lemma.

2.32. Lemma. $\rho_u(M_j \circ A, B) = \rho_u(A, B \circ M_j)$ where M_j is the operator of multiplication by a j-th coordinate u^j .

Proof. Using the identities:

$$L^h(u^j \circ a) = u^j L^h a + \sqrt{-1}\, h \partial_{\xi_j} L^{h-1} a$$

and

$$L^i(b \circ u^j) = u^j L^i b + \sqrt{-1}\, \partial_{\xi_j} L^{i-1}(i - L)b \, , \tag{20}$$

the following equalities are easily verified:

$$\sum_{h,i=0}^{\infty} \frac{1}{(1+h+i)!} L^h(u^j \circ a) \wedge L^i b = \sum_{h,i=0}^{\infty} \frac{1}{(1+h+i)!} \left[u^j + \frac{\sqrt{-1}(1+h)}{2+h+i} \partial_{\xi_j}\right] L^h a \wedge L^i b \tag{21}$$

$$\sum_{h,i=0}^{\infty} \frac{1}{(1+h+i)!} L^h a \wedge L^i(b \circ u^j) = \sum_{h,i=0}^{\infty} \frac{1}{(1+h+i)!} L^h a \wedge \left[u^j - \frac{\sqrt{-1}(1+h)}{2+h+i} \partial_{\xi_j}\right] L^i b . \tag{22}$$

Apply R_1 to both (21) and (22) and then evoke 2.11 to obtain

$$\rho_u(M_j \circ A, B) - \rho_u(A, B \circ M_j) = \sqrt{-1}\, \beta_n R_1 \left(\sum_{h,i=0}^{\infty} \frac{1}{(1+h+i)!} [u^j L^h a \wedge L^i b - L^h a \wedge u^j L^i b] \right) . \tag{23}$$

In view of $C^\infty(U)$-linearity of the $\underline{R}$-morphism (cf. 2.9) the right-hand side of (23) vanishes. □

2.33. As a corollary of Proposition 2.31 we obtain invariance of the form $\rho_u(A,B)$ with respect to gauge transformations (the gauge group is $GL_k(\mathbb{C}) \times GL_\ell(\mathbb{C})$; for the meaning of k and ℓ see 2.27).

2.34. To avoid complicated expressions we adopt the following notation. For a formal series $G \in C[[X,Y]]$ let $\underline{G} = G(1 \otimes L, L \otimes 1)$ denote the corresponding linear endomorphism of $P^\cdot \hat{\otimes} P^\cdot$. By $G_i (i = 1,2)$ we denote the partial derivatives $\partial G/\partial X$ and $\partial G/\partial Y$ respectively.

Finally, for $f \in \hat{P}^\cdot$, $\vec{f}$ will denote the operation on $\hat{P}^\cdot$ of composition with f on the left (in the sense of symbols), i.e. $\vec{f}(a) \equiv f \circ a$, $\overleftarrow{f}$ - the corresponding composition on the right, and the plain f will denote the ordinary commutative multiplication by f.

In this notation equalities (21) and (22) read as the following identities in $\mathrm{End}(P^\cdot \hat{\otimes} P^\cdot)$:

$$\underline{F} \cdot (\vec{u}^j \otimes 1) = (u^j \otimes 1)\underline{F} + (\sqrt{-1}\, \partial_{\xi_j} \otimes 1)\underline{F}_2$$

and

$$\underline{F} \cdot (1 \otimes \vec{u}^j) = (1 \otimes u^j)\underline{F} - (1 \otimes \sqrt{-1}\, \partial_{\xi_j})\underline{F}_2 \qquad (j = 1,\dots,n)$$

(F will always denote the series (16)).

Let $G = G(X,Y)$ be an arbitrary series and τ denote the transposition of factors in $P^\cdot \hat{\otimes} P^\cdot$. The following lemmas prove to be useful.

2.35. Lemma. One has

$$\tau \underline{G}_2 \tau = \underline{G}_1^\circ \quad \text{and} \quad \tau \underline{G}_1 \tau = \underline{G}_2^\circ$$

where $G^\circ(X,Y) \equiv G(Y,X)$.

2.36. Lemma. $\underline{G}(\overleftarrow{u}^j \otimes 1) = (u^j \otimes 1)G + (\sqrt{-1}\,\partial_{\xi_j} \otimes 1)(\underline{G}_2 - \underline{G})$, $(j=1,\dots,n)$. (Analogously for $\underline{G}(1 \otimes \overleftarrow{u}^j)$).

The former assertion reflects the fact that τ corresponds to the involution of $C[[X,Y]]$ which transposes X and Y. The identity of 2.36 is an immediate consequence of (20). □

2.37. Since the form $\rho_u(A,B)$, as was mentioned, depends at every point only on jets at that point of symbols of A and B it is clear that in order to examine the effect on ρ_u of an (orientation preserving) open embedding $V \to U$ it suffices to consider solely self-embeddings of the form $\exp(\eta)$ where η is a vector field(*).

In that case the functoriality of ρ_u would mean the identity

$$I_u(\eta;A,B) \equiv L_\eta \rho_u(A,B) - \rho_u([L_\eta,A],B) - \rho_u(A,[L_\eta,B]) = 0 \qquad (24)$$

to hold for an arbitrary pair of ψDOs A and B and any vector field η.

Actually, it turns out that $I_u(\eta;A,B)$ need not vanish, but, as we shall demonstrate below, it can be written as an explicit exact form:

$$I_u(\eta:A,B) = d\sigma_u(\eta;A,B) \qquad (25)$$

with $\sigma_u \in A_2(U)$ depending at each point on finite jets at that point of η and of symbols of A and B. It is not clear, however, whether it is possible to modify the form ρ_u in order that (24) be satisfied.

It should be added, perhaps, that namely the partial failing of functoriality prevented us from denoting $\rho_u(A,B)$ more naturally as $\mathrm{res}_{1,u}(A,B)$ or $\mathrm{res}_1(u;A,B)$, and calling it the *higher noncommutative residue* form.

(*) The instance of an orientation reversing embedding also reduces to the one because ρ_u is clearly functorial with respect to e.g. the involution $u^1 \to -u^1$, $u^i \to u^i$ $(i = 2,\dots,n)$.

2.38. Proposition. Equality (25) holds with

$$\sigma_u(\eta;A,B) = \sqrt{-1}\,\beta_n \sum_{j=1}^{n} i_{\partial/\partial u^j}\, R_1\phi^{\eta^j}\,(a \otimes b - b \otimes a)$$

where

$$\phi^f \equiv (f \otimes 1)\underline{F}_2 - \underline{F}_2(\overset{\leftarrow}{f} \otimes 1)\ ,$$

(F is the series (16) and $\eta \equiv \sum_{j=1}^{n} \eta^j\, \partial/\partial u^j$) .

The proof relies on the lemma which is of an independent interest.

2.39. Lemma. $R_1\underline{F}(a\circ\xi_j \otimes b - a\otimes\xi_j\circ b + b\circ a\otimes\xi_j) = \sqrt{-1}\, d i_{\partial/\partial u^j}\, R_1\underline{F}_2(a\otimes b)$

$(j = 1,\ldots,n)$. □

2.39^{bis}. Remark. The expression in brackets on the left is the boundary of the Hochschild 2-chain $a \otimes \xi_j \otimes b$ in the algebra of symbols. Since, however, $R_1\underline{F}$ is skew-symmetric it might be more proper to speak about the boundary of the *cyclic* 2-chain $a \otimes \xi_j \otimes b$:

$$R_1\underline{F}\partial^{cycl}(a \otimes \xi_j \otimes b) = \sqrt{-1}\, d i_{\partial/\partial u^j}\, R_1\underline{F}_2(a \otimes b)\ .$$

Proof of 2.39. Much as we did in the proof of 2.32 we establish the equalities

$$\underline{F}(\overset{\leftarrow}{\xi}_j \otimes 1) = (\xi_j \otimes 1)\underline{F} + (\sqrt{-1}\,\partial_{u^j} \otimes 1)\underline{F}_2 \tag{26}$$

and

$$\underline{F}(1 \otimes \overset{\rightarrow}{\xi}_j) = (1 \otimes \xi_j)\underline{F} + (1 \otimes \sqrt{-1}\,\partial_{u^j})(\underline{F}_1 - \underline{F})\ .$$

Since one has easily verifiable identity $F \equiv F_1 + F_2$ the latter equality can be written also as

$$\underline{F}(1 \otimes \overset{\rightarrow}{\xi}_j) = (1 \otimes \xi_j)\underline{F} - (1 \otimes \sqrt{-1}\,\partial_{u^j})\underline{F}_2\ . \tag{27}$$

We want to calculate $R_1\underline{F}(\overset{\leftarrow}{\xi}_j \otimes 1 - 1 \otimes \overset{\rightarrow}{\xi}_j)$. In order to do this let us notice first that:

$$R_1(\partial_{u^j} f\wedge g + f\wedge \partial_{u^j} g) = -R_1\underline{\partial}(f\wedge \xi_j \wedge g) - R_1(\{f,g\}\wedge \xi_j)$$

$$= dR_2(f\wedge\xi_j\wedge g) - i_{\partial/\partial u^j} R_0\{f,g\}$$

$$= d i_{\partial/\partial u^j} R_1(f\wedge g) - i_{\partial/\partial u^j} R_0\{f,g\},$$

and then recall (cf. diagram (14)) that $R_0\{.,.\} = \sqrt{-1}\, R_0 m(\underline{X} - \underline{Y})$.

Hence we obtain from (26) and (27) that

$$R_1\underline{F}(\vec{\xi}_j\otimes 1 - 1\otimes\vec{\xi}_j) = R_1\,(\xi_j\otimes 1 - 1\otimes\xi_j)\underline{F} + i_{\partial/\partial u^j} R_0 m(\underline{X} - \underline{Y})\underline{F}_2$$

$$+ \sqrt{-1}\, d i_{\partial/\partial u^j} R_1\underline{F}_2\ . \qquad (28)$$

We have another general equality (cf. 2.4)

$$R_1\cdot(\xi_j\otimes 1 - 1\otimes\xi_j)(f\otimes g) = -R_1(fg\otimes\xi_j) = -i_{\partial/\partial u^j} R_0(fg)$$

$$= -i_{\partial/\partial u^j} R_0 m(f\otimes g)\ .$$

In particular,

$$R_1\cdot(\xi_j\otimes 1 - 1\otimes\xi_j)\underline{F} + i_{\partial/\partial u^j} R_0 m(\underline{X}-\underline{Y})\underline{F}_2 = i_{\partial/\partial u^j} R_0 m(-\underline{F} + (\underline{X}-\underline{Y})\underline{F}_2)\ . \qquad (29)$$

Since

$$-F + (X-Y)F_2 = ((X-Y)F)_2 = (e^X - e^Y)_2 = -e^Y$$

and $R_0 m(e^L f\otimes g) = R_0(g\circ f)$ (cf. (15); the circle denotes the composition of symbols) we obtain that the right-hand side of (29) evaluated on the tensor $a\otimes b$ equals:

$$-i_{\partial/\partial u^j} R_0(b\circ a) = -R_1(b\circ a\otimes\xi_j).$$

Desired equality 2.39 follows then from (28) and from the following general lemma .

2.40. Lemma. The identity

$$R_1(\underline{G} - g_{00})(f \otimes \xi_j) = 0 \qquad (j = 1,\dots,n)$$

holds for arbitrary series $G = \sum_{k,\ell=0}^{\infty} g_{k\ell} X^k Y^\ell$ and $f \in \hat{P}^{\cdot}$.

Proof. One has obviously

$$(\underline{G} - g_{00})(f \otimes \xi_j) = \sum_{k=1}^{\infty} g_{k0} L^k f \otimes \xi_j .$$

Hence:

$$R_1(\underline{G} - g_{00})(f \otimes \xi_j) = i_{\partial/\partial u^j} R_0\left(\sum_{k=1}^{\infty} g_{k0} L^k f\right),$$

and the latter expression apparently vanishes (cf. 2.11). □

2.41. Proposition.

$$\rho_u(A \circ L_\eta, B) - \rho_u(A, L_\eta \circ B) + \rho_u(BA, L_\eta)$$

$$= \frac{1}{\sqrt{-1}} \beta_n d \sum_{j=1}^{n} i_{\partial/\partial u^j} R_1 \underline{F}_2(a \circ \eta^j \otimes b)$$

for an arbitrary vector field $\eta = \sum_{j=1}^{n} \eta^j \, \partial/\partial u^j$.

(This is a generalisation of 2.39).

Proof. To simplify notation we shall adopt in this and other proofs the summing convention with respect to repeating indices.

Direct computation using (17), Lemma 2.39 and Proposition 2.31 yields:

$$\rho_u(A \circ L_\eta, B) \overset{(17)}{=} \sqrt{-1}\ \beta_\eta R_1 \underline{F}(\sqrt{-1}\ a \circ (\eta^j \xi_j) \otimes b)$$

$$\overset{2.39}{=} \beta_n R_1 \underline{F}(-a \circ \eta^j \otimes \xi_j \circ b + b \circ a \otimes \eta^j \xi_j)$$

$$- \sqrt{-1}\ \beta_n d i_{\partial/\partial u^j} R_1 \underline{F}_2(a \circ \eta^j \otimes b) \qquad (30)$$

and the first term on the right-hand side of (30) equals

$$\rho_u(A, L_\eta \circ B) - \rho_u(BA, L_\eta) \ . \qquad \square$$

Before we prove Proposition 2.38, we need one more useful lemma.

2.42. Lemma. $\rho_u(L_\eta, C) = i_\eta \mathrm{res}_u C$.

Proof. Use Proposition 2.31 and Lemma 2.40 to obtain:

$$\rho_u(L_\eta, C) \overset{2.31}{=} \rho_u(L_{\partial/\partial u^j}, C \circ M_{\eta^j}) = -\beta_n R_1 \underline{F}(\xi_j \otimes c \circ \eta^j)$$

$$= \beta_n R_1 \underline{F}(c \circ \eta^j \otimes \xi_j) \overset{2.40}{=} \beta_n R_1(c \circ \eta^j \otimes \xi_j) = \beta_n i_{\partial/\partial u^j} R_0(c \circ \eta^j)$$

$$\overset{2.18}{=} \beta_n i_\eta R_0(c) \equiv i_\eta \mathrm{res}_u C \ . \qquad \square$$

Now the proof of Proposition 2.38 goes as follows.

The skew-symmetry of R_1 together with 2.35 and the identity $\underline{F} \equiv F_1 + F_2$ yield the equality

$$R_1 \underline{F}(f \otimes g) = R_1 \underline{F}_2(f \otimes g - g \otimes f) \ . \qquad (31)$$

Then 2.38 is verified by the following computation:

$$I_u(\eta; A, B)$$

$$\overset{2.25,2.42}{=} d i_\eta \rho_u(A,B) + \rho_u(L_\eta, [A,B]) - \rho_u([L_\eta, A], B) - \rho_u(A, [L_\eta, B])$$

$$\overset{2.41}{=} \sqrt{-1}\, \beta_n d i_\eta R_1 \underline{F}(a \otimes b) - \sqrt{-1}\, \beta_n d i_{\partial/\partial u^j}(R_1 \underline{F}_2(a \circ \eta^j \otimes b) - R_1 \underline{F}_2(b \circ \eta^j \otimes a))$$

$$\overset{(31)}{=} \sqrt{-1}\, \beta_n d i_{\partial/\partial u^j}(\eta^j R_1 \underline{F}_2(a \otimes b - b \otimes a) - R_1 \underline{F}_2(a \circ \eta^j \otimes b - b \circ \eta^j \otimes a))$$

$$\equiv \sqrt{-1}\, \beta_n d i_{\partial/\partial u^j} R_1 \phi^{\eta^j}(a \otimes b - b \otimes a) \ . \qquad \square$$

The following proposition establishes vanishing of $\sigma_u(\eta;\cdot,\cdot)$ for vector fields with constant and linear coefficients, and shows that this fails already for quadratic fields.

2.43. Proposition. a) The form $\sigma_u(\eta;A,B)$ vanishes for vector fields $\eta = \sum_{j=1}^{n} \eta^j \partial/\partial u^j$ with η^j's being linear functions of u.

b) For $\eta = u^p u^q \, \partial/\partial u^j$ $(j,p,q = 1,\dots,n;$ p and q may coincide) the actual value of $\sigma_u(\eta;A,B)$ is

$$\sigma_u(\eta;A,B) = \sqrt{-1}\; \beta_n{}^{i}{}_{\partial/\partial u^j}\; R_1(\partial^2_{\xi_p\xi_q} \otimes 1)\underline{F}_{12}(a \otimes b)$$

where $F_{12} \equiv \partial^2 F/\partial X \partial Y$.

Proof. a) We have to show that $R_1\phi^f(1-\tau)$ vanishes for f's which are linear functions. For $f = \text{const}$ even $\phi^f \equiv 0$, so there is nothing to do. For $f = u^j$ the actual value of ϕ^{u^j} is given by Lemma 2.36 (as applied to $G = F_2$) :

$$\phi^{u^j} = (\sqrt{-1}\,\partial_{\xi_j} \otimes 1)(\underline{F}_2 - \underline{F}_{22})$$

$(F_{22} \equiv \partial^2 F/\partial Y^2)$. Hence, by making use of 2.35, we obtain:

$$R_1\phi^{u^j}(1-\tau) = R_1(\sqrt{-1}\partial_{\xi_j}\otimes 1)(\underline{F}_2-\underline{F}_{22}) - R_1(\sqrt{-1}\,\partial_{\xi_j} \otimes 1)\tau(\underline{F}_1 - \underline{F}_{11})$$

$$= R_1(\sqrt{-1}\,\partial_{\xi_j}\otimes 1)(\underline{F}_2-\underline{F}_{22}) + R_1(1\otimes\sqrt{-1}\,\partial_{\xi_j})(\underline{F}_1-\underline{F}_{11})$$

$$\overset{2.11}{=} R_1(\sqrt{-1}\,\partial_{\xi_j}\otimes 1)(\underline{F}_2-\underline{F}_{22}-\underline{F}_1+\underline{F}_{11}) .$$

It can be easily verified that:

$$F_1 - F_2 = F_{11} - F_{22} \quad \text{in} \quad \mathbb{C}[[X,Y]], \tag{32}$$

and thereby we reach required conclusion.

b) Use Lemma 2.36 twice in order to obtain the general identity that is valid for an arbitrary series G:

$\underline{G}(\overset{\leftarrow}{u}^p\overset{\leftarrow}{u}^q\otimes 1)$

$= (u^p\otimes 1)\underline{G}(\overset{\leftarrow}{u}^q\otimes 1) + (\sqrt{-1}\partial_{\xi_p}\otimes 1)(\underline{G}_2-\underline{G})(\overset{\leftarrow}{u}^q\otimes 1)$

$= (u^pu^q\otimes 1)\underline{G} + [\sqrt{-1}(u^q\partial_{\xi_p}+u^p\partial_{\xi_q})\otimes 1](\underline{G}_2-\underline{G}) + (-\partial^2_{\xi_p\xi_q}\otimes 1)(\underline{G}_{22}-\underline{G}_2-\underline{G}_2+\underline{G})$.

In particular ϕ^f for $f = u^pu^q$ equals (one puts $G \equiv F_2$)

$$\phi^f = \left[\sqrt{-1}(u^q\partial_{\xi_p}+u^p\partial_{\xi_q})\otimes 1\right](\underline{F}_2-\underline{F}_{22}) + (\partial^2_{\xi_p\xi_q}\otimes 1)(\underline{F}_{222}-2\underline{F}_{22}+\underline{F}_2) \qquad (33)$$

and, by playing with τ and basic properties of R_1 (like 2.9 and 2.11), we get immediately

$$R_1\phi^f(1-\tau) = R_1([\sqrt{-1}(u^q\partial_{\xi_p}+u^p\partial_{\xi_q})\otimes 1](\underline{F}_2-\underline{F}_{22}-\underline{F}_1+\underline{F}_{11})$$
$$+(\partial^2_{\xi_p\xi_q}\otimes 1)(\underline{F}_{222}-2\underline{F}_{22}+\underline{F}_2+\underline{F}_{111}-2\underline{F}_{11}+\underline{F}_1)) \ . \qquad (34)$$

Let $F_1^{(k)} \equiv \partial^k F/\partial X^k$ and $F_2^{(k)} \equiv \partial^k F/\partial Y^k$. Note the identities:

$$F_1^{(k)} + F_2^{(k)} \equiv F - k\,\frac{F_1^{(k-1)}-F_2^{(k-1)}}{X-Y} \qquad (35)$$

and

$$F_{12} = \frac{F_1-F_2}{X-Y} \qquad (36)$$

(both easily verifiable).

From (35), (32) and (36) one easily gets that

$$F_{222} - 2F_{22} + F_2 + F_{111} - 2F_{11} + F_1 = F_{12} \ ,$$

and (34) turns, in view of (32), into

$$R_1\phi^f(1-\tau) = R_1(\partial^2_{\xi_p\xi_q}\otimes 1)\underline{F}_{12} \ . \qquad \square$$

2.44. Example. Part b) of Proposition 2.43 provides probably simplest instances when not only σ_u itself but also $I_u \equiv d\sigma_u$ do not vanish. Take e.g. operators with symbols

$$a(u,\xi) = \phi(\xi) \quad \text{and} \quad b(u,\xi) = u^j\psi(\xi) .$$

Then it can be easily verified that

$$I_u(u^p u^q \partial/\partial u^j;A,B) = \sqrt{-1}\ \beta_n d\, i_{\partial/\partial u^j} R_1(\partial^2_{\xi_p\xi_q} \otimes 1)\underline{F}_{12}(\phi \otimes u^j\psi)$$

$$= \frac{\sqrt{-1}\beta_n}{6}\, d(u^j i_{\partial/\partial u^j}\ R_1(\partial^2_{\xi_p\xi_q}\otimes 1)(\phi\otimes\psi)) ,$$

and the latter expression is a general *constant* form on U divisible by du^j.

2.45. Remark. Simply by iterating Lemma 2.36 one can generalise (33) to polynomial (and hence all smooth) f's in the following manner:

$$\phi^f = (\sqrt{-1}\ \frac{\partial f}{\partial u^p}\ \partial_{\xi_p} \otimes 1)(\underline{F}_2 - \underline{F}_{22}) + \left(\frac{1}{2}\ \frac{\partial^2 f}{\partial u^p \partial u^q}\ \partial^2_{\xi_p\xi_q} \otimes 1\right)(\underline{F}_{222} - 2\underline{F}_{22} + \underline{F}_2)$$

$$+ \text{ higher order terms}$$

(higher order - means terms which include higher than second derivative with respect to ξ's).

Therefore one can deduce exactly as in the proof of 2.43b) that

$$\sigma_u(f\partial/\partial u^j;A,B) = \sqrt{-1}\ \beta_n i_{\partial/\partial u^j}\ R_1\left(\frac{1}{2}\ \frac{\partial^2 f}{\partial u^p \partial u^q}\ \partial^2_{\xi_p\xi_q} \otimes 1\right)\underline{F}_{12}(a \otimes b)$$

$$+ \text{ 'higher order terms'}.$$

In particular, if one of the two operatores is a *differential* operator of order ≤ 2 the form $\sigma_u(\eta;A,B)$ vanishes for every vector field η.

2.46. Corollary. If one of the two ψDOs is a differential operator of order ≤ 2 the form $\rho_u(A,B)$ is functorial with respect to arbitrary morphisms $(V;\theta^{k'},\theta^{\ell'}) \to (U;\theta^{k},\theta^{\ell})$. □

2.47. (Comment on 2.46). Using Proposition 2.41 and Lemma 2.42 it is not difficult to obtain the *actual* value of $\rho_u(A,B)$. For example, for $A = L_{\eta_1} L_{\eta_2}$ Proposition 2.41 yields:

$$\rho_u(L_{\eta_1} L_{\eta_2},B) = \rho_u(L_{\eta_1}, L_{\eta_2}\circ B) + \rho_u(L_{\eta_2}, B\circ L_{\eta_1})$$

$$+ \frac{\beta_n}{\sqrt{-1}}\, di_{\partial/\partial u^j} R_1 \underline{F}_2(\sqrt{-1}\,\eta_1^i \xi_i \circ \eta_2^j \otimes b). \qquad (38)$$

Notice that

$$F_2 = \tfrac{1}{2}F \bmod (X\mathbb{C}[[X,Y]] + Y\mathbb{C}[[X,Y]]) .$$

Hence, by using 2.11 and 2.10 (twice) we can rewrite third term on he right-hand side of (38) as:

$$\frac{\beta_n}{2\sqrt{-1}}\, di_{\eta_2} R_1\underline{F}(\sqrt{-1}\eta_1^i\xi_i \otimes b) \equiv -\tfrac{1}{2} di_{\eta_2}\rho_u(L_{\eta_1},B) \overset{2.42}{=} \tfrac{1}{2}\, di_{\eta_1} i_{\eta_2} \mathrm{res}_u B$$

nd, finally, obtain from (38) the required expression:

$$\rho_u(L_{\eta_1} L_{\eta_2},B) = i_{\eta_1}\mathrm{res}_u(L_{\eta_2}\circ B) + i_{\eta_2}\mathrm{res}_u(B\circ L_{\eta_1}) + \tfrac{1}{2}\, di_{\eta_1} i_{\eta_2}\mathrm{res}_u B \qquad (39)$$

hose right-hand side is transparently functorial. We thereby eprove Corollary 2.46.

48. Real structures and adjoint operators. For completeness sake e shall determine the residue form of a complex conjugated A^c and djoint A^* operators.

The symbol of A^c is equal to $\overline{a(u,-\xi)}$. In particular,

$$\mathrm{res}_u A^c = \overline{\mathrm{res}_u A} .$$

In order to define A^* one has to specify hermitian metrics g_1 on θ^k and g_2 on θ^ℓ (viewed invariantly as $\mathbb{C}$-linear morphisms $\theta^k \to (\bar\theta^k)^*$ and $\theta^\ell \to (\bar\theta^\ell)^*$) and a positive density ν. Then A^* is defined by the requirement of fulfilling the equality

$$\int_U g_1(s,A^*t)\nu = \int_U g_2(As,t)\nu \tag{40}$$

for every $s \in C^\infty_{comp}(U,\theta^k)$ and $t \in C^\infty_{comp}(U;\theta^\ell)$.

It is clear from (40) that $A^* = (g_1 m)^{-1}\circ A^+ \circ (g_2 m)$ where $m = m(u) \equiv \nu/|du|$ and A^+ denotes the complex conjugated transpose of A. Thus, we obtain by using Lemma 2.18 that

$$res_u A^* = (g_1 m)^{-1}\cdot res_u A^+ \cdot (g_2 m) = g_1^{-1}\cdot res_u A^+ \cdot g_2 .$$

On the other hand, by using standard formula for the symbol of A^+ (cf. [9], Thm. 4.2 or [15], formula I.3.37) we obtain that

$$res_u A^+ = \beta_n \tau_* |Res_0 \sum_\alpha \frac{1}{\alpha!}\partial_\xi^\alpha D_u^\alpha (a^+)| \overset{2.2}{=} \beta_n \tau_* |Res_0 a^+| = (res_u A)^+ .$$

Thus we proved

<u>2.49.</u> <u>Lemma</u>. a) $res_u A^c = \overline{res_u A}$.

b) $res_u A^* = g_1^{-1}(res_u A)^+ g_2$. □

§3. <u>Noncommutative residue (global case)</u>

Throughout this section X denotes a fixed smooth open (i.e. without boundary but not necessarily compact) manifold, and E and F - two vector bundles on it of ranks equal to k and ℓ respectively.

<u>3.1.</u> <u>Lemma-Definition</u>. For an arbitrary ψDO $A : C^\infty_{comp}(X,E) \to C^\infty(X,F)$ there exists a unique density $res_X A \in A_0(X;\underline{Hom}(E,F))$ such that

$$\phi^* \mathrm{res}_X A = \mathrm{res}_X \phi^{\#} A$$

whenever $\phi : (U;\theta^k,\theta^\ell) \to (X;E,F)$ is a local coordinate patch (cf. 0.6).

This density will be called the (homomorphism valued) *residue density* (or *residue form*) of A.

(Lemma 3.1. is an immediate corollary of Lemma 2.4).

The basic properties of residue density are summed up in

3.2. Proposition

(I) (Linearity) For $A \in CL^\ell(X;E,F)$, $B \in CL^m(X;E,F)$ (with $\ell - m \in \mathbb{Z}$) and $\alpha,\beta \in \mathbb{C}$ one has

$$\mathrm{res}_X(\alpha A + \beta B) = \alpha\, \mathrm{res}_X A + \beta\, \mathrm{res}_X B$$

(II) (Functoriality) For any morphism $\psi : (Y;E',F') \to (X;E,F)$ (cf. 0.3) one has

$$\mathrm{res}_Y \psi^{\#} A = \psi^* \mathrm{res}_X A .$$

(III) (Locality) If at a neighbourhood of a point $x_0 \in X$ the complete symbol of A vanishes then $\mathrm{res}_{x_0} A = 0$.

(IV) If $\mathrm{ord}\, A < -\dim X$ or $\mathrm{ord}\, A \notin \mathbb{Z}$ then $\mathrm{res}_X A \equiv 0$.

(V) (Real structures) If E and F both admit real structures $\rho_E : E \to E$ and $\rho_F : F \to F$ respectively, and $A^c := \rho_F A \rho_E$ is the corresponding complex conjugated operator then

$$\mathrm{res}_X A^c = (\mathrm{res}_X A)^c \equiv \rho_F (\mathrm{res}_X A) \rho_E .$$

(VI) (Adjoints) $\mathrm{res}_X A^* = g_E^{-1} (\mathrm{res}_X A)^+ g_F$

where A^* is an operator adjoint to A with respect to hermitian metrics $g_E : E \to \bar{E}^*$ and

$g_F : F \to \bar{F}^*$ and a certain positive density on X (the sign $+$ denotes the complex conjugated transpose).

In particular, $\mathrm{res}_x A^*$ does not depend on the choice of volume density in question.

(VII) If $E = F$ and ∇_η denotes covariant differentiation of sections of E along a vector field η then

$$\mathrm{res}_x[\nabla_\eta, A] = L_\eta^{\tilde{\nabla}}\ \mathrm{res}_x A$$

where $L_\eta^{\tilde{\nabla}}$ denotes the "Lie derivative" on $A_0(X;\underline{\mathrm{End}}\ E)$ defined by the connection $\tilde{\nabla}$ on $\underline{\mathrm{End}}\ E$ induced by ∇.

<u>Comments on proof</u>: Assertions (I), (III) and (IV) follow directly from the definition; (II), (V) and (VI) follow from the corresponding local assertions: 2.20, 2.49 and from Lemma 3.1. Finally, (VII) is verified by a brief local computation using Lemma 2.2 (details are left as an easy exercise to the reader). □

<u>3.3. Corollary</u>. Assume $E = F$.

(a) If E admits a real structure and A^c is the corresponding complex conjugated operator then

$$\mathrm{tr}\ \mathrm{res}_x A^c = \overline{\mathrm{tr}\ \mathrm{res}_x A}$$

(the bar denotes the usual complex conjugation).

(b) $\mathrm{tr}\ \mathrm{res}_x A^* = \overline{\mathrm{tr}\ \mathrm{res}_x A}$.

(c) For an endomorphism $r : E \to E$ one has

$$\mathrm{tr}\ \mathrm{res}_x[A, r] \equiv 0 .$$

(d) $\mathrm{tr}\ \mathrm{res}_x[\nabla_\eta, A] = L_\eta\ \mathrm{tr}\ \mathrm{res}_x A$

(in particular, it does not depend on the connection). □

Parts (c) and (d) admit the following simple generalisation. Let $A : C^{\infty}_{comp}(X,E) \to C^{\infty}(X,F)$ be an arbitrary ψDO and $D : C^{\infty}(X,F) \to C^{\infty}(X,E)$ be a first order *differential* operator. For a density $\sigma \in A_0(X;\underline{Hom}(E,F))$ there is an invariantly defined *scalar* density $\langle L_D,\sigma\rangle \in A_0$ which is the image under the composition of sheaf-morphisms:

$$(T_X \otimes \underline{Hom}(F,E)) \otimes (A_{X,0} \otimes \underline{Hom}(E,F)) \xrightarrow{\text{contraction}} T_X \otimes A_{X,0} \xrightarrow{L} A_{X,0} \tag{1}$$

(L denotes Lie derivative) of $p_D \otimes \sigma$ where p_D is the principal symbol of D regarded as a vector field with coefficients in $\underline{Hom}(F,E)$.

A direct computation using Lemma 2.2 yields

3.4. <u>Proposition.</u> $\operatorname{tr} \operatorname{res}_x DA - \operatorname{tr} \operatorname{res}_x AD = \langle L_D, \operatorname{res}_x A\rangle$. □

(Notice that as in the scalar case $\langle L_D,\sigma\rangle$ is an explicit exact density:

$$\langle L_D,\sigma\rangle = d\langle i_D,\sigma\rangle \tag{2}$$

where $\langle i_D,\sigma\rangle$ is defined similarly to $\langle L_D,\sigma\rangle$ but with Lie derivative replaced in (1) by interior product.)

3.5. Let $A : C^{\infty}_{comp}(X,E) \to C^{\infty}(X,F)$ and $B : C^{\infty}_{comp}(X,F) \to C^{\infty}(X,E)$ be a pair of ψDOs similar to that considered in 2.24 (i.e. A or B, or both, are assumed to be proper ψDOs, and both are assumed to have integer orders (but also cf. 3.8 below)). We want to express $\operatorname{res}_x [A,B]$ as an explicit exact form.

Were A supported by the range of some local coordinate patch $\phi = (i : U \to X;r,s)$ (i.e. $A|_{X\setminus i(U)} \equiv 0$) we would have:

$$\operatorname{tr} \operatorname{res}_x[A,B] = i_! \rho_u(\phi^{\#} A,\phi^{\#} B) .$$

In general, one has to proceed as follows. It is always possible to represent A as a locally finite sum $\sum_{\chi} A_{\chi}$ (i.e. for any $x \in X$ the restriction of $\sum_{\chi} A_{\chi}$ to some neighbourhood of x has only finitely many non-zero entries) such that each A_{χ} is supported by the range of some local coordinates ϕ^{χ}. Then the bi-linearity and locality of the form ρ_u imply the equality:

$$\operatorname{tr} \operatorname{res}_x [A,B] = d \sum_{\chi} i_!^{\chi} \rho_{u_{(\chi)}} ((\phi^{\chi})^{\#} A_{\chi}, (\phi^{\chi})^{\#} B) . \tag{3}$$

Let us consider the expression under the sign of differential in (3) as an element of $A_1(X)/dA_2(X)$ and denote this element as

$$\bar{\rho}(A,B;\{A_{\chi}\},\{\phi^{\chi}\}) .$$

<u>3.6. Lemma</u>. $\bar{\rho}(A,B;\{A_{\chi}\},\{\phi^{\chi}\})$ depends neither on $\{\phi^{\chi}\}$, nor on the particular choice of the representation $A = \sum_{\chi} A_{\chi}$.

<u>Proof</u>. Independence of $\{\phi^{\chi}\}$ is clear from Proposition 2.31. In order to prove the second assertion we take such a partition of unity $\{f^{\lambda}\}$ that the sum $\sum_{\chi,\lambda,\mu} f^{\lambda} \circ A_{\chi} \circ f^{\mu}$ is locally finite. Then, clearly, we have

$$\sum_{\chi} i_!^{\chi} \rho_{u_{(\chi)}} ((\phi^{\chi})^{\#} A_{\chi}, (\phi^{\chi})^{\#} B) = \sum_{\lambda,\mu} \sum_{\chi} i_!^{\chi} \rho_{u_{(\chi)}} ((\phi^{\chi})^{\#} (f^{\lambda} \circ A_{\chi} \circ f^{\mu}), (\phi^{\chi})^{\#} B) . \tag{4}$$

All we need is to prove that for $A = 0$ the left-hand side of (4) belongs to dA_2 irrespective of the representation $\sum_{\chi} A_{\chi} = 0$. Actually, Proposition 2.38 implies that every summand under the sign of the first sum on the right-hand side of (4) is the differential of a form with support in $\operatorname{supp} f^{\lambda} \cap \operatorname{supp} f^{\mu}$. Hence, the right-hand side, and therefore the left-hand side of (4) belong to dA_2, as required. □

As the class $\bar{\rho}$ does not depend on extra choices we shall use notation $\bar{\rho}(A,B)$.

3.7. Proposition. $\operatorname{tr}\operatorname{res}_x[A,B] = d\bar{\rho}(A,B)$ where $\bar{\rho}(A,B)$ is the canonical element of $A_1(X)/dA_2(X)$ defined above. □

3.8. Remark. The assertion of Proposition 3.7 extends easily to the more general case of operators with arbitrary complex orders (cf. Remark 2.26).

An important property of the class $\bar{\rho}$ is stated in the following

3.9. Proposition. $\bar{\rho}(A_0A_1,A_2) - \bar{\rho}(A_0,A_1A_2) + \bar{\rho}(A_2A_0,A_1) \equiv 0$.

3.10. Remarks. 1) Since $\bar{\rho}$ is obviously invariant with respect to the 'stabilisation':

$$A \mapsto \begin{pmatrix} 0 & 0 \\ A & 0 \end{pmatrix}_{E\ \ F}, \qquad B \mapsto \begin{pmatrix} 0 & B \\ 0 & 0 \end{pmatrix}_{E\ \ F},$$

.9. is equivalent to:

$$\bar{\rho}\partial^{cycl}(A_0 \otimes A_1 \otimes A_2) = 0 ,$$

.e. to saying that $\bar{\rho}$ is a cyclic 1-cocycle with values in A_1/dA_2 n the algebra of symbols $CS^{\cdot}(X,E)$.

Thus Propositions 3.7 and 3.9 can be restated as the commutativity f the diagram:

$$\begin{array}{ccccc} CC_0(CS^{\cdot}(X,E)) & \xleftarrow{\partial} & CC_1(CS^{\cdot}(X,E)) & \xleftarrow{\partial} & CC_2(CS^{\cdot}(X,E)) \\ \downarrow{\scriptstyle \operatorname{tr}\operatorname{res}_x} & & \downarrow{\scriptstyle \bar{\rho}} & & \downarrow \\ A_0(x) & \xleftarrow{d} & A_1(X)/dA_2 & \longleftarrow & 0 \end{array} \qquad (5)$$

Diagram (5) is obviously an initial piece of the hypothetical noncommutative residue morphism. It is not clear yet how to construct the 'higher residues'. Quite probably, they arise in the setting f equivariant cyclic (or Chevalley) chain complexes. One can prove t least the existence of canonical homomorphisms in homology:

$HC_{\bullet}(CS^{\cdot}(X,E)) \to H(A_{\bullet}(X))$ and $H^{Lie}_{\bullet}(CS^{\cdot}(X,E))[-1] \to H(A_{\bullet}(X))$ (recall that $H_q(A_{\bullet}(X)) \simeq H_{q,c\ell}(X)$, i.e. the homology of X with closed supports), cf. [20].

2) There are several particular cases with a canonical choice of the representative for the class $\bar{\rho}$. This holds e.g. for *all* pairs (A,B) on a manifold with *linear* transition functions (arbitrary domains in $\mathbb{R}^n$, tori T^n etc.), as implied by Propositions 2.31 and 2.43a).

On the other hand it follows directly from formula (37) and its corollary (see 2.46) that the form $\rho_X(A,B;\{A_\chi\},\{\phi^\chi\}) \equiv \sum_\chi i_!^\chi \rho_{u_{(\chi)}}((\phi^\chi)^{\#}A_\chi,(\phi^\chi)^{\#}B)$ depends neither on particular representation $A = \sum_\chi A_\chi$ nor on local coordinates $\{\phi^\chi\}$ if one of the two operators is a differential one and of order ≤ 2. And this holds irrespective of geometry of an underlying manifold.

Proof of 3.9. The assertion reduces easily to the following local question. We want to prove for symbols of scalar ψDOs in a contractible domain U that

$$R_1\underline{F}(a^0\circ a^1\otimes a^2 - a^0\otimes a^1\circ a^2 + a^2\circ a^0\otimes a^1) = d\phi(a^0,a^1,a^2)$$

provided supp a^0 is compact. The form $\phi(a^0,a^1,a^2)$ is required to have support contained in supp a^0.

Let χ denote the form $R_1\underline{F}\partial^{cycl}(a^0\otimes a^1\otimes a^2)$. Clearly, supp $\chi \subset$ supp a^0, and $d\chi = 0$. The latter follows, in view of Lemma 2.25, from

$$d\chi = \frac{1}{\sqrt{-1}} R_0(\partial^{cycl})^2(a^0\otimes a^1\otimes a^2) = 0 .$$

Since $\chi \in A_{1,comp}(U)$, and U is contractible, there must be a form $\phi \in A_{2,comp}(U)$ with supp $\phi \subset$ supp χ such that $\chi = d\phi$. □

3.11. Identity 3.9 provides for a *differential* operator A a flexible way to produce a specific representative of the class

$\bar\rho(A,B)$. Indeed, represent A as a locally finite sum $\sum_\alpha A_1^\alpha \ldots A_{p(\alpha)}^\alpha$ where each A_j^α is of order ≤ 1. Then the recurrent use of Proposition 3.9 yields the equality:

$$\bar\rho(A,B) = \sum_\alpha \sum_{j=1}^{p(\alpha)} \bar\rho(A_j^\alpha, A_{j+1}^\alpha \ldots A_{p(\alpha)}^\alpha BA_1^\alpha \ldots A_{j-1}^\alpha) ,$$

and, as each class on the right possesses the canonical representative

$$\langle i_{A_j^\alpha}, \operatorname{res}_x (A_{j+1}^\alpha \ldots A_{p(\alpha)}^\alpha BA_1^\alpha \ldots A_{j-1}^\alpha)\rangle$$

(cf. Proposition 3.4 and (2)) we obtain the following

3.12. Proposition. For a differential operator

$$A = \sum_\alpha A_1^\alpha \ldots A_{p(\alpha)}^\alpha \tag{6}$$

the class mod $dA_2(X)$ of the form

$$\sum_\alpha \sum_{j=1}^{p(\alpha)} \langle i_{A_j^\alpha}, \operatorname{res}_x (A_{j+1}^\alpha \ldots A_{p(\alpha)}^\alpha BA_1^\alpha \ldots A_{j-1}^\alpha)\rangle \tag{7}$$

is equal to $\bar\rho(A,B)$.

In particular, it does not depend on the specific representation (6). □

3.13. Remark. As we already know (cf. Remark 3.10.2)) for an operator A of order ≤ 2 there is another canonical way to write down a form representing $\bar\rho(A,B)$. As demonstrates equality (39) of Section 2 the two differ!

The same equality also provides the simplest instance of dependence of the *actual* form (7) on the specific choice of the presentation (6). Assume e.g. both operators to be scalar ones, and let for A (that is our operator of the second order)

$$A = \sum_\alpha L_{\eta_1^\alpha} L_{\eta_2^\alpha} + \sum_\beta L_{\zeta^\beta} = \sum_\gamma L_{\eta_1^\gamma} L_{\eta_2^\gamma} + \sum_\delta L_{\zeta^\delta}$$

be its two different presentations (η's and ζ's denote vector fields). Then the difference between the two forms (7) is

$\frac{1}{2}d\{(\sum_{\alpha} i_{\eta_1^\alpha} i_{\eta_2^\alpha} - \sum_{\gamma} i_{\eta_1^\gamma} i_{\eta_2^\gamma})\mathrm{res}_x B\}$.

In the particular case of $A = L_{\eta_1} L_{\eta_2} = L_{\eta_2} L_{\eta_1} + L_{[\eta_1,\eta_2]}$ this reduces to $d i_{\eta_1} i_{\eta_2} \mathrm{res}_x B$.

3.14. Subresidue. Apart from its noncommutative residue density every ψDO $A : C^\infty_{comp}(X,E) \to C^\infty(X,F)$ determines yet another invariant local quantity which is a density with coefficients in $\Omega^1_X \otimes \underline{\mathrm{Hom}}(E,F)$. The relevant construction depends solely on A if $F = \theta^\ell$ is a trivial bundle. For general F one has, however, to specify also a connection ∇ on F.

Consider first the local case

$$X = U \subset \mathbb{R}^n, \quad E = \underline{\theta}^k, \quad F = \underline{\theta}^\ell \quad \text{and} \quad \nabla = \sum_{j=1}^n \Gamma_j du^j$$

where $\Gamma_j \in \mathrm{Mat}_\ell(C^\infty(U))$.

3.15. Definition.

$$\mathrm{sub\,res}^\nabla_u(A) = \sum_{j=1}^n du^j \otimes \left\{ \int_{|\xi|=1} (\hat\nabla_j a_{-n} + \sqrt{-1}\,\xi_j a_{-n-1}) |d\xi'| \right\} |du|$$

(we retain the notation of Definition 2.13; $\hat\nabla_j = \partial/\partial u^j + \Gamma_j \otimes \mathrm{id}_{(\theta^k)^*}$).

By definition $\mathrm{sub\,res}^\nabla_u(A)$ is a density on U with coefficients in the sheaf of matric-valued differential 1-forms. An alternative definition is:

$$\mathrm{sub\,res}^\nabla_u(A) = \beta_n \tau_* |\mathrm{Res}_0(d^{(\nabla)} + \alpha)a|$$

($d^{(\nabla)} \equiv \mathrm{id}_H \otimes d + \hat\nabla \wedge \mathrm{id}_\Omega$, where $H \equiv \theta^\ell \otimes (\theta^k)^*$, $\alpha \equiv i_\Xi \omega = \sum_{j=1}^n \xi_j du^j$ is the canonical 1-form on T^*U (cf. 1.3) and τ, as usual, denotes the projection $S^*U \to U$).

The quantity $\mathrm{sub\,res}^\nabla_u(A)$ will be called the *subresidue* of A (with respect to the connection ∇).

3.16. Lemma. Subresidue is functorial with respect to such morphisms $\phi = (f;r,s) : (V;\theta^{k'},\theta^{\ell}) \to (U;\theta^{k},\theta^{\ell})$ that s is an isomorphism.

3.17. Corollary. For an arbitrary ψDO $A : C^{\infty}_{comp}(X,E) \to C^{\infty}(X,F)$ and a connection ∇ on F there exists a unique section $\mathrm{subres}^{\nabla}_{x}(A)$ of the bundle $\Omega^1_X \otimes A_{X,0} \otimes \underline{\mathrm{Hom}}(E,F)$ with property that:

$$\phi^*\mathrm{subres}^{\nabla}_{x}(A) = \mathrm{subres}^{\phi^*\nabla}_{u}(\phi^{\#} A)$$

whenever ϕ is a local coordinate patch (as in 3.1).

Proof-Explanation. The subresidue is so defined that it corresponds under the natural isomorphism of sheaves

$\Omega^1_X \otimes A_{X,0} \otimes \underline{\mathrm{Hom}}(E,F) \simeq \underline{\mathrm{Hom}}(\mathcal{T}_X, A_{X,0} \otimes \underline{\mathrm{Hom}}(E,F))$ to $C^{\infty}(X)$-*linear* and transparently functorial homomorphism

$$\mathcal{T}X \ni \eta \mapsto \mathrm{res}_x(\nabla_\eta \circ A) \ . \qquad \square$$

When $E = F$ we obtain the global section $\mathrm{tr}\ \mathrm{subres}^{\nabla}_{x}(A)$ of the vector bundle $\Omega^1_X \otimes A_{X,0}$. The holomorphic realization of the latter and its $(n-1)$-st cohomology are the well known ingredients of deformation theory of complex structures. This coincidence is likely to indicate some deeper connection between ψDOs and variations of complex structures.

3.18. The relation to zeta function. Let X be closed (i.e. compact and without boundary) and $E = F$. Assume that $A : C^{\infty}(X,E) \to C^{\infty}(X,E)$ is an elliptic ψDO of positive order which admits complex powers A^{-s}_{θ} (the subscript θ means that the branch $\theta < \mathrm{Arg}\lambda < \theta + 2\pi$ of the argument is chosen). Let $A^{-s}_{\theta}(x,x)$ be the restriction to the diagonal of the Schwartz kernel of A^{-s}_{θ}. This is a density on X with coefficients in $\underline{\mathrm{End}}\ E$ which is defined only for $\mathrm{Re}\, s > \dim X/\mathrm{ord}\, A$. The classical nowadays theory of the zeta-function of an elliptic ψDO due to R.T. Seeley (cf. [14], and also [15] and [17]) asserts that the holomorphic in a half-plane function

$$s \mapsto A_\theta^{-s}(x,x) \qquad (\operatorname{Re} s > \dim X/\operatorname{ord} A)$$

with values in $A_0(X, \underline{\text{End}}\ E)$ admits an analytic continuation to a function $A_\theta^{(-s)}(x)$ that is meromorphic in the whole complex plane. The sole singularities of it are simple poles located at points of the arithmetic progression $s_j = (\dim X - j)/\operatorname{ord} A$ (except the origin where the function is always regular). The theory also gives explicit local formulae for all residues and the value at $s = 0$. Finally, if A happens to be a differential operator the residues at other negative integers vanish too and the theory also gives local formulae for values at those points.

Using the language of noncommutative residue the results of Seeley become extremely clear and the formulae - brief.

<u>Formulae for residues.</u> $$\operatorname{Res}_{s=s_j} A_\theta^{(-s)}(x) = \frac{\operatorname{res}_x(A_\theta^{-s_j})}{\operatorname{ord} A}. \tag{8}$$

In particular, $A_\theta^{(-s)}(x)$ can possess a pole exactly there where A_θ^{-s} can have a priori non-vanishing noncommutative residue form. For $s = 0$ we have

$$A_\theta^0 = I - P_0$$

(P_0 is the projector on the zero-eigenspace, it has a finite rank). Hence $\operatorname{res}_x A_\theta^0$ always vanishes identically. At other negative integers $s = -\ell$ one has $A = A \circ \ldots \circ A$ (ℓ times). Hence for a differential operator all A_θ^ℓ's ($\ell = 1,2,\ldots$) are also differential operators and $\operatorname{res} A_\theta^\ell$ vanishes identically.

<u>Formula for $A_\theta^{(0)}(x)$.</u> Let us consider first a local situation. For an elliptic ψDO $A : C^\infty_{\text{comp}}(U,\theta^k) \to C^\infty(U,\theta^k)$ let $b(\lambda) = \sum_{j=0}^{\infty} b_{-m-j}(\lambda)$ be defined by requiring that:

$$1^\circ \quad b_{-m-j}(u,t\xi,t^m\lambda) = t^{-m-j}b_{-m-j}(u,\xi,\lambda) \qquad (m = \text{ord } A)$$

$$2^\circ \quad (a-\lambda)\circ b(\lambda) = 1 \tag{9}$$

(condition 1° defines what is called sometimes a "symbol with a parameter". Notice that this is not a classical symbol which happens to depend on a parameter unless $m = 0$). Divide $b(\lambda)$ into the sum of its leading term $b_{-m}(\lambda)$ and the rest $b^-(\lambda)$, and suppose that for no $(u,\xi) \in T_0^*U$ there is an eigenvalue of the principal symbol $a_m(u,\xi)$ lying on the ray $\mathrm{Arg}\lambda = \theta$. Then the integral

$$\ell_\theta = \int_0^{\infty e^{i\theta}} b^-(\lambda)d\lambda \tag{10}$$

is well defined (notice that $\int_0^{\infty e^{i\theta}} b_{-m}(\lambda)d\lambda$ diverges). It follows easily from 1° that $\ell_\theta = \sum_{j=1}^{\infty} \ell_{\theta,-j}$ where each $\ell_{\theta,-j}$ is a homogeneous of order $-j$ matric function on T_0^*U. In order to indicate its dependence on A it will be also denoted ℓ_θ^A and called the logarithmic symbol" of A. It has obviously a different transformation rule from standard symbols (or even amplitudes) under local morphisms $\psi : (V;\theta^k) \to (U;\theta^k)$. It shares with them, however, the property that

$$\ell_\theta^{\psi^\# A} - (T^*\psi)^*(\ell_\theta^A) = \sum_{i=1}^{n} \partial_{\eta_i}\lambda_{(i)}$$

where $\lambda_{(i)} \in \hat{P}^\cdot(T_0^*V) \otimes \mathrm{Mat}_k(\mathbb{C})$ depend microlocally on the symbol of and on ψ, and $(\nu,\eta) \in T_0^*V$.

Thus Lemma 2.2 implies, as before (cf. 2.20), the functoriality of $_*|\mathrm{Res}_0\ell_\theta^A|$.

Lemma 3.19. The matric density $Z_{\theta,u}(A) := -\beta_n\tau_*|\mathrm{Res}_0\ell_\theta^A|$ is functorial with respect to morphisms $(V;\theta^k) \to (U;\theta^k)$. □

Corollary 3.20. For an arbitrary elliptic operator $: C^\infty_{comp}(X,E) \to C^\infty(X,E)$ with the property that no eigenvalue of its principal symbol lies on the ray $\operatorname{Arg}\lambda = \theta$ there is a well defined density $Z_{\theta,x}(A)$ with coefficients in End E such that

$$\phi^* Z_{\theta,x}(A) = Z_{\theta,u}(\phi^\# A)$$

for every local coordinate patch $\phi : (U;\theta^k) \to (V;E)$.

The density $Z_{\theta,x}(A)$ is, in particular, functorial with respect to global morphisms $\psi : (Y;F) \to (X;F)$. □

The promised formula for $A_\theta^{(0)}(x)$ reads then as

$$A_\theta^{(0)}(x) = \frac{Z_{\theta,x}(A)}{\operatorname{ord} A} - P_0(x,x) \tag{11}$$

where $P_0(x,x)$ is the restriction of the kernel of the projector P_0 to the diagonal.

3.21. Remark. Notice that $Z_{\theta,x}(A)$ is well defined even if $\operatorname{ord} A \le 0$. For an operator of zeroth order $b(\lambda)$ is simply a classical symbol of $(A - \lambda)^{-1}$ and $Z_{\theta,x}(A)$ turns to be equal

$$Z_{\theta,x}(A) = -\int_0^{\infty e^{i\theta}} \operatorname{res}_x (A - \lambda)^{-1} d\lambda. \tag{12}$$

This equality suggests that $Z_{\theta,x}(A)$ is roughly speaking the 'noncommutative residue of ≪log ∞_m≫-log A' where ∞_m is a figurative notation for the "infinite operator of order $m = \operatorname{ord} A$"; ℓ_θ^A is just the symbol of that non-existing operator with its divergent principal symbol thrown out.

It is easy to deduce from (12) that one has indeed the equality

$$\int_X \operatorname{tr} Z_{\theta,x}(A) = -\operatorname{res} \log^{(\theta)} A \tag{13}$$

for operators of zeroth order ($\log^{(\theta)} A$ is the logarithm of A defined by integration of $-\frac{\log\lambda}{2\pi i}(A - \lambda)^{-1} d\lambda$ along the closed contour in $\mathbb{C}\setminus\{\operatorname{Arg}\lambda = \theta\}$ enclosing the spectrum of the principal symbol of A). For more details see [19]. A simple comparison of (13) with (11) suggests that

$$\operatorname{ord} A(\zeta_\theta(0;A) + h_0(A)) \equiv \operatorname{ord} A\left[\int_X \operatorname{tr}(A_\theta^{(0)}(x) + P_0(x,x))\right]$$

should possess certain multiplicative properties (analogous to tr log functional on matrices). This is indeed the case (see [19]).

Finally, let us consider the case when A is a differential operator. Let $q \geq 1$ and $b^{(q)}(\lambda) := \sum_{j=q+1}^{\infty} b_{-j}(\lambda)$, then the integral

$$\ell^{(q),A} = \int_0^{\infty e^{i\theta}} \lambda^q b^{(q)}(\lambda)d\lambda \tag{14}$$

s well defined provided A is a differential operator. As before ne verifies easily that $z^{(q)}_{\theta,u}(A) := -\beta_n \tau_* |\operatorname{Res}_0 \ell_\theta^{(q),A}|$ gives rise to global functorial density $z^{(q)}_{\theta,x}(A)$. Then the formula for $A_\theta^{(-q)}(x)$ eads as:

$$A_\theta^{(-q)}(x) = \frac{z^{(q)}_{\theta,x}(A)}{\operatorname{ord} A}\,. \tag{15}$$

Because $\zeta_\theta(s;A) \equiv \int_X \operatorname{tr} A_\theta^{(-s)}(x)$, the integration of fibre races of (8), (11) and (15) yields the corresponding formulae for esidues and values of the zeta function of A.

.22. <u>Generalized zeta-functions.</u> The ordinary and more general eta-functions all arise as a special example of the following eneral picture. On the space $\coprod_{z\in\mathbb{C}} CL^z(X,E)$ there is a suitable opology making it a holomorphic fibre bundle over $A^1(\mathbb{C})$ so that $\Delta_E>^z \equiv (1+\Delta_E)^{z/2}$ provides a global invertible section of it Δ_E denotes the Laplacian with coefficients in E). For a holo- orphic section Φ let $\Phi(a;x,x)$ be the restriction of the Schwartz ernel of $\Phi(z)$ to the diagonal where it is defined (i.e. for ez $<$ -dim X). Then $\Phi(z;x,x)$ can be analytically continued to a eromorphic function $\Phi(z;x)$ with values in $A_{X,0}(\underline{\operatorname{End}}\,E)$. The sole ingularities are simple poles located at the points $z_j = j - \dim X$ $j = 0,1,\ldots$). The relevant residues are given by the simple formula

$$\operatorname{Res}_{z=z_j} \Phi(z;x) = -\operatorname{res}_x \Phi(z) \tag{16}$$

which is a generalization of (8). It follows from (16) that the residue of $\Phi(z;x)$ depends only on the *value* of section Φ at the relevant point and not, for instance, on the whole germ. It follows also that $\Phi(z;x)$ is regular at those points where $\Phi(z)$ happens to be a differential operator.

Formula (16) finds an immediate application to the zeta-functions $\zeta_\theta(s;Q|A)$ which are analytic continuations of the "Dirichlet series" $\operatorname{Tr} QA^{-s}$.

Sometimes it can be also possible to obtain formulae for values of $\Phi(z;x)$ at certain points (generalizing (11) and (15)). This happens, for instance, for the section $\Phi(z) = QA_\theta^w$ $(w \equiv (z - \operatorname{ord} Q)/\operatorname{ord} A)$ with Q being a differential operator. Let q be the symbol of Q and $b_Q^-(\lambda)$ be $q \circ b(\lambda)$ with first few divergent terms thrown out. Then we put (in similarity to (10))

$$\ell_\theta^{Q,A} = \int_0^{\infty e^{i\theta}} b_Q^-(\lambda)\,d\lambda \tag{17}$$

and

$$z_{\theta,u}(Q|A) = -\beta_n \tau_* |\operatorname{Res}_0 \ell_\theta^{Q,A}| \,. \tag{18}$$

Much as before we derive the existence of the global functorial density $z_{\theta,x}(Q|A)$, and the desired formula has the form:

$$\Phi(\operatorname{ord} Q;x) = \frac{z_{\theta,x}(Q|A)}{\operatorname{ord} A} \,. \tag{19}$$

For $Q = A^q$ $(q = 1,2,\dots)$ formula (19) is equivalent to Seeley's formula (14).

<u>3.23. Relation to the heat kernel expansions.</u> Let $D : C^\infty(X,E) \to C^\infty(X,F)$ be an arbitrary elliptic ψDO (of positive

order) and $D^* : C^\infty(X,F) \to C^\infty(X,E)$ be the corresponding adjoint operator (with respect to certain hermitian metrics on E and F, and a positive density on X). Put $\Box_E = D^*D$ and $\Box_F = DD^*$, and let $\zeta(s,x;\Box) := \mathrm{tr}\,\Box^{(-s)}(x)$ be the corresponding *local* zeta-function. Since one obviously has the equalities

$$D(\Box_E - \lambda)^{-1} = (\Box_F - \lambda)^{-1}D \quad \text{and} \quad (\Box_E - \lambda)^{-1}D^* = D^*(\Box_F - \lambda)^{-1} \qquad (20)$$

it follows easily from (8) and from Proposition 3.7 that

$$\mathrm{Res}_{s=s_j}(\zeta(s,x;\Box_E)-\zeta(s,x;\Box_F)) = \frac{1}{2\mathrm{ord}A}(\mathrm{tr}\ \mathrm{res}_x D^*\Box_F^{-s_j-1}D - \mathrm{tr}\ \mathrm{res}_x \Box_F^{-s_j-1}DD^*)$$

$$= \frac{1}{2\mathrm{ord}A}\, d\bar{\rho}(D^*,\Box_F^{-s_j-1}D)$$

or by interchanging D and D^*:

$$\mathrm{Res}_{s=s_j}(\zeta(s,x;\Box_E) - \zeta(s,x;\Box_F)) = \frac{1}{2\mathrm{ord}A}\, d\bar{\rho}(\Box_E^{-s_j-1}D^*,D) . \qquad (21)$$

If the order of A is not an integer we have to use $\bar{\rho}$ extended as mentioned in Remark 3.8).

Recall that the inverse Mellin transform translates the singularities of $\Gamma(s)\Box^{(-s)}(x)$ into the "high temperature" expansion

$$e^{-t\Box}(x) \sim \sum_{j=0}^{\infty} \alpha_j(x)t^{\frac{j-n}{2m}} + \sum_{q=1}^{\infty} \beta_q(x)t^q \log t \qquad (t \searrow 0) \qquad (22)$$

n = dim X, m = ord D) so that

$$\alpha_j(x) = \Gamma(s_j)\mathrm{Res}_{s=s_j}\Box^{(-s)}(x) \qquad (23)$$

$s_j = \frac{n-j}{2m}$; $j \neq n,\ n+2m,\ n+4m, \ldots)$,

$$\beta_q(x) = \frac{(-1)^{q-1}}{q!}\,\mathrm{Res}_{s=-q}\Box^{(-s)}(x) \qquad (q = 1,2,\ldots) \qquad (24)$$

nd

$$\alpha_0(x) = \Box^{(0)}(x) + P_0(x,x;\Box) \qquad (\text{cf. } (11)) \qquad (25)$$

depend locally on the symbol of $\Box$. By comparing (23) and (24) with (21) we obtain

3.24. Proposition. One has

$$\mathrm{tr}\ \alpha_j^E(x) - \mathrm{tr}\ \alpha_j^F(x) = \frac{\Gamma(s_j)}{2m}\, d\bar{\rho}\ (\Box_E^{-s_j-1} D^*, D) \tag{26}$$

$(s_j = \frac{n-j}{2m};\ j \neq n,\ n+2m,\ n+4m, \ldots)$, and

$$\mathrm{tr}\ \beta_q^E(x) - \mathrm{tr}\ \beta_q^F(x) = \frac{(-1)^{q-1}}{2m\ q!}\, d\bar{\rho}\ (\Box_E^{q-1} D^*, D) \ . \tag{27}$$

(Also to this situation extends the remark after (21)) .

3.25. Remark. If D is a differential operator all $\beta_q(x)$'s vanish, and then also

$$\alpha_{n+2mq}(x) \equiv \frac{(-1)^q}{q!}\ \Box^{(-q)}(x) \qquad (q = 1,2,\ldots)$$

depend locally on the symbol of $\Box$ (cf. (15)) .

Denote by d^* and $c^{(q)}$ the symbols of D^* and $\Box^q$ respectively. Then the second equality in (20) implies that

$$c_F^{(q)} \circ b(\lambda;\Box_F) = d \circ (c_E^{(q-1)} \circ b(\lambda;\Box_E) \circ d^*) \ .$$

On the other hand one has the obvious equality

$$c_E^{(q)} \circ b(\lambda;\Box_E) = (c_E^{(q-1)} \circ b(\lambda;\Box_E) \circ d^*) \circ d \ .$$

This yields (via formulae (17) - (19) applied to $A = \Box$ and $Q = \Box^q$)

$$\mathrm{tr}\ \alpha_{n+2mq}^E(u) - \mathrm{tr}\ \alpha_{n+2mq}^F(u) = \frac{(-1)^{q-1}\beta_n}{2mq!}\tau_*|\mathrm{Res}_0 \int_0^{-\infty} (\mathrm{tr}\{\tilde{b}^{(q-1)}(\lambda) \circ d\}^- - \mathrm{tr}\{d \circ \tilde{b}^{(q-1)}(\lambda)\}^-)d\lambda| \tag{28}$$

where $\tilde{b}^{(q-1)}(\lambda) \equiv c_E^{(q-1)} \circ b(\lambda;\Box_E) \circ d^*$ and the upper "minus" means the "non-divergent part of" obtained by throwing out terms $O(|\lambda|^{-1})$ and bigger (as $\lambda \to \infty$). Equality (28) suggests that some refinement of the constructions of Section 1 might lead also in this case to expressing the left-hand side of (28) as an explicit exact form. This purpose is served by the formalism of a suitably "twisted" symplectic residue which is sketched briefly below.

We are guided by the idea of entering with residue under the sign of integral $\int_0^\infty d\lambda$ (in formulae like 3.19 or (28)).

3.26. Λ-twisted symplectic residue (A sketch). Let us return to the situation of Section 1. We were dealing there with a symplectic cone Y^{2n} whose base was denoted by Z^{2n-1}. Let Λ be another space acted by the multiplicative group $\mathbb{R}_+^\times$. (Λ is assumed generally to be a manifold with boundary, but the boundary is allowed to have corners, etc.).

Let $\Phi^\cdot \subset C^\infty(Y,\Omega_\cdot(\Lambda)) \subset \Omega_\cdot(Y \times \Lambda)$ be a linear subspace which satisfies the following two conditions:

$$1)\quad d_\lambda \Phi^\cdot = 0 ,$$

$$2)\quad \Phi^\cdot = \bigoplus_{\ell \in \mathbb{Z}} \Phi^\ell \quad \text{and} \quad L_{\overline{\Xi}} = \ell \cdot \mathrm{id} \quad \text{on} \quad \Phi^\ell \tag{29}$$

d_λ denotes de Rham differential in the Λ-direction, and $\overline{\Xi}$ is the uler field determined by the product $\mathbb{R}_+^\times$-action on $Y \times \Lambda$) .

It should be clear that $\Phi^\cdot$ is closed with respect to L_f's for $\in P^\cdot$. Indeed, write $\overline{\Xi}$ as the sum $\Xi + \Psi$ where Ξ and Ψ are he corresponding Euler fields on Y and Λ. Then

$$L_{\overline{\Xi}} L_f = L_f L_{\overline{\Xi}} + L_{[\Xi,H_f]} \overset{1.7}{=} L_f (L_{\overline{\Xi}} + \ell_f - 1) .$$

In particular, $\Phi^\cdot$ is made a $P^\cdot$-module. Let $C_\cdot(P^\cdot;\Phi^\cdot)$ be he relevant chain complex. As before it is graded:

$$C_\cdot(P^\cdot;\Phi^\cdot) = \bigoplus_{k \in \mathbb{Z}} C_\cdot^{(k)}(P^\cdot;\Phi^\cdot) . \qquad \text{(cf. 1.17)}$$

Without loss of generality we can assume that $\Phi^\cdot$ is concentrated n a fixed codimension p, i.e. $\Phi^\cdot \subset C^\infty(Y,\Omega_p(\Lambda))$. Then a slight modification of 1.12-13 and (2) of Section 1 shows that

$$i_{f_1}\cdots i_{f_q}\left(\mu \wedge \phi + \frac{\omega^n}{n} \wedge i_\psi \phi\right) \qquad \left(\text{for } \ell_\phi + \sum_{i=1}^{q} \ell_{f_q} = q-n\right) \tag{30}$$

descends to a form $\bar{\mu}_{\phi;f_1,\ldots,f_q} \in \Omega_{p+q}(Y_\Lambda)$ where $Y_\Lambda := (Y \times \Lambda)/\mathbb{R}_+^\times$, and that the analogue of 1.16 holds for $\bar{\mu}$. Similarly, the collection of maps $\phi \otimes f_1 \wedge \ldots \wedge f_q \mapsto \bar{\mu}_{\phi;f_1,\ldots,f_q}$ gives rise to a morphism of chain complexes

$$\underline{\mathrm{Res}}^\Lambda : C_\cdot(P^\cdot;\Phi^\cdot) \to \Omega_\cdot(Y_\Lambda)[-p] . \tag{31}$$

The space Y_Λ is naturally fibred over Z. Any singular chain γ of the codimension p in Λ which is preserved by the multiplicative group defines a family Y_γ of chains in fibres of the projection $Y_\Lambda \to Z$. By γ_* we shall denote the map $\Omega_\cdot(Y_\Lambda)[-p] \to \Omega_\cdot(Z)$ which is the composition of the restriction to Y_γ with integration along fibres of $Y_\gamma \to Z$.

If γ is not closed then γ_* is not a morphism. It may happen, moreover, that γ is not even compact. Then γ_* is only a partial map defined on "γ-integrable" forms. For both reasons simultaneously some *exact* forms on Y_Λ need not remain exact when pushed down to Z. Precisely this phenomenon stands behind the fact that $\tilde{\alpha}_0 \equiv \mathrm{tr}\,\alpha_0^E(x) - \mathrm{tr}\,\alpha_0^F(x)$ may not be exact. We shall exploit it to propose in one of subsequent papers(*) a fresh approach to index theory.

Much as we did in 1.29 the above construction can be extended to the case of coefficients in a vector bundle on Z. The next step is to introduce twisted symplectic residue into the constructions of Section 2 (we omit the details).

(*) Say, in 'Chapter V'.

3.27. Example. Let Λ be a closed sector of the complex plane. For a fixed $m \in \mathbb{R}$ we define the action of $\mathbb{R}_+^\times$ on Λ as:

$$\chi_t^{(m)} : \lambda \mapsto t^m\lambda .$$

Put $Y = T_0^*X$ and $\Phi^\ell = \{\phi \in C^\infty(Y,\Omega^1_{hol}(\Lambda)) \mid (L_\Xi + mL_{|\lambda|}\frac{\partial}{\partial|\lambda|} - \ell)\phi = 0\}$. The ray $\operatorname{Arg}\lambda = \theta$ (contained in Λ) will play a role of γ. It is neither compact nor a cycle.

Suppose we have a Λ-elliptic ψDO $C^\infty_{comp}(X,E) \to C^\infty(X,E)$ of the order m (that will be assumed to be positive; recall that Λ-elliptic means that no eigenvalue of the principal symbol of A can lie in the sector Λ). Then for every local coordinate patch U the corresponding 'λ-symbol' $b(\lambda)$ (cf. (9)) is defined, and it should be clear that $\phi := b(\lambda)d\lambda$ determines an element of the formally completed space $\hat{\Phi}^\bullet$ (in fact, each $b_{-m-j}(\lambda)d\lambda$ belongs to $\Phi^{-j} \otimes \operatorname{Mat}_k(\mathbb{C})$ where $k = \operatorname{rank} E$). Since all $b_{-m-j}(\lambda)$'s (except $b_{-m}(\lambda)$) decrease for a fixed (u,ξ) as $O(|\lambda|^{-2})$, $\operatorname{Res}_0^\Lambda(b(\lambda)d\lambda)$ is apparently γ-integrable, and 3.19 can be rewritten as

$$Z_{\theta,u}(A) = -\beta_n\tau_*\gamma_*|\operatorname{Res}_0^\Lambda(b(\lambda)d\lambda)|. \tag{32}$$

Similarly, (18) can be rewritten as

$$Z_{\theta,u}(Q|A) = -\beta_n\tau_*\gamma_*|\operatorname{Res}_0^\Lambda(q \circ b(\lambda)\ d\lambda)|. \tag{33}$$

Notice that since q has no negative components, $\operatorname{Res}_0^\Lambda(q \circ b(\lambda)\ d\lambda)$ is still γ-integrable (the component of $q \circ b(\lambda)d\lambda$ of weight $-n$ behaves like $O(|\lambda|^{-2}d\lambda)$).

Finally, $Z_{\theta,u}^{(q)}(A)$ for a *differential* A can be also expressed as

$$Z_{\theta,u}^{(q)}(A) = -\beta_n\tau_*\gamma_*|\operatorname{Res}_0^\Lambda(\lambda^q b(\lambda)d\lambda)| \tag{34}$$

In this case the component of $\lambda^q b(\lambda)d\lambda$ of weight $-n$ behaves like $O|\lambda|^{-1-n-(m-1)q}d\lambda)$.

Now we can return to the differences $\tilde{\alpha}_{n+2mq}(x) \equiv \operatorname{tr} \alpha^{E}_{n+2mq}(x) - \operatorname{tr} \alpha^{F}_{n+2mq}(x)$ of the coefficients of two heat kernel expansions when D is a differential operator. We shall sketch below why for $q > 0$ $\tilde{\alpha}_{n+2mq}(x)$ can be expressed as fine explicit exact forms whereas for $q = 0$ this fails completely. First observe that for *all* q (i.e. $q = 0,1,2,\ldots$) the form

$$\tilde{\alpha}_{n+2mq}(u) \overset{(28)}{=} \frac{(-1)^{q-1}\beta_n}{2mq!} \tau_* \gamma_* | \operatorname{Res}^{\Lambda}_{0}\{\operatorname{tr}(\phi^{(q-1)} \circ d) - \operatorname{tr}(d \circ \phi^{(q-1)})\}| \quad (35)$$

($\phi^{(q-1)}$ denotes $\tilde{b}^{(q-1)}(\lambda)d\lambda \equiv (c_E^{(q-1)} \circ b(\lambda;\Box_E) \circ d^*)d\lambda$) can be written explicitly as an exact form on $(T_0^*U)_\Lambda$. This follows from the corresponding 'Λ-twisted version' of Proposition 2.28. That form for $q \geq 1$ is undoubtedly γ-integrable while for $q = 0$ (and for general D) it isn't. Notice also that there is no 'boundary term' in this case after the integration along γ was performed. This is related to the fact that $\lambda = 0$ is the fixed point of the $R_+^\times$-action on Λ and hence the Euler field Ψ vanishes at $\lambda = 0$ assuring thereby vanishing of (30) after restriction on $T_0^*U \times \{0\}$.

This gives the required representation locally. In order to obtain it globally one can use either the partition of unity argument (since $\phi^{(q-1)}$ and d in (35) can be regarded as being independent without destroying the mechanism of exactness described above or to try to establish the 'Λ-twisted' analogues of Proposition 3.7 and its more precise version for operators of order ≤ 2 (cf. Remark 3.10.2)). The details will be presented elsewhere.

3.28. Secondary classes. As is well known, for operators arising in Riemannian geometry such as Euler, signature and Cauchy-Riemann operators, or more physically significant Dirac operator, the differences $\tilde{\alpha}_j(x) \equiv \operatorname{tr} \alpha_j^E(x) - \operatorname{tr} \alpha_j^F(x)$ (for $j < n$) have in a number of cases tendency to vanish pointwise. This is called a local cancellation of divergencies (α_j's with $j < \dim X$ correspond to divergent terms in (22)), and is usually considered as a manifestation of some super-symmetries or integrability conditions (cf. e.g. local cancellation for the Cauchy-Riemann operator on a Kähler manifold [5],[7]. Such cancellations often rely on methods of Invariant Theory of classical groups (cf. [1], [6] and other papers by P.B. Gilkey).

In any case, whenever a local cancellation occurs Proposition 3.24 tells us that in place of the coefficient $\tilde{\alpha}_j(x)$ which has vanished arises a certain cohomology class (namely the class $\bar{\rho}(\square_E^{-s_j-1} D^*,D))$. In general, it seems to be no reason a priori for these classes to vanish (cf. Example 3.29 below) subject to the impression that the mechanism of local cancellations is rather some *extra* feature and probably should be regarded as being imposed from the outside on the fairly universal formalism of noncommutative residue.

The same applies to the differences $\tilde{\beta}_q(x) \equiv \mathrm{tr}\,\beta_q^E(x) - \mathrm{tr}\,\beta_q^F(x)$ in case they vanish.

.29. <u>Example</u>. Let $E = F$ and $D = A + \sqrt{-1}B$ where A and B are self-adjoint. Equality (27) yields

$$\tilde{\beta}_1(x) = \frac{1}{2m}\, d\bar{\rho}(D^*,D) = \frac{\sqrt{-1}}{m}\, d\bar{\rho}(A,B) \,. \tag{36}$$

Assume that $X = T^n$ is an n-dimensional torus and D is a ψDO with constant coefficients (in particular, E is assumed to be a trivial bundle). Then $\bar{\rho}(A,B)$ possesses a canonical representative (cf. Remark 3.10.2)) and this representative is clearly a constant form, and hence - closed. As we shall demonstrate below every constant (n-1)-form can be obtained in this way. Let e.g. A be the 1-st order operator

$$A = c\not{D} + \sum_{j=1}^{n} \alpha_j \frac{\partial}{\partial x^j} \tag{37}$$

where $\not{\partial} = -\sqrt{-1}\sum_{j=1}^{n}\gamma^j\partial/\partial x^j$ is the standard Dirac operator, α_j's are arbitrary complex scalars and c is a constant big enough to assure ellipticity of (37). As B we take an arbitrary scalar ψDO with principal symbol $|\xi|^{-n}$ (e.g. $B = (1+\Delta)^{-n/2}$). Note that B can be chosen so that it commute with A.

According to Proposition 3.12 and Definition 2.13 $\bar{\rho}(A,B)$ is represented by

$$\rho_x(A,B) = \sum_{j=1}^{n} tr(-\sqrt{-1}c\ \gamma^j + \alpha_j)\, i_{\partial/\partial x^j}\, res_x B$$

$$= \frac{d_n v_n}{(2\pi)^n}\sum_{j=1}^{n}(-1)^{j-1}\alpha_j|dx^1\wedge\ldots\wedge\widehat{dx^j}\wedge\ldots\wedge dx^n| \qquad (38)$$

if $n > 1$ (v_n is the volume of the unit sphere in $\mathbb{R}^n$ and $d_n = 2^{[n/2]}$ is the rank of the spinor bundle), and

$$\rho_x(A,B) = \frac{-\sqrt{-1}c+\alpha}{\pi}\,\frac{|dx|}{dx} \qquad (39)$$

if $n = 1$. (The asymmetry between the two cases is related to the fact that for $n = 2\nu+1$ and $\nu \geq 1$ the matrix $\gamma^{2\nu+1}$ is up to a power of $\sqrt{-1}$ the product of all γ^j's with $j \leq 2\nu$, and hence is traceless, whereas for $n = 1$, $\gamma^1 = 1$).

Equalities (38) and (39) show that every cohomology class from $H^{n-1}(T^n,\varepsilon)$ serves as the secondary class $\bar{\rho}(D^*,D)$ for a certain suitably rescaled and perturbed Dirac operator D. In the considered example not only all coefficients $\tilde{\alpha}_j(x)$'s and $\tilde{\beta}_q(x)$'s vanish but also $e^{-tD^*D} - e^{-tDD^*} \equiv 0$.

<u>3.30. Remark</u>. The above example shows even more. Recall that a ψDO $D \in CL^{\cdot}_{prop}(X,E)$ (for notation cf. 0.10) is called *normal* if $D^*\wedge D$ is a 2-cycle on the Lie algebra $\mathfrak{g} \equiv CL^{\cdot}_{prop}(X,E)$. According to diagram (5) $\bar{\rho}$ defines a map $H_2(\mathfrak{g}) \to H^{n-1}(X,\varepsilon_X)$. Example 3.29 shows that for $X = T^n$ and E - the (trivial) bundle of spinors this map is surjective and, moreover, that any class from

$H^{n-1}(X,\varepsilon_X)$ is, actually, the image of the cycle $D^* \wedge D$ where D is a normal elliptic ψDO of order 1. In [18] we shall demonstrate that at least the former assertion remains valid for arbitrary - bundle and closed manifold.

3.31. Let us return once more to differential operators arising in Riemannian geometry. In a 'non-integrable' situation even the coefficients $\tilde{\alpha}_j(x)$, $(j < n)$ corresponding to divergent terms of the high-temperature expansion may not vanish (e.g. for non-Kähler Hermitian manifolds, cf. e.g. [4], [5]). In certain cases, at least, methods of Invariant Theory of classical groups yield specific forms $q_j(x)$ such that $\tilde{\alpha}_j(x) = dq_j$ for all $j \neq \dim X$ (cf. [4],[7]). The order of considered operators rarely exceeds 2 (usually equals 1) Recall that for differential operators of order ≤ 2 we possess just not only the *class* $\bar{\rho}$ but also its *canonical* representative ρ_X (cf. Remark 3.13). In particular, the forms

$$\frac{\Gamma(s_j)}{2m}\,\rho_X(\Box_E^{-s_j-1} D^*,D) - q_j(x) \qquad (j \neq n, n+2m, n+4m)$$

are closed (as was pointed out cf. e.g. Example 3.27, it is possible to obtain analogues of the forms $\rho_X(\Box_E^{-s_j-1} D^*,D)$ also for $j = n+2m, n+4m, \ldots)$. It would be very interesting to know more about the associated cohomology classes.

References

[1] Atiyah, M.F., Bott, R., Patodi, V.K.: On the heat equation and the index theorem, Invent. math. 19, 279-330 (1973); Errata, *ibid.* 28, 277-280 (1975)

[2] Adler, M.: On a trace functional for formal pseudo-differential operators and the symplectic structure of the Korteweg-de Vries type equations. Invent. math. 50, 219-248 (1979)

[3] Bourbaki, N.: Variétés différentielles et analytiques, Par. 8 à 15. Paris: Hermann 1971

[4] Gilkey, P.B.: Curvature and the eigenvalues of the Dolbeault complex for Hermitian manifolds. Adv. Math. 21, 61-77 (1976)

[5] Gilkey, P.B.: Curvature and the eigenvalues of Dolbeault complex for Kähler manifolds. Adv. Math. 11, 311-325 (1973)

[6] Gilkey, P.B.: Curvature and the eigenvalues of the Laplacian for elliptic complexes. Adv. Math. 10, 344-382 (1973)

[7] Gilkey, P.B.: Local invariants of an embedded Riemannian manifold. Ann. Math. 102, 187-203 (1975)

[8] Guillemin, V.W.: A new proof of Weyl's formula on the asymptotic distribution of eigenvalues. Adv. Math. 55, 131-160 (1985)

[9] Hörmander, L.: Fourier integral operators, I. Acta Math. 127, 79-183 (1971)

[10] Manin, Yu. I.: Algebraic aspects of non-linear differential equations. Itogi Nauki i Tekhniki, ser. Sovremennyje Problemy Matematiki 11, 5-152 (1978), (Russian); J. Sov. Math. 11, 1-22 (1979) English

[11] Manin, Yu. I.: Gauge fields and complex geometry (in Russian). Moscow: Nauka 1984

[12] Manin, Yu. I., Radul A.O.: A supersymmetric extension of the Kadomtsev-Petviashvili hierarchy. Comm. Math. Phys. 98, 65-77 (1985)

[13] Penkov, I.B.: $\mathcal{D}$-modules on supermanifolds. Invent. math. 71, 501-512 (1983).

[14] Seeley, R.T.: Complex powers of an elliptic operator. Proc. Sympos. Pure Math. 10, 288-307, Amer. Math. Soc. 1967

[15] Shubin, M.A.: Pseudodifferential operators and spectral theory (in Russian). Moscow: Nauka 1978

[16] Wodzicki, M.: Commentary, in: Hermann Weyl's "Selected papers" (in Russian), ed. by V.I. Arnold and A.N. Parshin. Moscow: Nauka 1985

[17] Wodzicki, M.: Local invariants of spectral asymmetry. Invent. math. 75, 143-178 (1984)

[18] Wodzicki, M.: NC. Chapter II: The commutator structure of pseudodifferential operators (to appear)

[19] Wodzicki, M.: NC. Chapter III: Bi-multiplicative index formula and exotic log det (to appear)

[20] Wodzicki, M.: NC. Chapter IV: Homology of the algebras of differential operators and symbols (in preparation)

[21] Wodzicki, M.: Spectral asymmetry and local invariants (in Russian), Habilitation Thesis. Moscow: Steklov Math. Inst. 1984

Vol. 1117: D.J. Aldous, J.A. Ibragimov, J. Jacod, Ecole d'Été de Probabilités de Saint-Flour XIII - 1983. Édité par P.L. Hennequin. IX, 409 pages. 1985.

Vol. 1118: Grossissements de filtrations: exemples et applications. Seminaire, 1982/83. Edité par Th. Jeulin et M. Yor. V, 315 pages. 1985.

Vol. 1119: Recent Mathematical Methods in Dynamic Programming. Proceedings, 1984. Edited by I. Capuzzo Dolcetta, W.H. Fleming and T. Zolezzi. VI, 202 pages. 1985.

Vol. 1120: K. Jarosz, Perturbations of Banach Algebras. V, 118 pages. 1985.

Vol. 1121: Singularities and Constructive Methods for Their Treatment. Proceedings, 1983. Edited by P. Grisvard, W. Wendland and J.R. Whiteman. IX, 346 pages. 1985.

Vol. 1122: Number Theory. Proceedings, 1984. Edited by K. Alladi. VII, 217 pages. 1985.

Vol. 1123: Séminaire de Probabilités XIX 1983/84. Proceedings. Edité par J. Azéma et M. Yor. IV, 504 pages. 1985.

Vol. 1124: Algebraic Geometry, Sitges (Barcelona) 1983. Proceedings. Edited by E. Casas-Alvero, G.E. Welters and S. Xambó-Descamps. XI, 416 pages. 1985.

Vol. 1125: Dynamical Systems and Bifurcations. Proceedings, 1984. Edited by B.L.J. Braaksma, H.W. Broer and F. Takens. V, 129 pages. 1985.

Vol. 1126: Algebraic and Geometric Topology. Proceedings, 1983. Edited by A. Ranicki, N. Levitt and F. Quinn. V, 423 pages. 1985.

Vol. 1127: Numerical Methods in Fluid Dynamics. Seminar. Edited by F. Brezzi, VII, 333 pages. 1985.

Vol. 1128: J. Elschner, Singular Ordinary Differential Operators and Pseudodifferential Equations. 200 pages. 1985.

Vol. 1129: Numerical Analysis, Lancaster 1984. Proceedings. Edited by P.R. Turner. XIV, 179 pages. 1985.

Vol. 1130: Methods in Mathematical Logic. Proceedings, 1983. Edited by C.A. Di Prisco. VII, 407 pages. 1985.

Vol. 1131: K. Sundaresan, S. Swaminathan, Geometry and Nonlinear Analysis in Banach Spaces. III, 116 pages. 1985.

Vol. 1132: Operator Algebras and their Connections with Topology and Ergodic Theory. Proceedings, 1983. Edited by H. Araki, C.C. Moore, Ş. Strătilă and C. Voiculescu. VI, 594 pages. 1985.

Vol. 1133: K.C. Kiwiel, Methods of Descent for Nondifferentiable Optimization. VI, 362 pages. 1985.

Vol. 1134: G.P. Galdi, S. Rionero, Weighted Energy Methods in Fluid Dynamics and Elasticity. VII, 126 pages. 1985.

Vol. 1135: Number Theory, New York 1983–84. Seminar. Edited by D.V. Chudnovsky, G.V. Chudnovsky, H. Cohn and M.B. Nathanson. V, 283 pages. 1985.

Vol. 1136: Quantum Probability and Applications II. Proceedings, 1984. Edited by L. Accardi and W. von Waldenfels. VI, 534 pages. 1985.

Vol. 1137: Xiao G., Surfaces fibrées en courbes de genre deux. IX, 103 pages. 1985.

Vol. 1138: A. Ocneanu, Actions of Discrete Amenable Groups on von Neumann Algebras. V, 115 pages. 1985.

Vol. 1139: Differential Geometric Methods in Mathematical Physics. Proceedings, 1983. Edited by H. D. Doebner and J. D. Hennig. VI, 337 pages. 1985.

Vol. 1140: S. Donkin, Rational Representations of Algebraic Groups. VII, 254 pages. 1985.

Vol. 1141: Recursion Theory Week. Proceedings, 1984. Edited by H.-D. Ebbinghaus, G.H. Müller and G.E. Sacks. IX, 418 pages. 1985.

Vol. 1142: Orders and their Applications. Proceedings, 1984. Edited by I. Reiner and K. W. Roggenkamp. X, 306 pages. 1985.

Vol. 1143: A. Krieg, Modular Forms on Half-Spaces of Quaternions. XIII, 203 pages. 1985.

Vol. 1144: Knot Theory and Manifolds. Proceedings, 1983. Edited by D. Rolfsen. V, 163 pages. 1985.

Vol. 1145: G. Winkler, Choquet Order and Simplices. VI, 143 pages. 1985.

Vol. 1146: Séminaire d'Algèbre Paul Dubreil et Marie-Paule Malliavin. Proceedings, 1983–1984. Edité par M.-P. Malliavin. IV, 420 pages. 1985.

Vol. 1147: M. Wschebor, Surfaces Aléatoires. VII, 111 pages. 1985.

Vol. 1148: Mark A. Kon, Probability Distributions in Quantum Statistical Mechanics. V, 121 pages. 1985.

Vol. 1149: Universal Algebra and Lattice Theory. Proceedings, 1984. Edited by S. D. Comer. VI, 282 pages. 1985.

Vol. 1150: B. Kawohl, Rearrangements and Convexity of Level Sets in PDE. V, 136 pages. 1985.

Vol 1151: Ordinary and Partial Differential Equations. Proceedings, 1984. Edited by B.D. Sleeman and R.J. Jarvis. XIV, 357 pages. 1985.

Vol. 1152: H. Widom, Asymptotic Expansions for Pseudodifferential Operators on Bounded Domains. V, 150 pages. 1985.

Vol. 1153: Probability in Banach Spaces V. Proceedings, 1984. Edited by A. Beck, R. Dudley, M. Hahn, J. Kuelbs and M. Marcus. VI, 457 pages. 1985.

Vol. 1154: D.S. Naidu, A.K. Rao, Singular Pertubation Analysis of Discrete Control Systems. IX, 195 pages. 1985.

Vol. 1155: Stability Problems for Stochastic Models. Proceedings, 1984. Edited by V.V. Kalashnikov and V.M. Zolotarev. VI, 447 pages. 1985.

Vol. 1156: Global Differential Geometry and Global Analysis 1984. Proceedings, 1984. Edited by D. Ferus, R.B. Gardner, S. Helgason and U. Simon. V, 339 pages. 1985.

Vol. 1157: H. Levine, Classifying Immersions into $\mathbb{R}^4$ over Stable Maps of 3-Manifolds into $\mathbb{R}^2$. V, 163 pages. 1985.

Vol. 1158: Stochastic Processes – Mathematics and Physics. Proceedings, 1984. Edited by S. Albeverio, Ph. Blanchard and L. Streit. VI, 230 pages. 1986.

Vol. 1159: Schrödinger Operators, Como 1984. Seminar. Edited by S. Graffi. VIII, 272 pages. 1986.

Vol. 1160: J.-C. van der Meer, The Hamiltonian Hopf Bifurcation. VI, 115 pages. 1985.

Vol. 1161: Harmonic Mappings and Minimal Immersions, Montecatini 1984. Seminar. Edited by E. Giusti. VII, 285 pages. 1985.

Vol. 1162: S.J.L. van Eijndhoven, J. de Graaf, Trajectory Spaces, Generalized Functions and Unbounded Operators. IV, 272 pages. 1985.

Vol. 1163: Iteration Theory and its Functional Equations. Proceedings, 1984. Edited by R. Liedl, L. Reich and Gy. Targonski. VIII, 231 pages. 1985.

Vol. 1164: M. Meschiari, J.H. Rawnsley, S. Salamon, Geometry Seminar "Luigi Bianchi" II – 1984. Edited by E. Vesentini. VI, 224 pages. 1985.

Vol. 1165: Seminar on Deformations. Proceedings, 1982/84. Edited by J. Ławrynowicz. IX, 331 pages. 1985.

Vol. 1166: Banach Spaces. Proceedings, 1984. Edited by N. Kalton and E. Saab. VI, 199 pages. 1985.

Vol. 1167: Geometry and Topology. Proceedings, 1983–84. Edited by J. Alexander and J. Harer. VI, 292 pages. 1985.

Vol. 1168: S.S. Agaian, Hadamard Matrices and their Applications. III, 227 pages. 1985.

Vol. 1169: W.A. Light, E.W. Cheney, Approximation Theory in Tensor Product Spaces. VII, 157 pages. 1985.

Vol. 1170: B.S. Thomson, Real Functions. VII, 229 pages. 1985.

Vol. 1171: Polynômes Orthogonaux et Applications. Proceedings, 1984. Edité par C. Brezinski, A. Draux, A.P. Magnus, P. Maroni et A. Ronveaux. XXXVII, 584 pages. 1985.

Vol. 1172: Algebraic Topology, Göttingen 1984. Proceedings. Edited by L. Smith. VI, 209 pages. 1985.

Vol. 1173: H. Delfs, M. Knebusch, Locally Semialgebraic Spaces. XVI, 329 pages. 1985.

Vol. 1174: Categories in Continuum Physics, Buffalo 1982. Seminar. Edited by F.W. Lawvere and S.H. Schanuel. V, 126 pages. 1986.

Vol. 1175: K. Mathiak, Valuations of Skew Fields and Projective Hjelmslev Spaces. VII, 116 pages. 1986.

Vol. 1176: R.R. Bruner, J.P. May, J.E. McClure, M. Steinberger, H_∞ Ring Spectra and their Applications. VII, 388 pages. 1986.

Vol. 1177: Representation Theory I. Finite Dimensional Algebras. Proceedings, 1984. Edited by V. Dlab, P. Gabriel and G. Michler. XV, 340 pages. 1986.

Vol. 1178: Representation Theory II. Groups and Orders. Proceedings, 1984. Edited by V. Dlab, P. Gabriel and G. Michler. XV, 370 pages. 1986.

Vol. 1179: Shi J.-Y. The Kazhdan-Lusztig Cells in Certain Affine Weyl Groups. X, 307 pages. 1986.

Vol. 1180: R. Carmona, H. Kesten, J.B. Walsh, École d'Été de Probabilités de Saint-Flour XIV – 1984. Édité par P.L. Hennequin. X, 438 pages. 1986.

Vol. 1181: Buildings and the Geometry of Diagrams, Como 1984. Seminar. Edited by L. Rosati. VII, 277 pages. 1986.

Vol. 1182: S. Shelah, Around Classification Theory of Models. VII, 279 pages. 1986.

Vol. 1183: Algebra, Algebraic Topology and their Interactions. Proceedings, 1983. Edited by J.-E. Roos. XI, 396 pages. 1986.

Vol. 1184: W. Arendt, A. Grabosch, G. Greiner, U. Groh, H.P. Lotz, U. Moustakas, R. Nagel, F. Neubrander, U. Schlotterbeck, One-parameter Semigroups of Positive Operators. Edited by R. Nagel. X, 460 pages. 1986.

Vol. 1185: Group Theory, Beijing 1984. Proceedings. Edited by Tuan H.F. V, 403 pages. 1986.

Vol. 1186: Lyapunov Exponents. Proceedings, 1984. Edited by L. Arnold and V. Wihstutz. VI, 374 pages. 1986.

Vol. 1187: Y. Diers, Categories of Boolean Sheaves of Simple Algebras. VI, 168 pages. 1986.

Vol. 1188: Fonctions de Plusieurs Variables Complexes V. Séminaire, 1979–85. Edité par François Norguet. VI, 306 pages. 1986.

Vol. 1189: J. Lukeš, J. Malý, L. Zajíček, Fine Topology Methods in Real Analysis and Potential Theory. X, 472 pages. 1986.

Vol. 1190: Optimization and Related Fields. Proceedings, 1984. Edited by R. Conti, E. De Giorgi and F. Giannessi. VIII, 419 pages. 1986.

Vol. 1191: A.R. Its, V.Yu. Novokshenov, The Isomonodromic Deformation Method in the Theory of Painlevé Equations. IV, 313 pages. 1986.

Vol. 1192: Equadiff 6. Proceedings, 1985. Edited by J. Vosmansky and M. Zlámal. XXIII, 404 pages. 1986.

Vol. 1193: Geometrical and Statistical Aspects of Probability in Banach Spaces. Proceedings, 1985. Edited by X. Fernique, B. Heinkel, M.B. Marcus and P.A. Meyer. IV, 128 pages. 1986.

Vol. 1194: Complex Analysis and Algebraic Geometry. Proceedings, 1985. Edited by H. Grauert. VI, 235 pages. 1986.

Vol. 1195: J.M. Barbosa, A.G. Colares, Minimal Surfaces in $\mathbb{R}^3$. X, 124 pages. 1986.

Vol. 1196: E. Casas-Alvero, S. Xambó-Descamps, The Enumerative Theory of Conics after Halphen. IX, 130 pages. 1986.

Vol. 1197: Ring Theory. Proceedings, 1985. Edited by F.M.J. van Oystaeyen. V, 231 pages. 1986.

Vol. 1198: Séminaire d'Analyse, P. Lelong – P. Dolbeault – H. Skoda. Seminar 1983/84. X, 260 pages. 1986.

Vol. 1199: Analytic Theory of Continued Fractions II. Proceedings, 1985. Edited by W.J. Thron. VI, 299 pages. 1986.

Vol. 1200: V.D. Milman, G. Schechtman, Asymptotic Theory of Finite Dimensional Normed Spaces. With an Appendix by M. Gromov. VIII, 156 pages. 1986.

Vol. 1201: Curvature and Topology of Riemannian Manifolds. Proceedings, 1985. Edited by K. Shiohama, T. Sakai and T. Sunada. VI, 336 pages. 1986.

Vol. 1202: A. Dür, Möbius Functions, Incidence Algebras and Powe Series Representations. XI, 134 pages. 1986.

Vol. 1203: Stochastic Processes and Their Applications. Proceedings 1985. Edited by K. Itô and T. Hida. VI, 222 pages. 1986.

Vol. 1204: Séminaire de Probabilités XX, 1984/85. Proceedings. Edit par J. Azéma et M. Yor. V, 639 pages. 1986.

Vol. 1205: B.Z. Moroz, Analytic Arithmetic in Algebraic Number Fields VII, 177 pages. 1986.

Vol. 1206: Probability and Analysis, Varenna (Como) 1985. Semina Edited by G. Letta and M. Pratelli. VIII, 280 pages. 1986.

Vol. 1207: P.H. Bérard, Spectral Geometry: Direct and Invers Problems. With an Appendix by G. Besson. XIII, 272 pages. 1986.

Vol. 1208: S. Kaijser, J.W. Pelletier, Interpolation Functors an Duality. IV, 167 pages. 1986.

Vol. 1209: Differential Geometry, Peñíscola 1985. Proceeding Edited by A.M. Naveira, A. Ferrández and F. Mascaró. VIII, 306 page 1986.

Vol. 1210: Probability Measures on Groups VIII. Proceedings, 198 Edited by H. Heyer. X, 386 pages. 1986.

Vol. 1211: M.B. Sevryuk, Reversible Systems. V, 319 pages. 1986.

Vol. 1212: Stochastic Spatial Processes. Proceedings, 1984. Edit by P. Tautu. VIII, 311 pages. 1986.

Vol. 1213: L.G. Lewis, Jr., J.P. May, M. Steinberger, Equivari Stable Homotopy Theory. IX, 538 pages. 1986.

Vol. 1214: Global Analysis – Studies and Applications II. Edited Yu.G. Borisovich and Yu.E. Gliklikh. V, 275 pages. 1986.

Vol. 1215: Lectures in Probability and Statistics. Edited by G. del Pi and R. Rebolledo. V, 491 pages. 1986.

Vol. 1216: J. Kogan, Bifurcation of Extremals in Optimal Control. V 106 pages. 1986.

Vol. 1217: Transformation Groups. Proceedings, 1985. Edited by Jackowski and K. Pawalowski. X, 396 pages. 1986.

Vol. 1218: Schrödinger Operators, Aarhus 1985. Seminar. Edited E. Balslev. V, 222 pages. 1986.

Vol. 1219: R. Weissauer, Stabile Modulformen und Eisensteinreih III, 147 Seiten. 1986.

Vol. 1220: Séminaire d'Algèbre Paul Dubreil et Marie-Paule Mallia Proceedings, 1985. Edité par M.-P. Malliavin. IV, 200 pages. 1986.

Vol. 1221: Probability and Banach Spaces. Proceedings, 1985. Edi by J. Bastero and M. San Miguel. XI, 222 pages. 1986.

Vol. 1222: A. Katok, J.-M. Strelcyn, with the collaboration F. Ledrappier and F. Przytycki, Invariant Manifolds, Entropy Billiards; Smooth Maps with Singularities. VIII, 283 pages. 1986.

Vol. 1223: Differential Equations in Banach Spaces. Proceedin 1985. Edited by A. Favini and E. Obrecht. VIII, 299 pages. 1986.

Vol. 1224: Nonlinear Diffusion Problems, Montecatini Terme 19 Seminar. Edited by A. Fasano and M. Primicerio. VIII, 188 pa 1986.

Vol. 1225: Inverse Problems, Montecatini Terme 1986. Semi Edited by G. Talenti. VIII, 204 pages. 1986.

Vol. 1226: A. Buium, Differential Function Fields and Modul Algebraic Varieties. IX, 146 pages. 1986.

Vol. 1227: H. Helson, The Spectral Theorem. VI, 104 pages. 1986

Vol. 1228: Multigrid Methods II. Proceedings, 1985. Edited W. Hackbusch and U. Trottenberg. VI, 336 pages. 1986.

Vol. 1229: O. Bratteli, Derivations, Dissipations and Group Action C*-algebras. IV, 277 pages. 1986.

Vol. 1230: Numerical Analysis. Proceedings, 1984. Edited by Hennart. X, 234 pages. 1986.

Vol. 1231: E.-U. Gekeler, Drinfeld Modular Curves. XIV, 107 pa 1986.